# The Flight Navigator Handbook

U.S. Department
of Transportation

Federal Aviation
Administration

Equinoctial $\gamma$ or celestial equator

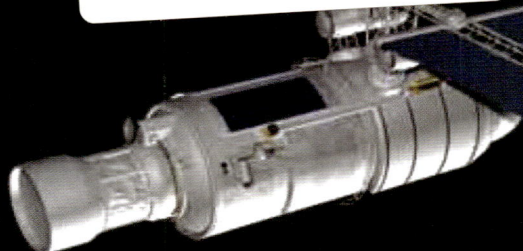

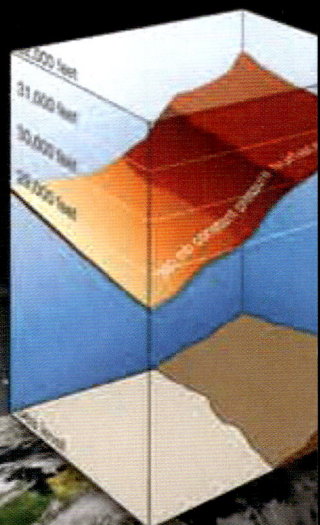

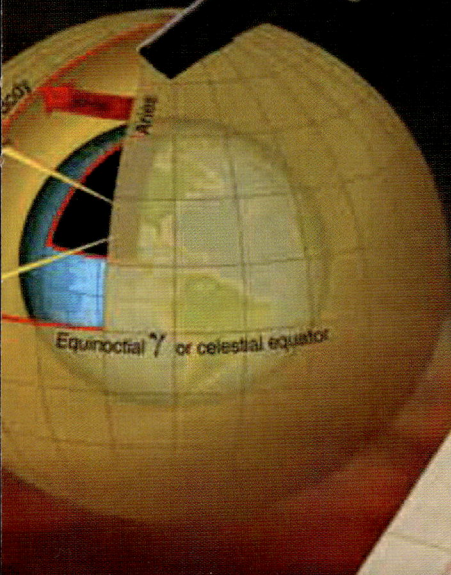

Step 5: Draw lop

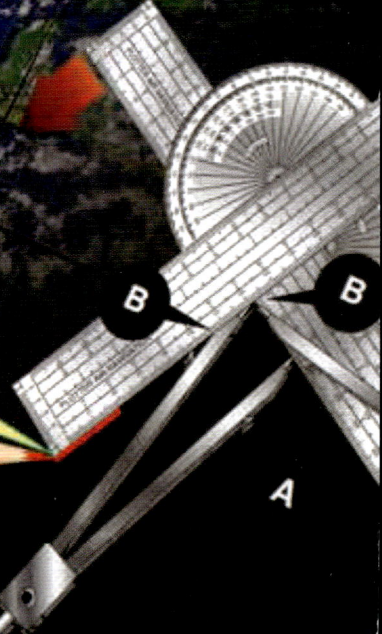

# Flight Navigator Handbook

U.S. Department of Transportation
**FEDERAL AVIATION ADMINISTRATION**
Flight Standards Service

Skyhorse Publishing

All inquiries should be addressed to Skyhorse Publishing, 307 West 36th Street, 11th Floor, New York, NY 10018.

Skyhorse Publishing books may be purchased in bulk at special discounts for sales promotion, corporate gifts, fund-raising, or educational purposes. Special editions can also be created to specifications. For details, contact the Special Sales Department, Skyhorse Publishing, 307 West 36th Street, 11th Floor, New York, NY 10018 or info@skyhorsepublishing.com.

Visit our website at www.skyhorsepublishing.com.

10 9 8 7 6 5 4 3 2 1

Library of Congress Cataloging-in-Publication Data is available on file.

ISBN: 978-1-62636-236-9

Printed in China

# Preface

The Flight Navigator Handbook provides information on all phases of air navigation. It is a source of reference for navigators and navigator students. This handbook explains how to measure, chart the earth, and use flight instruments to solve basic navigation problems. It also contains data pertaining to flight publications, preflight planning, in-flight procedures, and low altitude navigation. A listing of references and supporting information used in this publication is at Appendix A; mathematical formulas to use as an aid in preflight and in-flight computations are at Appendix B; chart and navigation symbols are at Appendix C.; and a Celestial Computation Sheet is at Appendix D.

Any time there is a conflict between the information in this handbook and specific information issued by an aircraft manufacturer, the manufacturer's data takes precedence over information in this handbook. Occasionally, the word "must" or similar language is used where the desired action is deemed critical. The use of such language is not intended to add to, interpret, or relieve a duty imposed by Title 14 of the Code of Federal Regulations (14 CFR).

It is essential for persons using this handbook to become familiar with and apply the pertinent parts of 14 CFR. The current Flight Standards Service (FSS) airman training and testing material and learning statements for all airman certificates and ratings can be obtained from www.faa.gov.

This handbook is available for download, in PDF format, from www.faa.gov.

This handbook is published by the United States Department of Transportation, Federal Aviation Administration, Airman Testing Standards Branch, AFS-630, P.O. Box 25082, Oklahoma City, OK 73125.

Comments regarding this publication should be sent, in email form, to the following address:

AFS630comments@faa.gov

# Table of Contents

## Chapter 16
## Navigation Systems ......................................... 16-1

## Appendix A
## References and Supporting Information .......... A-1

## Appendix B
## Mathematical Formulas ..................................... B-1

## Appendix C
## Chart and Navigation Symbols ........................ C-1

## Appendix D
## Celestial Computation Sheet ........................... D-1

## Glossary ................................................................ G-1

# Chapter 1

# Maps and Charts

## Introduction

Aviators use air navigation to determine where they are going and how to get there. This book serves as a reference for techniques and methods used in air navigation. In addition to this book, several other sources provide excellent references to methods and techniques of navigation:

- The Journal of the Institute of Navigation—published quarterly by The Institute of Navigation, 1800 Diagonal Road, Alexandria, Virginia, 22314, and covers the latest in navigation technology.

- The United States Observatory and United States Navy Oceanographic Office—publications include the Air Almanac, Nautical Almanac, NV Publication 9 (Volumes 1 and 2).

- The American Practical Navigator, SR Publication 249, Volumes 1 through 3, Sight Reduction Tables for Air Navigation and Catalog, and the NGA Public Sale Aeronautical Charts and Products—published by the National Geo-Spatial Intelligence Agency (NGA) (formally known as the National Imagery and Mapping Agency (NIMA)).

- Dutton's Navigation and Piloting—published by the Naval Institute Press, Annapolis, Maryland.

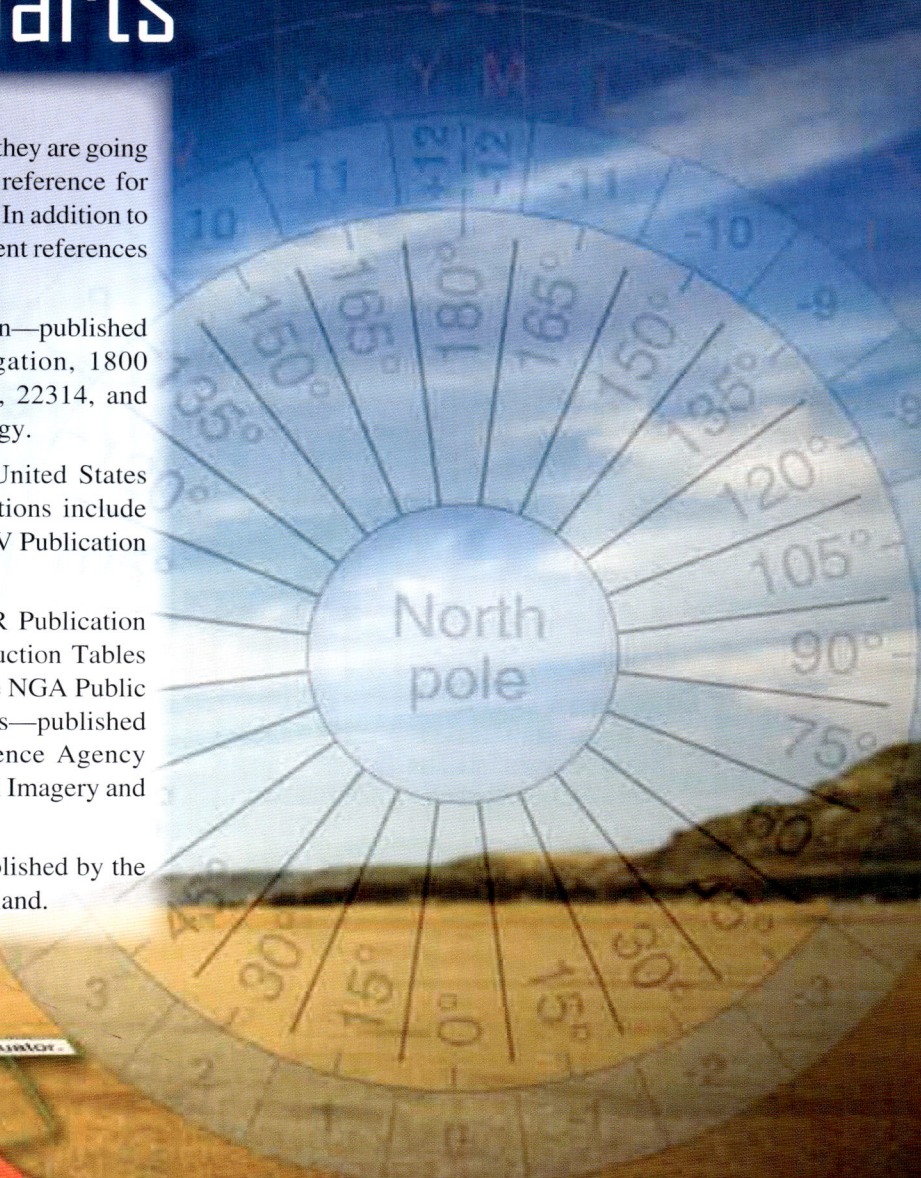

North pole

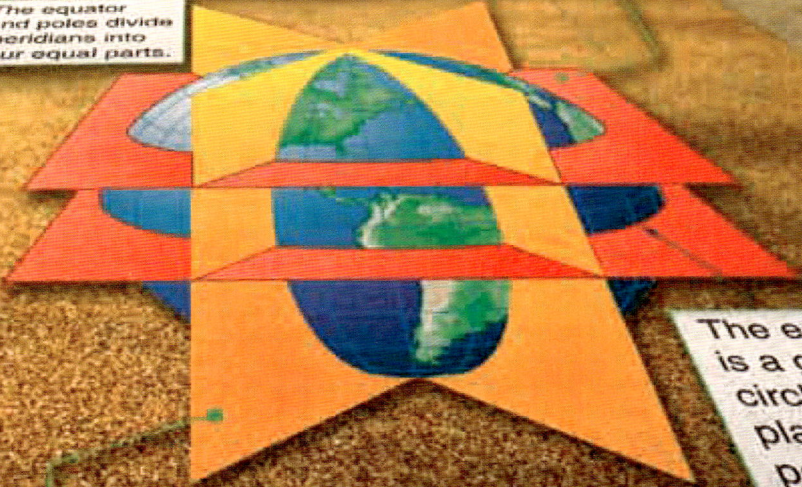

The plane of a parallel is parallel to the equator.

The equator and poles divide meridians into four equal parts.

The equator is a great circle whose plane is perpendi to the a

circles through the poles from meridians.

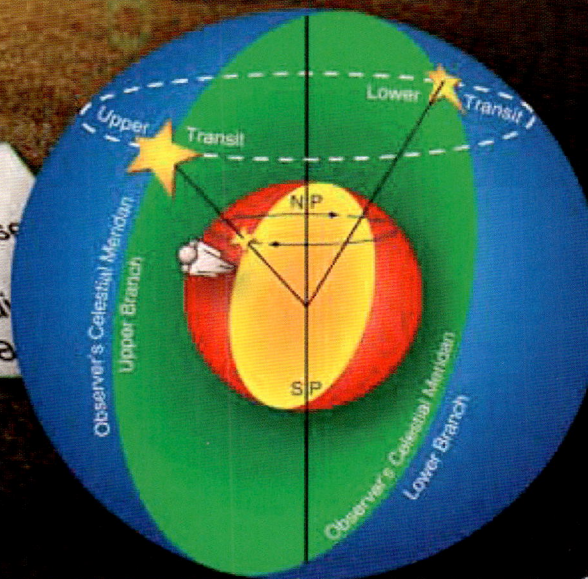

## Basic Terms

Basic to the study of navigation is an understanding of certain terms that could be called the dimensions of navigation. The navigator uses these dimensions of position, direction, distance, altitude, and time as basic references. A clear understanding of these dimensions as they relate to navigation is necessary to provide the navigator with a means of expressing and accomplishing the practical aspects of air navigation. These terms are defined as follows:

- Position—a point defined by stated or implied coordinates. Though frequently qualified by such adjectives as estimated, dead reckoning (DR), no wind, and so forth, the word position always refers to some place that can be identified. It is obvious that a navigator must know the aircraft's current position before being able to direct the aircraft to another position or in another direction.

- Direction—the position of one point in space relative to another without reference to the distance between them. Direction is not in itself an angle, but it is often measured in terms of its angular distance from a referenced direction.

- Distance—the spatial separation between two points, measured by the length of a line joining them. On a plane surface, this is a simple problem. However, consider distance on a sphere, where the separation between points may be expressed as a variety of curves. It is essential that the navigator decide exactly how the distance is to be measured. The length of the line can be expressed in various units (e.g., nautical miles (NM) or yards).

- Altitude—the height of an aircraft above a reference plane. Altitude can be measured as absolute or pressure. Absolute altitude is measured by a radar altimeter, and pressure altitude is measured from various datum planes. Compare with elevation, which is the height of a point or feature on the earth above a reference plane.

- Time—defined in many ways, but definitions used in navigation consist mainly of:
  1. Hour of the day.
  2. Elapsed interval.

- Methods of expression—the methods of expressing position, direction, distance, altitude, and time are covered fully in appropriate chapters. These terms, and others similar to them, represent definite quantities or conditions that may be measured in several different ways. For example, the position of an aircraft may be expressed in coordinates, such as a certain latitude and longitude. The position may also

be expressed as 10 miles south of a certain city. The study of navigation demands the navigator learn how to measure quantities, such as those just defined and how to apply the units by which they are expressed.

## The Earth

### Shape and Size

For most navigational purposes, the earth is assumed to be a perfect sphere, although in reality it is not. Inspection of the earth's crust reveals there is a height variation of approximately 12 miles from the top of the tallest mountain to the bottom of the deepest point in the ocean. A more significant deviation from round is caused by a combination of the earth's rotation and its structural flexibility. When taking the ellipsoidal shape of the planet into account, mountains seem rather insignificant. The peaks of the Andes are much farther from the center of the earth than Mount Everest.

Measured at the equator, the earth is approximately 6,378,137 meters in diameter, while the polar diameter is approximately 6,356,752.3142 meters. The difference in these diameters is 21,384.6858 meters, and this difference may be used to express the ellipticity of the earth. The ratio between this difference and the equatorial diameter is:

$$\text{Ellipticity} = \frac{21,384.6858}{6,378,137} = \frac{1}{298.257223}$$

Since the equatorial diameter exceeds the polar diameter by only 1 part in 298, the earth is nearly spherical. A symmetrical body having the same dimensions as the earth, but with a smooth surface, is called an ellipsoid. The ellipsoid is sometimes described as a spheroid, or an oblate spheroid.

In *Figure 1-1,* polar north (Pn), east (E), polar south (Ps), and west (W) represent the surface of the earth. Pn and Ps represent the axis of rotation. The earth rotates from west to east. All points in the hemisphere Pn, W, Ps approach the reader, while those in the opposite hemisphere recede from the reader. The circumference W-E is called the equator, which is defined as that imaginary circle on the surface of the earth whose plane passes through the center of the earth and is perpendicular to the axis of rotation.

### Great Circles and Small Circles

A great circle is defined as a circle on the surface of a sphere whose center and radius are those of the sphere itself. It is the largest circle that can be drawn on the sphere; it is the intersection with the surface of the earth of any plane passing through the earth's center. The arc of a great circle is the shortest distance between two points on a sphere, just as a straight line is the shortest distance between two points on a plane. On any sphere, an indefinitely large number of great

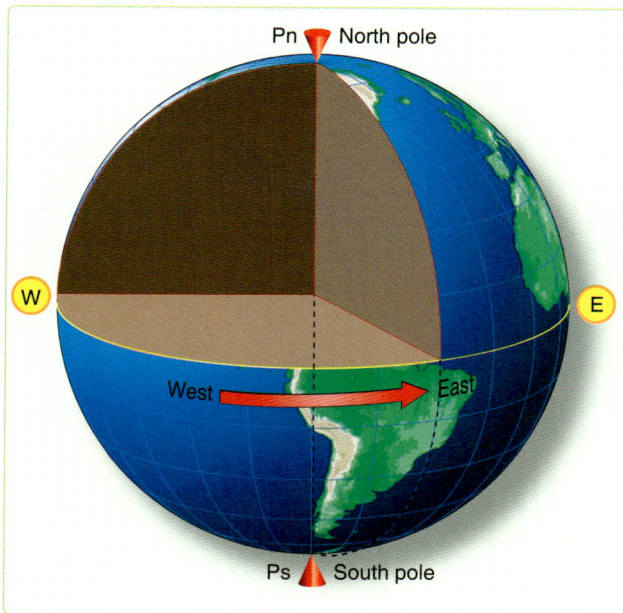

**Figure 1-1.** *Schematic representation of the earth showing its axis of rotation and equator.*

circles may be drawn through any point, though only one great circle may be drawn through any two points not diametrically opposite. Several great circles are shown in *Figure 1-2*.

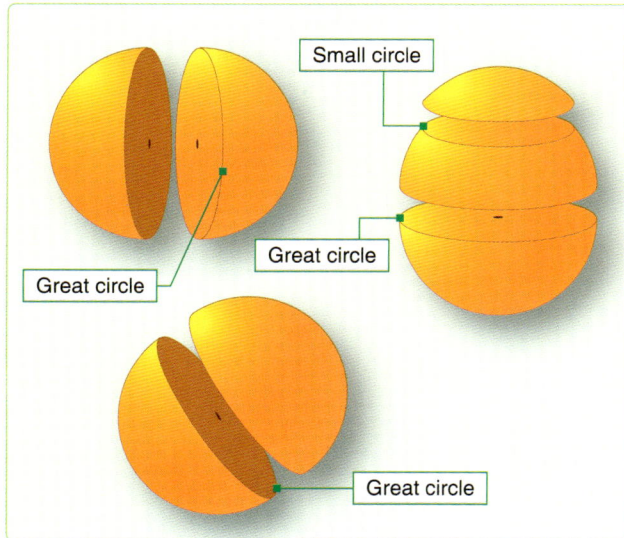

**Figure 1-2.** *A great circle is the largest circle in a sphere.*

Circles on the surface of the sphere, other than great circles, may be defined as small circles. A small circle is a circle on the surface of the earth whose center and/or radius are not that of the sphere. A set of small circles, called latitude, is discussed later. In summary, the intersection of a sphere and a plane is a great circle if the plane passes through the center of the sphere and a small circle if it does not.

## Latitude and Longitude

The nature of a sphere is such that any point on it is exactly like any other point. There is neither beginning nor ending as far as differentiation of points is concerned. In order that points may be located on the earth, some points or lines of reference are necessary so that other points may be located with regard to them. The location of New York City with reference to Washington, D.C. can be stated as a number of miles in a certain direction from Washington. Any point on the earth can be located in this manner.

## Imaginary Reference Lines

Such a system, however, does not lend itself readily to navigation, because it would be difficult to locate a point precisely in mid-ocean without any nearby geographic features to use for reference. We use a system of coordinates to locate positions on the earth by means of imaginary reference lines. These lines are known as parallels of latitude and meridians of longitude.

## Latitude

Once a day, the earth rotates on its north-south axis, which is terminated by the two poles. The equatorial plane is constructed at the midpoint of this axis at right angles to it. *[Figure 1-3]* A great circle drawn through the poles is called a meridian, and an infinite number of great circles may be constructed in this manner. Each meridian is divided into four quadrants by the equator and the poles. The circle is arbitrarily divided into 360°, and each of these quadrants contains 90°.

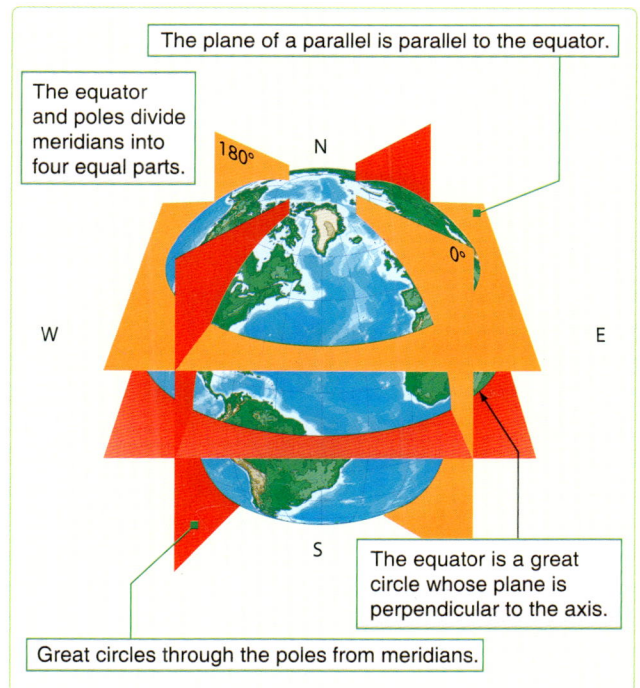

**Figure 1-3.** *Planes of the earth.*

Take a point on one of these meridians 30° N of the equator. Through this point passes a plane perpendicular to the north-south axis of rotation. This plane is parallel to the plane of the equator as shown in *Figure 1-3* and intersects the earth in a small circle called a parallel or parallel of latitude. The particular parallel of latitude chosen as 30° N, and every point on this parallel is at 30° N. In the same way, other parallels can be constructed at any desired latitude, such as 10°, 40°, etc.

Bear in mind that the equator is drawn as the great circle; midway between the poles and parallels of latitude are small circles constructed with reference to the equator. The angular distance measured on a meridian north or south of the equator is known as latitude *[Figure 1-4]* and forms one component of the coordinate system.

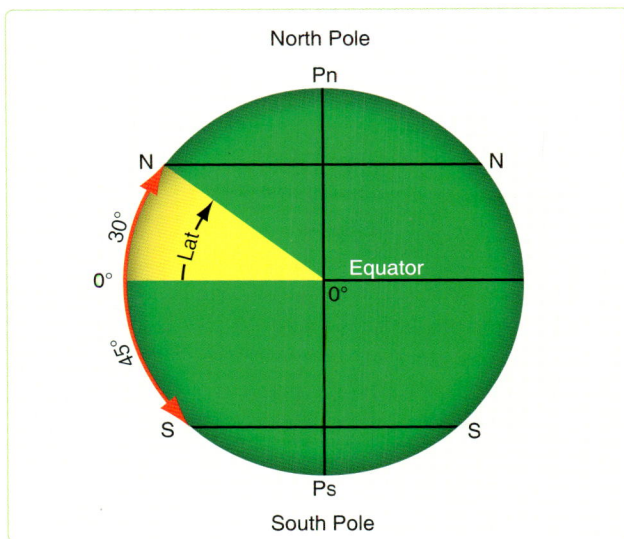

**Figure 1-4.** *Latitude as an angular measurement.*

## Longitude

The latitude of a point can be shown as 20° N or 20° S of the equator, but there is no way of knowing whether one point is east or west of another. This difficulty is resolved by use of the other component of the coordinate system, longitude, which is the measurement of this east-west distance. Longitude, unlike latitude, has no natural starting point for numbering. The solution has been to select an arbitrary starting point. A great many places have been used, but when the English-speaking people began to make charts, they chose the meridian through their principal observatory in Greenwich, England, as the origin for counting longitude. This Greenwich meridian is sometimes called the prime meridian, though actually it is the zero meridian. Longitude is counted east and west from this meridian through 180°. *[Figure 1-5]* Thus, the Greenwich meridian is the 0° longitude on one side of the earth and, after crossing the poles, it becomes the 180th meridian (180° east or west of the 0° meridian).

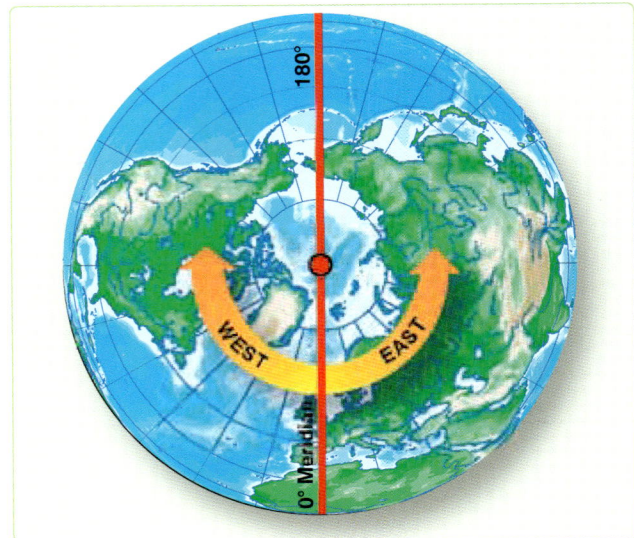

**Figure 1-5.** *Longitude is measured east and west of the Greenwich Meridian.*

In summary, if a globe has the circles of latitude and longitude drawn upon it according to the principles described, any point can be located on the globe using these measurements. *[Figure 1-6]*

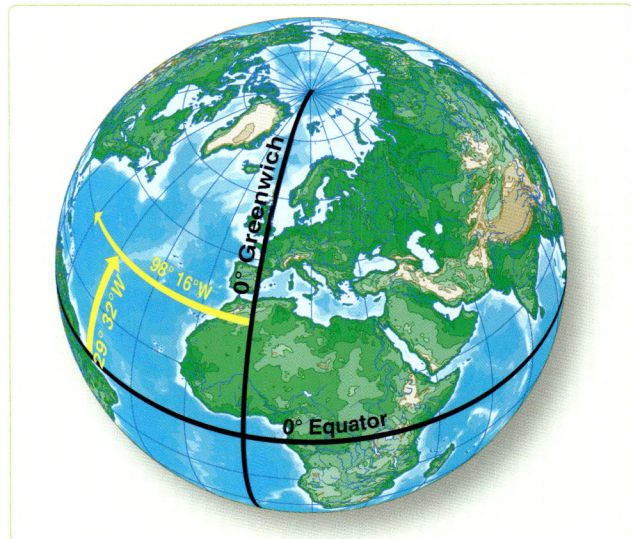

**Figure 1-6.** *Latitude is measured from the equator; longitude from the prime meridian.*

It is beneficial to point out here some of the measurements used in the coordinate system. Latitude is measured in degrees up to 90, and longitude is expressed in degrees up to 180. The total number of degrees in any one circle is always 360. A degree (°) of arc may be subdivided into smaller units by dividing each degree into 60 minutes (') of arc. Each minute may be further subdivided into 60 seconds (") of arc.

Measurement may also be expressed in degrees, minutes, and tenths of minutes.

A position on the surface of the earth is expressed in terms of latitude and longitude. Latitude is expressed as being either north or south of the equator, and longitude as either east or west of the prime meridian.

In actual practice, map production requires surveyors to measure the latitude and longitude of geographic objects in their area of interest. Local variation in the earth's gravity field can cause these measurements to be inconsistent. All coordinates from maps, charts, traditional surveys, and satellite positioning systems are tied to an individual mathematical model of the earth called a datum. Coordinates for a given point may differ between datums by hundreds of yards. In other words, latitude and longitude measured directly from observation of stars (called an astronomic coordinate) is consistent, but it may not match maps, charts, or surveyed points. The theoretical consistency of latitude and longitude is therefore not achievable in reality. Without knowledge of the datum used to establish a particular map or surveyed coordinate, the coordinate is suspect at best.

## Distance

Distance, as previously defined, is measured by the length of a line joining two points. The standard unit of distance for navigation is the nautical mile (NM). The NM can be defined as either 6,076 feet or 1 minute of latitude. Sometimes it is necessary to convert statute miles (SM) to NM and vice versa. This conversion is easily done with the following ratio:

$$\frac{\text{Number of statute miles}}{\text{Number of nautical miles}} = \frac{76}{66}$$

Closely related to the concept of distance is speed, which determines the rate of change of position. Speed is usually expressed in miles per hour (mph), this being either SM per hour or NM per hour. If the measure of distance is NM, it is customary to speak of speed in terms of knots. Thus, a speed of 200 knots and a speed of 200 NM per hour are the same thing. It is incorrect to say 200 knots per hour unless referring to acceleration.

## Direction

Remember, direction is the position of one point in space relative to another without reference to the distance between them. The time honored point system for specifying a direction as north, north-northwest, northwest, west-northwest, west, etc., is not adequate for modern navigation. It has been replaced for most purposes by a numerical system. *[Figure 1-7]* The numerical system divides the horizon into 360°, starting with north as 000° and continuing clockwise through east 090°, south 180°, west 270°, and back to north.

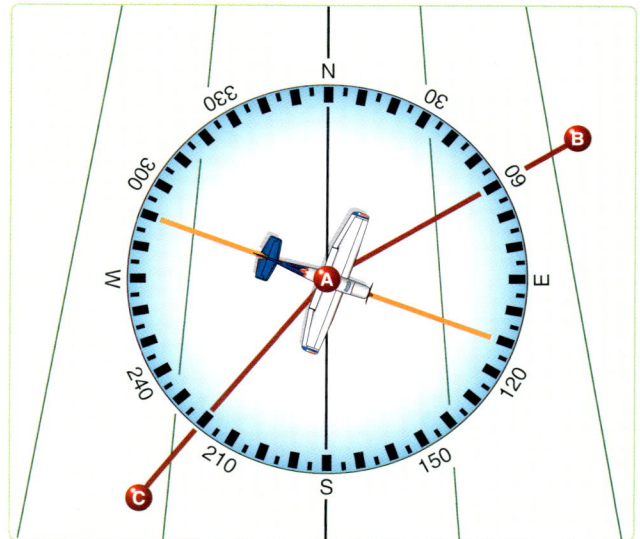

**Figure 1-7.** *Numerical system is used in air navigation.*

The circle, called a compass rose, represents the horizon divided into 360°. The nearly vertical lines in *Figure 1-7* are meridians drawn as straight lines with the meridian of position A passing through 000° and 180° of the compass rose. Position B lies at a true direction of 062° from A, and position C is at a true direction of 220° from A.

Since determination of direction is one of the most important parts of the navigator's work, the various terms involved should be clearly understood. Generally, in navigation, unless otherwise stated, directions are called true directions.

### Course

Course is the intended horizontal direction of travel. Heading is the horizontal direction in which an aircraft is pointed. Heading is the actual orientation of the longitudinal axis of the aircraft at any instant, while course is the direction intended to be made good. Track is the actual horizontal direction made by the aircraft over the earth.

### Bearing

Bearing is the horizontal direction of one terrestrial point from another. As illustrated in *Figure 1-8*, the direction of the island from the aircraft is marked by a visual bearing called the line of sight (LOS). Bearings are usually expressed in terms of one of two reference directions: true north (TN) or the direction in which the aircraft is pointed. If TN is the reference direction, the bearing is called a true bearing (TB). If the reference direction is the heading of the aircraft, the bearing is called a relative bearing (RB). *[Figure 1-9]*

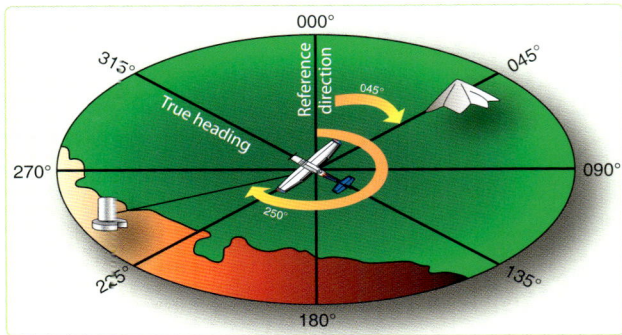

**Figure 1-8.** *Measuring true bearing from true north.*

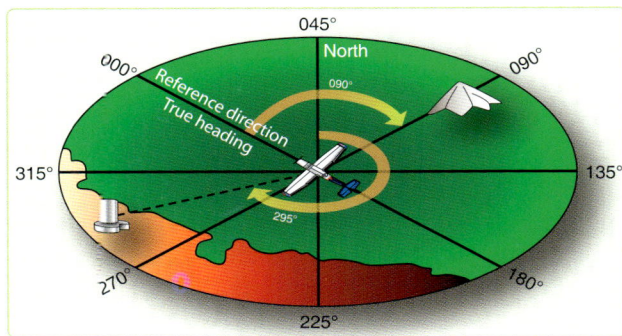

**Figure 1-9.** *Measuring relative bearing from aircraft heading.*

## Great Circle and Rhumb Line Direction

The direction of the great circle, shown in *Figure 1-10,* makes an angle of about 40° with the meridian near Washington, D.C., about 85° with the meridian near Iceland, and a still greater angle with the meridian near Moscow. In other words, the direction of the great circle is constantly changing as progress is made along the route and is different at every point along the great circle. Flying such a route requires constant change of direction and would be difficult to fly under ordinary conditions. Still, it is the most desirable route because it is the shortest distance between any two points.

A line that makes the same angle with each meridian is called a rhumb line. An aircraft holding a constant true heading would be flying a rhumb line. Flying this sort of path results in a greater distance traveled, but it is easier to steer. If continued, a rhumb line spirals toward the poles in a constant true direction but never reaches them. The spiral formed is called a loxodrome or loxodromic curve. *[Figure 1-11]*

Between two points on the earth, the great circle is shorter than the rhumb line, but the difference is negligible for short distances (except in high latitudes) or if the line approximates a meridian or the equator.

## Time

In celestial navigation, navigators determine the aircraft's position by observing the celestial bodies. The apparent position of these bodies changes with time. Therefore, determining the aircraft's position relies on timing the observation exactly. Time is measured by the rotation of the earth and the resulting apparent motions of the celestial bodies. This chapter considers several different systems of measurement, each with a special use. Before learning the various kinds of time, it is important to understand transit. Notice in *Figure 1-12* that the poles divide the observer's meridian into halves. The observer's position is in the upper branch. The lower branch is the opposite half. Every day, because of the earth's rotation, every celestial body transits the upper and lower branches of the observer's meridian. The first kind of time presented here is solar time.

## Apparent Solar Time

The sun as it is seen in the sky is called the true sun or the apparent sun. Apparent solar time is based upon the movement of the sun as it crosses the sky. A sundial accurately indicates apparent solar time. Apparent solar time is not useful, because the apparent length of day varies throughout the year. A timepiece would have to operate at different speeds to indicate correct apparent time. However, apparent time accurately indicates upper and lower transit. Upper transit occurs at noon; apparent time and lower transit at midnight apparent time. Difficulties in using apparent time led to the introduction of mean time.

**Figure 1-10.** *Great circle.*

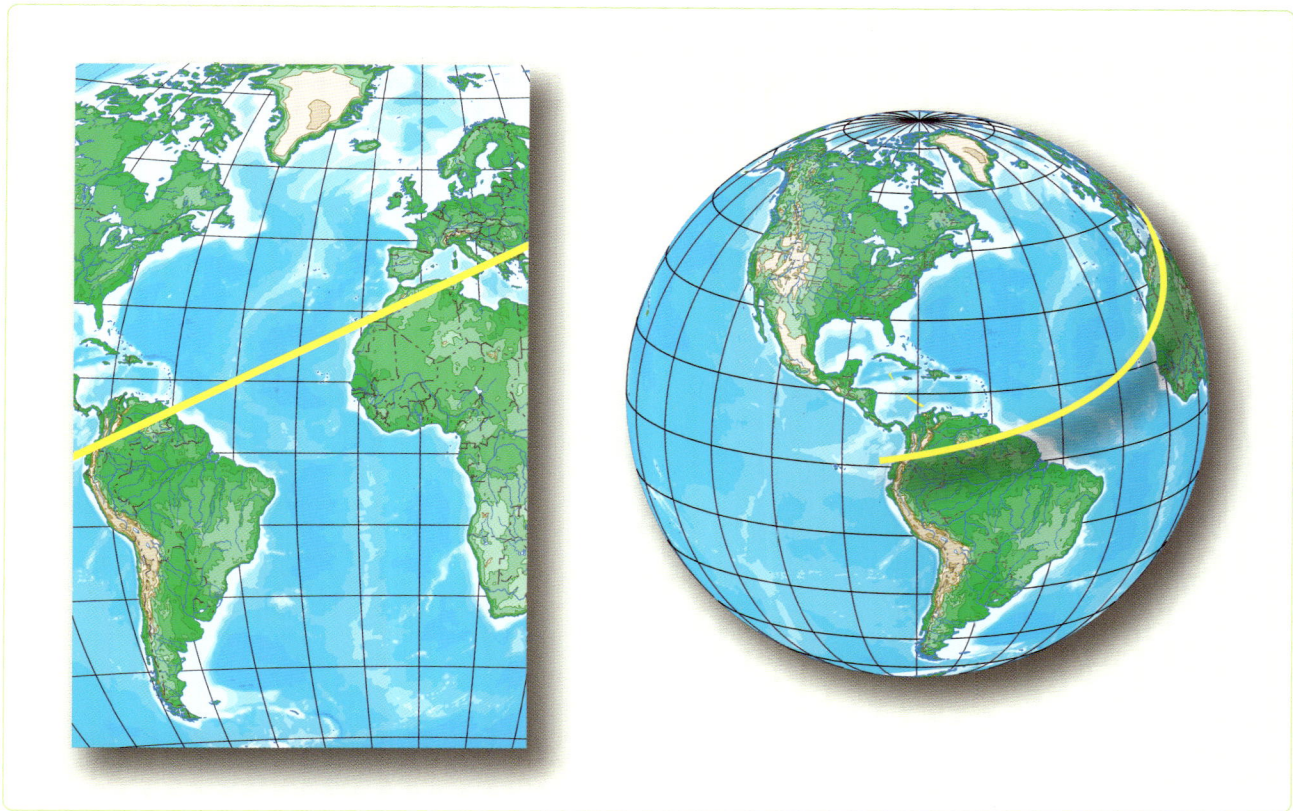

**Figure 1-11.** *A rhumb line or loxodrome.*

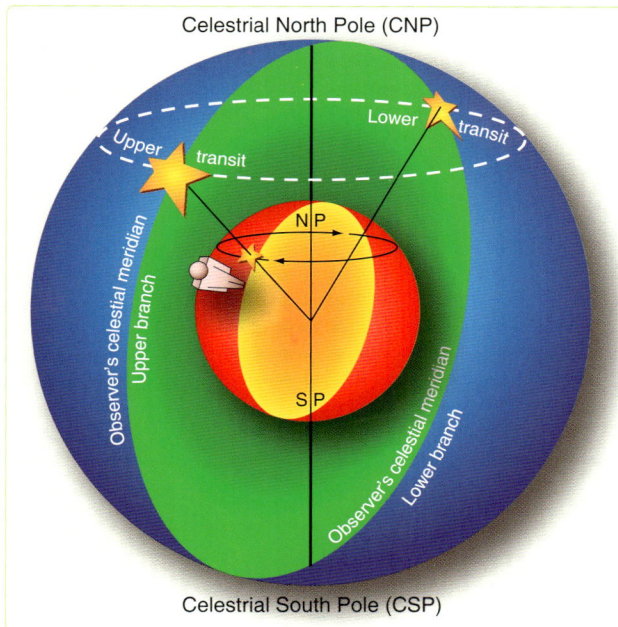

**Figure 1-12.** *Transit is caused by the earth's rotation.*

## Mean Solar Time

A mean day is an artificial unit of constant length, based on the average of all apparent solar days over a period of years. Time for a mean day is measured with reference to a fictitious body, the mean sun, so designed that its hour circle moves westward at a constant rate along the celestial equator. Time computed using the mean sun is called mean solar time. The coordinates of celestial bodies in the Air Almanac are tabulated in mean solar time, making it the time of primary interest to navigators. The difference in length between the apparent day (based upon the true sun) and the mean day (based upon the mean sun) is never as much as a minute. The differences are cumulative, however, so that the imaginary mean sun precedes or follows the apparent sun by approximately 15 minutes at certain times during the year.

## Greenwich Mean Time (GMT)

Greenwich Mean Time (GMT) is used for most celestial computations. GMT is mean solar time measured from the lower branch of the Greenwich meridian westward through 360° to the upper branch of the hour circle passing through the mean sun. *[Figure 1-13]* The mean sun transits the Greenwich meridian's lower branch at GMT 2400 (0000) each day and the upper branch at GMT 1200. The meridian at Greenwich is the logical selection for this reference, as it is the origin for the measurement of Greenwich hour angle (GHA) and the reckoning of longitude. Consequently, celestial coordinates and other information are tabulated in almanacs with reference to GMT. GMT is also called Zulu or Z time.

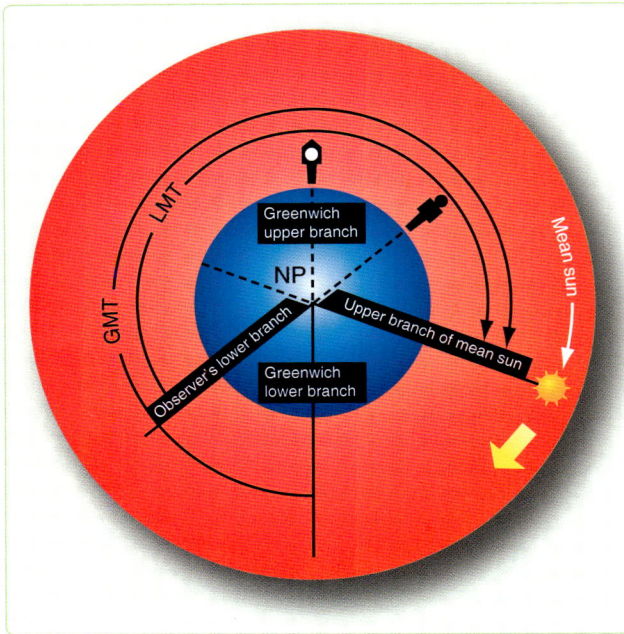

**Figure 1-13.** *Measuring Greenwich mean time.*

## Local Mean Time (LMT)

Just as GMT is mean solar time measured with reference to the Greenwich meridian, local mean time (LMT) is mean solar time measured with reference to the observer's meridian. LMT is measured from the lower branch of the observers meridian, westward through 360°, to the upper branch of the hour circle passing through the mean sun. *[Figure 1-13]* The mean sun transits the lower branch of the observer's meridian at LMT 0000 (2400) and the upper branch at LMT 1200. For an observer at the Greenwich meridian, GMT is LMT. Navigators use LMT to compute local sunrise, sunset, twilight, moonrise, and moonset at various latitudes along a given meridian.

## Relationship of Time and Longitude

The mean sun travels at a constant rate, covering 360° of arc in 24 hours. The mean sun transits the same meridian twice in 24 hours. The following relationships exists between time and arc:

| Time | Arc |
|---|---|
| 24 hours | 360° |
| 1 hour | 15° |
| 4 minutes | 1° |
| 1 minute | 15' |

Local time is the time at one particular meridian. Since the sun cannot transit two meridians simultaneously, no two meridians have exactly the same local time. The difference in time between two meridians is the time of the sun's passage

from one meridian to the other. This time is proportional to the angular distance between the two meridians. One hour is equivalent to 15°.

If two meridians are 30° apart, their time differs by 2 hours. The easternmost meridian has a later local time, because the sun has crossed its lower branch first; thus, the day is older there. These statements hold true whether referring to the apparent sun or the mean sun. *Figure 1-14* demonstrates that the sun crossed the lower branch of the meridian of observer 1 at 60° east longitude 4 hours before it crossed the lower branch of the Greenwich meridian (60 ÷ 15) and 6 hours before it crossed the lower branch of the meridian of observer 2 at 30° west longitude (90 ÷ 15). Therefore, the local time at 60° east longitude is later by the respective amounts.

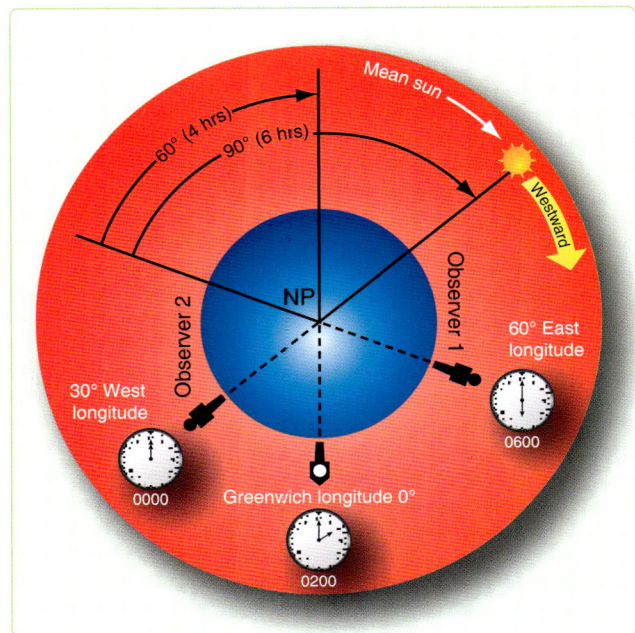

**Figure 1-14.** *Local time differences at different longitudes.*

## Standard Time Zone

The world is divided into 24 zones, each zone being 15° of longitude wide. Each zone uses the LMT of its central meridian. (A few areas of the world are further divided and use half-hour increments from GMT. Some notable examples include India, Bangladesh, Newfoundland, and parts of Australia and Thailand.) Since the Greenwich meridian is the central meridian for one of the zones, and each zone is 15° or 1 hour wide, the time in each zone differs from GMT by an integral number of hours. The zones are designated by numbers from 0 to 12 and –12, each indicating the number of hours that must be added or subtracted to local zone time (LZT) to obtain GMT. Since the time is earlier in the zones west of Greenwich, the numbers of these zones are plus; in those zones east of Greenwich, the numbers are minus. *[Figure 1-15]* Ground forces frequently refer to the zones by letters of the alphabet, and air forces use one of these letters (Z) for GMT. The zone boundaries have been modified to conform with geographical boundaries for greater convenience. For example, in case a zone boundary passed through a city, it would be impractical to use the time of one zone in one part of the city and the time of the adjacent zone in the other part. In some countries, which overlap two or three zones, one time is used throughout.

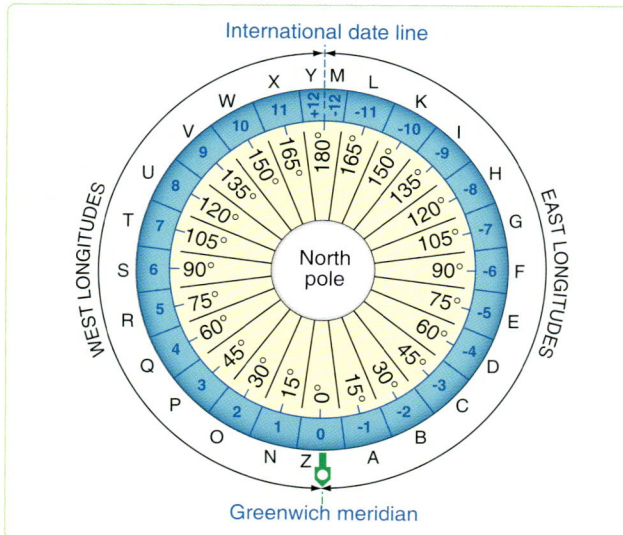

**Figure 1-15.** *Standard time zones.*

## Date Changes at Midnight

If travelling west from Greenwich around the world and setting a watch back an hour for each time zone, the watch would have been set back a total of 24 hours on arriving back at Greenwich, and the date would be 1 day behind. Conversely, traveling eastward, the watch would have been advanced a total of 24 hours, gaining a day.

To keep straight, a day must be added somewhere if going around the world to the west and a day must be lost if going around to the east. The 180° meridian is the international dateline where a day is gained or lost. The date line follows the meridian except where it detours to avoid eastern Siberia, the western Aleutian Islands, and several groups of islands in the South Pacific.

The local civil date changes at 2400 or midnight. Thus, the date changes as the mean sun transits the lower branch of the meridian. Consider the situation in another way. The hour circle of the mean sun is divided in half at the poles. On the half away from the sun (the lower branch), it is always midnight LMT. As the lower branch moves westward, it pushes the old date before it and drags the new date after it. *[Figure 1-16]* As the lower branch approaches the 180° meridian, the area of the old date decreases and the area of the new date increases. When the lower branch reaches the date line; that is, when the mean sun transits the Greenwich meridian, the old date is crowded out and the new date for that instant prevails in the world. Then, as the lower branch passes the date line, a newer date begins east of the lower

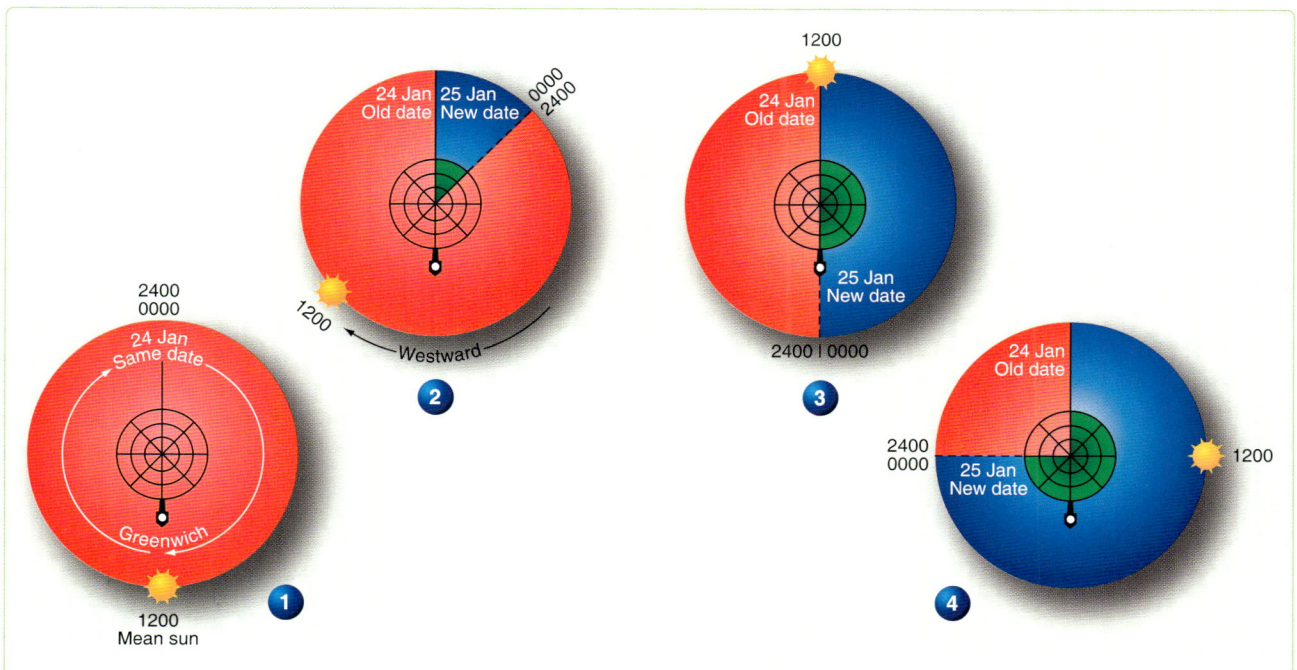

**Figure 1-16.** *Zone date changes.*

branch and the process starts all over again. The zone date changes at midnight zone time (ZT) or when the lower branch of the mean sun transits the central meridian of the zone.

## Time Conversion

Sometimes it is necessary to convert LMT time to GMT, or GMT to LMT. The Air Almanac contains a table for conversion of arc to time at a rate of 15° of arc per hour of time. *[Figure 1-17]* This conversion is only good for LMT to GMT, or GMT to LMT. ZT is influenced by daylight savings time and geographical boundaries. For example, to convert GMT to LMT at 126° –36' W:

126° 00 = 8 h 24 min 00s

36' = 02 min 24s

126° 36' = 8 h 26 min 24s

To derive LMT from GMT, subtract the time in the Western Hemisphere and add it in the Eastern Hemisphere. Do the opposite to convert LMT to GMT.

## Sidereal Time

Solar time is measured with reference to the true sun or the mean sun. Time may also be measured relative to a fixed point in space. Time measured with reference to the first point of Aries, which is considered stationary although it moves slightly, is sidereal or star time. The first point of Aries is defined as where the sun crosses the equator northbound on the first day of spring.

The sidereal day begins when the first point of Aries transits the upper branch of the observer's meridian. Local sidereal time (LST) is the number of hours that the first point of Aries has moved westward from the observers meridian. Expressed in degrees, it equals the local hour angle (LHA) of Aries. *[Figure 1-18]* LST at Greenwich is Greenwich sidereal time (GST), which is equivalent to the GHA of Aries. GST, or GHA of Aries, specifies the position of the stars with relation to the earth. Thus, a given star is in the same position relative to the earth at the same sidereal time each day.

## Number of Days in a Year

The earth revolves around the sun in a year. The number of days in the year equals the number of rotations of the earth during one revolution. The earth rotates eastward about 366.24 times during its yearly eastward revolution. The total effect of one revolution and 366.24 rotations is that the sun appears to revolve around the earth 365.24 times per year. Therefore, there are 365.24 solar days per year. Since the sidereal day is measured with reference to a fixed point, the length of the sidereal day is the period of the earth's rotation. Therefore, the number of sidereal days in the year is equal to the number of rotations per year, 366.24.

## Navigator's Use of Time

Navigators use three different kinds of time: GMT, LMT, and ZT. All three are based upon the motions of the fictitious mean sun. The mean sun revolves about the earth at the average rate of the apparent sun, completing one revolution in 24 hours. Time is based upon the motion of the sun relative to a given meridian. The time is 2400/0000 at lower transit and 1200 at upper transit. In GMT, the reference meridian is that of Greenwich; in LMT, the reference meridian is that of a given place; in ZT, the reference meridian is the standard meridian of a given zone.

The difference between two times equals the difference of longitude of their reference meridians expressed in time. GMT differs from ZT by the longitude of the zone's standard meridian; LMT differs from ZT by the difference of longitude between the zone's standard meridian and the meridian of the place. In interconverting ZT and GMT, the navigator uses the zone description. The zone difference is the time difference between its standard meridian and GMT, and it has a sign to indicate the correction to ZT to obtain GMT. The sign is plus (+) for west longitude and minus (–) for east longitude.

## Charts and Projections

There are several basic terms and ideas relative to charts and projections that the reader should be familiar with before discussing the various projections used in the creation of aeronautical charts.

1.  A map or chart is a small scale representation on a plane of the surface of the earth or some portion of it.

2.  The chart projection forms the basic structure on which a chart is built and determines the fundamental characteristics of the finished chart.

3.  There are many difficulties that must be resolved when representing a portion of the surface of a sphere upon a plane. Two of these are distortion and perspective.

    a.  Distortion cannot be entirely avoided, but it can be controlled and systematized to some extent in the drawing of a chart. If a chart is drawn for a particular purpose, it can be drawn in such a way as to minimize the type of distortion that is most detrimental to the purpose. Surfaces that can be spread out in a plane without stretching or tearing, such as a cone or cylinder, are called developable surfaces, and those like the sphere or spheroid that cannot be formed into a plane without distortion are called non-developable. *[Figure 1-19]* The problem of creating a projection lies in developing a method for transferring the meridians and parallels to the chart in a manner that preserves certain desired characteristics as

| ° | h | m | ° | h | m | ° | h | m | ° | h | m | ° | h | m | ° | h | m | m | s |
|---|---|---|---|---|---|---|---|---|---|---|---|---|---|---|---|---|---|---|---|
| 0 | 0 | 0 | 60 | 4 | 0 | 120 | 8 | 0 | 180 | 12 | 0 | 240 | 16 | 0 | 300 | 20 | 0 | 0 | 0 |
| 1 | 0 | 4 | 61 | 4 | 4 | 121 | 8 | 4 | 181 | 12 | 4 | 241 | 16 | 4 | 301 | 20 | 4 | 0 | 4 |
| 2 | 0 | 8 | 62 | 4 | 8 | 122 | 8 | 8 | 182 | 12 | 8 | 242 | 16 | 8 | 302 | 20 | 8 | 0 | 8 |
| 3 | 0 | 12 | 63 | 4 | 12 | 123 | 8 | 12 | 183 | 12 | 12 | 243 | 16 | 12 | 303 | 20 | 12 | 0 | 12 |
| 4 | 0 | 16 | 64 | 4 | 16 | 124 | 8 | 16 | 184 | 12 | 16 | 244 | 16 | 16 | 304 | 20 | 16 | 0 | 16 |
| 5 | 0 | 20 | 65 | 4 | 20 | 125 | 8 | 20 | 185 | 12 | 20 | 245 | 16 | 20 | 305 | 20 | 20 | 0 | 20 |
| 6 | 0 | 24 | 66 | 4 | 24 | 126 | 8 | 24 | 186 | 12 | 24 | 246 | 16 | 24 | 306 | 20 | 24 | 0 | 24 |
| 7 | 0 | 28 | 67 | 4 | 28 | 127 | 8 | 28 | 187 | 12 | 28 | 247 | 16 | 28 | 307 | 20 | 28 | 0 | 28 |
| 8 | 0 | 32 | 68 | 4 | 32 | 128 | 8 | 32 | 188 | 12 | 32 | 248 | 16 | 32 | 308 | 20 | 32 | 0 | 32 |
| 9 | 0 | 36 | 69 | 4 | 36 | 129 | 8 | 36 | 189 | 12 | 36 | 249 | 16 | 36 | 309 | 20 | 36 | 0 | 36 |
| 10 | 0 | 40 | 70 | 4 | 40 | 130 | 8 | 40 | 190 | 12 | 40 | 250 | 16 | 40 | 310 | 20 | 40 | 0 | 40 |
| 11 | 0 | 44 | 71 | 4 | 44 | 131 | 8 | 44 | 191 | 12 | 44 | 251 | 16 | 44 | 311 | 20 | 44 | 0 | 44 |
| 12 | 0 | 48 | 72 | 4 | 48 | 132 | 8 | 48 | 192 | 12 | 48 | 252 | 16 | 48 | 312 | 20 | 48 | 0 | 48 |
| 13 | 0 | 52 | 73 | 4 | 52 | 133 | 8 | 52 | 193 | 12 | 52 | 253 | 16 | 52 | 313 | 20 | 52 | 0 | 52 |
| 14 | 0 | 56 | 74 | 4 | 56 | 134 | 8 | 56 | 194 | 12 | 56 | 254 | 16 | 56 | 314 | 20 | 56 | 0 | 56 |
| 15 | 1 | 0 | 75 | 5 | 0 | 135 | 9 | 0 | 195 | 13 | 0 | 255 | 17 | 0 | 315 | 21 | 0 | 1 | 0 |
| 16 | 1 | 4 | 76 | 5 | 4 | 136 | 9 | 4 | 196 | 13 | 4 | 256 | 17 | 4 | 316 | 21 | 4 | 1 | 4 |
| 17 | 1 | 8 | 77 | 5 | 8 | 137 | 9 | 8 | 197 | 13 | 8 | 257 | 17 | 8 | 317 | 21 | 8 | 1 | 8 |
| 18 | 1 | 12 | 78 | 5 | 12 | 138 | 9 | 12 | 198 | 13 | 12 | 258 | 17 | 12 | 318 | 21 | 12 | 1 | 12 |
| 19 | 1 | 16 | 79 | 5 | 16 | 139 | 9 | 16 | 199 | 13 | 16 | 259 | 17 | 16 | 319 | 21 | 16 | 1 | 16 |
| 20 | 1 | 20 | 80 | 5 | 20 | 140 | 9 | 20 | 200 | 13 | 20 | 260 | 17 | 20 | 320 | 21 | 20 | 1 | 20 |
| 21 | 1 | 24 | 81 | 5 | 24 | 141 | 9 | 24 | 201 | 13 | 24 | 261 | 17 | 24 | 321 | 21 | 24 | 1 | 24 |
| 22 | 1 | 28 | 82 | 5 | 28 | 142 | 9 | 28 | 202 | 13 | 28 | 262 | 17 | 28 | 322 | 21 | 28 | 1 | 28 |
| 23 | 1 | 32 | 83 | 5 | 32 | 143 | 9 | 32 | 203 | 13 | 32 | 263 | 17 | 32 | 323 | 21 | 32 | 1 | 32 |
| 24 | 1 | 36 | 84 | 5 | 36 | 144 | 9 | 36 | 204 | 13 | 36 | 264 | 17 | 36 | 324 | 21 | 36 | 1 | 36 |
| 25 | 1 | 40 | 85 | 5 | 40 | 145 | 9 | 40 | 205 | 13 | 40 | 265 | 17 | 40 | 325 | 21 | 40 | 1 | 40 |
| 26 | 1 | 44 | 86 | 5 | 44 | 146 | 9 | 44 | 206 | 13 | 44 | 266 | 17 | 44 | 326 | 21 | 44 | 1 | 44 |
| 27 | 1 | 48 | 87 | 5 | 48 | 147 | 9 | 48 | 207 | 13 | 48 | 267 | 17 | 48 | 327 | 21 | 48 | 1 | 48 |
| 28 | 1 | 52 | 88 | 5 | 52 | 148 | 9 | 52 | 208 | 13 | 52 | 268 | 17 | 52 | 328 | 21 | 52 | 1 | 52 |
| 29 | 1 | 56 | 89 | 5 | 56 | 149 | 9 | 56 | 209 | 13 | 56 | 269 | 17 | 56 | 329 | 21 | 56 | 1 | 56 |
| 30 | 2 | 0 | 90 | 6 | 0 | 150 | 10 | 0 | 210 | 14 | 0 | 270 | 18 | 0 | 330 | 22 | 0 | 2 | 0 |
| 31 | 2 | 4 | 91 | 6 | 4 | 151 | 10 | 4 | 211 | 14 | 4 | 271 | 18 | 4 | 331 | 22 | 4 | 2 | 4 |
| 32 | 2 | 8 | 92 | 6 | 8 | 152 | 10 | 8 | 212 | 14 | 8 | 272 | 18 | 8 | 332 | 22 | 8 | 2 | 8 |
| 33 | 2 | 12 | 93 | 6 | 12 | 153 | 10 | 12 | 213 | 14 | 12 | 273 | 18 | 12 | 333 | 22 | 12 | 2 | 12 |
| 34 | 2 | 16 | 94 | 6 | 16 | 154 | 10 | 16 | 214 | 14 | 16 | 274 | 18 | 16 | 334 | 22 | 16 | 2 | 16 |
| 35 | 2 | 20 | 95 | 6 | 20 | 155 | 10 | 20 | 215 | 14 | 20 | 275 | 18 | 20 | 335 | 22 | 20 | 2 | 20 |
| 36 | 2 | 24 | 96 | 6 | 24 | 156 | 10 | 24 | 216 | 14 | 24 | 276 | 18 | 24 | 336 | 22 | 24 | 2 | 24 |
| 37 | 2 | 28 | 97 | 6 | 28 | 157 | 10 | 28 | 217 | 14 | 28 | 277 | 18 | 28 | 337 | 22 | 28 | 2 | 28 |
| 38 | 2 | 32 | 98 | 6 | 32 | 158 | 10 | 32 | 218 | 14 | 32 | 278 | 18 | 32 | 338 | 22 | 32 | 2 | 32 |
| 39 | 2 | 36 | 99 | 6 | 36 | 159 | 10 | 36 | 219 | 14 | 36 | 279 | 18 | 36 | 339 | 22 | 36 | 2 | 36 |
| 40 | 2 | 40 | 100 | 6 | 40 | 160 | 10 | 40 | 220 | 14 | 40 | 280 | 18 | 40 | 340 | 22 | 40 | 2 | 40 |
| 41 | 2 | 44 | 101 | 6 | 44 | 161 | 10 | 44 | 221 | 14 | 44 | 281 | 18 | 44 | 341 | 22 | 44 | 2 | 44 |
| 42 | 2 | 48 | 102 | 6 | 48 | 162 | 10 | 48 | 222 | 14 | 48 | 282 | 18 | 48 | 342 | 22 | 48 | 2 | 48 |
| 43 | 2 | 52 | 103 | 6 | 52 | 163 | 10 | 52 | 223 | 14 | 52 | 283 | 18 | 52 | 343 | 22 | 52 | 2 | 52 |
| 44 | 2 | 56 | 104 | 6 | 56 | 164 | 10 | 56 | 224 | 14 | 56 | 284 | 18 | 56 | 344 | 22 | 56 | 2 | 56 |
| 45 | 2 | 0 | 105 | 7 | 0 | 165 | 11 | 0 | 225 | 15 | 0 | 285 | 19 | 0 | 345 | 23 | 0 | 2 | 0 |
| 46 | 2 | 4 | 106 | 7 | 4 | 166 | 11 | 4 | 226 | 15 | 4 | 286 | 19 | 4 | 346 | 23 | 4 | 2 | 4 |
| 47 | 2 | 8 | 107 | 7 | 8 | 167 | 11 | 8 | 227 | 15 | 8 | 287 | 19 | 8 | 347 | 23 | 8 | 2 | 8 |
| 48 | 2 | 12 | 108 | 7 | 12 | 168 | 11 | 12 | 228 | 15 | 12 | 288 | 19 | 12 | 348 | 23 | 12 | 2 | 12 |
| 49 | 2 | 16 | 109 | 7 | 16 | 169 | 11 | 16 | 229 | 15 | 16 | 289 | 19 | 16 | 349 | 23 | 16 | 2 | 16 |
| 50 | 3 | 20 | 110 | 7 | 20 | 170 | 11 | 20 | 230 | 15 | 20 | 290 | 19 | 20 | 350 | 23 | 20 | 3 | 20 |
| 51 | 3 | 24 | 111 | 7 | 24 | 171 | 11 | 24 | 231 | 15 | 24 | 291 | 19 | 24 | 351 | 23 | 24 | 3 | 24 |
| 52 | 3 | 28 | 112 | 7 | 28 | 172 | 11 | 28 | 232 | 15 | 28 | 292 | 19 | 28 | 352 | 23 | 28 | 3 | 28 |
| 53 | 3 | 32 | 113 | 7 | 32 | 173 | 11 | 32 | 233 | 15 | 32 | 293 | 19 | 32 | 353 | 23 | 32 | 3 | 32 |
| 54 | 3 | 36 | 114 | 7 | 36 | 174 | 11 | 36 | 234 | 15 | 36 | 294 | 19 | 36 | 354 | 23 | 36 | 3 | 36 |
| 55 | 3 | 40 | 115 | 7 | 40 | 175 | 11 | 40 | 235 | 15 | 40 | 295 | 19 | 40 | 355 | 23 | 40 | 3 | 40 |
| 56 | 3 | 44 | 116 | 7 | 44 | 176 | 11 | 44 | 236 | 15 | 44 | 296 | 19 | 44 | 356 | 23 | 44 | 3 | 44 |
| 57 | 3 | 48 | 117 | 7 | 48 | 177 | 11 | 48 | 237 | 15 | 48 | 297 | 19 | 48 | 357 | 23 | 48 | 3 | 48 |
| 58 | 3 | 52 | 118 | 7 | 52 | 178 | 11 | 52 | 238 | 15 | 52 | 298 | 19 | 52 | 358 | 23 | 52 | 3 | 52 |
| 59 | 3 | 56 | 119 | 7 | 56 | 179 | 11 | 56 | 239 | 15 | 56 | 299 | 19 | 56 | 359 | 23 | 56 | 3 | 56 |

The above table is for converting expressions in arc to their equivalent in time; its main use in this almanac is for conversion of longitude to LMT (added if west, subtracted if east) to give GMT, or vice versa, particularly in the case of sunrise or sunset.

**Figure 1-17.** *Air almanac conversion of arc to time.*

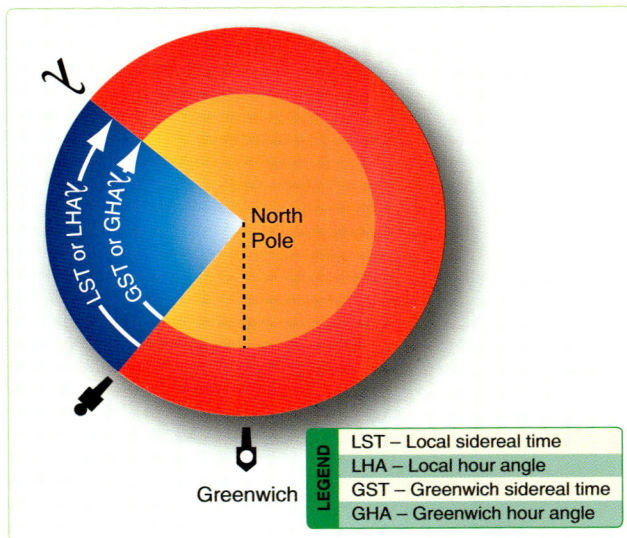

**Figure 1-18.** *Greenwich sidereal time.*

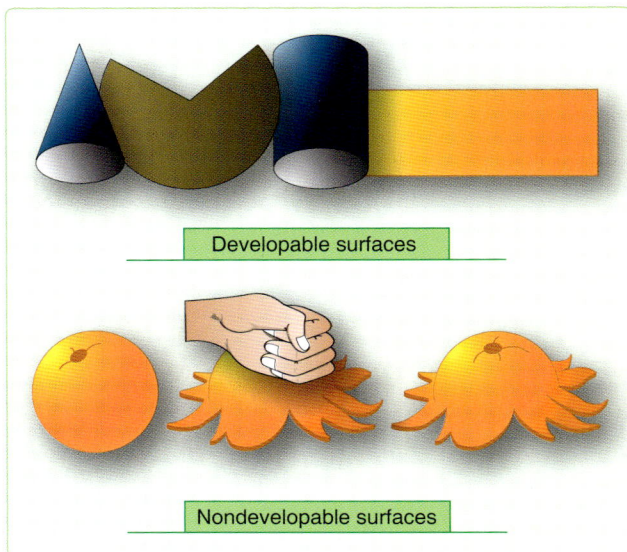

**Figure 1-19.** *Developable and nondevelopable surfaces.*

nearly as possible. The methods of projection are either mathematical or perspective.

b. The perspective or geometric projection consists of projecting a coordinate system based on the earth-sphere from a given point directly onto a developable surface. The properties and appearance of the resultant map depends upon two factors: the type of developable surface and the position of the point of projection.

4. The mathematical projection is derived analytically to provide certain properties or characteristics that cannot be arrived at geometrically. Consider some of the choices available for selecting projections that best accommodate these properties and characteristics.

## Choice of Projection

The ideal chart projection would portray the features of the earth in their true relationship to each other; that is, directions would be true and distances would be represented at a constant scale over the entire chart. This would result in equality of area and true shape throughout the chart. Such a relationship can only be represented on a globe. On a flat chart, it is impossible to preserve constant scale and true direction in all directions at all points, nor can both relative size and shape of the geographic features be accurately portrayed throughout the chart. The characteristics most commonly desired in a chart projection are conformality, constant scale, great circles as straight lines, rhumb lines as straight lines, true azimuth, and geographic position easily located.

## Conformality

Conformality is very important for air navigation charts. For any projection to be conformal, the scale at any point must be independent of azimuth. This does not imply, however, that the scales at two points at different latitudes are equal. It means the scale at any given point is, for a short distance, equal in all directions. For conformality, the outline of areas on the chart must conform in shape to the feature being portrayed. This condition applies only to small and relatively small areas; large land masses must necessarily reflect any distortion inherent in the projection. Finally, since the meridians and parallels of earth intersect at right angles, the longitude and latitude lines on all conformal projections must exhibit this same perpendicularity. This characteristic facilitates the plotting of points by geographic coordinates.

## Constant Scale

The property of constant scale throughout the entire chart is highly desirable, but impossible to obtain as it would require the scale to be the same at all points and in all directions throughout the chart.

## Straight Line

The rhumb line and the great circle are the two curves that a navigator might wish to have represented on a map as straight lines. The only projection that shows all rhumb lines as straight lines is the Mercator. The only projection that shows all great circles as straight lines is the gnomonic projection. However, this is not a conformal projection and cannot be used directly for obtaining direction or distance. No conformal chart represents all great circles as straight lines.

## True Azimuth

It would be extremely desirable to have a projection that showed directions or azimuths as true throughout the chart. This would be particularly important to the navigator, who must determine from the chart the heading to be flown. There is no chart projection representing true great circle direction along a straight line from all points to all other points.

## Coordinates Easy to Locate

The geographic latitudes and longitudes of places should be easily found or plotted on the map when the latitudes and longitudes are known.

## Chart Projections

Chart projections may be classified in many ways. In this book, the various projections are divided into three classes according to the type of developable surface to which the projections are related. These classes are azimuthal, cylindrical, and conical.

### *Azimuthal Projections*

An azimuthal, or zenithal projection, is one in which points on the earth are transferred directly to a plane tangent to the earth. According to the positioning of the plane and the point of projection, various geometric projections may be derived. If the origin of the projecting rays (point of projection) is the center of the sphere, a gnomonic projection results. If it is located on the surface of the earth opposite the point of the tangent plane, the projection is a stereographic, and if it is at infinity, an orthographic projection results. *Figure 1-20* shows these various points of projection.

### *Gnomonic Projection*

All gnomonic projections are direct perspective projections. Since the plane of every great circle cuts through the center of the sphere, the point of projection is in the plane of every great circle. This property then becomes the most important and useful characteristic of the gnomonic projection. Each and every great circle is represented by a straight line on the projection. A complete hemisphere cannot be projected onto this plane because points 90° from the center of the map project lines parallel to the plane of projection. Because the gnomonic is nonconformal, shapes or land masses are distorted, and measured angles are not true. At only one point, the center of the projection, are the azimuths of lines true. At this point, the projection is said to be azimuthal. Gnomonic projections are classified according to the point of tangency of the plane of projection. A gnomonic projection is polar gnomonic when the point of tangency is one of the poles,

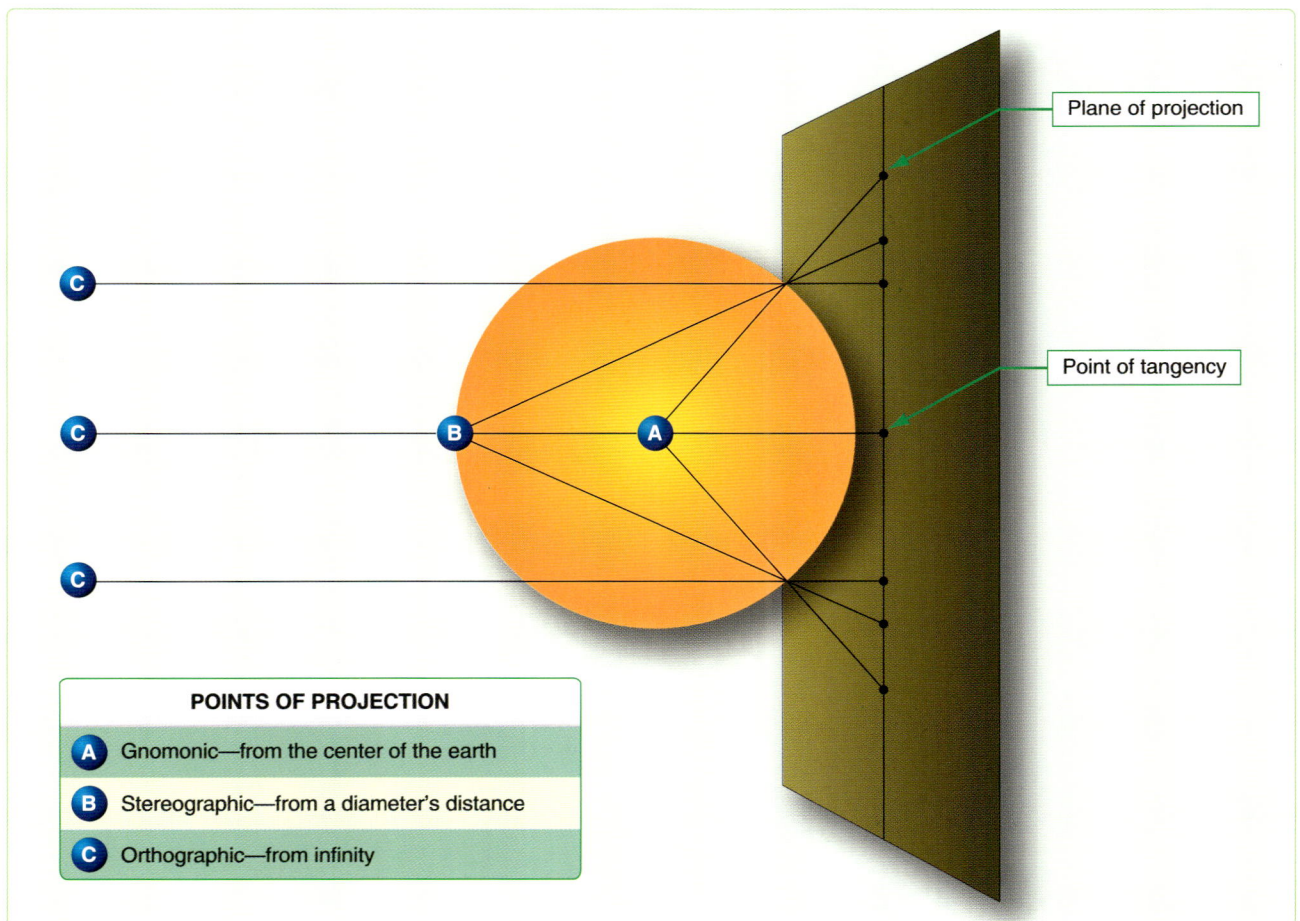

Figure 1-20. *Azimuthal projections.*

equatorial gnomonic when the point of tangency is at the equator and any selected meridian. *[Figure 1-21]*

## *Stereographic Projection*

The stereographic projection is a perspective conformal projection of the sphere. The term oblique stereographic is applied to any stereographic projection where the center of the projection is positioned at any point other than the geographic poles or the equator. If the center is coincident with one of the poles of the reference surface, the projection is called polar stereographic. The illustration in *Figure 1-21* shows both gnomonic and stereographic projections. If the center lies on the equator, the primitive circle is a meridian, which gives the name meridian stereographic or equatorial stereographic.

## *Cylindrical Projections*

The only cylindrical projection used for navigation is the Mercator, named after its originator, Gerhard Mercator (Kramer), who first devised this type of chart in the year 1569. The Mercator is the only projection ever constructed that is conformal and, at the same time, displays the rhumb line as a straight line. It is used for navigation, for nearly all atlases (a word coined by Mercator), and for many wall maps.

Imagine a cylinder tangent to the equator, with the source of projection at the center of the earth. It would appear much like the illustration in *Figure 1-22,* with the meridians being straight lines and the parallels being unequally spaced circles around the cylinder. It is obvious from *Figure 1-22* that those parts of the terrestrial surface close to the poles could not be projected unless the cylinder was tremendously long, and the poles could not be projected at all.

On the earth, the parallels of latitude are perpendicular to the meridians, forming circles of progressively smaller diameters as the latitude increases. On the cylinder, the parallels of latitude are shown perpendicular to the projected meridians but, since the diameter of a cylinder is the same at any point along the longitudinal axis, the projected parallels are all the same length. If the cylinder is cut along a vertical line (a meridian) and spread flat, the meridians appear as equally spaced, vertical lines, and the parallels as horizontal lines, with distance between the horizontal lines increasing with distance away from the false (arbitrary) meridian.

The cylinder may be tangent at some great circle other than the equator, forming other types of cylindrical projections. If the cylinder is tangent at a meridian, it is a transverse cylindrical projection; if it is tangent at any point other than the equator or a meridian, it is called an oblique cylindrical projection. The patterns of latitude and longitude appear quite different on these projections because the line of tangency and the equator no longer coincide.

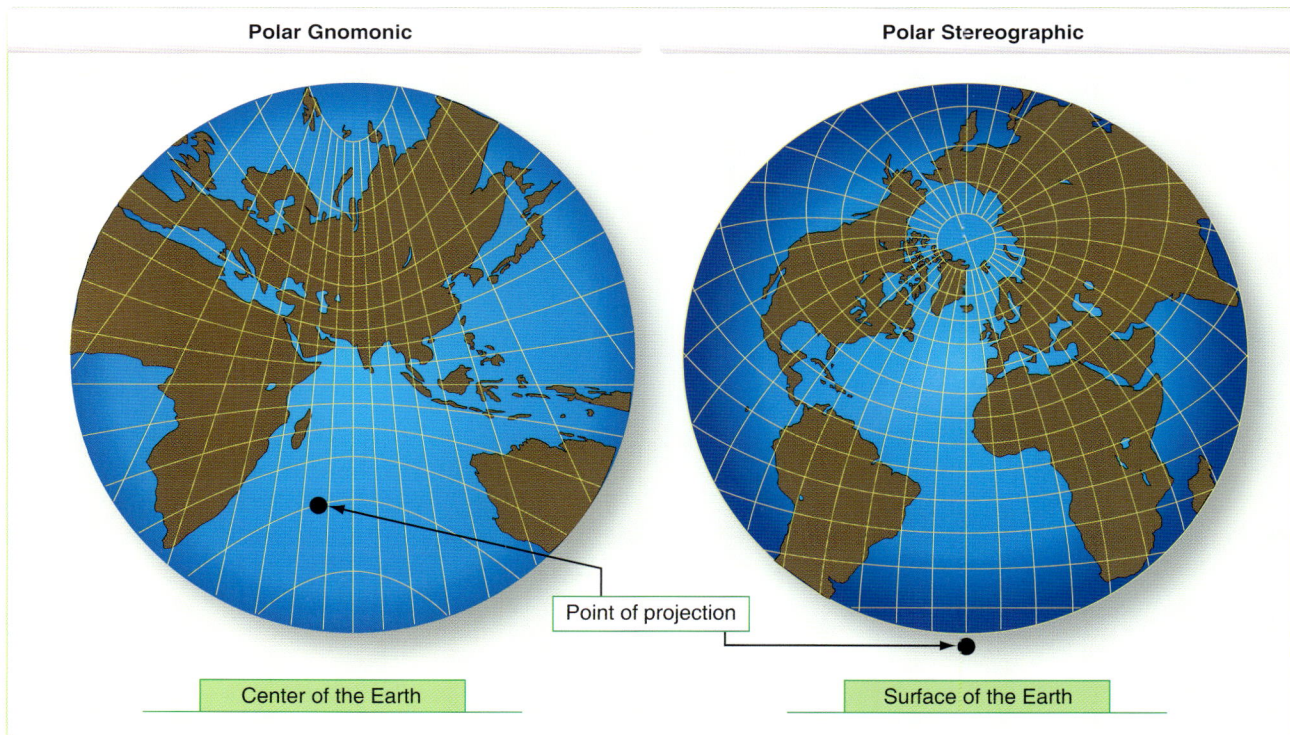

**Figure 1-21.** *Polar gnomonic and stereographic projections.*

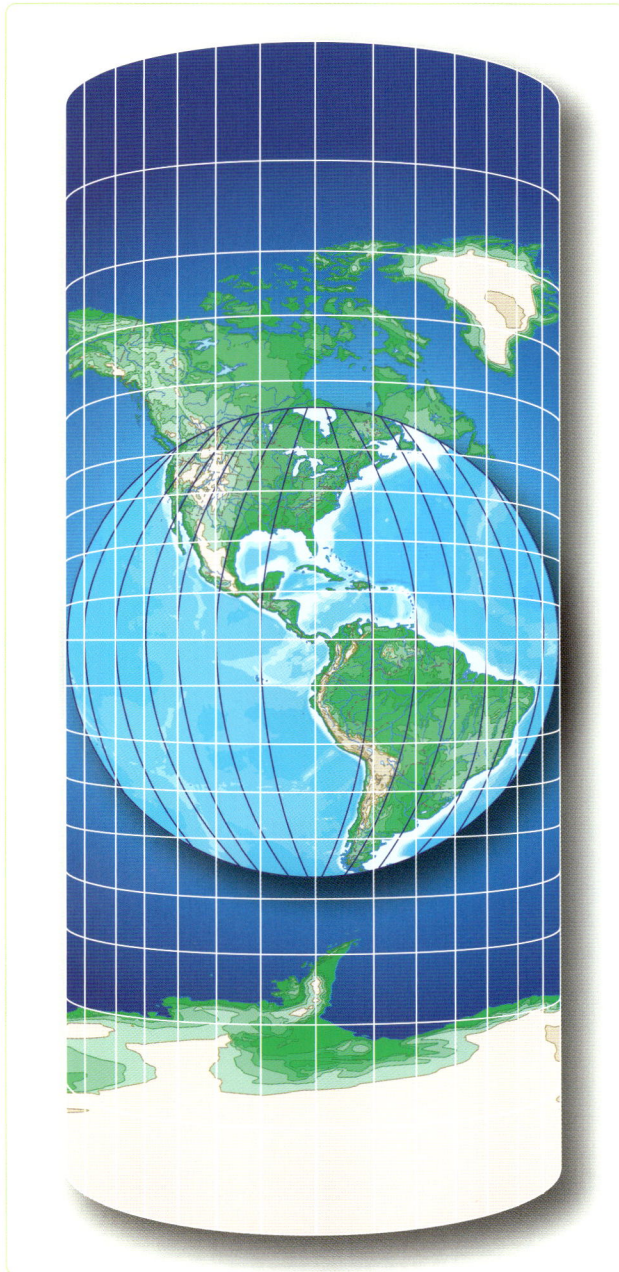

**Figure 1-22.** *Cylindrical projection.*

## *Mercator Projection*

The Mercator projection is a conformal, nonperspective projection; it is constructed by means of a mathematical transformation and cannot be obtained directly by graphical means. The distinguishing feature of the Mercator projection among cylindrical projections is: At any latitude the ratio of expansion of both meridians and parallels is the same, thus, preserving the relationship existing on the earth. This expansion is equal to the secant of the latitude, with a small correction for the ellipticity of the earth. Since expansion

is the same in all directions and since all directions and all angles are correctly represented, the projection is conformal.

Rhumb lines appear as straight lines and their directions can be measured directly on the chart. Distance can also be measured directly, but not by a single distance scale on the entire chart, unless the spread of latitude is small. Great circles appear as curved lines, concave to the equator or convex to the nearest pole. The shapes of small areas are very nearly correct, but are of increased size unless they are near the equator. *[Figure 1-23]* The Mercator projection has the following disadvantages:

1. Measuring large distances accurately is difficult.

2. Must apply conversion angle to great circle bearing before plotting.

3. Is useless above 80° N or below 80° S since the poles cannot be shown.

The transverse or inverse Mercator is a conformal map designed for areas not covered by the equatorial Mercator.

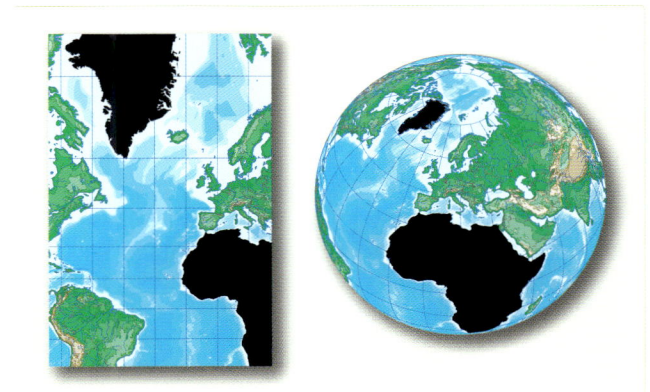

**Figure 1-23.** *Mercator is conformal but not equal area.*

With the transverse Mercator, the property of straight meridians and parallels is lost, and the rhumb line is no longer represented by a straight line. The parallels and meridians become complex curves and, with geographic reference, the transverse Mercator is difficult to use as a plotting chart. The transverse Mercator, though often considered analogous to a projection onto a cylinder, is in reality a nonperspective projection, constructed mathematically. This analogy, however, does permit the reader to visualize that the transverse Mercator shows scale correctly along the central meridian, which forms the great circle of tangency. *[Figure 1-24]* In effect, the cylinder has been turned 90° from its position for the ordinary Mercator, and a meridian, called the central meridian, becomes the tangential great circle. One series of NGA charts using this type of projection places the cylinder tangent to the 90° E–90° W longitude.

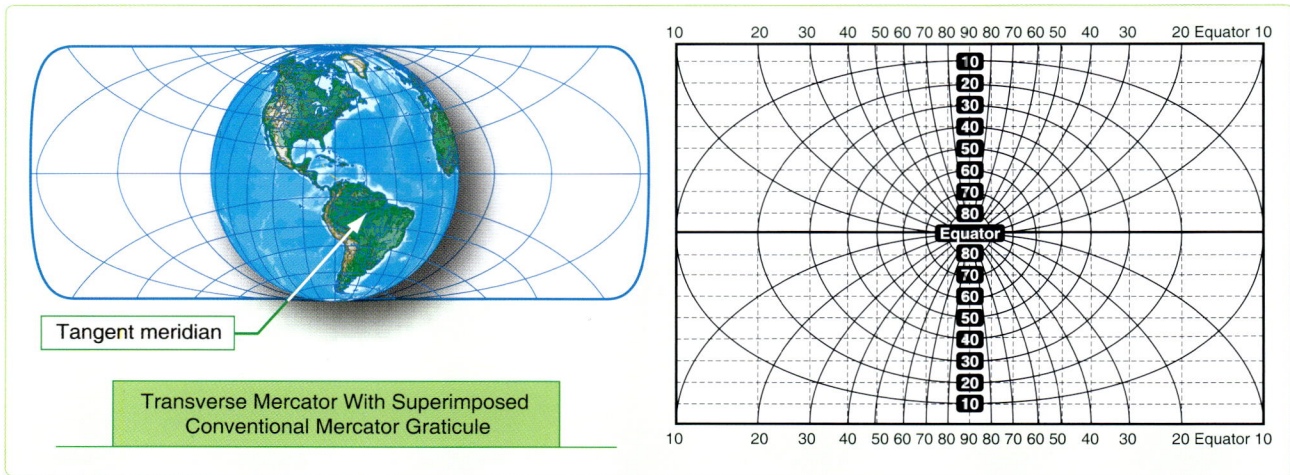

**Figure 1-24.** *Transverse cylindrical projection—cylinder tangent at the poles.*

These projections use a fictitious graticule similar to, but offset from, the familiar network of meridians and parallels. The tangent great circle is the fictitious equator. Ninety degrees from it are two fictitious poles. A group of great circles through these poles and perpendicular to the tangent constitutes the fictitious meridians, while a series of lines parallel to the plane of the tangent great circle forms the fictitious parallels.

On these projections, the fictitious graticule appears as the geographical one ordinarily appearing on the equatorial Mercator. That is, the fictitious meridians and parallels are straight lines perpendicular to each other. The actual meridians and parallels appear as curved lines, except the line of tangency. Geographical coordinates are usually expressed in terms of the conventional graticule. A straight line on the transverse Mercator projection makes the same angle with all fictitious meridians, but not with the terrestrial meridians. It is, therefore, a fictitious rhumb line.

The appearance of a transverse Mercator using the 90° E–90° W meridian as a reference or fictitious equator is shown in *Figure 1-24*. The dotted lines are the lines of the fictitious projection. The N–S meridian through the center is the fictitious equator, and all other original meridians are now curves concave on the N–S meridian with the original parallels now being curves concave to the nearer pole. To straighten the meridians, use the graph in *Figure 1-25* to extract a correction factor that mathematically straightens the longitudes.

## Conic Projections

There are two classes of conic projections. The first is a simple conic projection constructed by placing the apex of the cone over some part of the earth (usually the pole) with the cone tangent to a parallel called the standard parallel and projecting the graticule of the reduced earth onto the cone.

*[Figure 1-26]* The chart is obtained by cutting the cone along some meridian and unrolling it to form a flat surface. Notice, in *Figure 1-27,* the characteristic gap appears when the cone is unrolled. The second is a secant cone, cutting through the earth and actually contacting the surface at two standard parallels as shown in *Figure 1-28*.

## *Lambert Conformal (Secant Cone)*

The Lambert conformal conic projection is of the conical type in which the meridians are straight lines that meet at a common point beyond the limits of the chart and parallels are concentric circles, the center of each being the point of intersection of the meridians. Meridians and parallels intersect at right angles. Angles formed by any two lines or curves on the earth's surface are correctly represented. The projection may be developed by either the graphic or mathematical method. It employs a secant cone intersecting the spheroid at two parallels of latitude, called the standard parallels, of the area to be represented. The standard parallels are represented at exact scale. Between these parallels, the scale factor is less than unity and, beyond them, greater than unity. For equal distribution of scale error (within and beyond the standard parallels), the standard parallels are selected at one-sixth and five-sixths of the total length of the segment of the central meridian represented. The development of the Lambert conformal conic projection is shown by *Figure 1-29*.

The chief use of the Lambert conformal conic projection is in mapping areas of small latitudinal width but great longitudinal extent. No projection can be both conformal and equal area but, by limiting latitudinal width, scale error is decreased to the extent the projection gives very nearly an equal area representation in addition to the inherent quality of conformality. This makes the projection very useful for aeronautical charts. Some of the advantages of the Lambert conformal conic projection are:

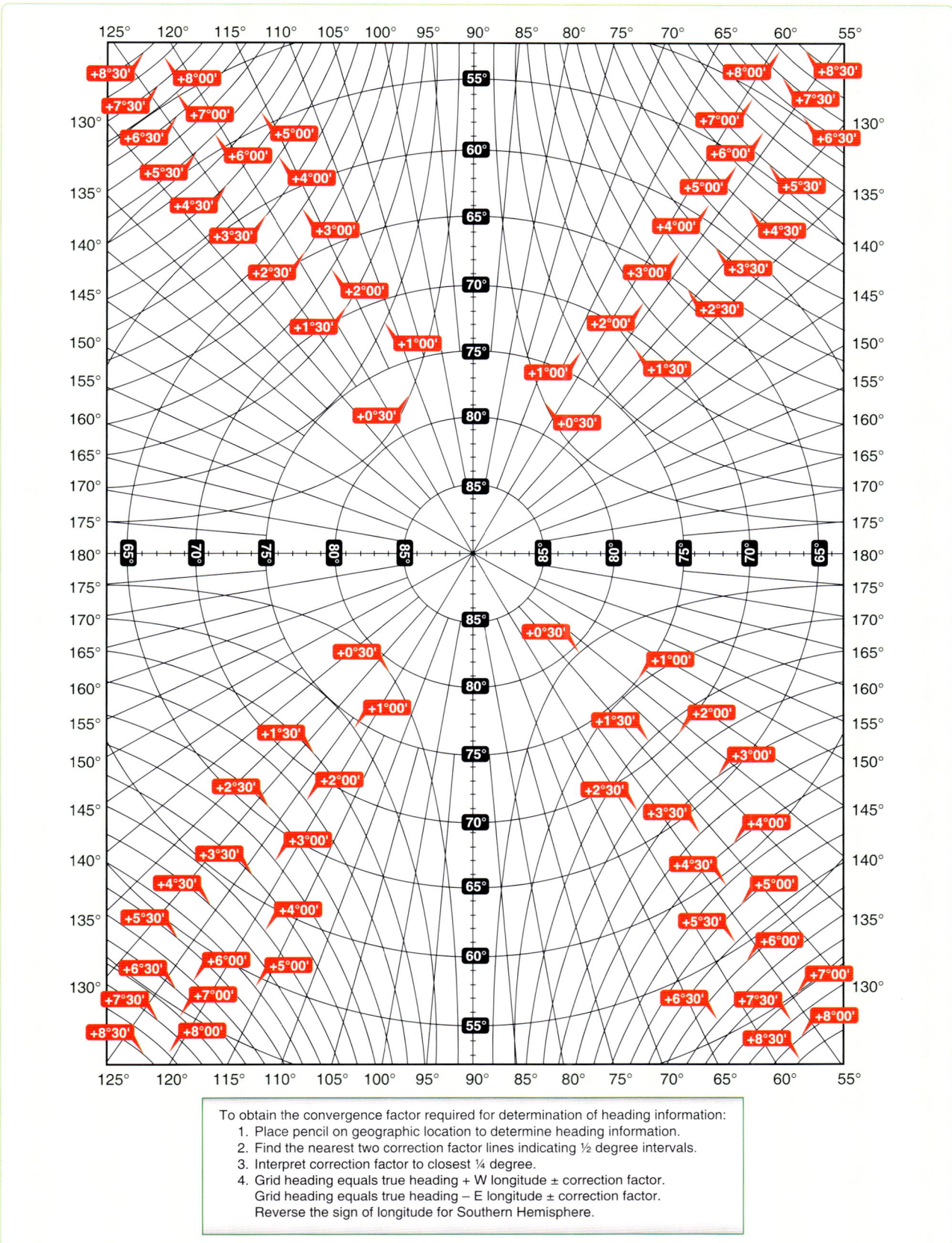

**Figure 1-25.** *Transverse Mercator convergence graph.*

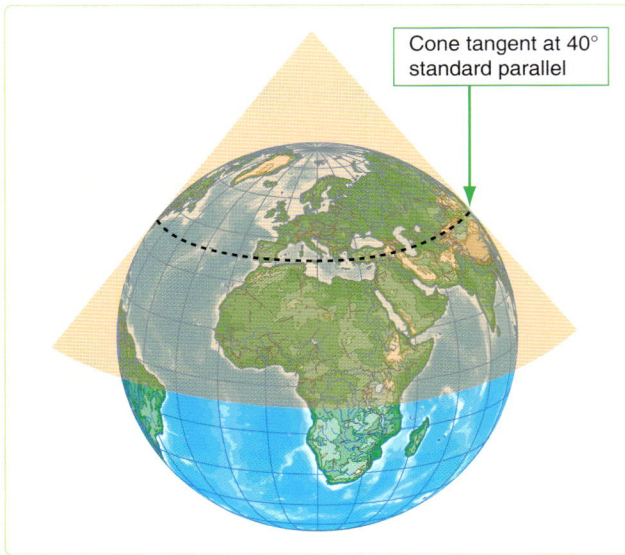

Figure 1-26. *Simple conic projection.*

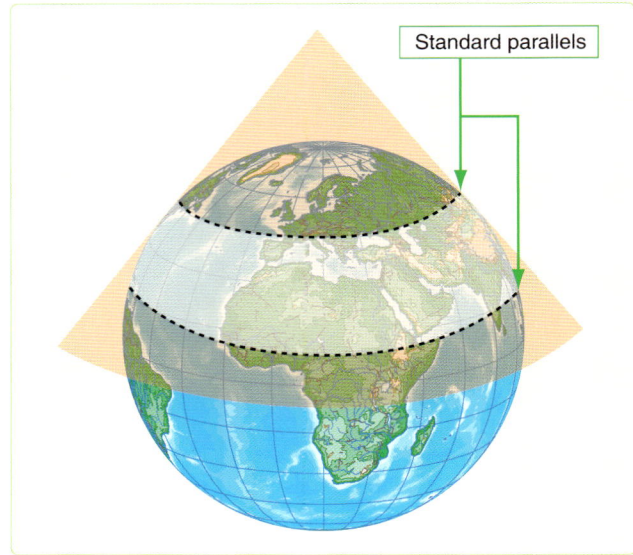

Figure 1-28. *Conic projection using secant cone.*

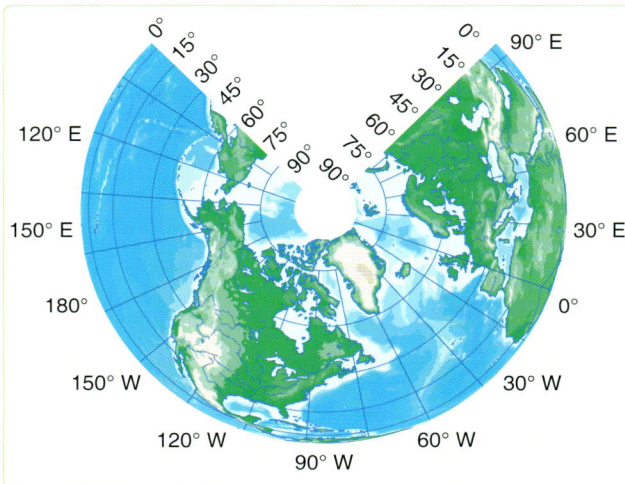

Figure 1-27. *Simple conic projection of northern hemisphere.*

1. Conformality.

2. Great circles are approximated by straight lines (actually concave toward the midparallel).

3. For areas of small latitudinal width, scale is nearly constant. For example, the United States may be mapped with standard parallels at 33° N and 45° N with a scale error of only 2 percent for southern Florida. The maximum scale error between 30°30' N and 47°30' N is only one-half of 1 percent.

4. Positions are easily plotted and read in terms of latitude and longitude. Construction is relatively simple.

5. Its two standard parallels give it two lines of strength (lines along which elements are represented true to shape and scale).

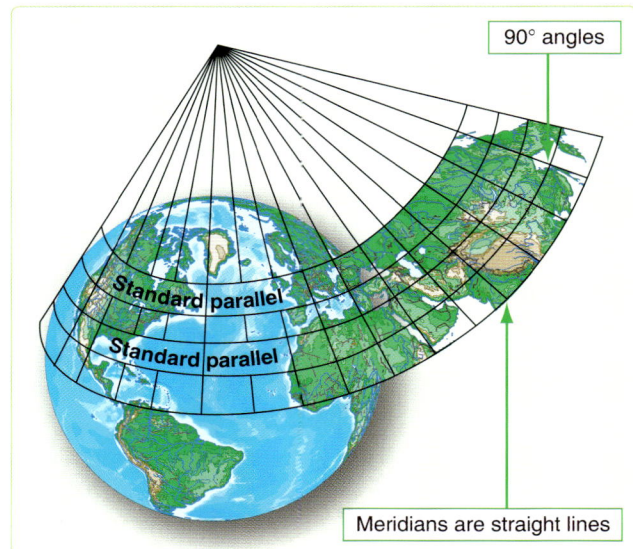

Figure 1-29. *Lambert conformal conic projection.*

6. Distance may be measured quite accurately. For example, the distance from Pittsburgh to Istanbul is 5,277 NM; distance as measured by the graphic scale on a Lambert projection (standard parallels 36° N and 54° N) without application of the scale factor is 5,258 NM; an error of less than 0.4 percent.

Some of the chief limitations of the Lambert Conformal conic projection are:

1. Rhumb lines are curved lines that cannot be plotted accurately.

2. Maximum scale increases as latitudinal width increases.

3.  Parallels are curved lines (arcs of concentric circles).

4.  Continuity of conformality ceases at the junction of two bands, even though each is conformal. If both have the same scale along their standard parallels, the common parallel (junction) has a different radius for each band and does not join perfectly.

## Constant of the Cone

Most conic charts have the constant of the cone (convergence factor) computed and listed on the chart somewhere in the chart margin.

## Convergence Angle (CA)

The convergence angle (CA) is the actual angle on a chart formed by the intersection of the Greenwich meridian and another meridian; the pole serves as the vertex of the angle. CAs, like longitudes, are measured east and west from the Greenwich meridian.

## Convergence Factor (CF)

A chart's convergence factor (CF) is a decimal number that expresses the ratio between meridional convergence as it actually exists on the earth and as it is portrayed on the chart. When the CA equals the number of the selected meridian, the chart CF is 1.0. When the CA is less than the number of the selected meridian, the chart CF is proportionately less than 1.0. The subpolar projection illustrated in *Figure 1-30* portrays the standard parallels, 37° N and 65° N. It presents 360° of the earth's surface on 282.726° of paper. Therefore, the chart has a CF of 0.78535 (282.726° divided by 360° equals 0.78535). Meridian 90° W forms a west CA of 71° with the Greenwich meridian.

Express as a formula:

CF × longitude = CA

0.78535 × 90° W = 71° west CA

Approximate a chart's CF on subpolar charts by drawing a straight line covering 10 lines of longitude and measuring the true course at each end of the line, noting the difference between them, and dividing the difference by 10. NOTE: The quotient represents the chart's CF.

## Aeronautical Charts

An aeronautical chart is a pictorial representation of a portion of the earth's surface upon which lines and symbols in a variety of colors represent features or details seen on the earth's surface. In addition to ground image, many additional symbols and notes are added to indicate navigational aids (NAVAID) and data necessary for air navigation. Properly used, a chart is a vital adjunct to navigation; improperly used, it may become a hazard. Without it, modern navigation would never have reached its present state of development. Because of their great importance, the navigator must be thoroughly familiar with the wide variety of aeronautical charts and understand their many uses.

### Lambert Conformal

Aeronautical charts are produced on many different types of projections. Since the demand for variety in charts is so great and the properties of the projections vary greatly, there is no one projection satisfying all navigation needs. The projection that most nearly answers all of the navigator's problems is the Lambert conformal, and this projection is the one most widely used for aeronautical charts. An aeronautical chart of some projection and scale can be obtained for any portion of the earth.

### Datums

Maps made by a given country traditionally use the datum created by that country. There may be as many as a thousand of these various datums in use throughout the world. Inherent problems result from over a hundred countries using widely different methods and standards to measure coordinate systems. When added to the effects of local variations in topography and the gravity field, systems are created that differ substantially from each other. These individualized datums are classified as local or regional.

The Department of Defense adopted a datum in 1987 called World Geodetic System 84 (WGS 84). This global datum is a system that models the entire planet, instead of one small piece. WGS 84 is used by NGA for production of almost all new maps and charts. The purpose of such a system is to minimize the confusion created by the proliferation of local

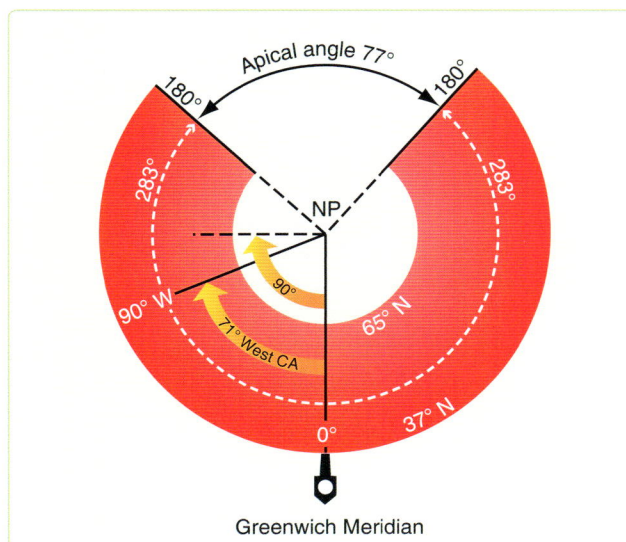

**Figure 1-30.** *A Lambert Conformal, convergence factor 0.78535.*

datums. As long as all coordinates are stated in WGS 84, combat interoperability problems are minimized. In addition, WGS 84 positions may be computed from global positioning system (GPS) equipment to an extreme level of precision by NGA surveyors, well under half a meter anywhere in the world. Widespread use of WGS 84 virtually eliminates problems due to different datums.

It is important to realize that every coordinate is related to a specific datum. A latitude and longitude extracted from a WGS 84 chart is still a WGS 84 coordinate, and an MGRS point pulled from that same chart is also WGS 84. However, a ground survey of that same point could have established a local datum coordinate that is different from the map derived one by as much as a half mile. Always use the same datum throughout a mission, or serious positional errors are possible.

## Scale

Obviously, charts are much smaller than the area they represent. The ratio between any given unit of length on a chart and the true distance it represents on the earth is the scale of the chart. The scale varies, and may vary greatly from one part of the chart to another. Charts are made to various scales for different purposes. If a chart is to show the whole world and yet not be too large, it must be drawn to small scale. If a chart is to show much detail, it must be drawn to a large scale; then it shows a smaller area than does a chart of the same size drawn to a small scale. Remember, large area, small scale; small area, large scale.

## Aeronautical Chart

The scale of a chart may be given by a simple statement, such as 1 inch equals 10 miles. This means a distance of 10 miles on the earth's surface is shown 1 inch long on the chart. On aeronautical charts, the scale is indicated in one of two ways: representative fraction or graphic scale.

## Representative Fraction

The scale may be given as a representative fraction, such as 1:500,000 or 1/500,000. This means one of any unit on the chart represents 500,000 of the same unit on the earth. For example, 1 inch on the chart represents 500,000 inches on the earth. A representative fraction can be converted into a statement of miles to the inch. Thus, if the scale is 1:1,000,000, 1 inch on the chart stands for 1,000,000 inches or 1,000,000 divided by (6,076 × 12) equaling about 13.7 NM. Similarly, if the scale is 1:500,000, 1 inch on the chart represents about 6.86 NM. Thus, the larger the denominator of the representative fraction, the smaller the scale.

## Graphic Scale

The graphic scale may be shown by a graduated line. It usually is found printed along the border of a chart. Take a measurement on the chart and compare it with the graphic scale of miles. The number of miles the measurement represents on the earth may be read directly from the graphic scale on the chart. The distance between parallels of latitude also provides a convenient scale for distance measurement. One degree of latitude always equals 60 NM and 1 minute of latitude always equals 1 NM.

## Types of Charts

Aeronautical charts are differentiated on a functional basis by the type of information they contain. Navigation charts are grouped into three major types: general purpose, special purpose, and plotting. The name of the chart is a reasonable indication of its intended use. A Minimal Flight Planning Chart is primarily used in minimal flight planning techniques; a Jet Navigation Chart has properties making it adaptable to the speed, altitude, and instrumentation of jet aircraft. In addition to the specific type of information contained, charts vary according to the amount of information displayed. Charts designed to facilitate the planning of long distance flights carry less detail than those required for navigation en route. Local charts present great detail.

## Standard Chart Symbols

Symbols are used for easy identification of information portrayed on aeronautical charts. While these symbols may vary slightly between various projections, the amount of variance is slight and, once the basic symbol is understood, variations of it are easy to identify. A chart legend is the key to explaining the meaning of the relief, culture, hydrography, vegetation, and aeronautical symbols. *[Figure 1-31]*

## Relief (Hypsography)

Chart relief shows the physical features related to the differences in elevation of land surface. These include features, such as mountains, hills, plateaus, plains, depressions, etc. Standard symbols and shading techniques are used in relief portrayal on charts; these include contours, spot elevations and variations in tint, and shading to represent shadows.

## Contour Lines

A contour line is a line connecting points of equal elevation. *Figure 1-32* shows the relationship between contour lines and terrain. Notice on steep slopes the contours are close together and on gentle slopes they are farther apart. The interval of the contour lines usually depends upon the scale of the chart and the terrain depicted. In *Figure 1-32*, the contour interval is 1,000 feet. Depression contours are regular contour lines with spurs or ticks added on the down slope side.

## Spot Elevations

Spot elevations are the height of a particular point of terrain above an established datum, usually sea level.

## ELEVATIONS IN FEET

ALL ELEVATION VALUES
(AERONAUTICAL, RELIEF AND
HYDROGRAPHIC) ARE
BASED ON MEAN SEA LEVEL

# LEGEND

## RELIEF PORTRAYAL

Elevations are in feet. HIGHEST TERRAIN
elevation is **14433** feet
located at *39°07' 106°27'W*

## TERRAIN CHARACTERISTIC TINTS

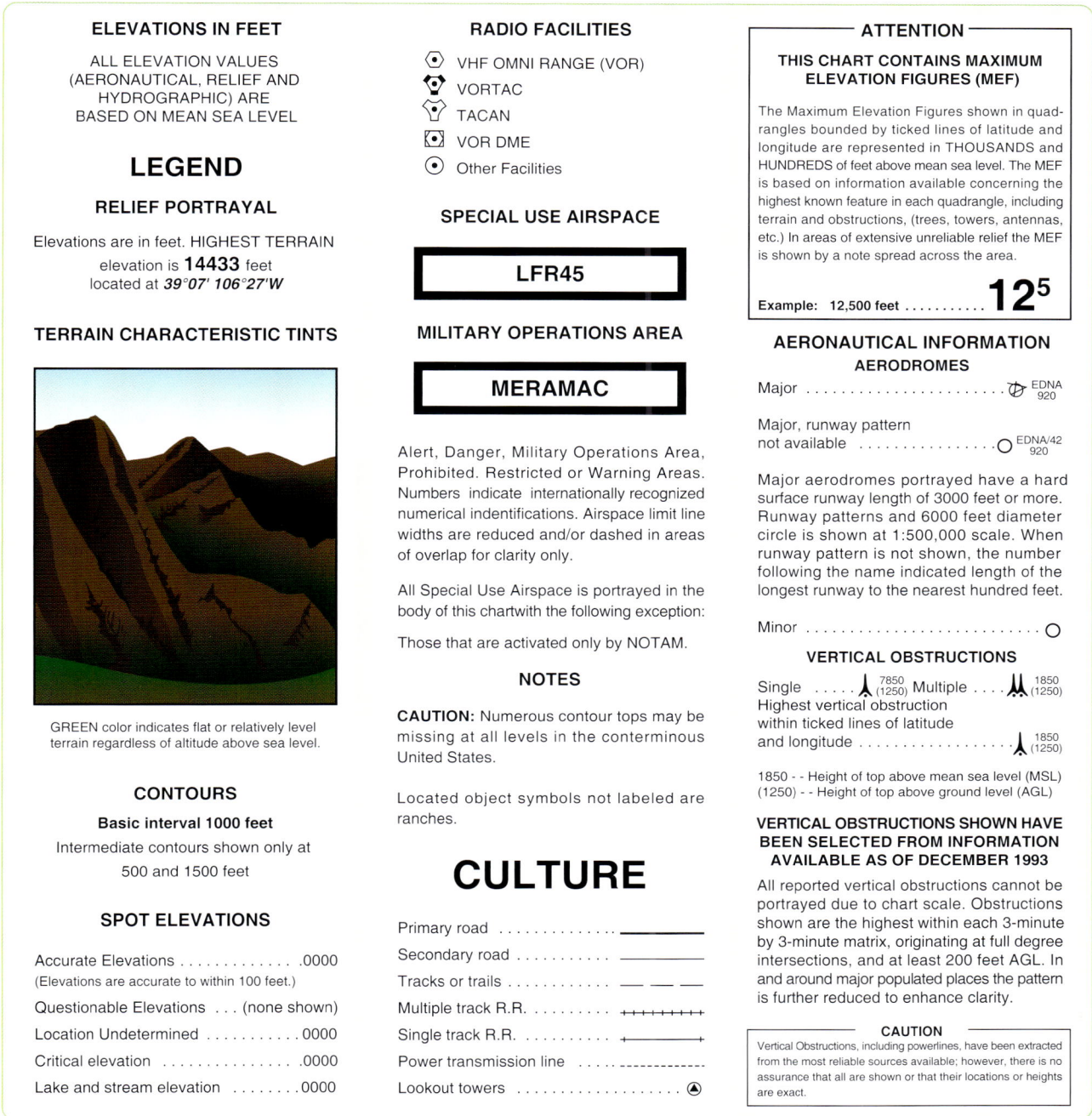

GREEN color indicates flat or relatively level
terrain regardless of altitude above sea level.

## CONTOURS

**Basic interval 1000 feet**

Intermediate contours shown only at
500 and 1500 feet

## SPOT ELEVATIONS

| | |
|---|---|
| Accurate Elevations | .0000 |

(Elevations are accurate to within 100 feet.)

| | |
|---|---|
| Questionable Elevations | (none shown) |
| Location Undetermined | 0000 |
| Critical elevation | 0000 |
| Lake and stream elevation | 0000 |

## RADIO FACILITIES

- ⊙ VHF OMNI RANGE (VOR)
- ⬡ VORTAC
- ⬠ TACAN
- ⊡ VOR DME
- ⊙ Other Facilities

## SPECIAL USE AIRSPACE

**LFR45**

## MILITARY OPERATIONS AREA

**MERAMAC**

Alert, Danger, Military Operations Area,
Prohibited. Restricted or Warning Areas.
Numbers indicate internationally recognized
numerical indentifications. Airspace limit line
widths are reduced and/or dashed in areas
of overlap for clarity only.

All Special Use Airspace is portrayed in the
body of this chartwith the following exception:

Those that are activated only by NOTAM.

## NOTES

**CAUTION:** Numerous contour tops may be
missing at all levels in the conterminous
United States.

Located object symbols not labeled are
ranches.

# CULTURE

| | |
|---|---|
| Primary road | ─────── |
| Secondary road | ─────── |
| Tracks or trails | ── ── ── |
| Multiple track R.R. | ++++++++ |
| Single track R.R. | ──┼──┼── |
| Power transmission line | ───────── |
| Lookout towers | ⬤ |

## AERONAUTICAL INFORMATION
### AERODROMES

| | |
|---|---|
| Major | ⊕ EDNA 920 |
| Major, runway pattern not available | ○ EDNA/42 920 |

Major aerodromes portrayed have a hard
surface runway length of 3000 feet or more.
Runway patterns and 6000 feet diameter
circle is shown at 1:500,000 scale. When
runway pattern is not shown, the number
following the name indicated length of the
longest runway to the nearest hundred feet.

| | |
|---|---|
| Minor | ○ |

### VERTICAL OBSTRUCTIONS

| | |
|---|---|
| Single | ⋏ 7850 (1250) |
| Multiple | ⋔ 1850 (1250) |

Highest vertical obstruction
within ticked lines of latitude
and longitude ⋏ 1850 (1250)

1850 - - Height of top above mean sea level (MSL)
(1250) - - Height of top above ground level (AGL)

**VERTICAL OBSTRUCTIONS SHOWN HAVE
BEEN SELECTED FROM INFORMATION
AVAILABLE AS OF DECEMBER 1993**

All reported vertical obstructions cannot be
portrayed due to chart scale. Obstructions
shown are the highest within each 3-minute
by 3-minute matrix, originating at full degree
intersections, and at least 200 feet AGL. In
and around major populated places the pattern
is further reduced to enhance clarity.

### CAUTION

**Figure 1-31.** *Sample chart legend.*

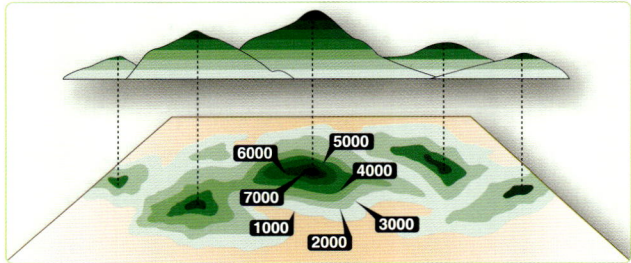

**Figure 1-32.** *Contour lines.*

## Gradient Tints

The relief indicating contours is further emphasized on charts by a system of gradient tints. They are used to designate areas within certain elevation ranges by different color tints.

## Shading

Perhaps the most obvious portrayal of relief is supplied by graduated shading applied to the southeastern side of elevated terrain and the northwestern side of depressions. This shading simulates the shadows cast by elevated features, lending a sharply defined, three-dimensional effect.

## Cultural Features

All structural developments appearing on the terrain are known as cultural features. Three main factors govern the amount of detail given to cultural features: the scale of the chart, the use of the chart, and the geographical area covered. Populated places, roads, railroads, installations, dams, bridges, and mines are some of the many kinds of cultural features portrayed on aeronautical charts. The true representative size and shape of larger cities and towns are shown. Standardized coded symbols and type sizes are used to represent smaller population centers. Some symbols denoting cultural features are usually keyed in a chart legend. However, some charts use pictorial symbols which are self-explanatory.

## Hydrography

In this category, aeronautical charts depict oceans, coast lines, lakes, rivers, streams, swamps, reefs, and numerous other hydrographic features. Open water may be portrayed by tinting or vignetting, or may be left blank.

## Vegetation

Vegetation is not shown on most small scale charts. Forests and wooded areas in certain parts of the world are portrayed on some medium scale charts. On some large scale charts, park areas, orchards, hedgerows, and vineyards are shown. Portrayal may be by solid tint, vignette, or supplemented vignette.

## Aeronautical Information

In the aeronautical category, coded chart symbols denote airfields, radio aids to navigation, commercial broadcasting stations, Air Defense Identification Zones (ADIZ), compulsory corridors, restricted airspace, warning notes, lines of magnetic variation, and special navigational grids. Some aeronautical information is subject to frequent change.

For economy of production, charts are retained in stock for various periods of time. To keep the charts current, only the stable kinds of information are printed on navigation charts.

NGA produces and distributes all aeronautical charts and Flight Information Publication (FLIP) documents. A summary of the typical charts is in *Figure 1-33*. Requisitions should indicate item identification and terminology for each item requested as listed in the catalog. List aeronautical charts by series in numerical or alphabetical sequence; list FLIP documents by type (en route, planning, terminal), title, and geographic area of coverage. Contact NGA or its squadrons and detachments for technical assistance in preparing statements of requirements. Addresses are listed in the NGA Catalog of Maps, Charts, and Related Products.

## Chapter Summary

This chapter discussed the basic terminology used for reading maps and charts which are used throughout the book. An introduction to map reading is explained by defining concepts, such as the shape and size of the earth, longitude, latitude, and rhumb line. Other major topics discussed in the chapter include time, charts, and projections. Aeronautical charts are explained in detail discussing the different types, the features of each, and how to acquire the necessary maps needed for specific flights.

| Name and Code | Projection | Scale and Coverage | Description | Purpose |
|---|---|---|---|---|
| USAF Jet Navigation Charts (JNG) | Charts 4, 5, 6, &7 transverse Mercator. All others, Lambert. | 1:2,000,000 Northern Hemisphere | Charts show all pertinent hydrographic and cultural features. Complete transportation network in areas surrounding cities. A maximum of radar-significant detail as required for long-range navigation, and a suitable aeronautical information overprint, including runway pattern. | For preflight planning and enroute navigation by long-range, high-speed aircraft; for radar, celestial, and grid navigation. This series can also be used for navigation at medium speeds and altitudes. |
| Operational Navigation Charts (ONG) | 0°–80° Lambert conformal conic 80°–90° polar stereographic | 1:1,000,000 Worldwide | Charts show all types of hydrographic and cultural features. All important NAVAIDs and air facilities are included. | Standard series of aeronautical charts designed for military low-altitude navigation. Also used for planning intelligence briefings, plotting, and wall displays. |
| Tactical Pictage Charts (TPC) | 0°–80° Lambert conformal conic 80°–90° polar stereographic | 1:500,000 Worldwide | Charts show all types of hydrographic and cultural features. All important NAVAIDs and air facilities are included. | To provide charts with detailed ground features significant to visual and radar low level navigation for immediate ground or chart orientation at predetermined checkpoints. |
| Joint Operations Graphic Air Series 1501AIR | Latitude 84°N to 80°S. Transverse Mercator latitude 84° to 90°N and 80° to 90°S, Polar Stereographic | 1:250,000 Worldwide | Charts show all types of hydrographic and cultural features. All important NAVAIDs and air facilities are included. | To provide Army, Navy, and Air Force with a common large-scale graphic. Used for tactical air operations, close air support and interdiction at low altitudes. Used for preflight planning and inflight navigation for short-range flights using DR and visual pilotage. Also used for operational planning and intelligence briefing. |
| Air Navigation Charts (V 30) | Mercator, except Antarctic-Polar Stereographic | 1:2,188,800 Worldwide with relating arctic coverage | Charts contain essential topographical, hydrographical, and aeronautical information. Primary roads and railroads are shown. | Basic long-range air navigation plotting series designed primarily for use in larger aircraft. |
| Special Navigation Charts (VS) | Mercator | 1:750,000 to 1:4,000,000 Special Areas | Charts show essential hydrographic and cultural features. Roads and major railroads are shown. Important air facilities included. | Designed for general air navigation. |
| Gnomonic Tracking Charts (GT) | Gnomonic | Various scales from North Pole to about 40° South latitude | Majority are two-color outline charts with blue for graticule and buff for all land areas. Majority of charts are overprinted with special airways and air communication services. | Series of plotting charts suitable for accurate tracking of aircraft by electronic devices and small scale plotting charts for accurate great circle courses. |
| USN/USAF Plotting Sheets (VP 30) | Mercator | 1:2,187,400 70° North to 70° South latitude | Projection graticule only. | To provide uncluttered universal plotting sheets of suitable scope for long-range DR and celestial navigation for selected bands of latitudes. |
| USAF Global Navigation and Planning Charts (GNC) | Polar regions Transverse Mercator Lower latitudes Lambert conformal conic | 1:5,000,000 Worldwide | Show spot evaluations, major cities, roads, principal hydrography and stable aeronautical information. | Long distance operational planning. Also suitable for long-range inflight navigation at high altitudes and speeds. |

**Figure 1-33.** *Summary of typical charts.*

## Chapter 2

# Flight Planning

## Introduction

Before boarding an aircraft, a navigator must thoroughly plan the flight. A well-planned flight provides a professional atmosphere, enhancing safety and accomplishment of flight objectives. Also, adequate flight planning prior to flight can avoid unnecessary in-flight problems. This chapter describes the navigator's flight planning. It begins with a discussion of air traffic control (ATC) systems, followed by a brief description of publications, with which a professional navigator should be familiar. All phases of ground planning are discussed, from chart selection to arrival study.

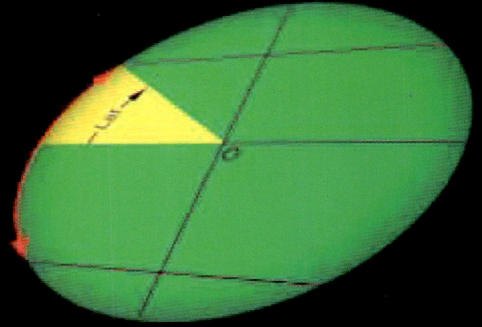

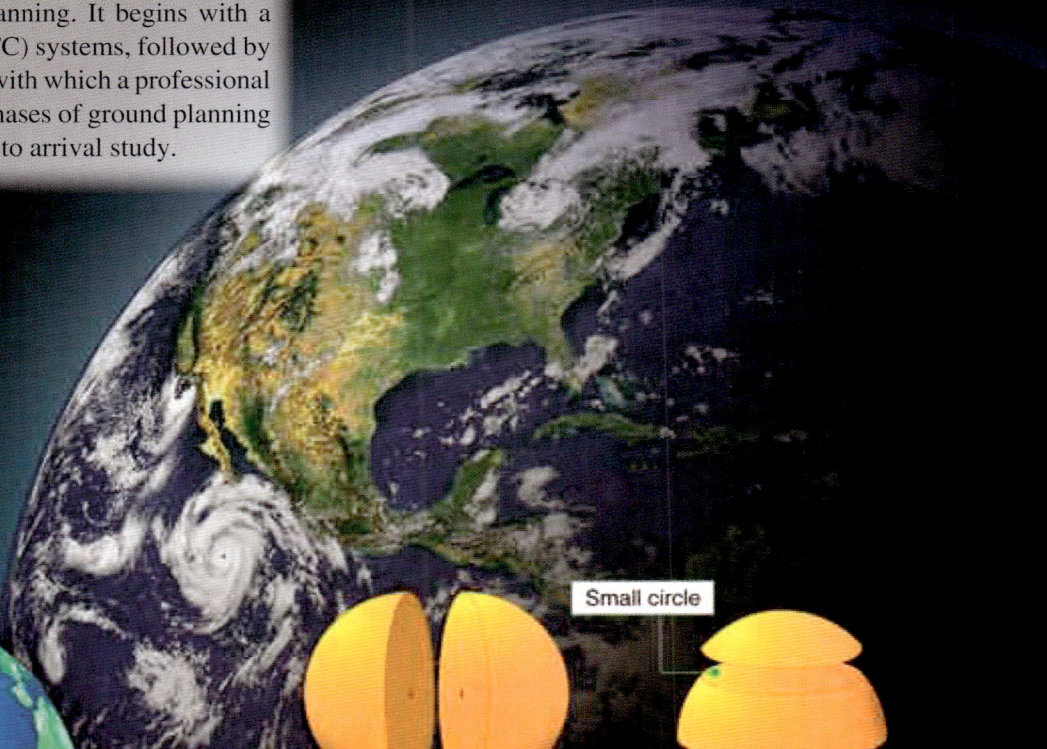

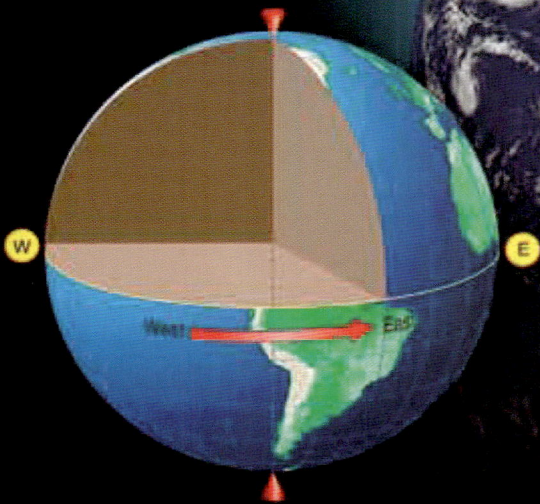

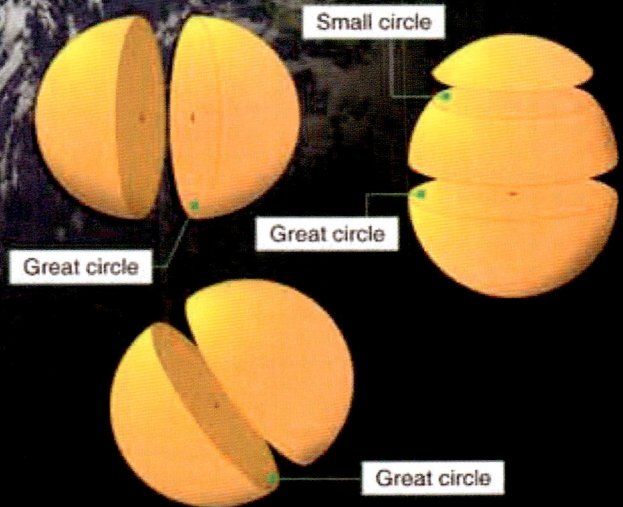

Small circle

Great circle

Great circle

Great circle

San Francisco
Reno
Elko
Rock Springs
Salt Lake City
Rawlins
Cheyenne
North Platte
Omaha
Iowa City
Chicago
Bryan
Cleveland
Bellefonte
New York

# Air Traffic Control (ATC) System

Most nations of the world today have established airspace, air traffic units, and air traffic services to promote a safe, orderly, and expeditious flow of traffic. Furthermore, in the interest of standardization, many nations are establishing systems according to the standards and recommended practices adopted by the International Civil Aviation Organization (ICAO). Navigators must understand what these air traffic services are and how they can be used.

- Air Traffic Service—a general term referring to any of the following services:

    - Air Traffic Control (ATC)—a service provided by ground agencies to prevent collisions and to expedite and maintain an orderly flow of traffic. ATC includes such services as area and en route control, approach control, and tower control. It is used primarily under instrument flight rules (IFR).

    - Advisory Service—provided to give air information that is useful for the safe and effective conduct of flight. This service is usually associated with the visual flight rules (VFR) environment and includes such services as weather conditions, location of known traffic, status of navigational aids (NAVAID), and status of airports and facilities.

    - Alerting Service—a service provided to notify applicable organizations regarding aircraft in need of search and rescue aid and to assist such organizations as required.

    - Airspace—when it has been determined that air traffic services are to be provided, portions of the airspace are designed in relation to the air traffic services that are required. Consult Flight Information Publication (FLIP) for an in-depth explanation of airspace.

    - Air Traffic Service Units—provide the air traffic service within defined airspace.

    - Air Route Traffic Control Centers (ARTCC)— provides ATC to IFR flights within controlled airspace.

    - Approach Control—provides ATC to aircraft arriving at or departing from one or more airports.

    - Airport Control Tower—provides ATC service for airport traffic.

    - Flight Service Station (FSS)—operated by the Federal Aviation Administration (FAA) to provide flight assistance service.

    - International Civil Aviation Organization (ICAO)—establishes international rules for ATC,

the ICAO was formed in April 1947. ICAO is affiliated with the United Nations as a specialized international body dealing with aviation matters. The member states (refer to flight information publications (FLIP) General Planning (GP)) of the ICAO subscribe to ICAO rules and procedures. These rules and procedures are used except for national deviations, which are usually filed with ICAO. Since standardization in ICAO is based upon the same technical principles and policies which are in actual effect in the continental United States (CONUS), American airmen can fly all major routes following the same general rules of the air, and governed by the same traffic control service with which they are familiar at home.

- Federal Aviation Administration (FAA)— the United States is a member of ICAO and follows ICAO standards. Deviations from ICAO standards are filed with ICAO. The FAA is responsible for air traffic services in the United States and its possessions according to the Federal Aviation Act of 1958, which consolidated all air traffic regulatory agencies under the control of the FAA. Some of the responsibilities of the FAA include:

    1. Operates the ATC system within the United States airspace.

    2. Establishes and ensures compliance with Title 14 of the Code of Federal Regulations (14 CFR), which is binding on the entire aviation community.

    3. Issues certificates to aircrew members, maintenance personnel, and control tower operators.

    4. Investigates aircraft accidents.

    5. Maintains communication stations and conducts flight checks on NAVAIDs.

# Flight Planning Publications and Charts

## Flight Information Publications (FLIP)

Complete aeronautical information concerning air traffic systems is published in FLIP. Published by the National Geo-Spatial Intelligence Agency (NGA), FLIP are divided into three phases of flight: planning, en route operations, and terminal operations. The en route and terminal phase publications have been divided into the following areas:

1. CONUS;

2. Alaska;

3. Canada and North Atlantic;

4. Caribbean and South America;

5. Europe, North Africa, and Middle East;

6. Africa;

7. Pacific, Australasia, and Antarctica; and

8. Eastern Europe and Asia.

### General Planning (GP)

This document is revised every 32 weeks with planning change notices (PCN) issued at the 16-week midpoint. Urgent change notices (UCN) are issued as required. The FLIP GP document contains information that is applicable worldwide. It is supplemented by the information published in seven Area Publication Sections.

### Area Planning (AP/1, 2, and 3)

Located behind GP in the FLIP Planning document binder, these publications contain planning and procedure information for a specific geographical area. Area Planning Documents 1, 2, and 3 are respectively North and South America, Europe-Africa-Middle East, and Pacific-Australasia-Antarctica.

### Area Planning (AP/1A, 2A, and 3A)

Located behind their respective Area Planning documents, these publications contain a tabulation of all prohibited, restricted, danger, warning, and alert areas. In addition, they contain intensive student jet training areas, military training areas, and known parachute jumping areas within their specific geographical area.

### Area Planning (AP/1B)

Located behind AP/1A in the FLIP Planning document, AP/1B contains information relative to military training routes in North and South America, including IFR and VFR military training routes.

### Planning Change Notices (PCN)

Planning change notices (PCN) are in textual form and are used to update the FLIP Planning document.

### Flight Information Handbook (FIH)

The fight informaton handbook (FIH) contains information for in-flight use. Sections include emergency procedures, national and international flight data and procedures, meteorological information, conversion tables and frequency pairings, standard time signals, and FLIP/Notices to Airmen (NOTAM) abbreviations and codes.

### FLIP En Route Charts

Charts portray airway systems, radio aids to navigation, airports, airspace divisions, and other aeronautical data for IFR operations. FLIP En Route Charts are divided into high altitude (18,000 feet mean sea level (MSL) through FL450) for use in the jet route system, and low altitude (1,200 feet above the surface up to but not including 18,000 feet MSL) for use in the airway systems. Packets of low and high altitude charts are available for each geographic area: CONUS; Alaska; Canada and North Atlantic; Caribbean and South America; Europe, North Africa, and Middle East; Africa; Pacific, Australasia, and Antarctica; and Eastern Europe and Asia.

### FLIP En Route Supplements

A FLIP En Route Supplement is published for each geographical areas. Each supplement contains an airport or facility directory, en route procedures, special notices, and other textual data required to support FLIP En Route Charts. In the United States, there are two supplements. One supplement is designed for IFR operations and contains IFR airport and facility directory, special notices, and procedures required to support the FLIP En Route and Area Charts. The other supplement is designed for VFR operations and contains a listing of selected VFR airports with sketches and an IFR or VFR city and airport cross-reference listing. In all other FLIP areas, airport sketches are published for a limited number of selected airports and are provided with a separate section of the FLIP En Route Supplement. Airport sketch details include airport identification, city name, distance, direction, and elevation, as well as a diagram of each airport.

### Area and Terminal Area Charts

These charts are large-scale graphics of selected terminal areas. In the United States, area charts are provided primarily as area enlargements; in foreign areas, the terminal area charts are published primarily to provide arrival and departure routings. The area and terminal charts are printed on the same size sheet as the FLIP En Route Charts (that is, the terminal or area sheet contains several terminal or area charts) and are distributed with the en route FLIP.

### Approaches and Departure Procedures

FLIP terminal instrument approach procedures and departure procedures (DP) plates are divided into low altitude approaches (approaches initiated below 18,000 feet MSL), and high altitude approaches (approaches initiated normally at or above 18,000 feet MSL, such as high performance aircraft). Each instrument approach procedure shows an airport sketch, with additional data if necessary, for an approach under IFR conditions.

### Terminal Change Notices (TCN)

Terminal Change Notices (TCNs) contain revisions to approach procedures and are published normally at the midpoint of the FLIP terminal booklets. The changes may be in textual or graphic form. In the United States, area TCNs revise only the low altitude approaches; however, in the Europe, North Africa, Middle East area, and Pacific,

Australasia, and Antarctica areas, TCNs revise both low and high altitude approaches. In the other four FLIP areas, TCNs are not published and Notice to Airman (NOTAM) must be consulted for changes to approach procedures.

## Standard Terminal Arrival Route (STAR)

Standard Terminal Arrival Route (STARs) contain preplanned IFR ATC arrival routes and are published in graphic and/or textual form. STARs provide transition from the en route structure to a fix or point from which an approach can be made. In Alaska, Pacific, Australasia, and Antarctica areas, STAR information is contained in the FLIP terminal booklets. In the United States, STARs are published in a bound booklet with civilian DPs.

## Notice to Airman (NOTAM)

A NOTAM is a message requiring expeditious and wide dissemination by telecommunication means. NOTAMs provide information that is essential to all personnel concerned with flight operations. NOTAM information is normally in the form of abbreviations or a NOTAM code. The FIH contains an alphabetical list of these abbreviations.

# Flight Planning

In the air, there is little time for lengthy processes of reasoning. Decisions must be made quickly and accurately; therefore, careful planning is essential to any flight. A smooth, successful flight requires a careful step-by-step plan that can be followed from takeoff to landing.

## Route Determination

When planning a route to be flown, many factors enter into consideration. The route may be dictated by operational requirements of the flight; it may be a preplanned route, or the navigator may have the prerogative of selecting the route to be flown. In any case, definite factors affect route selection and the navigator must be aware of them.

In most cases, a direct route is usually best because it conserves both time and fuel. Such things as airways, routing, high terrain, and bad weather, however, can affect this. The direction of prevailing winds can affect route selection, because the proper use of a jet stream often decreases total flying time, even though a direct route is not flown.

## Chart Selection

Once a route is established, navigation charts appropriate to the intended flightpath should be selected. Correct selection depends on distance to be flown, airspeeds, methods of navigation, and chart accuracy.

## Total Distance to Fly

A great circle is the shortest distance between two points. Considerable distance can be saved by flying a great circle course, particularly on long-range flights in polar latitudes. A straight line on a gnomonic chart represents a great circle course. One way to flight plan a great circle course is to plot the entire route on a gnomonic chart, then transfer coordinates to charts more suitable for navigation, such as a Transverse Mercator. Select coordinates at intervals of approximately 300 nautical miles (NM). Once the route is plotted on the navigational chart, record true courses and distances for each leg of the flight on the flight plan.

## Chart and Methods of Navigation

The method of navigation is determined by flight requirements and the route/area that is used. Select charts for the flight that are best suited to the navigational techniques chosen. For example, radar flights require charts with representative terrain and cultural returns for precision fixing and grid flights require charts with a grid overlay. Charts produced by NGA are shown in *Figure 2-1*.

| National Geo-Spatial Intelligence Agency | | |
|---|---|---|
| 1501 AIR | Joint Operations Graphics | 1: 250,000 |
| | Sectional Aeronautical Charts | 1: 500,000 |
| PC | Pilotage Charts (Small Size) | 1: 500,000 |
| TPC | Tactical Pilotage Charts | 1: 500,000 |
| ONC | Operational Navigational Charts | 1: 1,000,000 |
| WAC | World Aeronautical Charts | 1: 1,000,000 |
| JNC | Jet Navigation Charts | 1: 2,000,000 |
| JNU | Universal Jet Navigation Charts | 1: 2,000,000 |
| CEC | Continental Entry Charts | 1: 2,000,000 |
| JNCA | Jet Navigation Charts | 1: 3,000,000 |
| GNC | Global Navigation Planning Charts | 1: 5,000,000 |
| NASC | Antarctic Strip Charts | Various |

**Figure 2-1.** *National Geo-Spatial Intelligence Agency charts.*

## Scale

The scale of charts used for navigation varies inversely with the speed of the aircraft. For example, Jet Navigation Charts (JNC) have a small scale and contain features appropriate for high speed navigation. Navigation at slower speeds requires large scale charts providing more detailed coverage.

## NOTAMs

Interim aeronautical flight information changes are disseminated by NOTAMs until the change is provided in all pertinent FLIPs. NOTAMs also provide the most current information on restrictions to flight, reliability of airport facilities and services, en route hazards, radio aids, etc.

## Airways

### Types and Use of Airways

Airways are corridors established by a national government within its airspace to facilitate the navigation and control of air traffic under IFR conditions. Usually, an airway is 10 statute miles wide and follows a route over the ground defined by radio NAVAIDs.

Generally, there are many different airways within a country as evidenced by those established in the United States. In the United States, as well as in other countries, there are two sets of airways (one for low altitudes and one for high altitudes.) To distinguish one airway from another, each has its own designator, such as V (low altitude) and J (high altitude). These designators simplify the preparation of a flight plan and improve the communication between aircrews and air traffic controllers.

### Alternate Airfield

An alternate airfield is where an aircraft intends to land if weather conditions prevent landing at a scheduled destination. Occasionally, an airfield may also be identified as an alternate for takeoff purposes. This procedure is at the direction of company procedures and operations specifications that authorize the use of lower minimums for takeoff than for landing.

### Emergency Airfields

During flight planning, select certain airfields along the planned flight route as possible emergency landing areas and then annotate these airfields on the charts for quick reference. Consider the following factors when selecting an emergency airfield: type of aircraft, weather conditions, runway length, runway weight-bearing capacity, runway lighting, and radio NAVAIDs. The NOTAMs for these airfields should be checked prior to flight.

### Highest Obstruction

After the route has been determined, the navigator should study the area surrounding the planned route and annotate the highest obstruction (terrain or cultural). The distance within which the highest obstruction is annotated is in accordance with governing or local directives. The highest obstruction is taken into consideration when determining the minimum en route altitude (MEA) and in emergency procedures discussion.

### Special Use Airspace

When determining the flight planned route, the locations of special use airspace has to be considered. The best way to find the locations of the areas is by checking an en route chart. After the route is determined, any special use airspace that may be close enough to the route of flight to cause concern (as per governing directives) should be annotated on the chart with pertinent information. Annotate time and days of operation, effective altitudes, and any restriction applicable to that area. These areas, when annotated on the chart, assist the navigator with in-flight changes and prevent planning a route of flight that cannot be flown.

## Flight Plans

### IFR and VFR Flight Plans

Flight plans are documents filed by pilots, or a Flight Dispatcher with the local Civil Aviation Authority (e.g., FAA in the United States), prior to departure. *[Figure 2-2]* They generally include basic information, such as departure and arrival points, estimated time en route, alternate airports in case of bad weather, type of flight (whether IFR or VFR), pilot's name, and number of people onboard. In most countries, flight plans are required for flights under IFR. Under VFR, they are optional unless crossing national borders; however, they are highly recommended, especially when flying over inhospitable areas, such as water, as they provide a way of alerting rescuers if the flight is overdue. For IFR flights, flight plans are used by ATC to initiate tracking and routing services. For VFR flights, their only purpose is to provide needed information should search and rescue operations be required.

### International Flight Plans

Flight plans are required for all flights into international and foreign airspace. The standard flight plan form is the FAA Form 7233-4, available at most U.S. Flight Service Stations (FSSs). *[Figure 2-3]* Flight plans must be transmitted to and should be received by ATC authorities in each ATC Region to be entered at least 2 hours prior to entry, unless otherwise stated in the various country requirements. It is extremely important that, when filing flight plans in countries outside the U.S., inquiry be made by the pilot as to the method used for subsequent transmission of flight plan information to pertinent en route and destination points and of the approximate total elapsed time applicable to such transmissions.

### En Route Fuel

En route fuel is determined with a fuel graph, such as the one depicted in *Figure 2-4*. Each type aircraft has a series of fuel graphs based on: aircraft gross weight, pressure or density altitude, true airspeed (TAS) or Mach number and, on some aircraft, the aerodynamic drag of external stores. En route fuel is computed in a manner that takes into account the worst fuel consumption situation, such as the lowest cruise altitude and highest airspeed. Most fuel graphs are designed for standard day conditions, so temperature deviation has to be considered. En route fuel can be calculated from the start descent point or initial approach fix (IAF).

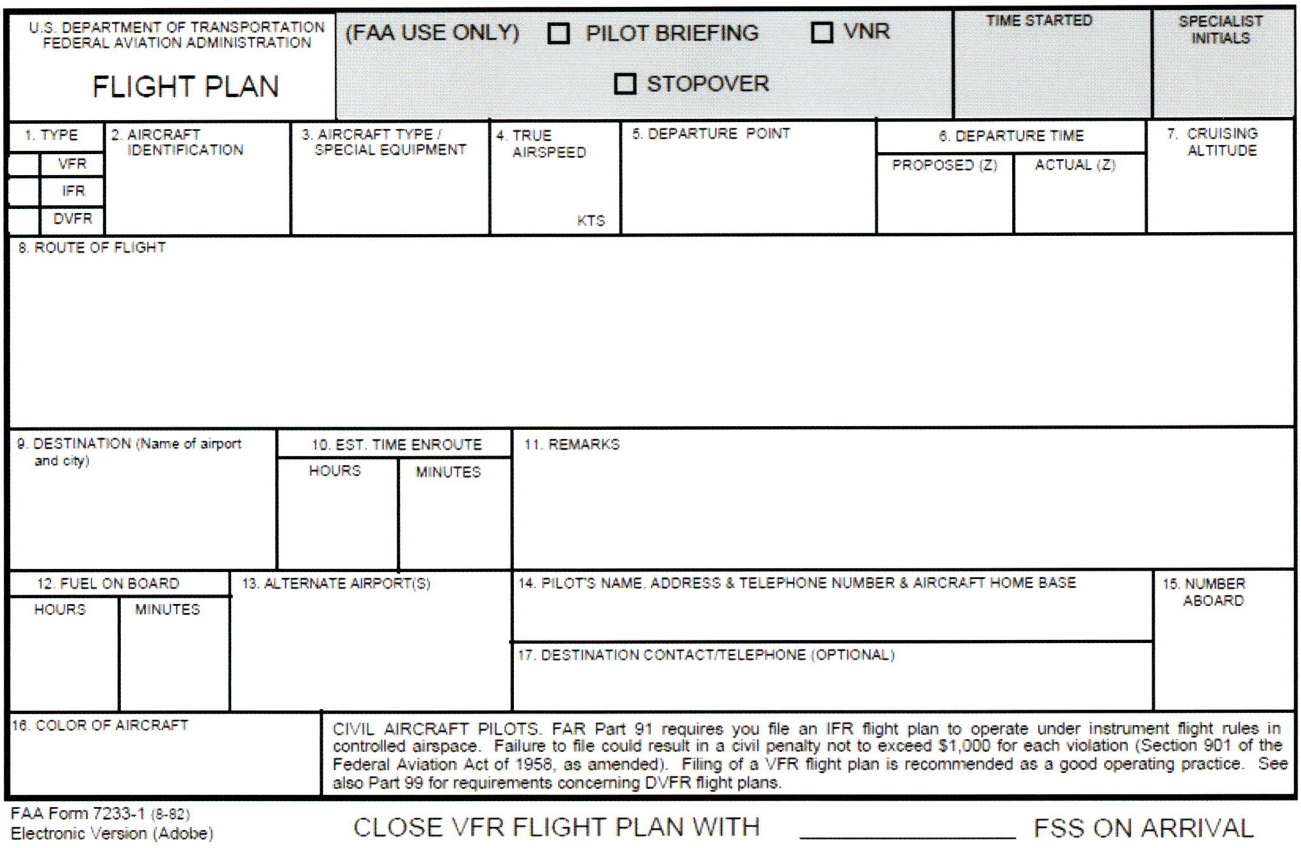
U.S. DEPARTMENT OF TRANSPORTATION
FEDERAL AVIATION ADMINISTRATION

FLIGHT PLAN

(FAA USE ONLY) ☐ PILOT BRIEFING ☐ VNR

☐ STOPOVER

TIME STARTED

SPECIALIST INITIALS

1. TYPE
VFR
IFR
DVFR

2. AIRCRAFT IDENTIFICATION

3. AIRCRAFT TYPE / SPECIAL EQUIPMENT

4. TRUE AIRSPEED
KTS

5. DEPARTURE POINT

6. DEPARTURE TIME
PROPOSED (Z)
ACTUAL (Z)

7. CRUISING ALTITUDE

8. ROUTE OF FLIGHT

9. DESTINATION (Name of airport and city)

10. EST. TIME ENROUTE
HOURS
MINUTES

11. REMARKS

12. FUEL ON BOARD
HOURS
MINUTES

13. ALTERNATE AIRPORT(S)

14. PILOT'S NAME, ADDRESS & TELEPHONE NUMBER & AIRCRAFT HOME BASE

15. NUMBER ABOARD

17. DESTINATION CONTACT/TELEPHONE (OPTIONAL)

16. COLOR OF AIRCRAFT

CIVIL AIRCRAFT PILOTS. FAR Part 91 requires you file an IFR flight plan to operate under instrument flight rules in controlled airspace. Failure to file could result in a civil penalty not to exceed $1,000 for each violation (Section 901 of the Federal Aviation Act of 1958, as amended). Filing of a VFR flight plan is recommended as a good operating practice. See also Part 99 for requirements concerning DVFR flight plans.

FAA Form 7233-1 (8-82)
Electronic Version (Adobe)

CLOSE VFR FLIGHT PLAN WITH ______________ FSS ON ARRIVAL

**Figure 2-2.** *FAA Flight Plan Form 7233-1 (8-82).*

### Fuel Reserve

Fuel reserve is the quantity of fuel carried in excess of flight requirements if the flight is completed as planned.

### En Route Plus Reserve

Add en route time and reserve time together to obtain the en route plus reserve time.

### Alternate Fuel

The fuel to the alternate is based on the fuel flow for the gross weight of the aircraft at destination, TAS, and altitude flown to the alternate. Some flight manuals include graphs designed for computing fuel to the alternate, but the fuel can also be computed by adding the time to the alternate and to the en route time. This time is then used to extract the total fuel required from takeoff to alternate. En route fuel is then subtracted from this to obtain the fuel to the alternate. A standard fuel amount may be added to allow for a missed approach at the original destination.

### Holding Fuel

Adverse weather, air traffic, or aircraft malfunction in the terminal area may force the aircraft to hold in the local area for a period of time before landing. The amount of holding fuel is based on any planned delays according to applicable directives.

### Approach and Landing Fuel

Approach and landing fuel is the fuel required from the terminal fix to the runway. This is computed for a prescribed amount of time (usually 15 minutes). The amount of fuel needed for approach and landing varies with the aircraft.

### Total Takeoff or Flaps Up Fuel

Total takeoff, or flaps up, is the cumulative total fuel from takeoff or flaps up that is required for en route, reserve, alternate, holding, and approach and landing.

### Taxi and Runup Fuel

Taxi and runup fuel is the fuel needed for taxiing, engine runup, and acceleration to takeoff speed. It is usually a predetermined value for each type of aircraft.

### Required Ramp Fuel

Required ramp fuel is the amount of fuel required at engine start to complete the flight.

### Actual Ramp Fuel

Actual ramp fuel is the fuel onboard prior to engine start.

## International Flight Plan

U S Department of Transportation
**Federal Aviation Administration**

PRIORITY      ADDRESSEE(S)

**<=FF**

<=

FILING TIME      ORIGINATOR

<=

SPECIFIC IDENTIFICATION OF ADDRESSEE(S) AND / OR ORIGINATOR

3 MESSAGE TYPE     7 AIRCRAFT IDENTIFICATION     8 FLIGHT RULES     TYPE OF FLIGHT

**<=(FPL**     —     —     —     **<=**

9 NUMBER     TYPE OF AIRCRAFT     WAKE TURBULENCE CAT.     10 EQUIPMENT

—     /     —     /     **<=**

13 DEPARTURE AERODROME     TIME

—     **<=**

15 CRUISING SPEED     LEVEL     ROUTE

—

<=

TOTAL EET

16 DESTINATION AERODROME     HR   MIN     ALTN AERODROME     2ND ALTN AERODROME

<=

18 OTHER INFORMATION

—

<=

SUPPLEMENTARY INFORMATION (NOT TO BE TRANSMITTED IN FPL MESSAGES)

19     ENDURANCE               EMERGENCY RADIO

    HR   MIN     PERSONS ON BOARD     UHF    VHF    ELBA

— **E/**     **P/**     **R/**

SURVIVAL EQUIPMENT           JACKETS

    POLAR DESERT MARITIME JUNGLE     LIGHT FLUORES UHF VHF

    /     /

DINGHIES

NUMBER CAPACITY COVER     COLOR

**D** /     <=

AIRCRAFT COLOR AND MARKINGS

**A/**

REMARKS

**N** /     <=

PILOT-IN-COMMAND

**C/**     )<=

FILED BY     ACCEPTED BY     ADDITIONAL INFORMATION

FAA Form 7233-4 (7-93)

**Figure 2-3.** *FAA Form 7233-4, International Flight Plan (7-93).*

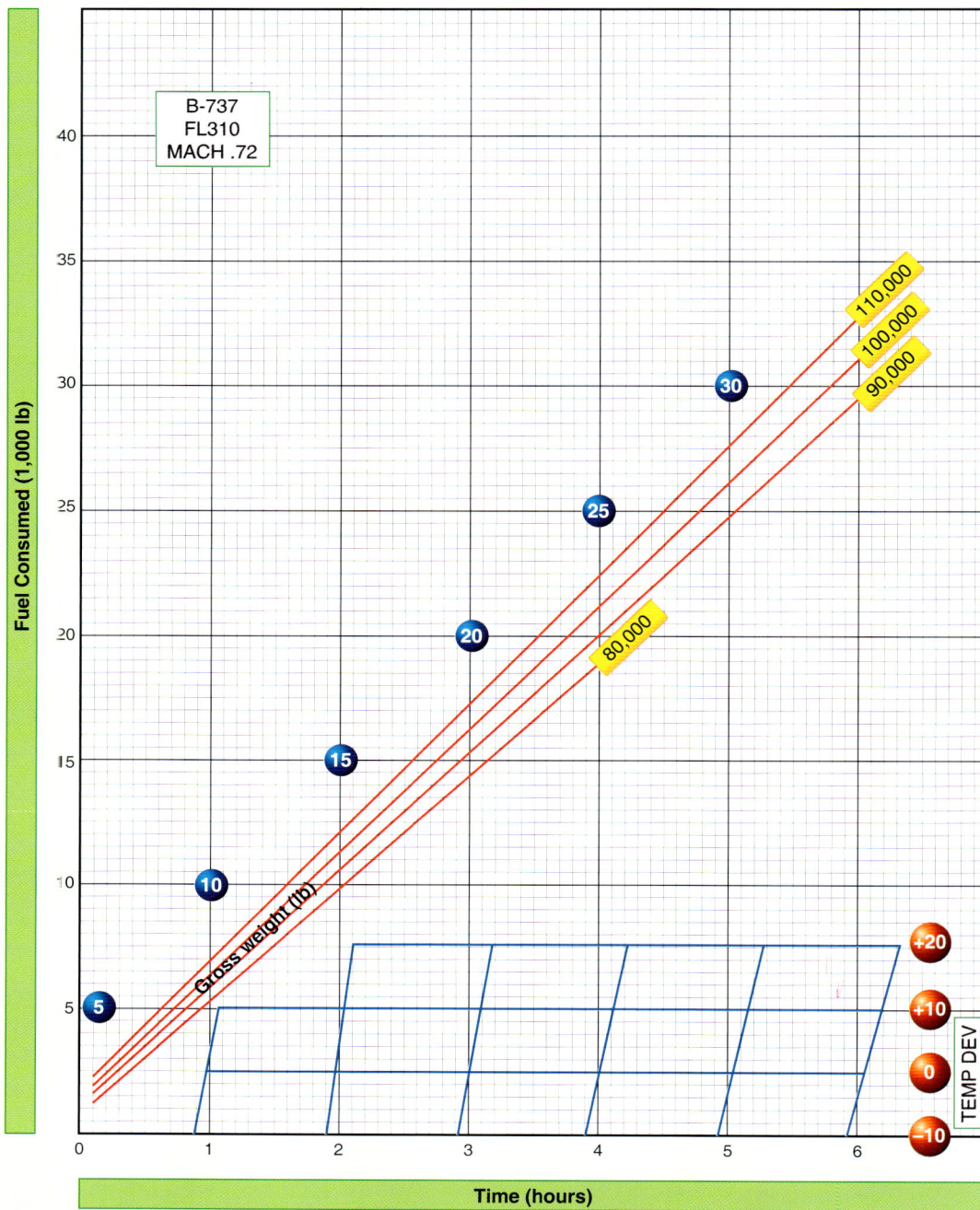

**Figure 2-4.** *Fuel planning graph.*

## Unidentified Extra Fuel

Additional fuel over and above that required by the flight plan is referred to as unidentified extra fuel. It is the difference between required ramp fuel and actual ramp fuel.

## Burnoff Fuel

Burnoff is the planned amount of fuel to be used after takeoff. This value subtracted from takeoff gross weight is equal to the approximate aircraft gross weight at landing.

## Range Control Graph

The range control graph portrays planned, minimum, and actual fuel consumption. *[Figure 2-5]* It is used to flight plan fuel consumption and serves as an in-flight worksheet for comparing actual and planned fuel consumption. The range control graph can be constructed with information taken from a completed flight plan, such as *Figure 2-2* and the applicable fuel planning graph in *Figure 2-4*. After calculating the required fuel at checkpoints along the route, the fuel remaining (vertical) is plotted against time remaining (horizontal). The planned fuel consumption is then plotted on the graph along with the minimum required fuel line. In-flight fuel readings are taken periodically and plotted on the graph to determine the fuel consumption in relation to that planned.

The planned line is determined by calculating the fuel remaining and time remaining at predetermined points in the flight and then plotting these points on the graph and connecting them with a line. The minimum line is determined by adding up all fuel required as a minimum at the destination (reserve, alternate, approach, etc.) and plotting it on the zero time remaining line. The difference between the minimum fuel required and the planned fuel on the zero time remaining line is then plotted below each of the predetermined fuel remaining points on the planned line. The points are connected with a line that represents the minimum required fuel line. This line is used to determine whether or not to continue the flight.

In-flight fuel readings are obtained and plotted against time remaining to determine fuel status. These plotted points are then connected with a dotted line that represents the actual fuel consumption. The trend of the in-flight fuel readings indicates actual fuel consumption and is used to make flight decisions with regard to fuel.

## Equal Time Point (ETP)

The equal time point (ETP) is that point along the route (normally one with an extended overwater leg) from which it takes the same amount of time to return to departure (or the last suitable airfield prior to beginning the overwater leg of the flight) as it would to continue to destination (or the first suitable airfield for landing). *[Figure 2-6]* The ETP is not necessarily the midpoint in time from departure to destination. Its location is somewhere near the midpoint of the route (between suitable airfields), and it is dependent upon the wind factors.

A wind factor (WF) is a headwind or tailwind component, computed at planned altitude between suitable airfields by comparing the average groundspeed (GS) to the average TAS. To do this, algebraically subtract the TAS from the GS. A WF with a negative value is a headwind; positive is a tailwind. When computing ETP, obtain a WF for each half of the route. Wind factors may play a major role in determining whether or not a destination can be reached. The overall or total wind factor (TFW) is the average of $WF_1$ and $WF_2$ and is computed using the formula ($WF_1 + WF_2$) divided by 2. An ETP is computed using the following formula:

$$\text{Total Distance} = T$$
$$(WF_2 - WF_1) + (2X \text{ TAS}) (60 \text{ min})$$

Total distance is the distance in NMs from the last suitable airfield to the first suitable airfield, measured along the route of flight. $WF_2$ and $WF_1$ are wind factors for the second and first halves of the route segment, respectively. T is the time remaining in minutes from the ETP to the first suitable airfield. This time can be converted to distance by applying the GS for the second half of the route segment. The distance can then be measured uptrack and the ETP plotted on the chart. The time should be plotted on the range control graph with a vertical line that crosses both the planned and minimum lines. If the first suitable airfield is not the planned landing airfield, then the time should be added between the first suitable airfield and the landing airfield to determine the ETP.

## Endurance

Endurance is the length of time an aircraft can remain airborne, not including minimum required fuel. Endurance can be computed by taking the last plotted fuel reading and following a line parallel to the fuel remaining lines in the direction of increasing time remaining until intercepting the minimum line. This point and its corresponding time remaining represent the endurance at the time of the fuel reading that is being used. Endurance is critical in making in-flight diversion decisions.

# Route Study

## Flight Planning

During flight planning, crewmembers should conduct a route study. For the navigator, a route study encompasses three phases of flight: takeoff and climb, cruise, and approach and landing.

## Fuel Planning

| Flight Phase | Time Accum Time | Time Remaining | T/O Fuel Weight A | Fuel Consumed B | Fuel Remaining A – B |
|---|---|---|---|---|---|
| Taxi and Run up | ▓ | 2+51 | ▓ | ▓ | ▓ |
| Climb | +19 / +19 | 2+32 | 36.0 | 3.5 | 32.5 |
| Cruise 1 | +38 / +37 | 1+54 | 36.0 | 6.5 | 29.5 |
| Cruise 2 | +38 / 1+35 | 1+16 | 36.0 | 10.1 | 26.9 |
| Cruise 3 | +38 / 2+13 | +38 | 36.0 | 13.4 | 22.6 |
| Cruise 4 | +38 / 2+51 | ▓ | 36.0 | 16.7 | 19.3 |
| Long Range Alternate | | ▓ | | | ▓ |

| | | | |
|---|---|---|---|
| Operating Wt: 66,400 | Climb Temp Dev: +5 | ACFT Call UGM: Gator 79 | |
| Cargo/Pax Wt: 4,600 | Cruise Temp Dev: +1 | ACFT Serial No: 1404 | |
| Ramp Fuel Wt: 37,000 | Takeoff Wt: 107,000 | Zulu Date: 2 Jul 79 | |
| Ramp Gross Wt: 106,000 | Pg: 4-6 | Time | Fuel |
| 1. Enroute | | 2+51 | 16.7 |
| 2. Reserve | | +20 | 4.0 |
| 3. Enroute Plus Reserve | | 3+11 | 20.7 |
| 4. Alternate (and Missed Approach) | | NR | NR |
| 5. Holding | | — | — |
| 6. Approach/Landing | | +15 | 1.0 |
| 7. Identified Extra | | ▓ | — |
| 8. Total (3+4+5+6+7) Takeoff/Flaps Up | | 3+26 | 21.7 |
| 9. Taxi and Run up (Acceleration) | | ▓ | 1.0 |
| 10. Required Ramp | | ▓ | 22.7 |
| 11. Actual Ramp | | ▓ | |
| 12. Unidentified Extra | | ▓ | |
| 13. Dest Plus Reserve (2+4+5) | | ▓ | 5.0 |

## EPT Summary

| Wind Factor: | Total: | 1st Half: | 2nd Half: |
|---|---|---|---|
| | -2 | -5 | +3 |

$$\frac{\text{TOTAL DISTANCE} \ (\ 1186\ )}{(WF_2 - WF_1) + 2 \ (TAS) \ (\ 860\ )} = T \ \frac{(\ 83\ )}{\Delta} \text{ MIN}$$

$$\frac{(+3 - (-5)) + (2 \times 426)}{}$$

$$\frac{8 + 852}{860}$$

## Clearances

ATC Clearance

Departure Clearance

Enroute Clearance

Arrival Clearance

## Position Reporting

| | Present | | | | |
|---|---|---|---|---|---|
| | Position | | | | |
| | Time | | | | |
| | Altitude | | | | |
| | Next | | | | |
| | Position | | | | |
| | Time | | | | |

**Figure 2-5.** *Range control graph.*

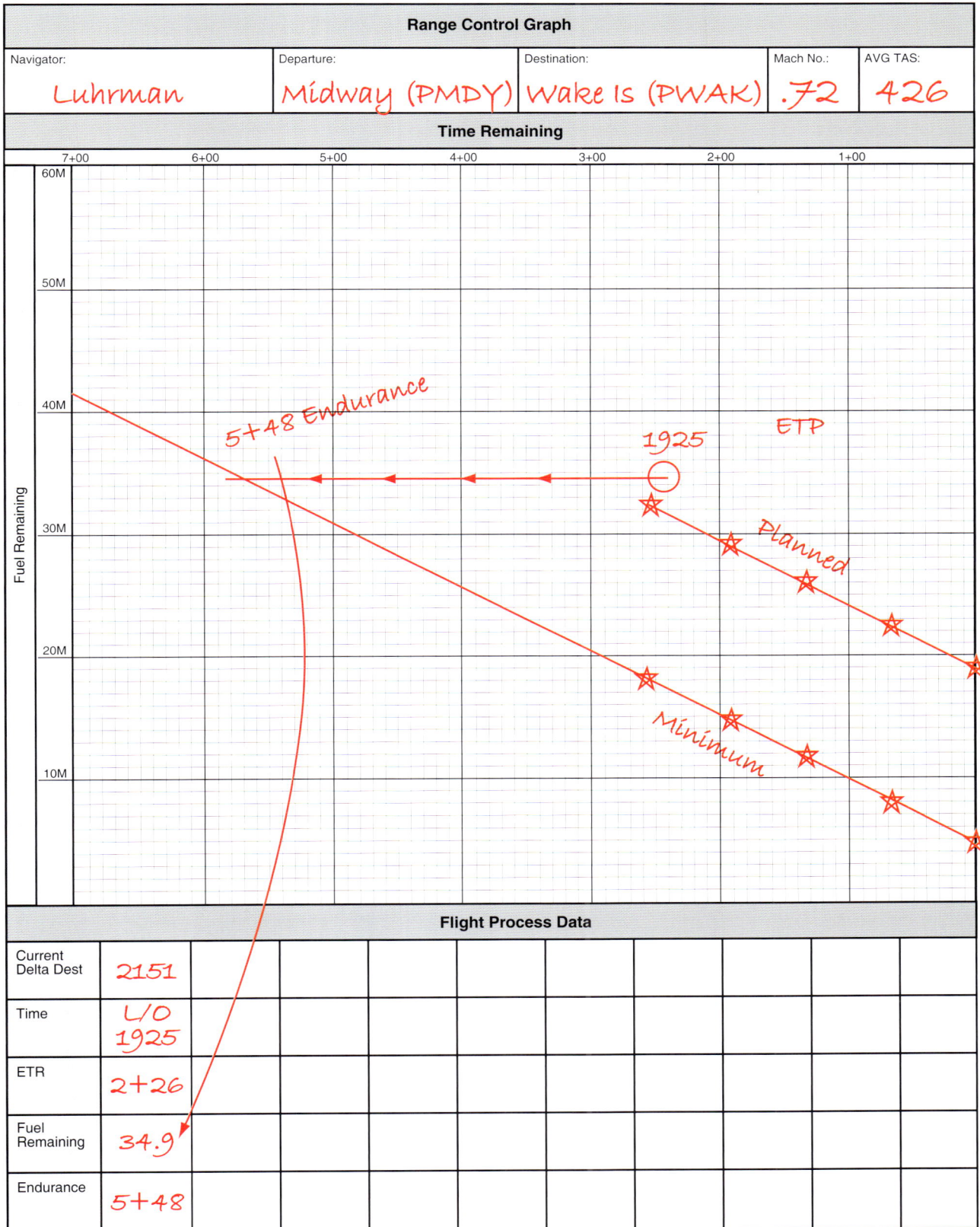

**Figure 2-5.** *Range control graph (continued).*

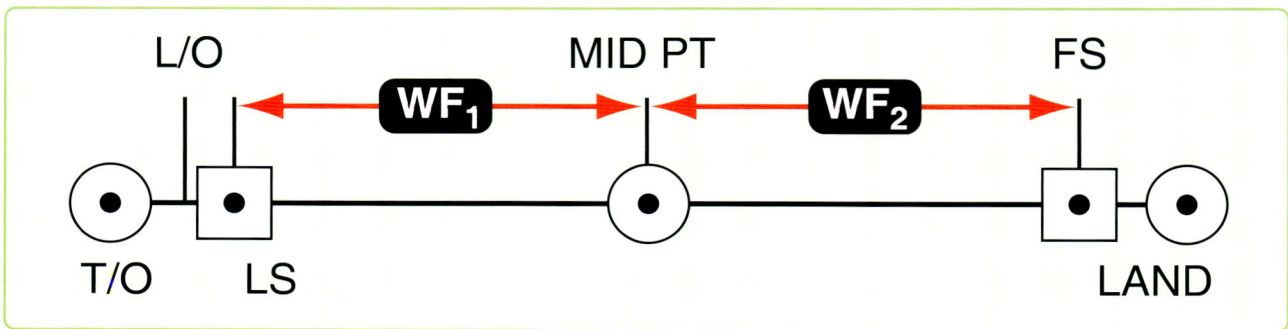

**Figure 2-6.** *Equal time point.*

### Takeoff and Climb

Before completing pre-flight planning, be familiar with the departure procedure filed. Check the NAVAIDs, such as tactical air navigation (TACAN), very high frequency (VHF) omnidirectional range (VOR), and automatic direction finder (ADF) to be used on the departure procedure. Note magnetic headings, radials, or bearings and altitude restrictions.

#### Duties

During takeoff and climb, the navigator's duties include monitoring the departure procedure, copying clearances, and ensuring applicable altitude restrictions and terrain clearance are maintained.

#### Estimated Times of Arrival (ETA)

During climb-out, controlling authorities may request estimated time of arrivals (ETAs), and generally it is the navigator's duty to compute these ETAs. Use the best source of groundspeed to compute ETAs. As a backup, use the inertial navigation system (INS), flight-planned GS, or TACAN distance measuring equipment (DME) change for groundspeed. When a controlled time of arrival is required, the navigator should compute an IAS or TAS for the pilot to maintain when approaching cruise altitude.

### Cruise

#### Navigator's Duties

While the primary duty is to monitor and direct the progress of the aircraft, the navigator must meet many associated requirements, such as completing the log, filling out forms, working controlled ETAs, and analyzing the information received from the navigation equipment.

#### Log

The navigator's log is usually the only record of the aircraft's actual position at any given time during the flight. For this reason, it must be accurate and complete. Log procedures vary between organizations; however, the basic log requirements and purpose remain the same—to keep an accurate record

of data for the navigator's reference and debriefing purposes and to serve as a worksheet for the navigator. Generally, required items for log entry are all information necessary to reconstruct the flight.

#### Celestial Precomps

Like logs, celestial precomp forms also vary between organizations. Like the log, the computational format may vary; however, the celestial computations themselves are essentially the same.

### Approach and Landing

#### Standard Approaches

The descent portion of the flight is similar to the climb portion. Instrument approach plates are established for almost all airfields of any significance in the world. The published approaches are normally flight-checked for safety of flight; if not, they are appropriately annotated. The navigator must make certain that the route affords adequate terrain clearance given by the approach control. Because of congested air traffic, approaches must be followed precisely. The navigator should monitor the aircraft position and altitude during descent and advise the pilot of any deviations.

#### Airborne Radar Approach (ARA)

In-flight duties for the navigator are many and diverse. As it is true that pilots earn their money on takeoffs and landings, navigators earn their money during the cruise portion, but they are also responsible for monitoring the departure and approach. Since crew safety and flight accomplishment are affected by each crewmember's performance, the success of any flight depends in part on the navigator's competence. Conscientious performance of in-flight duties can avert embarrassing and dangerous situations for the entire crew.

## Chapter Summary

This chapter discussed the importance of flight planning and understanding the resources available to the navigator. ATC systems are introduced along with all of the publications associated with flight planning that navigators should be familiar with. Ground planning is discussed in detail along with chart definitions and route study, which encompasses flight planning, take-off and arrival study.

# Chapter 3
# Basic Instruments

## Introduction

Instruments mechanically measure physical quantities or properties with varying degrees of accuracy. Much of a navigator's work consists of applying corrections to the indications of various instruments and interpreting the results. Therefore, navigators must be familiar with the capabilities and limitations of the instruments available to them.

A navigator obtains the following flight information from basic instruments: direction, altitude, temperature, airspeed, drift, and groundspeed (GS). Some of the basic instruments are discussed in this chapter. The more complex instruments that make accurate and long distance navigation possible are discussed in later chapters.

| Compass: Magnetic | | | |
|---|---|---|---|
| Swung: 12 APR, 95 | | By: TTD | |
| To Fly | Steer | To Fly | Steer |
| N | 001 | 180 | 179 |
| 15 | 016 | 195 | 194 |
| 30 | 131 | 210 | 209 |
| 45 | 046 | 225 | 224 |
| 60 | 062 | 240 | 238 |
| 75 | 077 | 255 | 253 |
| 90 | 092 | 270 | 268 |
| 105 | 107 | 285 | 283 |
| 120 | 122 | 300 | 298 |
| 135 | 135 | 315 | 314 |
| 150 | 149 | 330 | 330 |
| 165 | 164 | 345 | 346 |

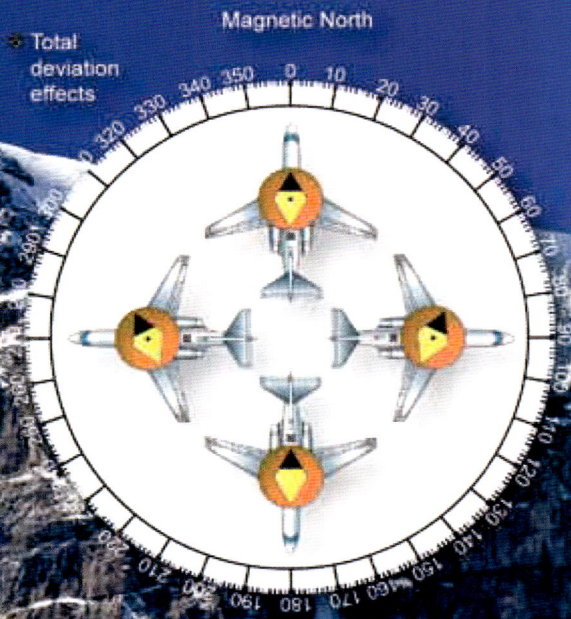

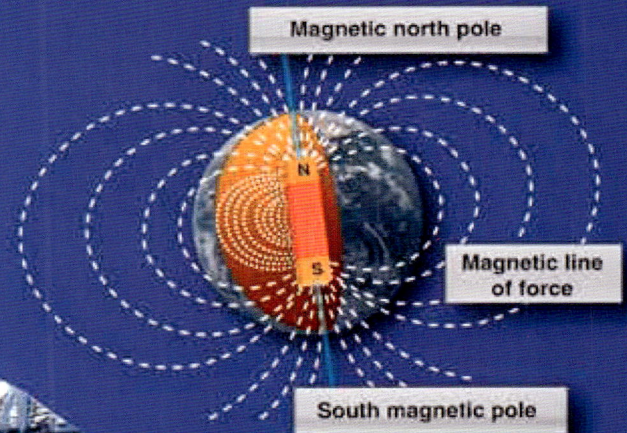

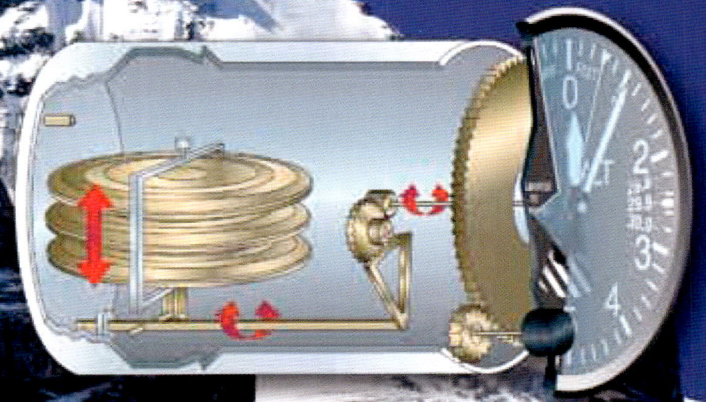

Magnetic north pole

Magnetic line of force

South magnetic pole

Magnetic North

Total deviation effects

# Direction

## Basic Instruments

The navigator must have a fundamental background in navigation to ensure accurate positioning of the aircraft. Dead reckoning (DR) procedures aided by basic instruments give the navigator the tools to solve the three basic problems of navigation: position of the aircraft, direction to destination, and time of arrival. Using only a basic instrument, such as the compass and drift information, you can navigate directly to any place in the world. Various fixing aids, such as celestial and radar, can greatly improve the accuracy of basic DR procedures. This chapter discusses the basic instruments used for DR and then reviews the mechanics of DR, plotting, wind effect, and computer solutions.

Directional information needed to navigate is obtained by use of the earth's magnetic lines of force. A compass system uses a device that detects and converts the energy from these lines of force to an indicator reading. The magnetic compass operates independently of the aircraft electrical systems. Later developed compass systems require electrical power to convert these lines of force to an aircraft heading.

## Earth's Magnetic Field

The earth has some of the properties of a bar magnet; however, its magnetic poles are not located at the geographic poles, nor are the two magnetic poles located exactly opposite each other as on a straight bar. The north magnetic pole is located approximately at 73° N latitude and 100° W longitude on Prince of Wales Island. The south magnetic pole is located at 68° S latitude and 144° E longitude on Antarctica.

The earth's magnetic poles, like those of any magnet, can be considered to be connected by a number of lines of force. These lines result from the magnetic field that envelops the earth. They are considered to be emanating from the south magnetic pole and terminating at the north magnetic pole. [Figure 3-1]

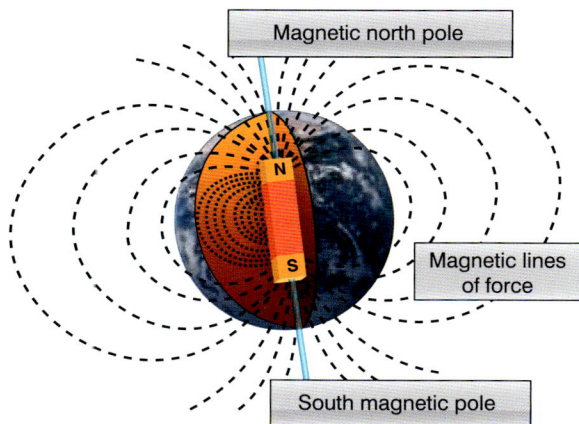

**Figure 3-1.** *Earth's magnetic field.*

The force of the magnetic field of the earth can be divided into two components: the vertical and the horizontal. The relative intensity of these two components varies over the earth so that, at the magnetic poles, the vertical component is at maximum strength and the horizontal component is minimum strength. At approximately the midpoint between the poles, the horizontal component is at maximum strength and the vertical component is at minimum strength. Only the horizontal component is used as a directive force for a magnetic compass. Therefore, a magnetic compass loses its usefulness in an area of weak horizontal force, such as the area around the magnetic poles. The vertical component causes the end of the needle nearer to the magnetic pole to tip as the pole is approached. [Figure 3-1] This departure from the horizontal is called magnetic dip.

## Compasses

A compass may be defined as an instrument that indicates direction over the earth's surface with reference to a known datum. Various types of compasses have been developed, each of which is distinguished by the particular datum used as the reference from which direction is measured. Two basic types of compasses are in current use: the magnetic and gyrocompass.

The magnetic compass uses the lines of force of the earth's magnetic field as a primary reference. Even though the earth's field is usually distorted by the pressure of other local magnetic fields, it is the most widely used directional reference. The gyrocompass uses as its datum an arbitrary fixed point in space determined by the initial alignment of the gyroscope axis. Compasses of this type are widely used today and may eventually replace the magnetic compass entirely.

## Magnetic Compass

The magnetic compass indicates direction in the horizontal plane with reference to the horizontal component of the earth's magnetic field. This field is made up of the earth's field in combination with other magnetic fields in the vicinity of the compass. These secondary fields are caused by the presence of ferromagnetic objects.

Magnetic compasses may be divided into two classes:

1. The direct-indicating magnetic compass in which the measurement of direction is made by a direct observation of the position of a pivoted magnetic needle; and

2. The remote-indicating gyro-stabilized magnetic compass.

Magnetic direction is sensed by an element located at positions where local magnetic fields are at a minimum, such as the vertical stabilizer and wing tips. The direction is then transmitted electrically to repeater indicators on the instrument panels.

### Direct-Indicating Magnetic Compass

Basically, the magnetic compass is a magnetized rod pivoted at its middle, with several features incorporated to improve its performance. One type of direct-indicating magnetic compass, the B-16 compass (often called the whiskey compass), is illustrated in *Figure 3-2*. It is used as a standby compass in case of failure of the electrical system that operates the remote compasses. It is a reliable compass and gives good navigational results if used carefully.

#### Magnetic Variation and Compass Errors

The earth's magnetic poles are joined by irregular curves called magnetic meridians. The angle between the magnetic meridian and the geographic meridian is called the magnetic variation. Variation is listed on charts as east or west. When variation is east, magnetic north (MN) is east of true north (TN). Similarly, when variation is west, MN is west of TN. *[Figure 3-3]* Lines connecting points having the same magnetic variation are called isogonic lines. *[Figure 3-4]* Compensate for magnetic variation to convert a compass direction to true direction.

Compass error is caused by nearby magnetic influences, such as magnetic material in the structure of the aircraft and its electrical systems. These magnetic forces deflect a compass needle from its normal alignment. The amount of such deflection is called deviation which, like variation, is labeled "east" or "west" as the north-seeking end of the compass is deflected east or west of MN, respectively.

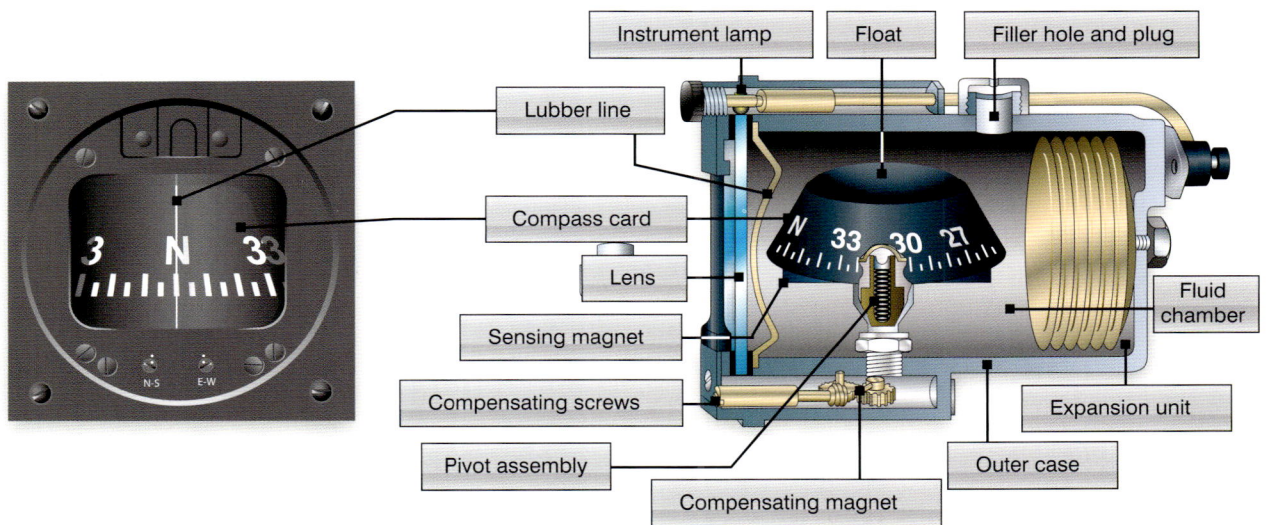

**Figure 3-2.** *Magnetic compass.*

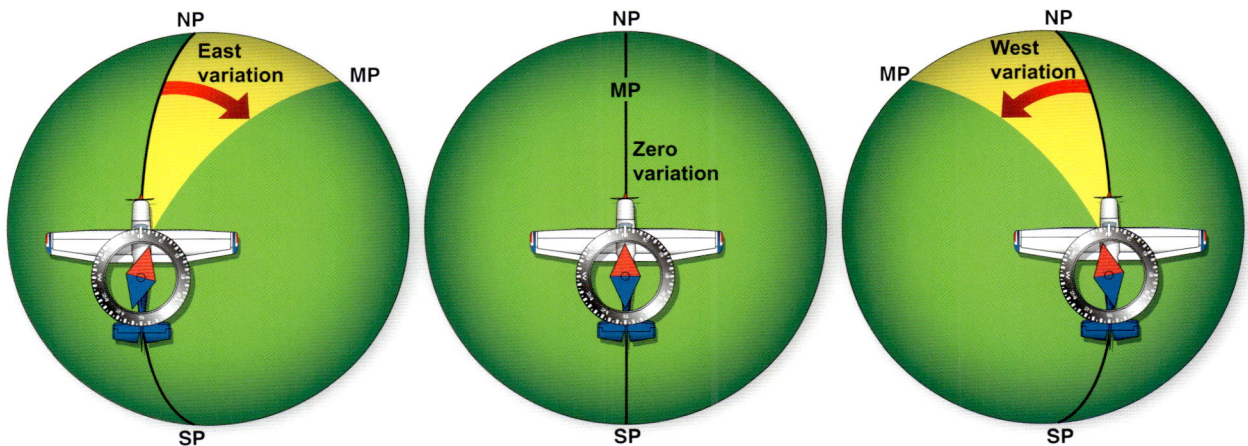

**Figure 3-3.** *Effects of variation.*

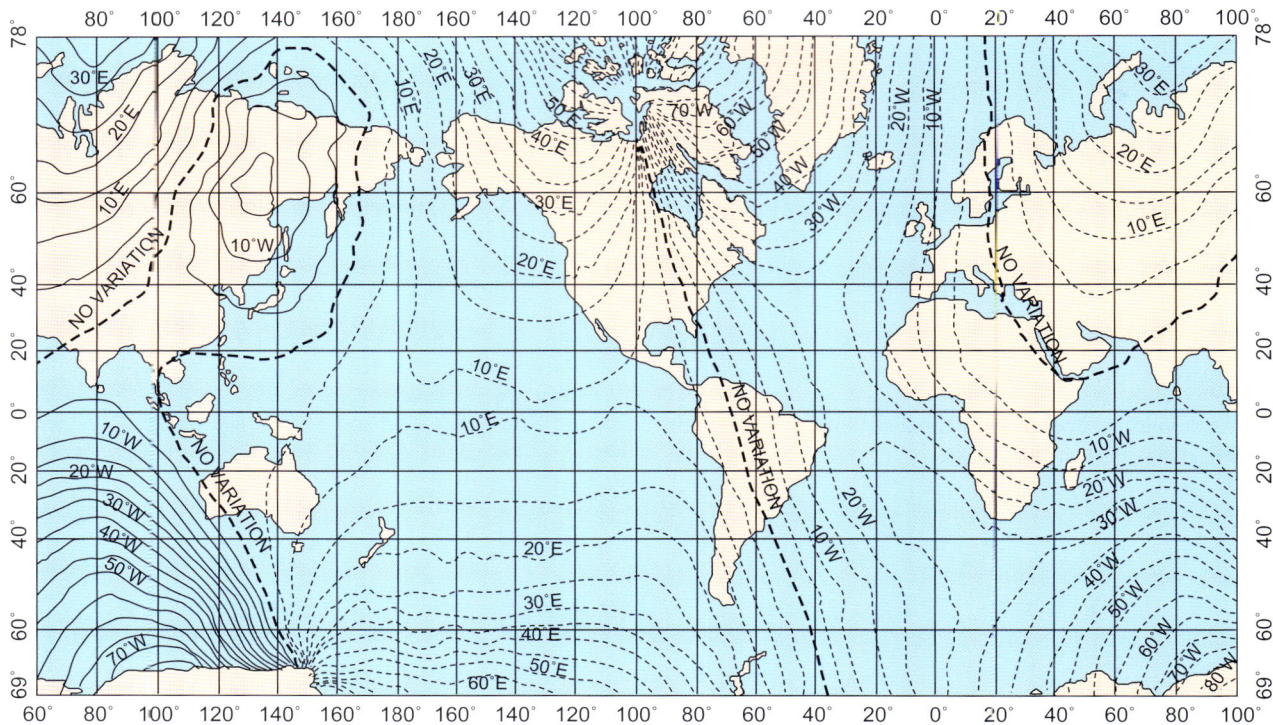

**Figure 3-4.** *Isogonic lines show same magnetic variation.*

The correction for variation and deviation is usually expressed as a plus or minus value and is computed as a correction to true heading (TH). If variation or deviation is east, the sign of the correction is minus; if west, the sign is plus. A rule of thumb for this correction is easily remembered as east is least and west is best.

Aircraft headings are expressed as TH or magnetic headings (MH). If the heading is measured in relation to geographical north, it is a TH. If the heading is in reference to MN, it is a MH; if it is in reference to the compass lubber line, it is a compass heading (CH). CH corrected for variation and deviation is TH. MH corrected for variation is TH.

This relationship is best expressed by reference to the navigator's log, where the various headings and corrections are listed as TH, variation (var), MH, deviation (dev), and CH. *[Figure 3-5]* Thus, if an aircraft is flying in an area where the variation is 10° E and the compass has a deviation of 3° E, the relationship would be expressed as follows, assuming a CH of 125°:

TH var MH dev CH
138 − 10 = 128 − 3 = 125

*Variation*

Variation has been measured throughout the world and the values found have been plotted on charts. Isogonic lines are

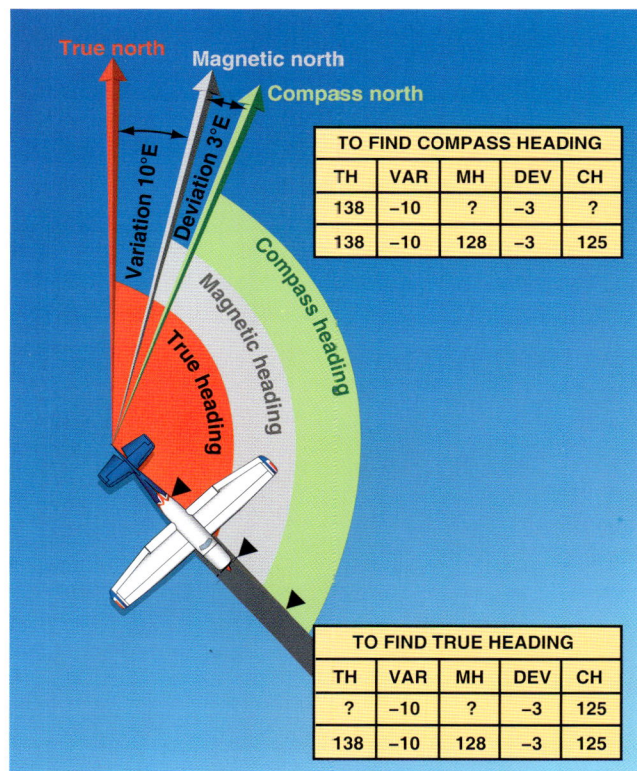

| TO FIND COMPASS HEADING | | | | |
| --- | --- | --- | --- | --- |
| TH | VAR | MH | DEV | CH |
| 138 | −10 | ? | −3 | ? |
| 138 | −10 | 128 | −3 | 125 |

| TO FIND TRUE HEADING | | | | |
| --- | --- | --- | --- | --- |
| TH | VAR | MH | DEV | CH |
| ? | −10 | ? | −3 | 125 |
| 138 | −10 | 128 | −3 | 125 |

**Figure 3-5.** *Find true heading by working backwards.*

3-4

printed on most charts used in aerial navigation so that, if the aircraft's approximate position is known, the amount of variation can be determined by visual interpolation between the printed lines. At high altitudes, these values can be considered quite realistic. Conversely, at low altitudes, these magnetic values are less reliable because of local anomalies.

Variation changes slowly over a period of years and the yearly amount of such change is printed on most charts. Variation is also subject to small diurnal (daily) changes that may generally be neglected in air navigation.

## Deviation

Because deviation depends upon the distribution of magnetic forces in the aircraft itself, it must be obtained individually for each magnetic compass on each aircraft. The process of determining deviation, known as compass swinging, should be discussed in the technical order for each compass.

Deviation changes with heading are shown in *Figure 3-6*. The net result of all magnetic forces of the aircraft (those forces excluding the earth's field) is represented by a dot located just behind the wings of the aircraft. If the aircraft is headed toward MN, the dot attracts one pole of the magnetic compass (in this case, the South Pole) but, on this heading, does not change its direction. The only effect is to amplify the directive force of the earth's field. If the aircraft heads toward magnetic east, the dot is now west of the compass, and attracts the South Pole of the compass, causing easterly deviation. *Figure 3-6* also shows that the deviation is zero on a south heading, and westerly when the aircraft is heading west.

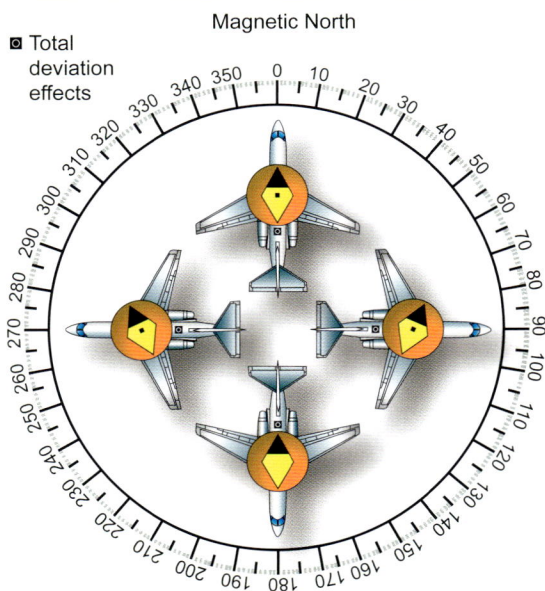

**Figure 3-6.** *Deviation changes with heading.*

Deviation can be reduced (but not eliminated) in some direct-indicating magnetic compasses by adjusting the small compensating magnets in the compass case. Remaining deviation is referred to as residual deviation and can be determined by comparison with true values. This residual deviation is recorded on a compass correction card showing actual deviation on various headings or the compass headings. From the compass correction card illustrated in *Figure 3-7*, the navigator knows that to fly a magnetic heading (MH) of 270°, the pilot must steer a CH of 268°.

| Compass: Magnetic | | | |
|---|---|---|---|
| Swung: 12 APR 95 | | By: TTD | |
| To Fly | Steer | To Fly | Steer |
| N | 001 | 180 | 179 |
| 15 | 016 | 195 | 194 |
| 30 | 131 | 210 | 209 |
| 45 | 046 | 225 | 224 |
| 60 | 062 | 240 | 238 |
| 75 | 077 | 255 | 253 |
| 90 | 092 | 270 | 268 |
| 105 | 107 | 285 | 283 |
| 120 | 122 | 300 | 298 |
| 135 | 135 | 315 | 314 |
| 150 | 149 | 330 | 330 |
| 165 | 164 | 345 | 346 |

**Figure 3-7.** *Compass correction card.*

### Errors in Flight

Unfortunately, deviation is not the only error of a magnetic compass. Additional errors are introduced by the motion of the aircraft itself. These errors have minimal effect on the use of magnetic compasses and come into play normally during turns or changes in speed. They are mentioned only to bring awareness of the limitations of the basic compass. Although a basic magnetic compass has some shortcomings, it is simple and reliable. The compass is very useful to both the pilot and navigator and is carried on all aircraft as an auxiliary compass. Because compass systems are dependent upon the electrical system of the aircraft, a loss of power means a loss of the compass system. For this reason, an occasional check on the standby compass provides an excellent backup to the main systems.

## Remote-Indicating Gyro-Stabilized Magnetic Compass System

A chief disadvantage of the simple magnetic compass is its susceptibility to deviation. In remote-indicating gyro-stabilized compass systems, this difficulty is overcome by locating the compass direction-sensing device outside magnetic fields created by electrical circuits in the aircraft. This is done by installing the direction-sensing device in a remote part of the aircraft, such as the outer extremity of a wing or vertical stabilizer. Indicators of the compass system can then be located throughout the aircraft without regard to magnetic disturbances.

Several kinds of compass system are used in aircraft systems. All include the following five basic components: remote compass transmitter, directional gyro (DG), amplifier, heading indicators, and slaving control. Though the names of these components vary among systems, the principle of operation is identical for each. Thus, the N-1 compass system shown in *Figure 3-8* can be considered typical of all such systems.

The N-1 compass system is designed for airborne use at all latitudes. It can be used either as a magnetic-slaved compass or as a DG. In addition, the N-1 generates an electric signal that is used as an azimuth reference by the autopilot, the radar system, the navigation and bombing computers, and various compass cards.

### Remote Compass Transmitter

The remote compass transmitter is the magnetic-direction sensing component of the compass system when the system is in operation as a magnetic-slaved compass. The transmitter is located as far from magnetic disturbances of the aircraft as possible, usually in a wing tip or the vertical stabilizer. The transmitter senses the horizontal component of the earth's magnetic field and electrically transmits it to the master indicator. The compensator, an auxiliary unit of the remote compass transmitter, is used to eliminate most of the magnetic deviation caused by the aircraft electrical equipment and ferrous metal when a deviation-free location for the remote compass transmitter is not available.

### Directional Gyro (DG)

The DG is the stabilizing component of the compass system when the system is in magnetic-slaved operation. When the compass system is in DG operation, the gyro acts as the directional reference component of the system.

### Amplifier

The amplifier is the receiving and distributing center of the compass system. Azimuth correction and leveling signals originating in the components of the system are each received, amplified, and transmitted by separate channels in the amplifier. Primary power to operate the compass is fed to the amplifier and distributed to the systems components.

### Master Indicator

The master indicator is the heading-indicating component of the compass system. The mechanism in the master indicator integrates all data received from the directional gyro and the remote compass transmitter, corrects the master indicator heading pointer for azimuth drift of the DG due to the earth's rotation, and provides takeoff signals for operating remote indicators, radar, navigation computers, and directional control of the autopilot.

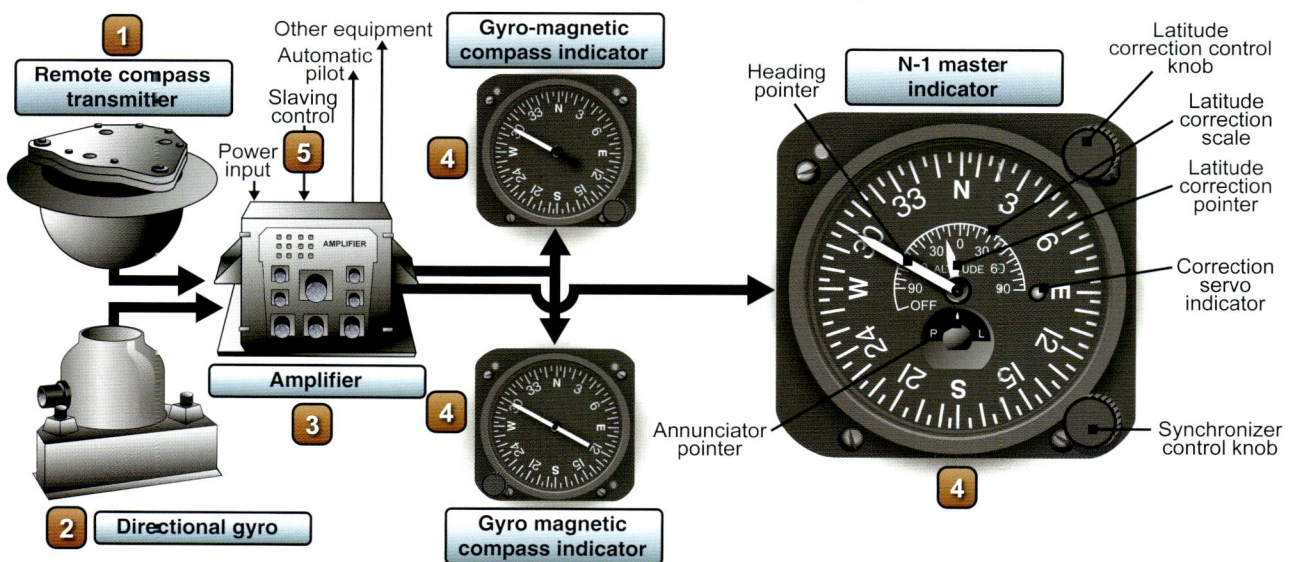

**Figure 3-8.** *N-1 compass system components.*

The latitude correction control provides a means for selecting either magnetic-slaved operation or DG operation of the compass system, as well as the proper latitude correction rate. The latitude correction pointer is mechanically connected to the latitude correction control knob and indicates the latitude setting on the latitude correction scale at the center of the master indicator dial face.

The synchronizer control knob at the lower right-hand corner of the master indicator face provides a means of synchronizing the master indicator heading pointer with the correct MH when the system is in magnetic-slaved operation. It also provides a means of setting the master indicator heading pointer on the desired gyro heading reference when the system is in DG operation. The annunciator pointer indicates the direction in which to rotate the synchronizer control knob to align the heading pointer with the correct MH.

### Gyro-Magnetic Compass Indicators

The gyro-magnetic compass indicators are remote-reading, movable dial compass indicators. They are intended for supplementary use as directional compass indicators when used with the compass system. The indicators duplicate the azimuth heading of the master indicator heading pointer. A setting knob is provided at the front of each indicator for rotating the dial 360° in either direction without changing the physical alignment of the pointer.

### Slaving Control

The slaving control is a gyro control rate switch that reduces errors in the compass system during turns. When the aircraft turns at a rate of 23° or more per minute, the slaving control prevents the remote compass transmitter signal from being transmitted to the compass system during magnetic-slaved operation. It also interrupts leveling action in the DG when the system is in magnetic-slaved or DG operation.

### Gyro Basics

Any spinning body exhibits gyroscopic properties. A wheel designed and mounted to use these properties is called a gyroscope or gyro. Basically, a gyro is a rapidly rotating mass that is free to move about one or both axes perpendicular to the axis of rotation and to each other. The three axes of a gyro (spin, drift, and topple) shown in *Figure 3-9* are defined as follows:

1.  In a DG, the spin axis or axis of rotation is mounted horizontally;

2.  The topple axis is that axis in the horizontal plane that is 90° from the spin axis;

3.  The drift axis is that axis 90° vertically from the spin axis.

Gyroscopic drift is the horizontal rotation of the spin axis about the drift axis. Topple is the vertical rotating of the spin

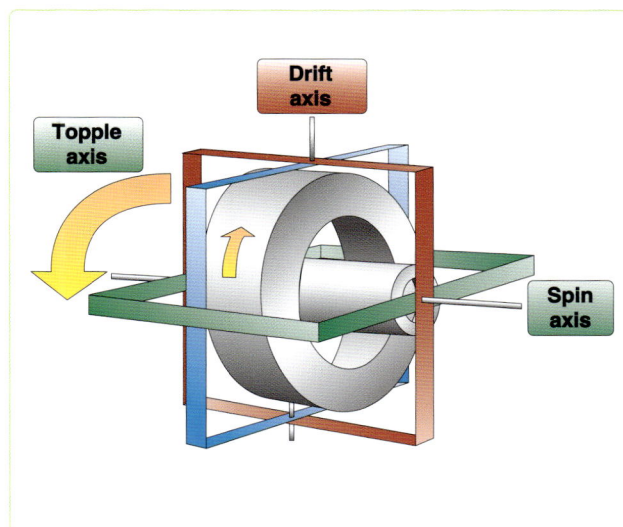

**Figure 3-9.** *Gyroscope axes.*

axis about the topple axis. These two component drifts result in motion of the gyro called precession.

A freely spinning gyro tends to maintain its axis in a constant direction in space, a property known as rigidity in space or gyroscopic inertia. Thus, if the spin axis of a gyro were pointed toward a star, it would keep pointing at the star. Actually, the gyro does not move, but the earth moving beneath it gives it an apparent motion. This apparent motion is called apparent precession. *[Figure 3-10]* The magnitude of apparent precession is dependent upon latitude. The horizontal component, drift, is equal to 15° per hour times the sine of the latitude, and the vertical component, topple, is equal to 15° per hour times the cosine of the latitude.

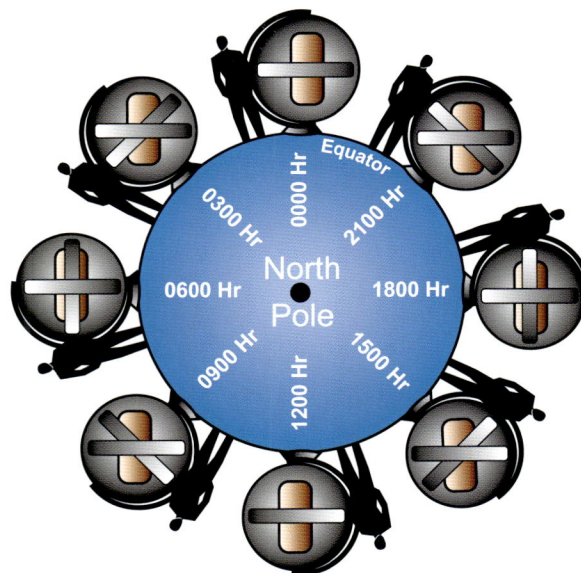

**Figure 3-10.** *Apparent precession.*

These computations assume the gyro is stationary with respect to the earth. If the gyro is to be used in a high-speed aircraft, however, it is readily apparent that its speed with respect to a point in space may be more or less than the speed of rotation of the earth. If the aircraft in which the gyro is mounted is moving in the same direction as the earth, the speed of the gyro with respect to space is greater than the earth's speed. The opposite is true if the aircraft is flying in a direction opposite to that of the earth's rotation. This difference in the magnitude of apparent precession caused by transporting the gyro over the earth is called transport precession.

A gyro may precess because of factors other than the earth's rotation. When this occurs, the precession is labeled real precession. When a force is applied to the plane of rotation of a gyro, the plane tends to rotate, not in the direction of the applied force, but 90° around the spin axis from it. This torquing action, shown in *Figure 3-11*, may be used to control the gyro by bringing about a desired reorientation of the spin axis, and most DGs are equipped with some sort of device to introduce this force. However, friction within the bearings of a gyro may have the same effect and cause a certain amount of unwanted precession. Great care is taken in the manufacture and maintenance of gyroscopes to eliminate this factor as much as possible, but, as yet, it has not been possible to eliminate it entirely. Precession caused by the mechanical limitations of the gyro is called real or induced precession. The combined effects of apparent precession, transport precession, and real precession produce the total precession of the gyro. The properties of the gyro that most concern the navigator are rigidity and precession. By understanding these two properties, the navigator is well equipped to use the gyro as a reliable steering guide.

## Directional Gyro (DG)

The discussion thus far has been of a universally mounted gyro, free to turn in the horizontal or vertical or any component of these two. This type of gyro is seldom, if ever, used as a DG. When the gyro is used as a steering instrument, it is restricted so that the spin axis remains parallel to the surface of the earth. Thus, the spin axis is free to turn only in the horizontal plane (assuming the aircraft normally flies in a near-level attitude), and only the horizontal component (drift) affects a steering gyro. In the terminology of gyro steering, precession always means the horizontal component of precession.

The operation of the instrument depends upon the principle of rigidity in space of the gyroscope. Fixed to the plane of the spin axis is a circular compass card, similar to that of the magnetic compass. Since the spin axis remains rigid in space, the points on the card hold the same position in space relative to the horizontal plane. The case, to which the lubber line is attached, simply revolves about the card.

It is important at this point to understand that the numbers on the compass card have no meaning within themselves, as on the magnetic compass. The fact that the gyro may indicate 100° under the lubber line is not an indication that

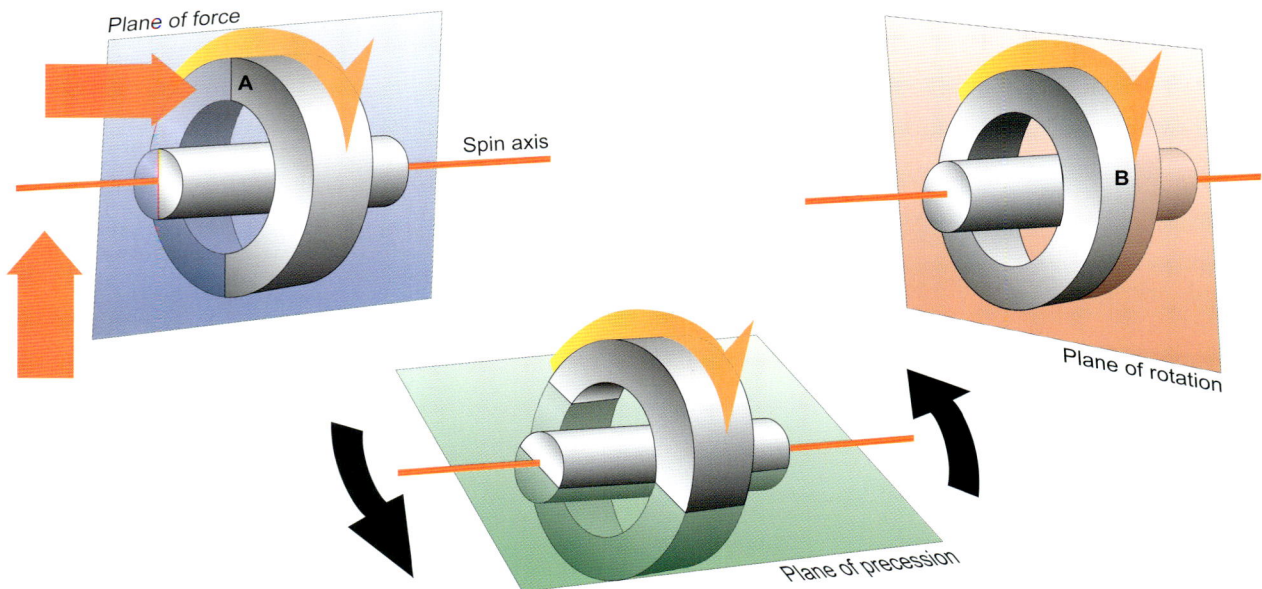

**Figure 3-11.** *By applying an upward pressure on the gyro spin axis, a deflective force is applied to the rim of the gyro at point A (plane of force). The resultant force is 90° ahead in the direction of rotation to point B (plane of rotation), which causes the gyro to precess (plane of precession).*

the instrument is actually oriented to magnetic north (MN), or any other known point. To steer by the gyro, the navigator must first set it to a known direction or point. Usually, this is MN or geographic north, though it can be at any known point. If, for example, MN is set as the reference, all headings on the gyro read relative to the position of the magnetic poles.

The actual setting of the initial reference heading is done by using the principle discussed earlier of torque application to the spinning gyro. By artificially introducing precession, the navigator can set the gyro to whatever heading is desired and can reset it at any time, by using the same technique.

## Gyrocompass Errors

The major error affecting the gyro and its use as a steering instrument is precession. Apparent precession causes an apparent change of heading equal to 15° per hour times the sine of the latitude. Real precession, caused by defects in the gyro, may occur at any rate, but is typically very small in current gyros. Apparent precession is a known value depending upon location and can be compensated for. In some of the more complex gyro systems, apparent precession is compensated for by setting in a constant correction equal to, and in the opposite direction to, the precession caused by the earth's rotation.

## Altitude and Altimeters

Altitude may be defined as a vertical distance above some point or plane used as a reference. Knowledge of the aircraft altitude is imperative for terrain clearance, aircraft separation, and a multitude of operational reasons. There are as many kinds of altitude as there are reference planes from which to measure them. Only six concern the navigator: indicated altitude, calibrated altitude, pressure altitude (PA), density altitude (DA), true altitude (TA), and absolute altitude. There are two main types of altimeters: the pressure altimeter, which is installed in every aircraft, and the absolute or radar altimeter. To understand the pressure altimeter's principle of operation, a knowledge of the standard datum plane is essential.

## Standard Datum Plane

The standard datum plane is a theoretical plane where the atmospheric pressure is 29.92 inches of mercury ("Hg) and the temperature is 15 °C. The standard datum plane is the zero elevation level of an imaginary atmosphere known as the standard atmosphere. In the standard atmosphere, pressure is 29.92 "Hg at 0 feet and decreases upward at the standard pressure lapse rate. The temperature is 15 °C at 0 feet and decreases at the standard temperature lapse rate. Both the pressure and temperature lapse rates are given in *Figure 3-12.* The standard atmosphere is theoretical. It was derived by averaging the readings taken over a period of many years. The list of altitudes and their corresponding

| Altitude (feet) | Standard Pressure (millibars) | Standard Pressure (inches of mercury) | Standard Temperature (°C) | Standard Temperature (°F) |
|---|---|---|---|---|
| 60,000 | 71.7 | 2.12 | −56.5 | −69.7 |
| 59,000 | 45.2 | 2.22 | −56.5 | −69.7 |
| 58,000 | 79.0 | 2.33 | −56.5 | −69.7 |
| 57,000 | 82.8 | 2.45 | −56.5 | −69.7 |
| 56,000 | 86.9 | 2.57 | −56.5 | −69.7 |
| 55,000 | 91.2 | 2.69 | −56.5 | −69.7 |
| 54,000 | 95.7 | 2.83 | −56.5 | −69.7 |
| 53,000 | 100.4 | 2.96 | −56.5 | −69.7 |
| 52,000 | 105.3 | 3.11 | −56.5 | −69.7 |
| 51,000 | 110.5 | 3.26 | −56.5 | −69.7 |
| 50,000 | 116.0 | 3.42 | −56.5 | −69.7 |
| 49,000 | 121.7 | 3.59 | −56.5 | −69.7 |
| 48,000 | 127.7 | 3.77 | −56.5 | −69.7 |
| 47,000 | 134.0 | 3.96 | −56.5 | −69.7 |
| 46,000 | 140.6 | 4.15 | −56.5 | −69.7 |
| 45,000 | 147.5 | 4.35 | −56.5 | −69.7 |
| 44,000 | 154.7 | 4.57 | −56.5 | −69.7 |
| 43,000 | 162.4 | 4.79 | −56.5 | −69.7 |
| 42,000 | 170.4 | 5.04 | −56.5 | −69.7 |
| 41,000 | 178.7 | 5.28 | −56.5 | −69.7 |
| 40,000 | 187.5 | 5.54 | −56.5 | −69.7 |
| 39,000 | 196.8 | 5.81 | −56.5 | −69.7 |
| 38,000 | 206.5 | 6.10 | −56.5 | −69.7 |
| 37,000 | 216.6 | 6.40 | −56.5 | −69.7 |
| 36,000 | 227.3 | 6.71 | −56.3 | −69.4 |
| 35,000 | 238.4 | 7.04 | −54.3 | −65.8 |
| 34,000 | 250.0 | 7.38 | −52.4 | −62.2 |
| 33,000 | 262.0 | 7.74 | −50.4 | −58.7 |
| 32,000 | 274.5 | 8.11 | −48.4 | −55.1 |
| 31,000 | 287.4 | 8.49 | −46.4 | −51.6 |
| 30,000 | 300.9 | 8.89 | −44.4 | −48.0 |
| 29,000 | 314.8 | 9.30 | −42.5 | −44.4 |
| 28,000 | 329.3 | 9.72 | −40.5 | −40.9 |
| 27,000 | 344.3 | 10.17 | −38.5 | −37.3 |
| 26,000 | 359.9 | 10.63 | −36.5 | −33.7 |
| 25,000 | 376.0 | 11.10 | −34.5 | −30.2 |
| 24,000 | 392.7 | 11.60 | −32.5 | −26.6 |
| 23,000 | 410.0 | 12.11 | −30.6 | −23.0 |
| 22,000 | 427.9 | 12.64 | −28.6 | −19.5 |
| 21,000 | 446.4 | 13.18 | −26.6 | −15.9 |
| 20,000 | 465.6 | 13.75 | −24.6 | −12.3 |
| 19,000 | 485.5 | 14.34 | −22.6 | −8.8 |
| 18,000 | 506.0 | 14.94 | −20.7 | −5.2 |
| 17,000 | 527.2 | 15.57 | −18.7 | −1.6 |
| 16,000 | 549.2 | 16.22 | −16.7 | 1.9 |
| 15,000 | 571.8 | 16.89 | −14.7 | 5.5 |
| 14,000 | 595.2 | 17058 | −12.7 | 9.1 |
| 13,000 | 619.4 | 18.29 | −10.8 | 12.6 |
| 12,000 | 644.4 | 19.03 | −8.8 | 16.2 |
| 11,000 | 670.2 | 19.79 | −6.8 | 19.8 |
| 10,000 | 696.8 | 20.58 | −4.8 | 23.3 |
| 9,000 | 724.3 | 21.39 | −2.8 | 26.9 |
| 8,000 | 752.6 | 22.22 | −0.8 | 30.5 |
| 7,000 | 781.8 | 23.09 | 1.1 | 34.0 |
| 6,000 | 812.0 | 23.98 | 3.1 | 37.6 |
| 5,000 | 843.1 | 24.90 | 5.1 | 41.2 |
| 4,000 | 875.1 | 25.84 | 7.1 | 44.7 |
| 3,000 | 908.1 | 26.82 | 9.1 | 48.6 |
| 2,000 | 942.1 | 27.82 | 11.0 | 51.9 |
| 1,000 | 977.2 | 28.86 | 13.0 | 55.4 |
| Sea level | 1,013.2 | 29.92 | 15.0 | 59.0 |

**Figure 3-12.** *Standard lapse rate table.*

values of temperature and pressure given in the table were determined by these averages. The height of the aircraft above the standard datum plane (29.92 "Hg and 15 °C) is the PA. *[Figure 3-13]*

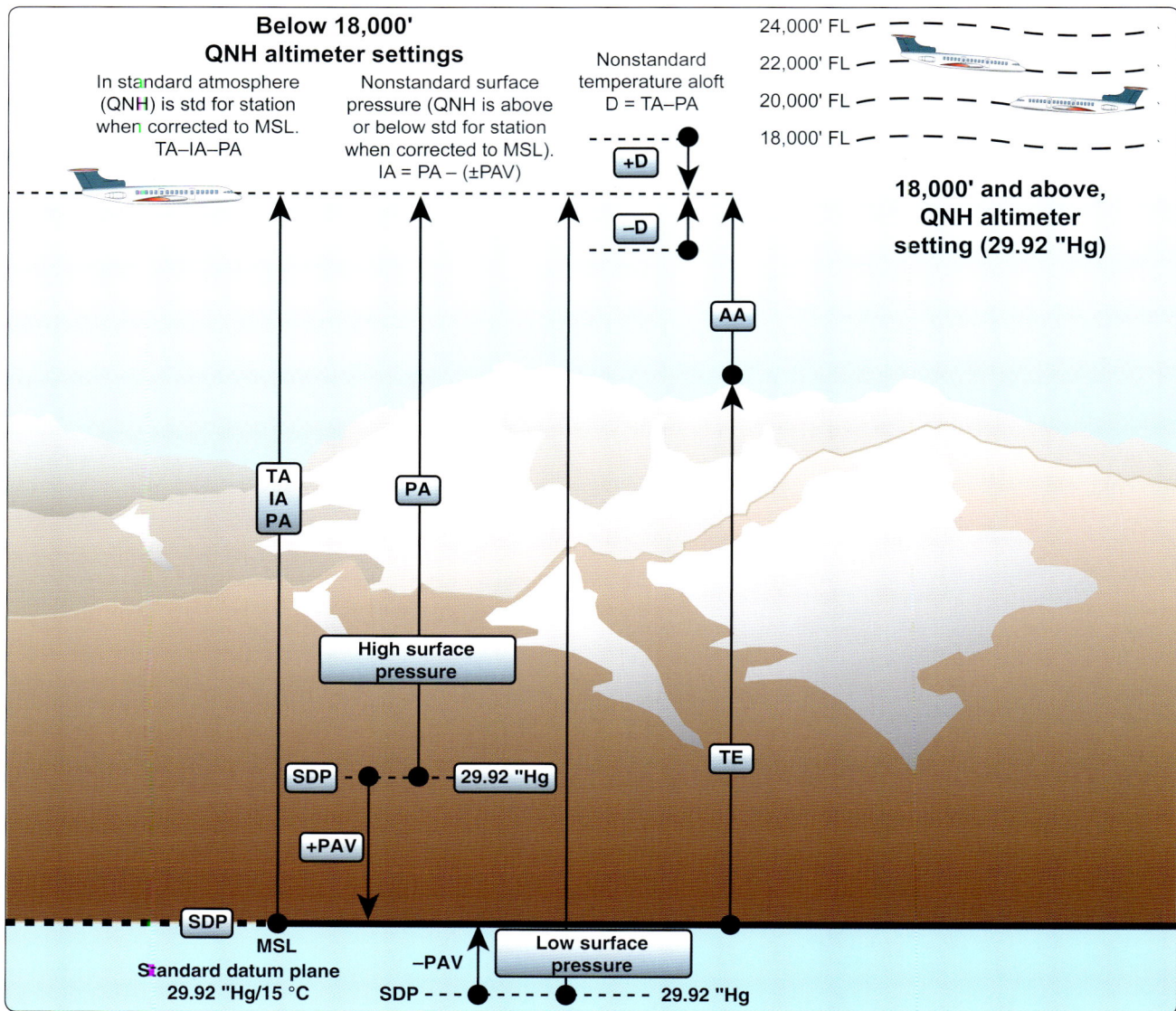

**Figure 3-13.** *Depiction of altimetry terms.*

## Pressure Altimeter Principles of Operation

The pressure altimeter is an aneroid barometer calibrated to indicate feet of altitude instead of pressure. As shown in *Figure 3-14*, the pointers are connected by a mechanical linkage to a set of aneroid cells. These aneroid cells expand or contract with changes in barometric pressure. In this manner, the cells assume a particular thickness at a given pressure level and, thereby, position the altitude pointers accordingly. On the face of the altimeter is a barometric scale that indicates the barometric pressure (expressed in inches of mercury) of the point or plane from which the instrument is measuring altitude. Turning the barometric pressure set knob on the altimeter manually changes this altimeter setting on the barometric scale and results in simultaneous movement of the altitude pointers to the corresponding altitude reading. Like all measurements, an altitude reading is meaningless if the point from which it starts is unknown. The face of the pressure altimeter supplies both values. The position of the pointers indicates the altitude in feet, and the barometric pressure appearing on the barometric scale is that of the reference plane above which the measurement is made.

### Altimeter Displays

#### Counter-Pointer Altimeter

The counter-pointer altimeter has a two-counter digital display unit located in the 9 o'clock position of the dial. The counter indicates altitude in 1,000 foot increments from zero to 80,000 feet. *[Figure 3-15]* A single conventional pointer indicates 100s of feet on the fixed circular scale. It makes one complete revolution per 1,000 feet of altitude change and, as it passes through the 900- to 1,000-foot area of the dial,

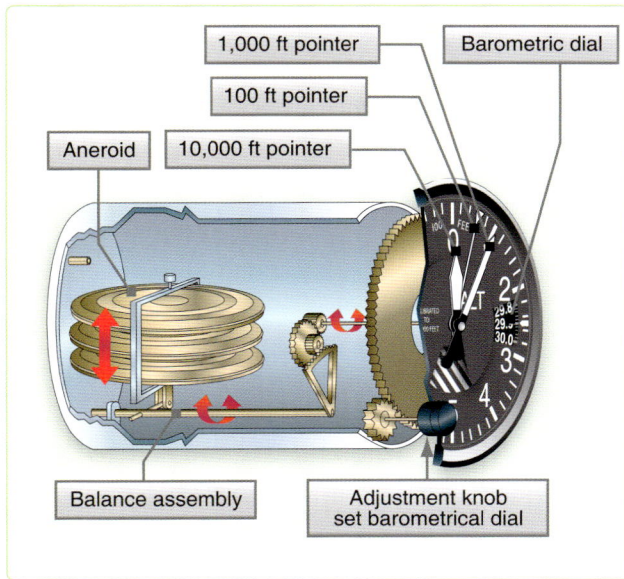

**Figure 3-14.** *Altimeter mechanical linkage.*

**Figure 3-15.** *Counter-pointer.*

the 1,000-foot counter is actuated. The shaft of the 1,000-foot counter in turn actuates the 10,000-foot counter at each 10,000-feet of altitude change. To determine the indicated altitude, first read the 1,000-foot counter and then add the 100-foot pointer indication.

It is possible to misinterpret the counter-pointer altimeter by 1,000 feet immediately before or after the 1,000-foot counter moves. This error is possible because the 1,000-foot counter changes when the 100-foot pointer is between the 900- and 1,000-foot position.

*Counter-Drum-Pointer Altimeter*
Aside from the familiar circular scale and 100-foot pointer, the counter-drum-pointer presentation differs somewhat in appearance from the present three-pointer altimeter. Starting at the left of the instrument illustrated in *Figure 3-16* and

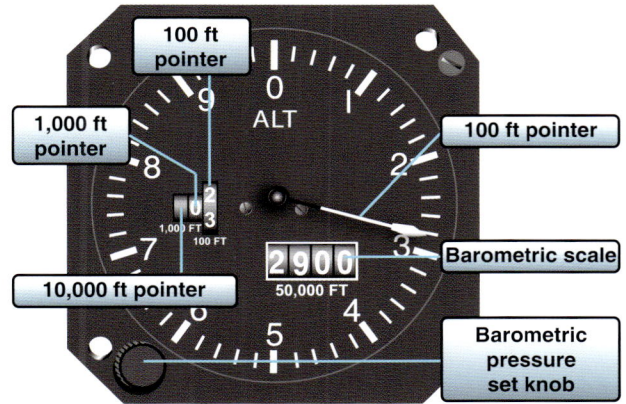

**Figure 3-16.** *Counter-drum pointer altimeter.*

reading from left to right, there are two counter windows and one drum window (white). The numerals presented in the counter windows indicate 10,000s and 1,000s of feet, respectively. The drum window numbers always follow the pointer number, thereby indicating 100s of feet.

Two methods may be used to read indicated pressure altitude on the counter-drum-pointer altimeter:

1. Read the counter-drum window, without referring to the 100-foot pointer, as a direct digital readout of both thousands and hundreds of feet;

2. Read the two counter indications, without referring to the drum, and then add the 100-foot pointer indication. The 100 foot pointer serves as a precise readout of values less than 100 feet.

The differential air pressure that is used to operate the counter-drum-pointer altimeter is processed by an altitude transducer where it is converted to electrical signals that drive the indicator. The transducer is also used to send digital signals to a transponder for purposes of automatic altitude reporting to Air Route Traffic Control Centers (ARTCC). A standby system is available for use if an electrical malfunction occurs. In the standby system, the altimeter receives static air pressure directly from the pitot-static system. When the instrument is operating in the standby system, the word STANDBY appears on the instrument face. A switch in the upper right-hand corner of the instrument is provided to return the instrument to its normal mode of operation. This switch may also be used to manually place the instrument in the STANDBY mode.

# Altimeter Errors

The pressure altimeter is subject to certain errors that fall in five general categories.

## Mechanical Error

A mechanical error is caused by misalignment of gears and levers that transmit the aneroid cell expansion and contraction to the pointers of the altimeter. This error is not constant and must be checked before each flight by the setting procedure.

## Scale Error

A scale error is caused by irregular expansion of the aneroid cells and is recorded on a scale correction card maintained for each altimeter in the instrument maintenance shop.

## Installation or Position Error

An installation, or position, error is caused by the airflow around the static ports. This error varies with the type of aircraft, airspeed, and altitude. The magnitude and direction of this error can be determined by referring to the performance data section in the aircraft technical order. An altimeter correction card is installed in some aircraft that combines the installation or position and scale errors. The card indicates the amount of correction required at different altitudes and airspeeds. Installation, or position, error may be considerable at high speeds and altitudes. Apply the corrections as outlined in the technical order or on the altimeter correction card.

## Reversal Error

A reversal error is caused by inducing false static pressure in the static system. It normally occurs during abrupt or large pitch changes. This error appears on the altimeter as a momentary indication in the opposite direction.

## Hysteresis Error

A hysteresis error is a lag in altitude indication caused by the elastic properties of the material within the altimeter. This occurs after an aircraft has maintained a constant altitude for an extended period of time and then makes a large, rapid altitude change. After a rapid descent, altimeter indications are higher than actual. This error is negligible during climbs and descents at slow rate or after maintaining a new altitude for a short period of time.

## Setting the Altimeter

The barometric scale is used to set a reference plane into the altimeter. Rotating the barometric pressure set knob increases or decreases the scale reading and the indicated altitude. Each .01 change on the barometric scale is equal to 10 feet of altitude. The majority of altimeters have mechanical stops at or just beyond the barometric scale limits (28.10 to 31.00). Attempting to adjust outside this range may cause damage to the instrument. Altimeters not equipped with mechanical stops near the barometric scale limits can be set with a 10,000 foot error. Therefore, when setting the altimeter, ensure the 10,000 foot pointer is reading correctly. Check the altimeter for accuracy before every flight.

To check and set the altimeter:

1. Set the current altimeter setting on the barometric scale.

2. Check the altimeter at a known elevation and note the error in feet.

3. Set the reported altimeter setting on the barometric scale and compare the indicated altitude to the elevation of a known cockpit.

## Nonstandard Atmospheric Effects

The altimeter setting is a correction for nonstandard surface pressure only. Atmospheric pressure is measured at each station and the value obtained is corrected to sea level according to the surveyed field elevation. Thus, the altimeter setting is the computed sea level pressure and should be considered valid only in close proximity to the station and the surface. It does not reflect nonstandard temperature nor distortion of atmospheric pressure at higher altitudes.

## Types of Altitude
### Indicated Altitude

Indicated altitude is the value of altitude that is displayed on the pressure altimeter.

### Calibrated Altitude

Calibrated altitude is indicated altitude corrected for installation or position error.

### Pressure Altitude (PA)

The height above the standard datum plane (29.92 "Hg and 15 °C) is PA. [Figure 3-13]

### Density Altitude (DA)

Density is mass per unit volume. The density of the air varies with temperature and with height. Warm air expands and is less dense than cold air. Normally, the higher the PA, the less dense the air becomes. The density of the air can be expressed in terms of the standard atmosphere. DA is the PA corrected for temperature in the density altitude window of the DR computer. This calculation converts the density of the air to the standard atmospheric altitude having the same density. DA is used in performance data and true airspeed (TAS) calculations.

## True Altitude (TA)

TA is the actual vertical distance above mean sea level (MSL), measured in feet. It can be determined by two methods:

1. Set the local altimeter setting on the barometric scale of the pressure altimeter to obtain the indicated true altitude (ITA). The ITA can then be resolved to TA by use of the DR computer. [Figure 3-17]

2. Measure altitude over water with an absolute altimeter.

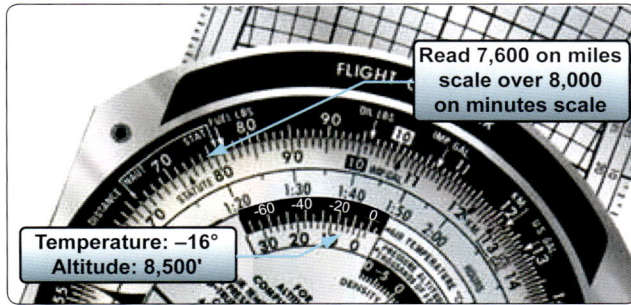

**Figure 3-17.** *Finding true altitude.*

The height above the terrain is called absolute altitude. It is computed by subtracting terrain elevation from TA, or it can be read directly from a radar altimeter.

The two altitudes most commonly accomplished on the computer are TA and DA. Nearly all DR computers have a window by which DA can be determined; however, be certain that the window is labeled density altitude.

### True Altitude (TA) Determination

In the space marked FOR ALTITUDE COMPUTATIONS are two scales: a centigrade scale in the window and a PA scale on the upper disk. When a PA is placed opposite the temperature at that height, all values on the outer (miles) scale are equal to the corresponding values on the inner (minutes) scale, increased or decreased by 2 percent for each 5.5 °C that the actual temperature differs from the standard temperature at that PA, as set in the window. Although the PA is set in the window, the ITA is used on the inner (minutes) scale for finding the TA, corrected for difference in temperature lapse rate.

PA = 8,500 feet
ITA = 8,000 feet
Air Temperature = –16 °C

Place PA (8,500 feet) opposite the temperature (–16) on the FOR ALTITUDE COMPUTATIONS scale. Opposite the ITA (8,000 feet) on the inner scale, read the TA (7,600 feet) on the outer scale. [Figure 3-l7]

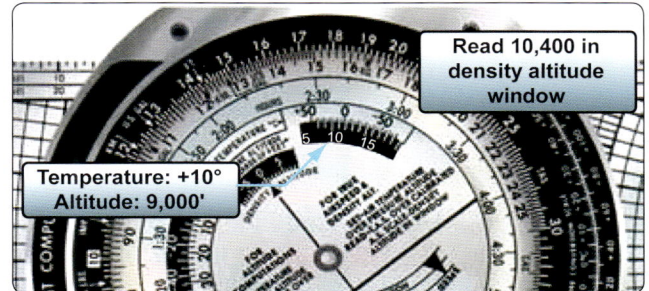

**Figure 3-18.** *Finding density altitude.*

### Density Altitude (DA) Determination

DA determination on the computer is accomplished by using the window just above FOR AIRSPEED AND DENSITY ALTITUDE COMPUTATIONS and the small window just above that marked DENSITY ALTITUDE.

PA = 9,000 feet
Air temperature = 10 °C

Place pressure altitude (9,000 feet) opposite air temperature (10) in window marked FOR AIRSPEED AND DENSITY ALTITUDE, read DA (10,400 feet). [Figure 3-18]

## Absolute Altimeter

Accurate absolute altitude is an important requisite for navigation, as well as for safe aviating. It is particularly important in pressure pattern navigation. Absolute altitude may be computed from the PA readings if the position of the aircraft is known, but the results are often inaccurate. Under changing atmospheric conditions, corrections applied to PA readings to obtain TAs are only approximate. In addition, any error made in determining the terrain elevations results in a corresponding error in the absolute altitude.

### Radar Altimeter High-Level

A typical high-level radar altimeter is designed to indicate absolute altitude of the aircraft up to 50,000 feet above the terrain, land, or water. This altimeter does not warn of approaching obstructions, such as mountains, because it measures altitude only to a point directly below the aircraft. [Figure 3-19]

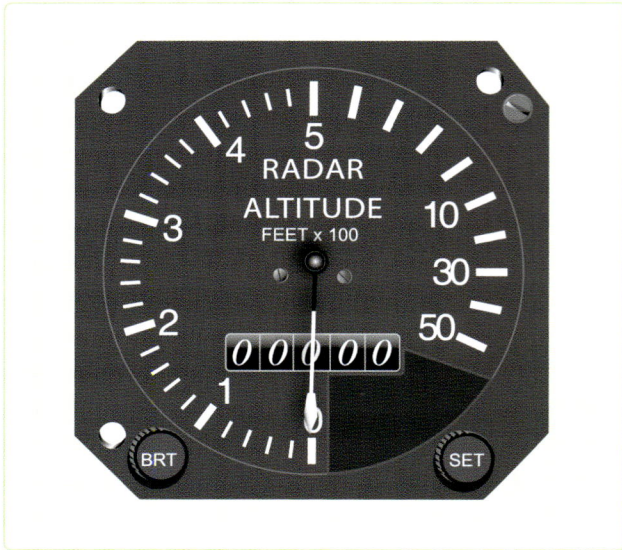

Figure 3-19. *High-level radar altimeter.*

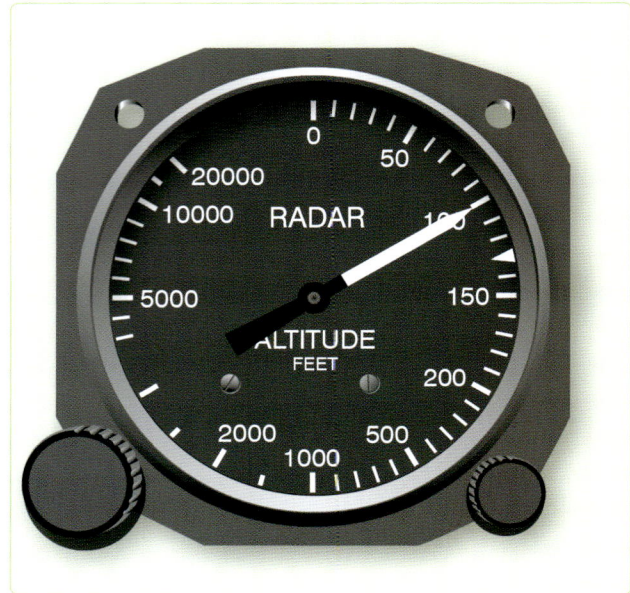

Figure 3-20. *Low-level radar altimeter.*

A typical set consists of the radar receiver-transmitter, height indicator, and antenna. The transmitter section of the receiver-transmitter unit develops recurring pulses of radio frequency (RF) energy that are delivered to the transmitter antenna located either flush mounted or on the underside of the aircraft. The transmitter antenna radiates the pulsed energy downward to reflect off the earth and return to the receiver antenna on the aircraft. The time consumed between transmission and reception of the RF pulse is determined only by the absolute altitude of the aircraft above the terrain since the radio wave velocity is constant.

The receiver antenna delivers the returned pulse to the receiver section of the receiver-transmitter unit where it is amplified and detected for presentation on the indicator unit. The radar altimeter indicator displays absolute altitude, which is used in pressure pattern navigation, terrain clearance, or as a backup for the PA.

### Low-Level Radar Altimeter

This type of altimeter provides a dial or digital indication of the altitude of the aircraft above the terrain. It is designed to eliminate the necessity of adding antennas or any other equipment external to the surface of the aircraft. This equipment may also be used in conjunction with automatic pilot or other devices requiring altitude limit data. *[Figure 3-20]*

Systems vary widely, but typically include a receiver-transmitter, height indicator, and electronic control amplifier. The height indicator contains the only operating control on the equipment. This instrument normally gives altitude readings up to 35,000 feet. If the instrument has an analog scale, the markings are usually logarithmic, graduated for

the low altitude portion of its range. A variable altitude limit indicator system is included to provide an indication of flight below a preset altitude.

To operate the equipment, turn the ON-LIMIT control to on. After warmup, the terrain clearance of the aircraft within the range of 0–20,000 is read directly from the single pointer on the indicator. *[Figure 3-20]* This pointer can be preset to any desired altitude by the ON-LIMIT control and is used as a reference for flying at fixed altitudes. The altitude can be maintained by observing the position of the pointer with respect to the small triangular marker instead of the actual altitude scale. In addition, a red light on the front of the indicator lights up when the aircraft is at or below the preset altitude. To turn off the equipment, it is only necessary to turn off the ON-LIMIT control on the indicator.

## Temperature

Determination of correct temperature is necessary for accurate computation of airspeed and altitude. Temperature, airspeed, and altitude are all closely interrelated, and the navigator must be familiar with each in order to work effectively and accurately.

### Temperature Gauges

The temperature gauge that is most commonly used in aircraft employs a bimetallic element. The instrument is a single unit consisting of a stainless steel stem that projects into the airstream and a head that contains the pointer and scale. The sensitive element in the outside end of the stem is covered by a radiation shield of brightly polished metal to cut down the amount of heat that the element might absorb by direct radiation from the sun. The bimetallic element (called the sensitive element) is so named because it consists of two strips of different metal alloys welded together. When the element is heated, one alloy expands more rapidly than the other, causing this element, which is shaped like a coil spring, to turn. This, in turn, causes the indicator needle to move on the pointer dial. Temperatures between –60 °C and +50 °C can be measured on this type of gauge.

## Temperature Scales

In the United States, temperature is usually expressed in terms of the Fahrenheit scale (°F). In aviation, temperature is customarily measured on the centigrade, or Celsius (°C), scale. Although aircraft thermometers are usually calibrated in °C, it is sometimes necessary to interconvert Fahrenheit and Centigrade temperatures. The following formulas may be used:

$$°F = (1.8 \times °C) + 32$$
$$°C = (°F – 32°) \div 1.8$$

Temperature error is the total effect of scale error and heat of compression error. Scale error is simply an erroneous reading of the pointer under standard conditions. It is difficult for a crewmember to evaluate this error without sensitive testing equipment. With this in mind, the reading of the indicator is considered correct and is called indicated air temperature (IAT).

The second error, heat of compression error, causes the instrument to read too warm. Heating occurs at high speeds from friction and the compression of air on the forward edge of the temperature probe. Thus, the IAT is always corrected by a minus correction factor to produce true air temperature (TAT). Heat of compression increases with TAS. The TAT can be obtained from the aircraft flight manual.

# Airspeed

Airspeed is the speed of the aircraft in relation to the air mass surrounding that aircraft.

## Pitot-Static System

Accurate airspeed measurement is obtained by means of a pitot-static system. The system consists of:

1. A tube mounted parallel to the longitudinal axis of the aircraft in an area that is free of turbulent air generated by the aircraft, and

2. A static source that provides still, or undisturbed, air pressure.

Ram and static pressures may be taken from a single pitot-static tube or from completely separate sources. A pitot-static tube usually has a baffle plate *[Figure 3-21]* to reduce turbulence and to prevent rain, ice, and dirt from entering

the tube. There may be one or more drain holes in the bottom of the tube to dispose of condensed moisture. A built-in electrical heating element, controlled by a switch inside the aircraft, prevents the formation of ice in the tube.

Reasonable care should be taken with the pitot-static system to ensure continuous, reliable service. The drain holes should be checked periodically to ensure they are not clogged. At the completion of each flight, a cover is placed over the intake end of the tube to prevent foreign objects and moisture from collecting in the tube.

## Principles of Operation of Airspeed Indicators

The heart of the airspeed indicator is a diaphragm that is sensitive to pressure changes. *Figure 3-21* shows it located inside the indicator case and connected to the ram air source in the pitot tube. The indicator case is sealed airtight and connected to the static pressure source. The differential pressure created by the relative effects of the impact and static pressures on the diaphragm causes it to expand or contract. As the speed of the aircraft increases, the impact pressure increases, causing the diaphragm to expand. Through mechanical linkage, the expansion is displayed as an increase in airspeed. This principle is used in the IAS meter, the TAS meter, and the Machmeter.

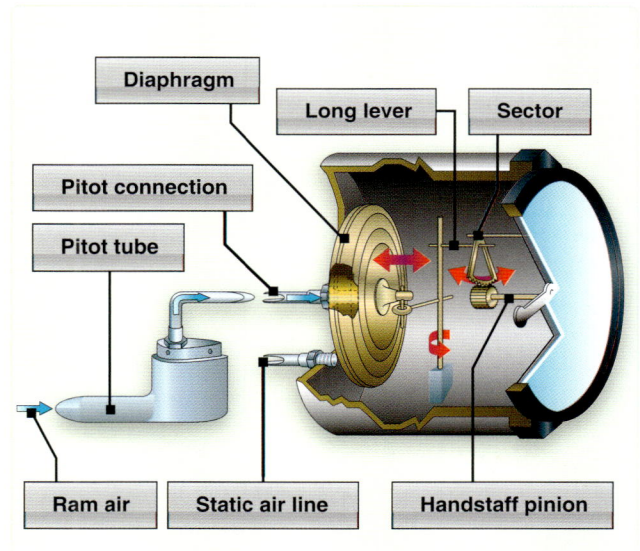

**Figure 3-21.** *Expansion and contraction of the diaphragm is transmitted to the pointer of the airspeed indicator.*

## Airspeed Definitions

There are many reasons for the difference between IAS and TAS. Some of the reasons are the error in the mechanical makeup of the instrument, the error caused by incorrect installation, and the fact that density and pressure of the atmosphere vary from standard conditions.

## Indicated Airspeed (IAS)

IAS is the uncorrected reading taken from the face of the indicator. It is the airspeed that the instrument shows on the dial.

## Basic Airspeed (BAS)

Basic airspeed (BAS) is the IAS corrected for instrument error. Each airspeed indicator has its own characteristics that cause it to differ from any other airspeed indicator. These differences may be caused by slightly different hairspring tensions, flexibility of the diaphragm, accuracy of the scale markings, or even the effect of temperature on the different metals in the indicator mechanism. The effect of temperature introduces an instrument error due to the variance in the coefficient of expansion of the different metals comprising the working mechanisms. This error can be removed by the installation of a bimetallic compensator within the mechanical linkage. This bimetallic compensator is installed and properly set at the factory, thereby eliminating the temperature error within the instrument. The accuracy of the airspeed indicator is also affected by the length and curvature of the pressure line from the pitot tube. These installation errors must be corrected mathematically. Installation, scale, and instrument errors are all combined under one title called instrument error. Instrument error is factory-determined to be within specified tolerances for various airspeeds. It is considered negligible or is accounted for in technical order tables and graphs.

## Calibrated Airspeed (CAS)

Calibrated airspeed (CAS) is basic airspeed corrected for pitot-static error or attitude of the aircraft. The pitot-static system of a moving aircraft has some error. Minor errors are found in the pitot section of the system. The major difficulty is encountered in the static pressure section. As the flight attitude of the aircraft changes, the pressure at the static inlets changes. This is caused by the airstream striking the inlet at an angle. Different types and locations of installations cause different errors. It is immaterial whether the status source is located in the pitot-static head or at some flush mounting on the aircraft. This error is essentially the same for all aircraft of the same model, and a correction can be computed by referring to tables in the appendix of the flight manual.

## Equivalent Airspeed (EAS)

Equivalant airspeed (EAS) is CAS corrected for compressibility. Compressibility becomes noticeable when the airspeed is great enough to create an impact pressure that causes the air molecules to be compressed within the impact chamber of the pitot tube. The amount of the compression is directly proportionate to the impact pressure. As the air is compressed, it causes the dynamic pressure to be greater than it should be. Therefore, the correction is a negative value.

The correction for compressibility error can be determined by referring to the performance data section of the aircraft flight manual or by using the F-correction factor on the DR computer.

## Density Airspeed (DAS)

Density airspeed (DAS) is calibrated airspeed corrected for PA and TAT. Pitot pressure varies not only with airspeed but also with air density. As the density of the atmosphere decreases with height, pitot pressure for a given airspeed must also decrease with height. Thus, an airspeed indicator operating in a less dense medium than that for which it was calibrated indicates an airspeed lower than true speed. The higher the altitude, the greater the discrepancy. The necessary correction can be found on the DR computer. Using the window on the computer above the area marked FOR AIRSPEED DENSITY ALTITUDE COMPUTATIONS, set the PA against the TAT. Opposite the CAS on the minutes scale, read the DAS on the miles scale. At lower airspeeds and altitudes, DAS may be taken as true airspeed with negligible error. However, at high speeds and altitudes, this is no longer true and compressibility error must be considered. (Compressibility error is explained in the equivalent airspeed section.) When DA is multiplied by the compressibility factor, the result is true airspeed.

## True Airspeed (TAS)

TAS is equivalent airspeed that has been corrected for air density error. By correcting EAS for TAT and PA, the navigator compensates for air density error and computes an accurate value of TAS. The TAS increases with altitude when the IAS remains constant. When the TAS remains constant, the IAS decreases with altitude. CAS and EAS can be determined by referring to the performance data section of the aircraft flight manual.

### Computing True Airspeed
### ICE-T Method

To compute TAS using the ICE-T method on the DR computer, solve, for each type of airspeed, in the order of I, C, E, and T; that is, change IAS to CAS, change CAS to EAS, and change EAS to TAS. This process is illustrated by the following sample problem. (Refer to definitions as necessary.)

Given:
PA = 30.000'
Temperature = –37 °C
IAS = 253 knots
Flight Manual Correction Factor = 2 knots

Find: CAS, EAS, and TAS

CAS is determined by algebraically adding to IAS the correction factor taken from the chart in the flight manual. (This correction is insignificant at low speeds but can be higher than 10 knots near Mach 1.) To correct CAS to EAS, use the chart on the slide of the computer entitled F-CORRECTION FACTORS FOR TAS. *[Figure 3-22]* Enter the chart with CAS and PA. The F factor is .96. Multiply CAS by .96 or take 96 percent of 255 knots. To do this, place 255 knots on the inner scale under the 10 index on the outer scale. Locate 96 on the outer scale and read EAS on the inner scale: 245 knots.

Now, we need to correct EAS for temperature and altitude to get TAS. As shown in *Figure 3-22,* in the window marked FOR AIRSPEED AND DENSITY ALTITUDE COMPUTATIONS, place temperature over PA. Locate the EAS of 245 knots on the inner scale and read TAS on the outer scale. The TAS is 408 knots.

### Alternate TAS Method

There is an alternate method of finding TAS when given CAS. The instructions for alternate solution are printed on the computer directly below the F factor table (Multiply F factor by TAS obtained with computer to obtain TAS corrected for compressibility). Mathematically, the answer should be the same regardless of the procedure being used, but the ICE-T method is used most often because the computation can be worked backwards from TAS. If the desire is to maintain a constant TAS, determine what IAS to fly by working the ICE-T method in reverse (also known as reverse ICE-T). *[Figure 3-23]*

### Machmeters

Machmeters indicate the ratio of aircraft speed to the speed of sound at any particular altitude and temperature during flight. It is often necessary to convert TAS to a Mach number or vice versa. Instructions are clearly written on the computer in the center portion of the circular slide rule. Locate the window marked FOR AIRSPEED AND DENSITY ALTITUDE COMPUTATIONS and rotate the disk until the window points to the top of the computer (toward the l0 index on the outer scale). Within the window is an arrow entitled MACH NO. INDEX. *[Figure 3-24]* To obtain TAS from a given Mach number, set air temperature over the MACH NO. INDEX and, opposite the Mach number on the MINUTES scale, read the TAS on the outer scale.

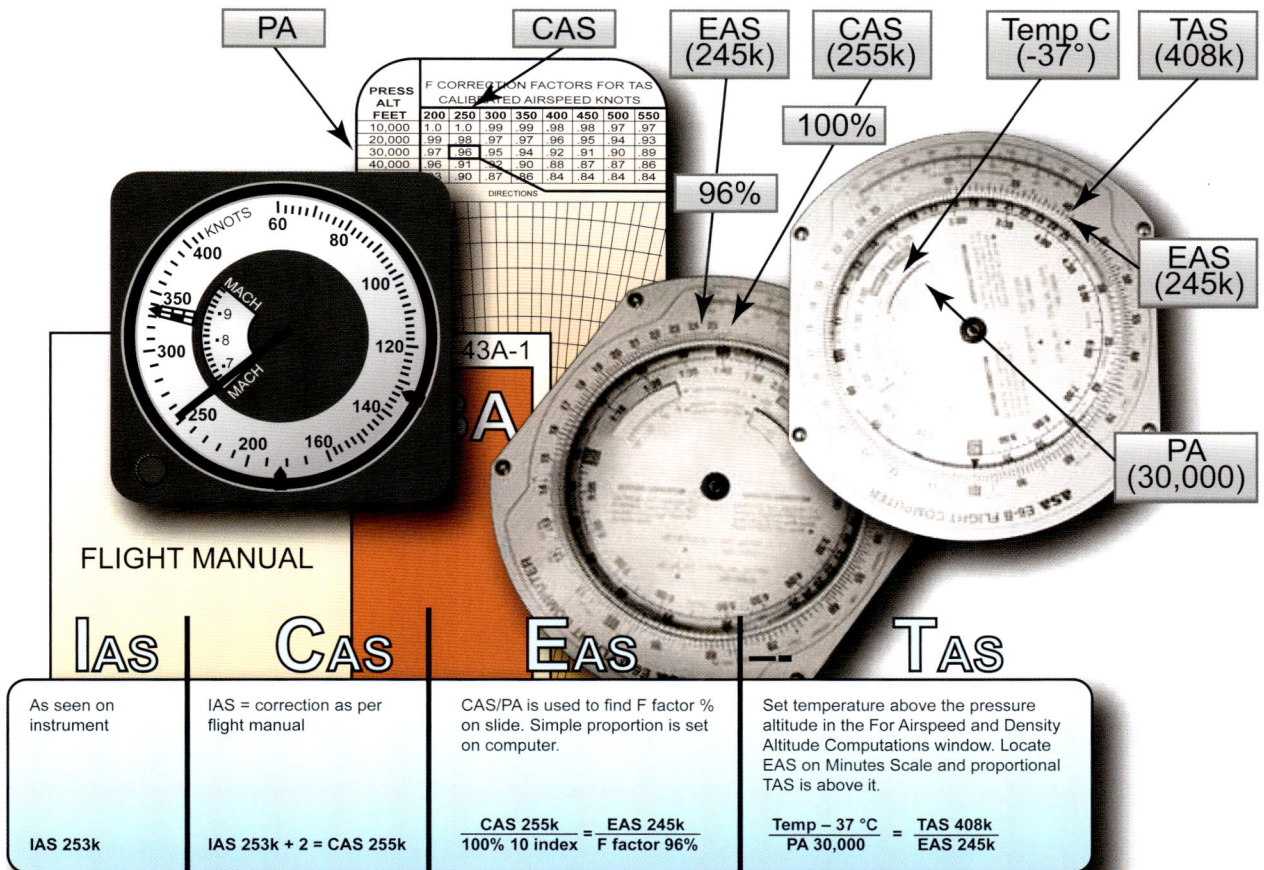

Figure 3-22. *ICE-T method.*

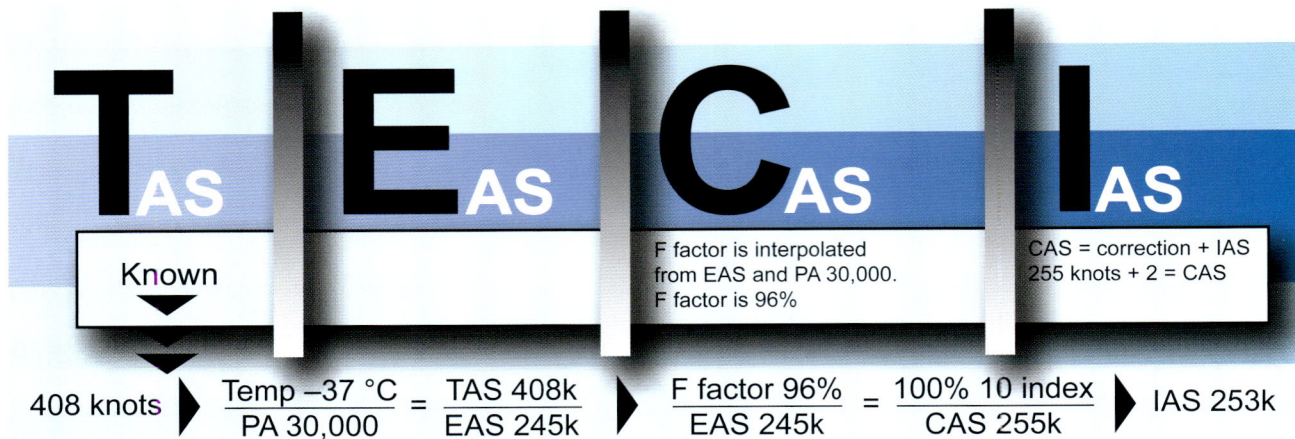

Figure 3-23. *ICE-T in reverse.*

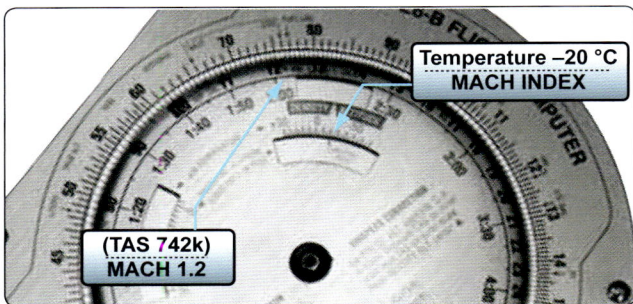

Figure 3-24. *Finding true airspeed from Mach number.*

Example: If you are planning to maintain Mach 1.2 on a cross-country flight, place the air temperature at flight altitude over the MACH NO. INDEX. Read the TAS on the outer scale opposite 1.2 on the inner scale. If the temperature is –20 °C, the TAS is 742 knots.

## Airspeed Indicators

The combined airspeed-Mach indicator, shown in *Figure 3-25*, is usually found in high-performance aircraft or where instrument panel space is limited. It simultaneously displays IAS, indicated Mach number, and maximum allowable airspeed. It contains a differential pressure diaphragm and two aneroid cells. The diaphragm drives the airspeed-Mach pointer. One aneroid cell rotates the Mach scale, permitting IAS and Mach number to be read simultaneously. The second aneroid cell drives the maximum allowable airspeed pointer. This pointer is preset to the aircraft's maximum IAS. Unlike the maximum IAS and unlike the maximum allowable airspeed, Mach number increases with altitude. An airspeed marker set knob positions a movable airspeed marker. This marker serves as a memory reference for desired airspeed.

## Air Data Computer

The air data computer is an electro-pneumatic unit that uses pitot and static pressures and total air temperature to compute outputs for various systems. These output parameters of voltage and resistance represent functions of altitude, Mach number, TAS, computed airspeed, and static air temperature. Air data computer outputs are used with the flight director computers, automatic flight controls, cabin pressurization equipment, and normal basic indicators. The air data computer provides extreme accuracy and increased reliability.

## Doppler

Doppler radar provides the navigator with continuous, instantaneous, and accurate readings of groundspeed (GS) and drift angle in all weather conditions, both over land and water. It does this automatically with equipment that is of practical size and weight. Its operation makes use of the Doppler effect.

Figure 3-25. *Combined airspeed Mach indicator.*

Two basic Doppler radar systems exist: the four-beam and the three-beam. Both types use either continuous-wave (CW) or pulse-wave (PW) transmission. CW transmission requires one antenna for transmission and a second antenna for reception. Both systems use an X-shaped beam configuration. Groundspeed is computed by comparing Doppler shift between front and rear beams, and drift angle is computed by comparing the shift between the left and right beams.

Doppler is not the only source of drift angle and GS. The same basic information is available on virtually all inertial navigation systems (INS), and now global positioning system (GPS) computers can also give accurate information under a wider range of conditions.

## Chapter Summary

While the operation of some of the equipment in this chapter seems crude and primitive compared to state-of-the-art navigation computers currently in use, these instruments still have two very important uses—they act as crosschecks to ensure the computers are functioning properly, and they become your primary means of navigation when the computers are not.

## Chapter 4

# Dead Reckoning

## Basic Navigation

### Dead Reckoning (DR)

Having discussed the basic instruments available to the navigator, this chapter reviews the mechanics of dead reckoning (DR) procedures, plotting, determining wind effect, and MB-4 computer solutions. Using basic skills in DR procedures, a navigator can predict aircraft positions in the event more reliable navigation equipment is unavailable or not operative. Therefore, a good foundation in DR is imperative for the navigator.

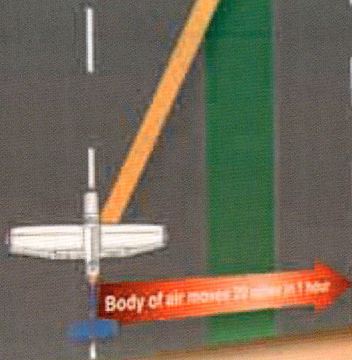

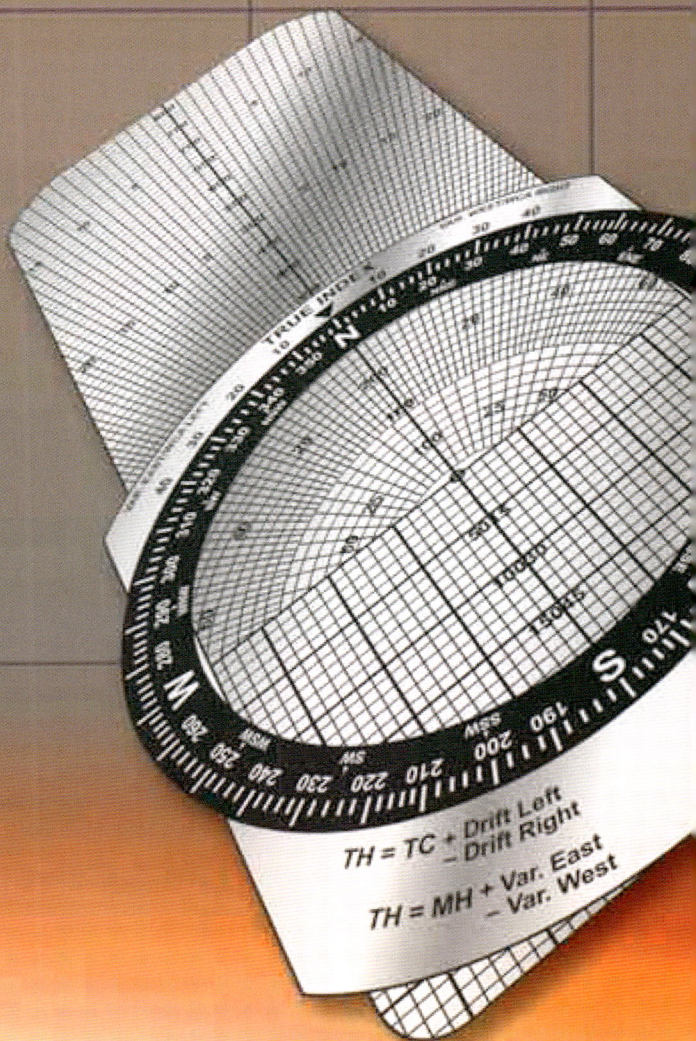

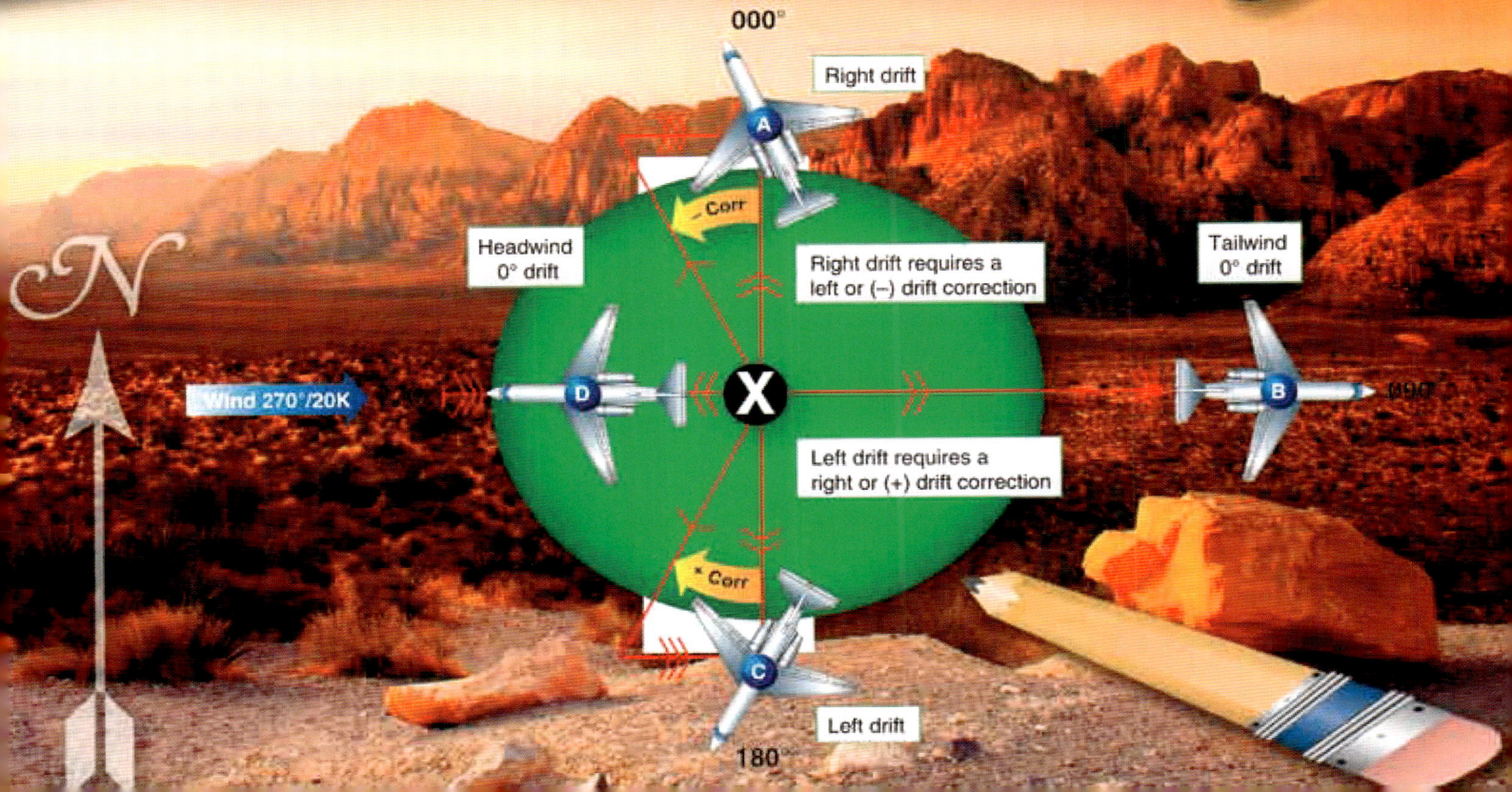

20 NM

Body of air moves 20 miles in 1 hour

$$TH = TC + \text{Drift Left}/- \text{Drift Right}$$

$$TH = MH + \text{Var. East}/- \text{Var. West}$$

000°

Right drift

A

Headwind
0° drift

– Corr

Right drift requires a
left or (–) drift correction

Tailwind
0° drift

N

Wind 270°/20K

D

X

B

090°

Left drift requires a
right or (+) drift correction

+ Corr

C

Left drift

180°

## Plotting

Chart work should be an accurate and graphic picture of the progress of the aircraft from departure to destination and, with the log, should serve as a complete record of the flight. Thus, it also follows that the navigator must be familiar with and use accepted standard symbols and labels on charts. *[Figure 4-1]* See Appendix A for additional chart and navigation symbols.

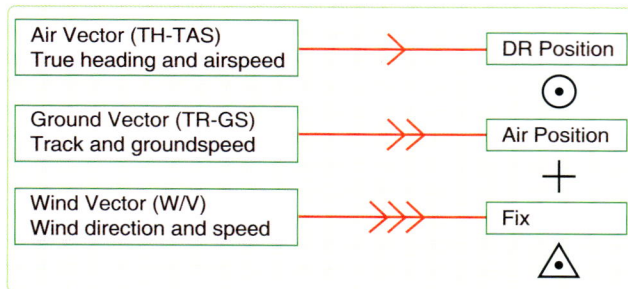

**Figure 4-1.** *Standard plotting symbols.*

## Terms

Several terms have been mentioned in earlier portions of this handbook. Precise definitions of these terms must now be understood before the mechanics of chart work are learned.

- True Course (TC)—the intended horizontal direction of travel over the surface of the earth, expressed as an angle measured clockwise from true north (000°) through 360°.

- Course Line—the horizontal component of the intended path of the aircraft comprising both direction and magnitude or distance.

- Track—the horizontal component of the actual path of the aircraft over the surface of the earth track may, but very seldom does, coincide with the TC or intended path of the aircraft. The difference between the two is caused by an inability to predict perfectly all inflight conditions.

- True Heading (TH)—the horizontal direction in which an aircraft is pointed. More precisely, it is the angle measured clockwise from true north through 360° to the longitudinal axis of the aircraft. The difference between track and TH is caused by wind and is called drift.

- Groundspeed (GS)—the speed of the aircraft over the ground. It may be expressed in nautical miles (NM), statute miles (SM), or kilometers (km) per hour, but, a navigator uses NM per hour (knots).

- True Airspeed (TAS)—the rate of motion of an aircraft relative to the airmass surrounding it. Since the airmass is usually in motion in relation to the ground, airspeed and GS seldom are the same.

- Dead Reckoning Position (DR Position)—a point in relation to the earth established by keeping an accurate account of time, GS, and track since the last known position. It may also be defined as the position obtained by applying wind effect to the TH and TAS of the aircraft.

- Fix—a position determined from terrestrial, electronic, or astronomical data.

- Air Position (AP)—the location of the aircraft in relation to the airmass surrounding it. TH and TAS are the components of the vector used to establish an AP.

- Most Probable Position (MPP)—a position determined with partial reference to a DR position and partial reference to a fixing aid.

## Plotting Equipment

A fine-tipped pencil, a good pair of dividers, and a plotter are imperative for accurate chart work.

### *Dividers*

Dividers should be manipulated with one hand, leaving the other free to use the plotter, pencil, or chart as necessary. Some navigation dividers have a tension screw that can be adjusted to prevent the dividers from becoming either too stiff or too loose for convenient use. Adjust the points of the dividers to approximately equal length. A small screwdriver, required for these adjustments, should be a part of the navigator's equipment.

### *Plotter*

A common plotter is shown in *Figure 4-2*. This plotter is a semicircular protractor with a straight edge attached to it. A small hole at the base of the protractor portion indicates the center of the arc of the angular scale. Two complete scales cover the outer edge of the protractor and are graduated in degrees. An abbreviated inner scale measures the angle from the vertical. *[Figures 4-3 and 4-4]* The angle measured is the angle between the meridian and the straight line. The outer scale is used to read all angles between north through east to south, and the inner scale is used to read all angles between south through west to north.

## Plotting Procedures for Mercator Charts
### *Preparation*

Many charts and plotting sheets are printed on the Mercator projection. Before starting any plot, note the scale and projection of the chart and check the date to make sure that it is the latest edition. The latitude scale is used to represent NM. The longitude scale should never be used to measure distance. Some charts carry a linear scale in the margin, and, where present, it indicates that the same scale may be used anywhere on the chart.

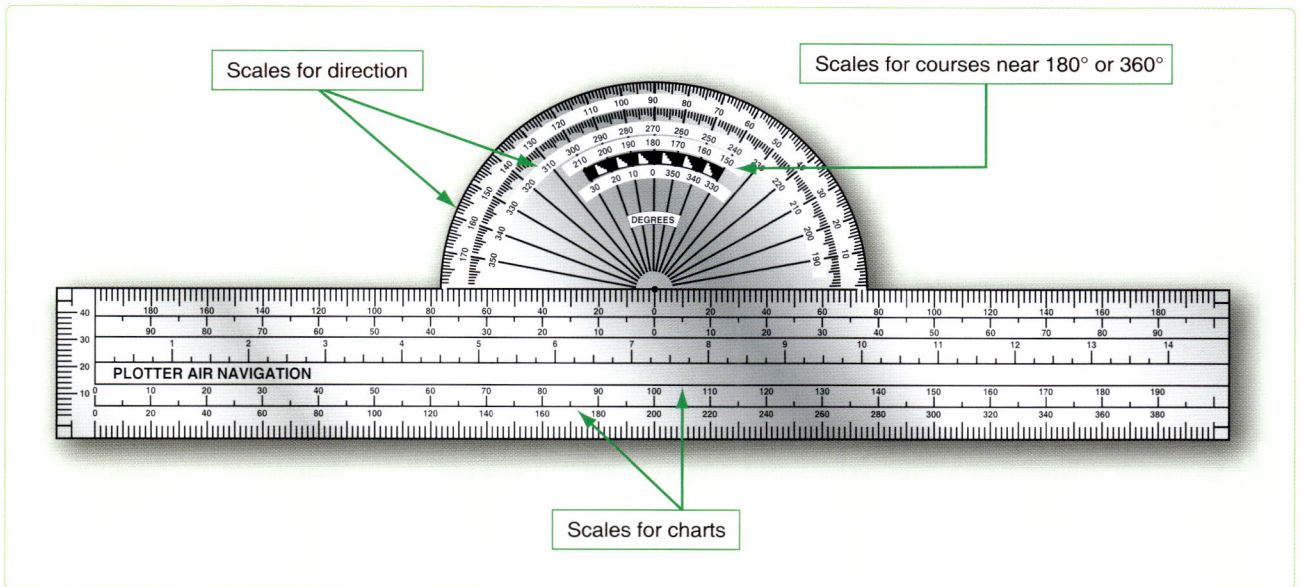

**Figure 4-2.** *Plotter.*

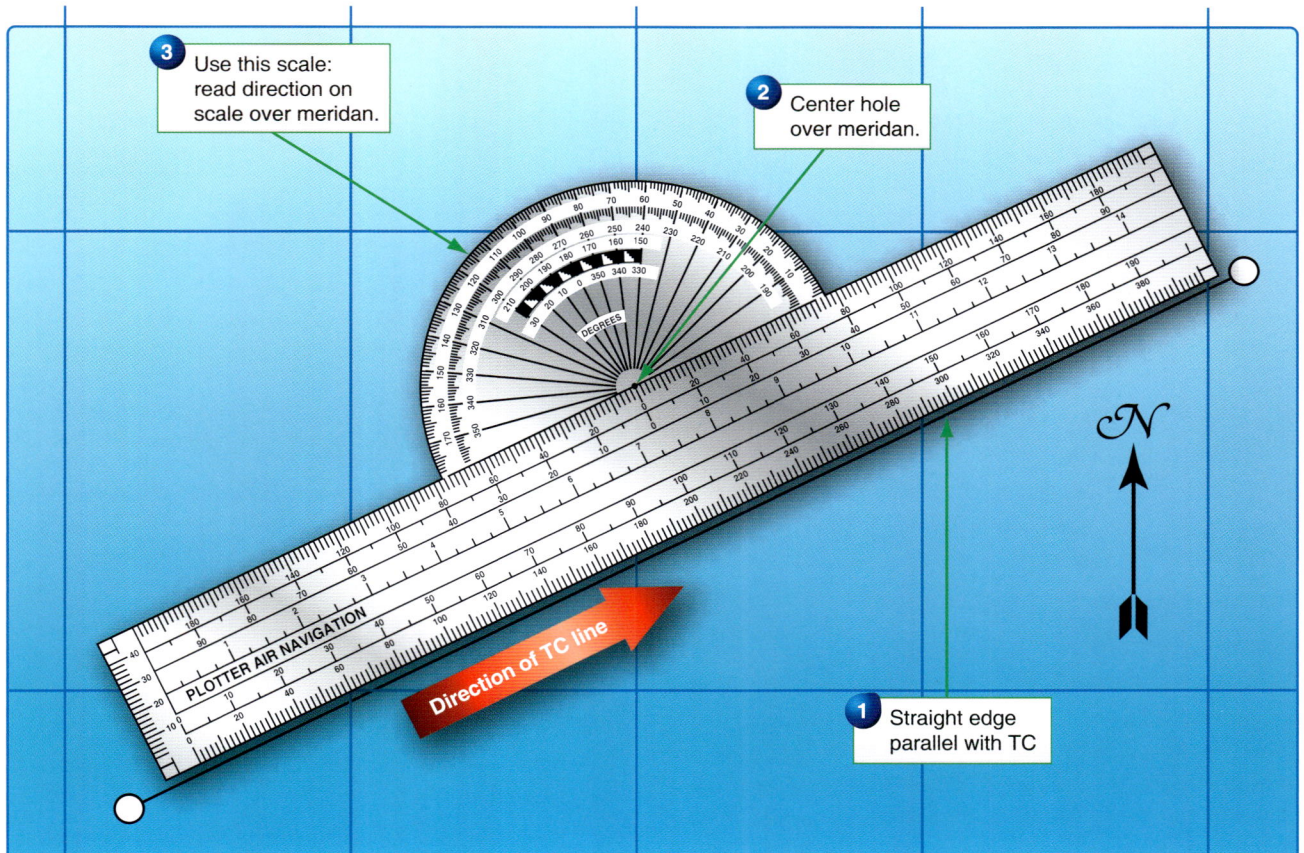

**Figure 4-3.** *Measuring true course.*

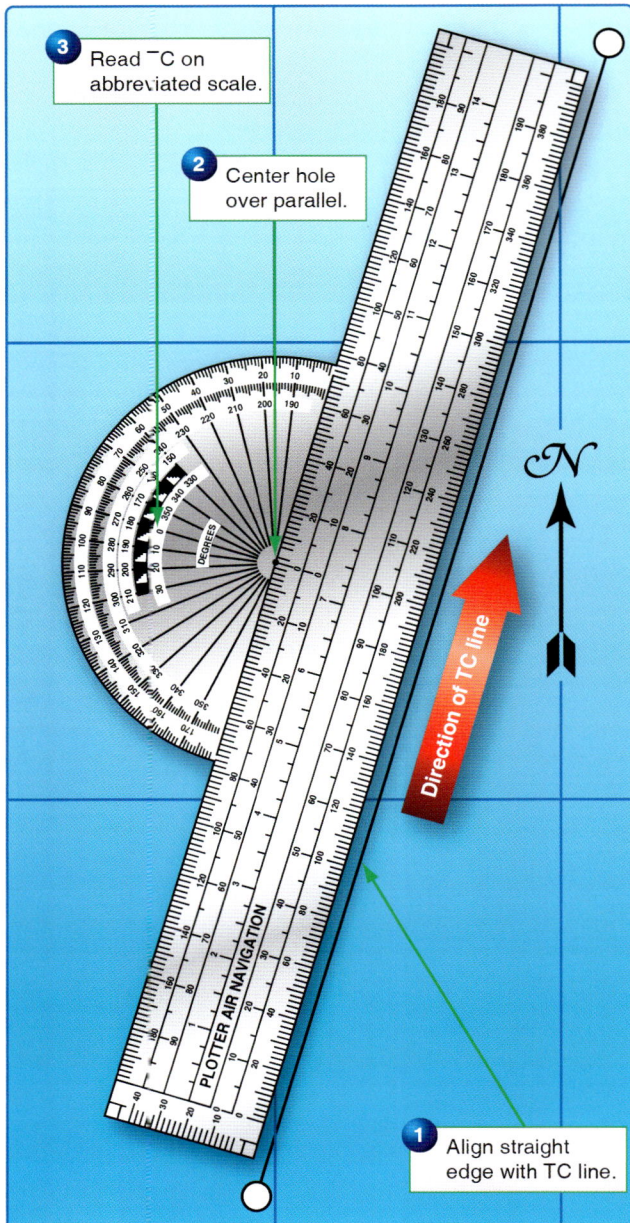

**Figure 4-4.** *Measuring true course near 180° or 360°.*

## Plotting Positions

On most Mercator charts, the spacing between meridians and parallels is widely spaced, necessitating the use of dividers. There are several methods by which positions can be plotted on Mercator charts. *[Figure 4-5]* Place the straight edge of the plotter in a vertical position at the desired longitude. Set the dividers to the desired number of minutes of latitude. Hold one point against the straight edge on the parallel of latitude corresponding to the whole degree of latitude given. Let the other point also rest against the straight edge and lightly prick the chart. This marks the desired position. In measuring the latitude and longitude of a position already plotted, reverse the procedure.

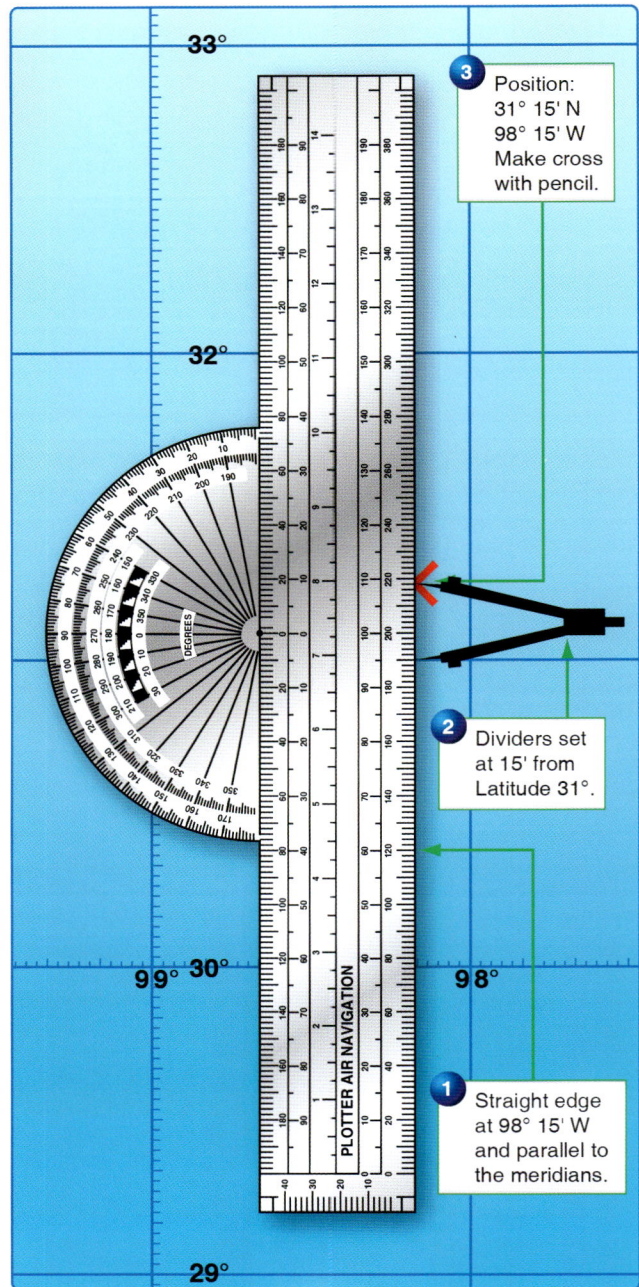

**Figure 4-5.** *Plotting positions on a Mercator.*

## Plotting and Measuring Courses

Step 1—Plot departure and destination on the chart. *[Figure 4-6]*

Step 2—Draw the course line between the two points. If they are close together, the straight edge of the plotter can be used. If they are far apart, two plotters can be used together or a longer straight edge can be used. If none of these methods is adequate, fold the edge of the charts so that the fold connects the departure and destination points, and make a series of pencil marks along the edge. A plotter or straight

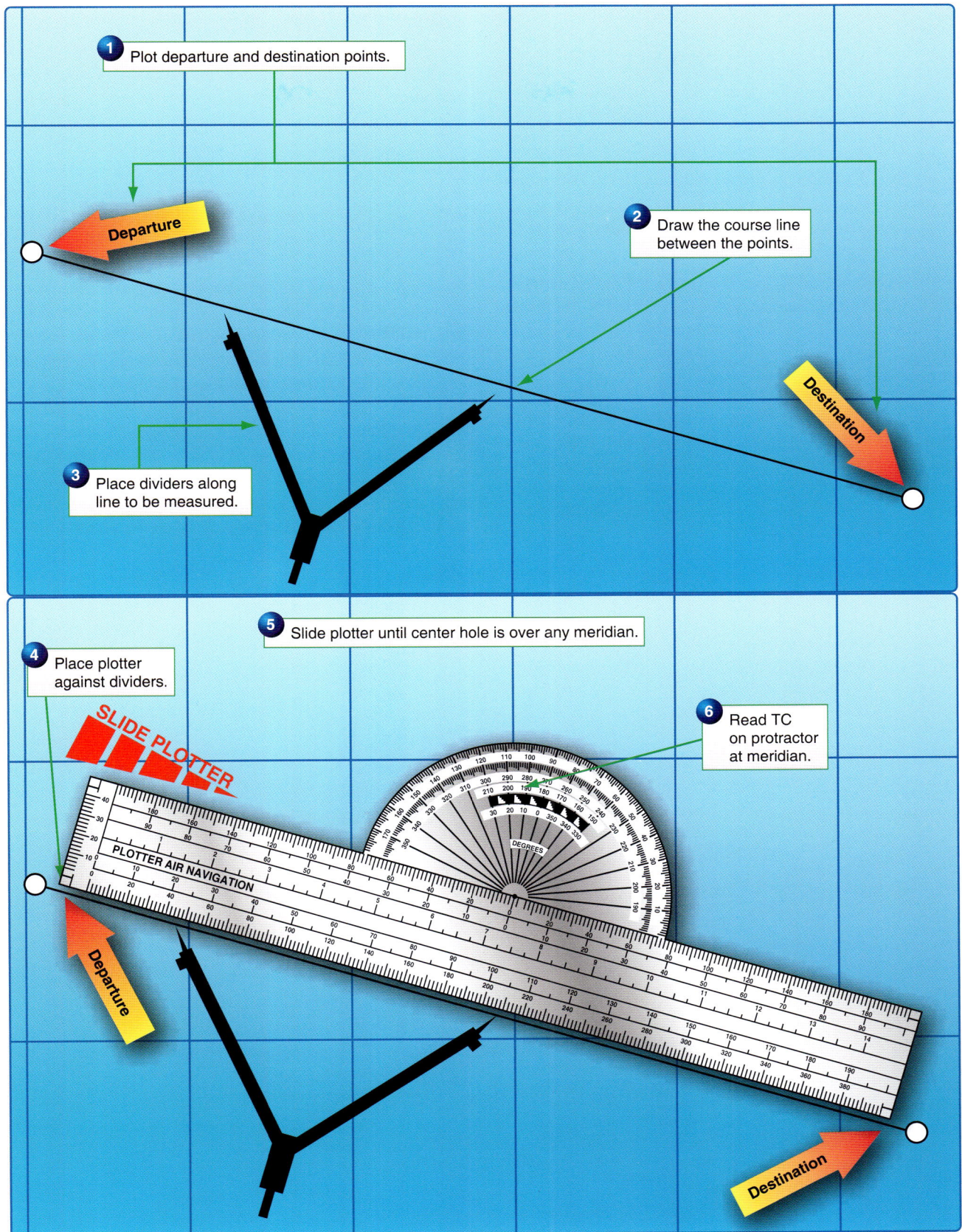

**Figure 4-6.** *Reading direction of a course line.*

1. Plot departure and destination points.

2. Draw the course line between the points.

3. Place dividers along line to be measured.

Departure

Destination

4. Place plotter against dividers.

5. Slide plotter until center hole is over any meridian.

6. Read TC on protractor at meridian.

SLIDE PLOTTER

PLOTTER AIR NAVIGATION

DEGREES

Departure

Destination

edge can then be used to connect the points where the chart is unfolded. After the course line has been plotted, the next step is to determine its direction.

Step 3—Place the points of the dividers, or a pencil, anywhere along the line to be measured.

Step 4—Place the plotter against the dividers.

Step 5—Slide the plotter until the center hole is over the midmeridian. Make a mental estimate of the approximate direction to avoid obtaining a reciprocal course.

Step 6—Read TC on the protractor at the meridian. Using the midmeridian gives an average TC for the leg.

### Plotting Course from Given Position

A course from a given position can be plotted quickly in the following manner. Place the point of a pencil on the position and slide the plotter along this point, rotating it as necessary, until the center hole and the figure on the protractor representing the desired direction are lined up with the same meridian. Hold the plotter in place and draw the line along the straight edge. [Figure 4-7]

### Measuring Distance

One of the disadvantages of the Mercator chart is the lack of a constant scale. If the two points between which the distance is to be measured are approximately in a north-south direction, and the total distance between them can be spanned, the distance can be measured on the latitude scale opposite the midpoint. However, the total distance between any two points that do not lie approximately north or south of each other should not be spanned unless the distance is short.

In the measurement of long distances, select a midlatitude lying approximately halfway between the latitudes of the two points. By using dividers set to a convenient, reasonably short distance, such as 60 NM picked off at the midlatitude scale, determine an approximate distance by marking off units along the line to be measured. [Figure 4-8]

The scale at the midlatitude is accurate enough if the course line does not cover more than 5° of latitude (somewhat less in high latitudes). If the course line exceeds this amount or if it crosses the equator, divide it into two or more legs and measure the length of each leg with the scale of its own midlatitude.

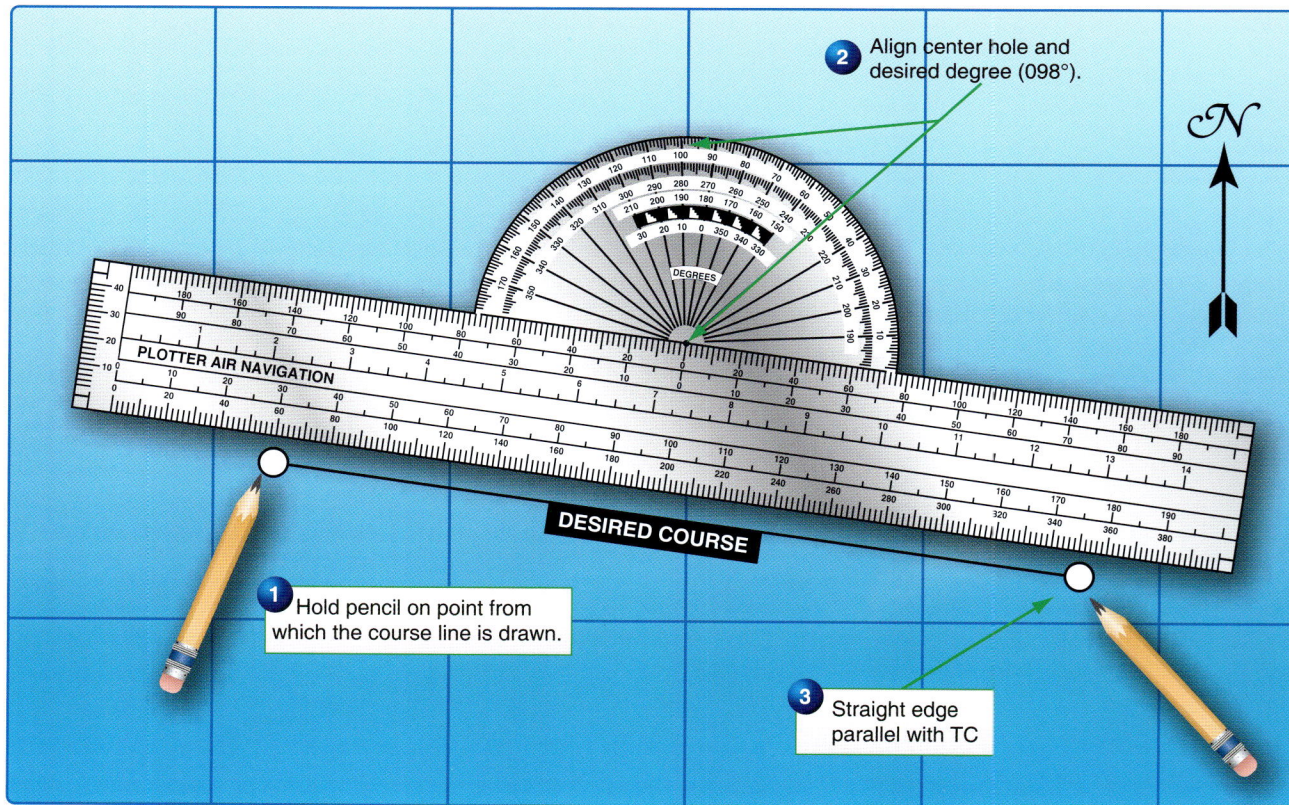

**Figure 4-7.** *Plotting course from given direction.*

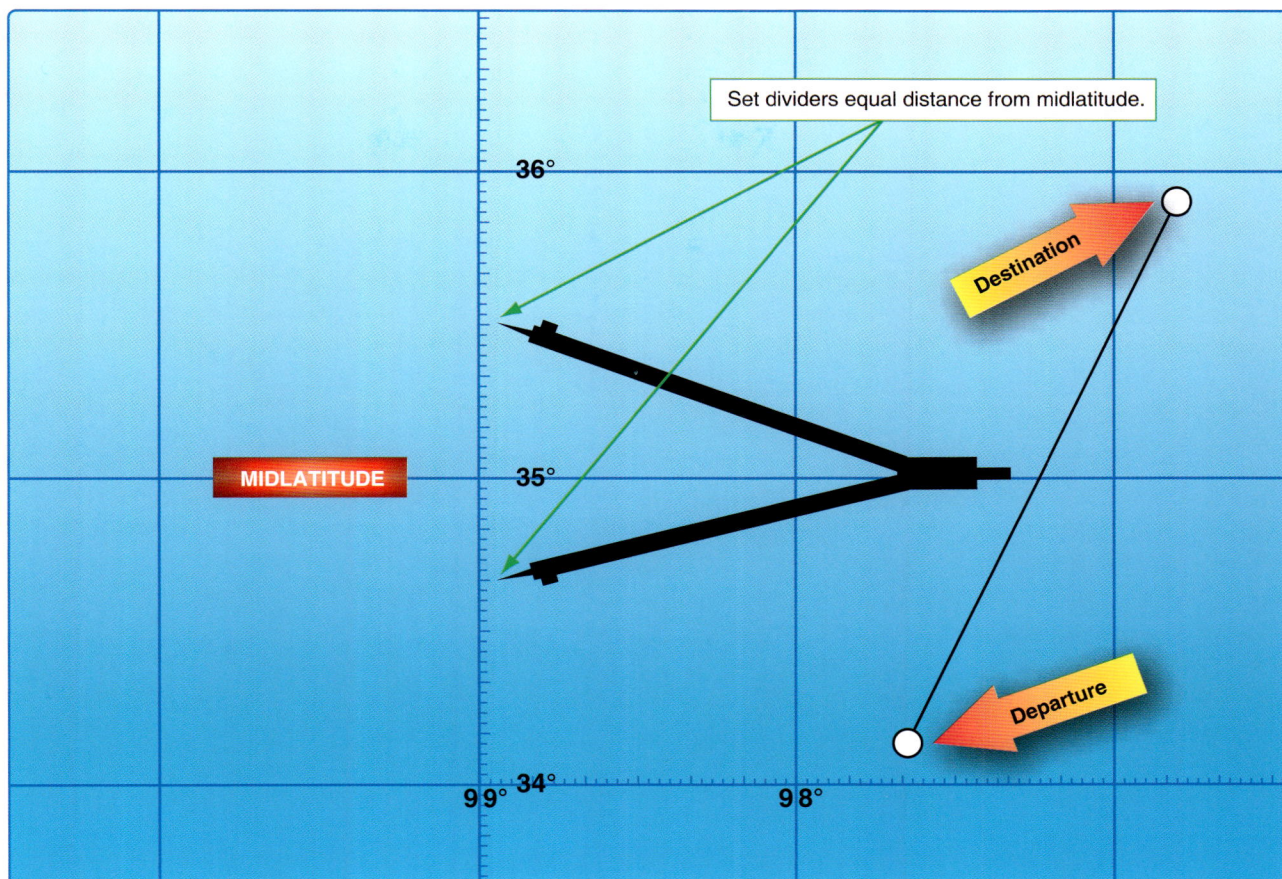

**Figure 4-8.** *Midlatitude scale.*

## Plotting Procedures for Lambert Conformal and Gnomonic Charts

### Plotting Positions

On a Lambert conformal chart, the meridians are not parallel as on a Mercator chart. Therefore, plotting a position by the method described under Mercator charts may not be accurate. On small scale charts, or where there is marked convergence, the plotter should intersect two graduated parallels of latitude at the desired longitude rather than parallel to the meridian. Then, mark off the desired latitude with dividers. On a large scale chart, the meridians are so nearly parallel that this precaution is unnecessary. The scale on all parts of a Lambert conformal chart is essentially constant. Therefore, it is not absolutely necessary to pick off minutes of latitude near any particular parallel except in the most precise work.

### Plotting and Measuring Courses

Any straight line plotted on a Lambert conformal chart is approximately an arc of a great circle. On long distance flights, this feature is advantageous since the great circle course line can be plotted as easily as a rhumb line on a Mercator chart.

However, for shorter distances where the difference between the great circle and rhumb line is negligible, the rhumb line is more desirable because a constant heading can be held. For such distances, the approximate direction of the rhumb line course can be found by measuring the great circle course at midmeridian. *[Figure 4-9]* In this case, the track is not quite the same as that indicated by the course line drawn on the chart, since the actual track (a rhumb line) appears as a curve convex to the equator on a Lambert conformal chart, while the course line (approximately a great circle) appears as a straight line. Near midmeridian, the two have approximately the same direction (except for very long distances) along an oblique course line. *[Figure 4-10]*

For long distances involving great circle courses, it is not possible to change heading continually, as is necessary when following a great circle exactly, and it is customary to divide the great circle into a series of legs, each covering about 5° of longitude. The direction of the rhumb line connecting the ends of each leg is found at its midmeridian.

4-7

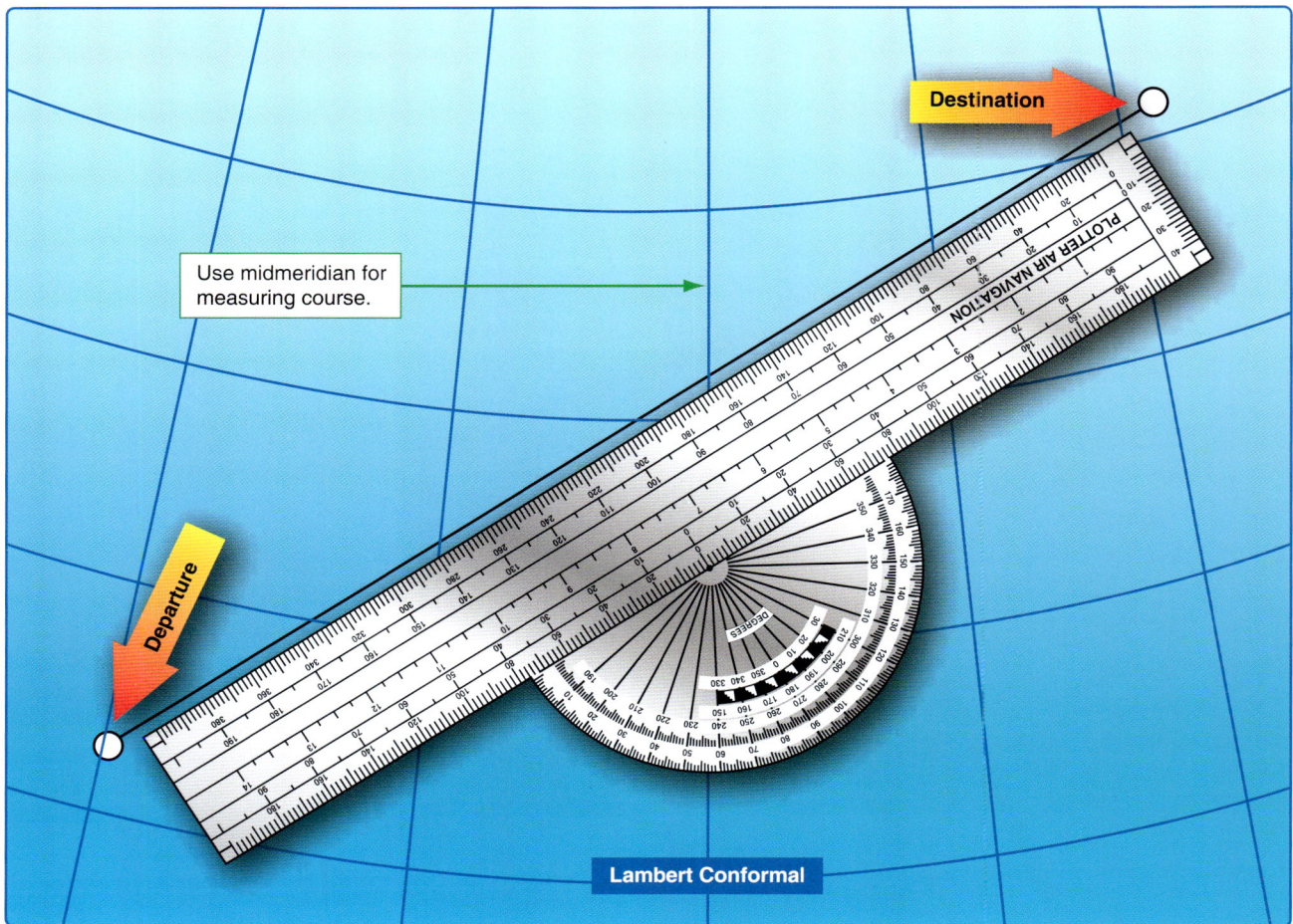

**Figure 4-9.** *Use midmeridian to measure course on a Lambert conformal.*

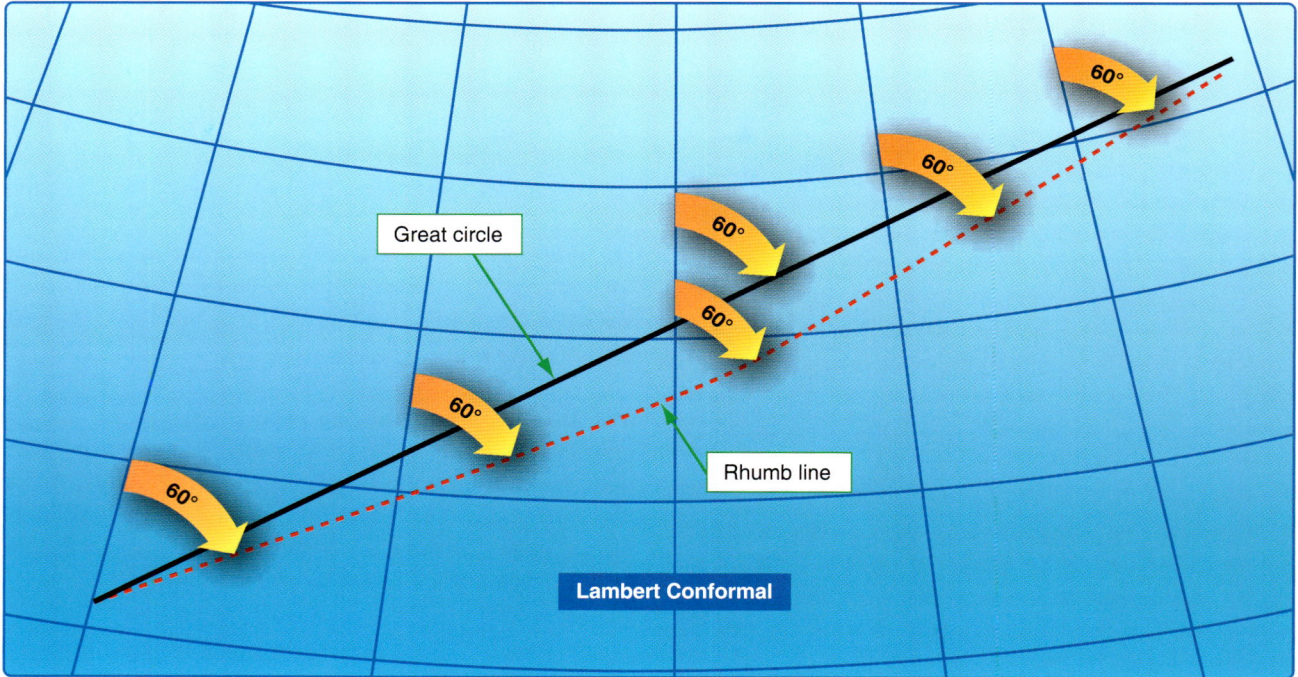

**Figure 4-10.** *At meridian, rhumb line and great circle have approximately the same direction.*

## Measuring Distance

As previously stated, the scale on a Lambert conformal chart is practically constant, making it possible to use any part of a meridian graduated in minutes of latitude to measure NM.

## Plotting on a Gnomonic Chart

Gnomonic charts are used mainly for planning great circle routes. Since any straight line on a gnomonic chart is an arc of a great circle, a straight line drawn from the point of departure to destination gives a great circle route. Once obtained, this great circle route is transferred to a Mercator chart by breaking the route into segments. *[Figure 4-11]*

## *Plotting Hints*

The following suggestions should prove helpful in developing good plotting procedures:

1. Measure all directions and distances carefully. Double-check all measurements, computations, and positions.

2. Avoid plotting unnecessary lines. If a line serves no purpose, erase it.

3. Keep plotting equipment in good working order. If the plotter is broken, replace it. Keep sharp points on dividers. Use a sharp pencil and an eraser that does not smudge.

4. Draw light lines at first, as they may have to be erased. When the line has been checked and proven to be correct, then darken it if desired.

5. Label lines and points immediately after they are drawn. Use standard labels and symbols. Letter the labels legibly. Be neat and exact.

## DR Computer

Almost any type of navigation requires the solution of simple arithmetical problems involving time, speed, distance, fuel consumption, and so forth. In addition, the effect of the wind on the aircraft must be known; therefore, the wind must be computed. To solve such problems quickly and with reasonable accuracy, various types of computers have been devised. The computer described in this pamphlet is simply a combination of two devices: a circular slide rule for the solution of arithmetical problems *[Figure 4-12]*, and a specially designed instrument for the graphical solution of the wind problem. *[Figure 4-13]*

The slide rule is a standard device for the mechanical solution of various arithmetical problems. Slide rules operate on the basis of logarithms. Slide rules are either straight or circular; the one on the DR computer is circular.

The slide rule face of the computer consists of two flat metallic disks, one of which can be rotated around a common center. These disks are graduated near their edges with adjacent logarithmic scales to form a circular slide rule approximately equivalent to a straight, 12-inch slide rule. Because the outer scale usually represents a number of miles and the inner scale represents a number of minutes, they are called the miles scale and the minutes or time scale, respectively. *[Figure 4-12]*

The numbers on each scale represent the printed figure with the decimal point moved any number of places to the right or left. For example, the numbers on either scale can represent 1.2, 12, 120, 1200, etc. Since speed (or fuel consumption) is expressed in miles (or gallons or pounds) per hour (60 minutes), a large black arrow marked speed index is placed at the 60-minute mark.

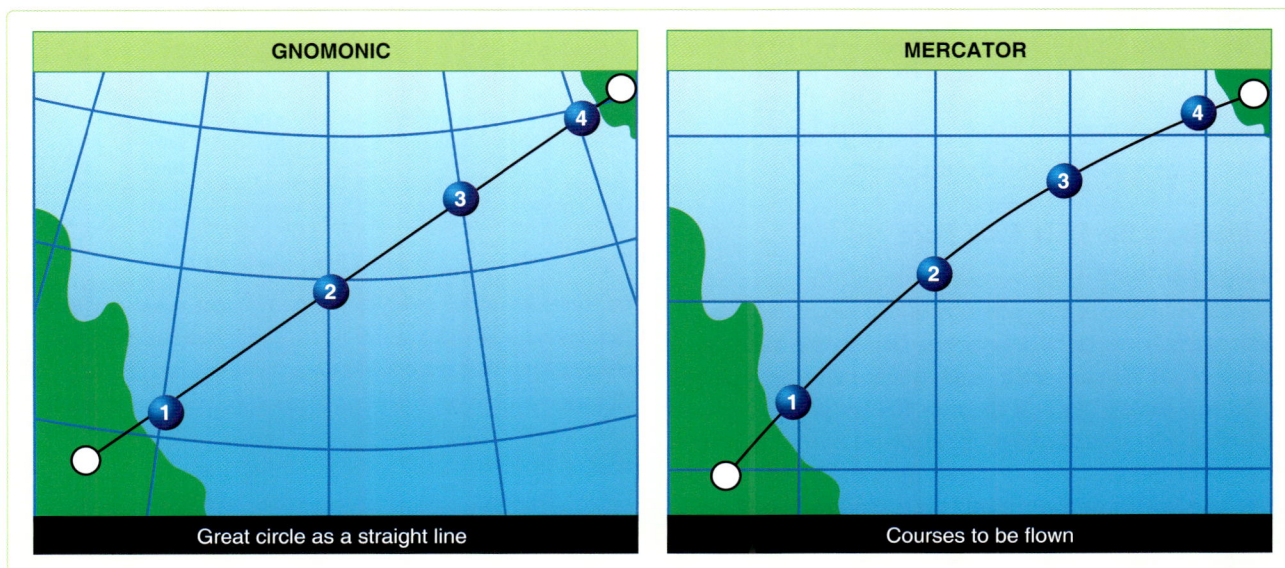

**Figure 4-11.** *Transferring great circle route from gnomonic to mercator chart.*

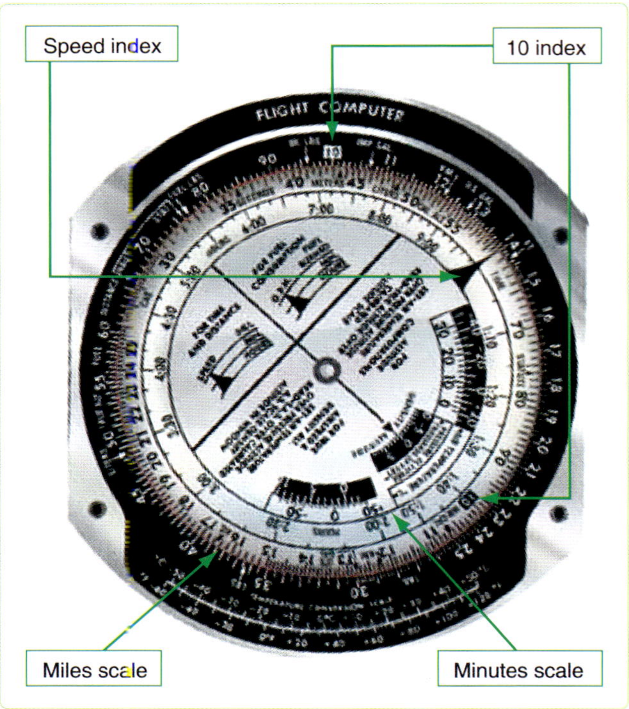

**Figure 4-12.** *Dead reckoning computer slide rule face.*

Graduations of both scales are identical. The graduations are numbered from l0 to 100, and the unit intervals decrease in size as the numbers increase in size. Not all unit intervals are numbered. The first element of skill in using the computer is a sure knowledge of how to read the numbers.

### Reading the Slide Rule Face

The unit intervals that are numbered present no difficulty. The problem lies in giving the correct values to the many small lines that come between the numbered intervals. There are no numbers given between 25 and 30 as shown in *Figure 4-14,* for example, but it is obvious that the larger intermediate divisions are 26, 27, 28, and 29. Between 25 and (unnumbered) 26, there are five smaller divisions, each of which would, therefore, be .2 of the larger unit. A mental estimate aids in placing the decimal point.

### Problems on the Slide Rule Face

*Simple Proportion*

The slide rule face of the computer is so constructed that any relationship between two numbers, one on the miles scale and one on the minutes scale, holds true for all other numbers on the two scales. Thus, if the two 10s are placed opposite each other, all other numbers are identical around the circle. If 20 on the minutes scale is placed opposite 10 on the miles scale, all numbers on the minutes scale are double of those on the miles scale. This feature allows one to supply the fourth term of any mathematical proportion. Thus, the unknown in the equation could be solved on the computer by setting 18

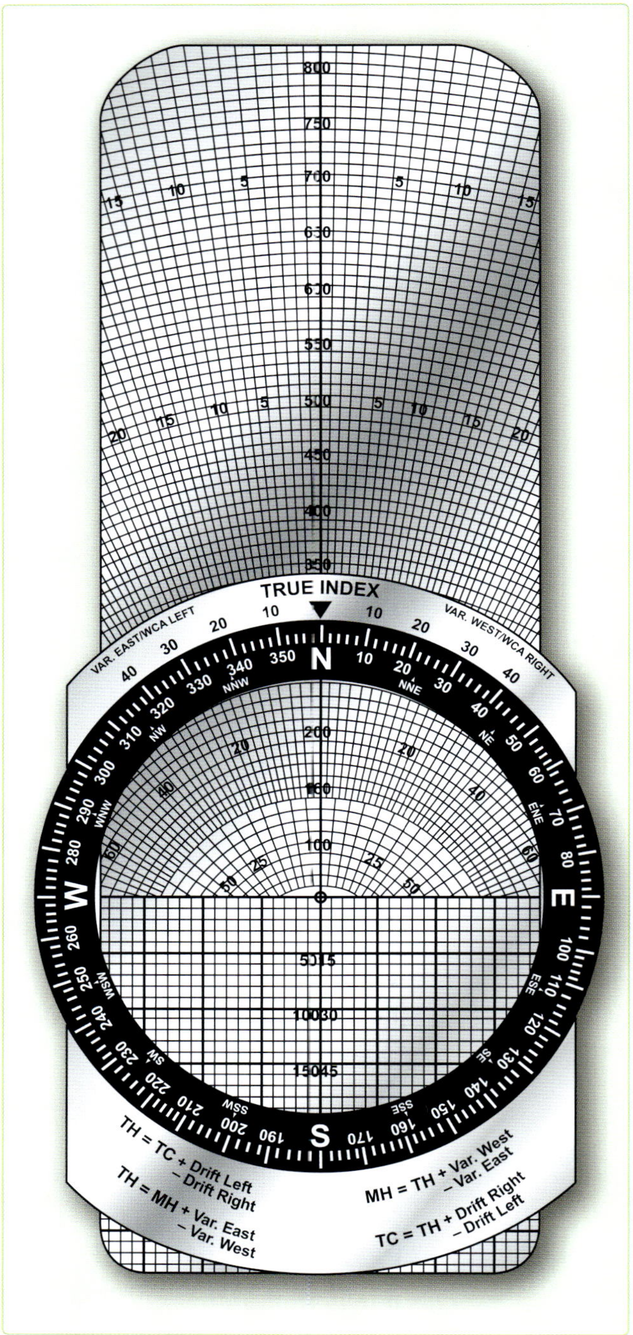

**Figure 4-13.** *Dead reckoning computer wind face.*

on the miles scale over 45 on the minutes scale and reading the answer (32) above the 80 on the minutes scale. It is this relationship that makes possible the solution of time-speed-distance problems. This can also be solved algebraically:

$$45X = \frac{18 \times 80}{18 \times 80}$$
$$X = 45$$

**A** Not all unit intervals are numbered. Those not numbered must be counted.

**B** For example, 83 is found as follows:

1. First find nearest smaller numbered graduation (80).

2. Then locate nearest larger numbered graduation (90).

3. Observe the unit division markers between these two; finally, pick up the third to locate 83.

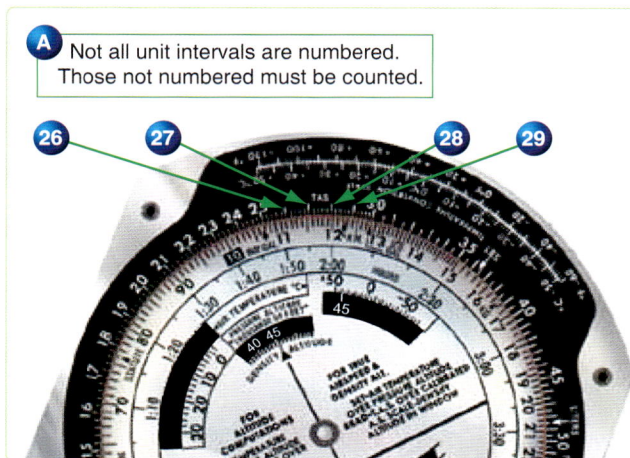

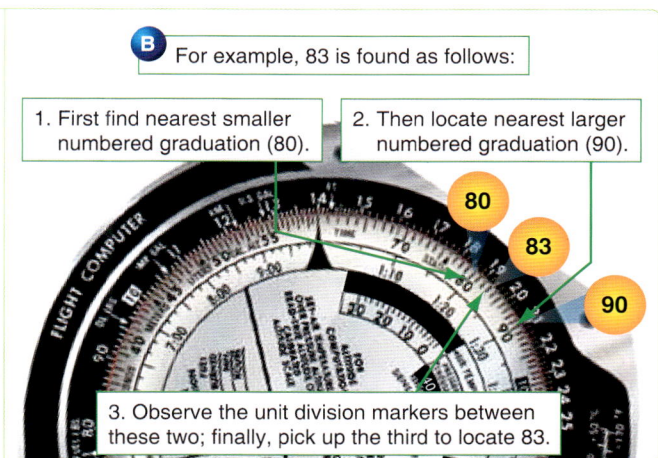

**Figure 4-14.** *Reading the slide rule face.*

*Time, Speed, and Distance*

An aircraft has traveled 24 miles in 8 minutes. How many minutes are required to travel 150 miles? This is a simple proportion which can be written as:

$$\frac{24}{8} = \frac{150}{1}$$

Setting the 24 over the 8 on the computer as illustrated in *Figure 4-15* and reading under the 150, the answer is 50 minutes.

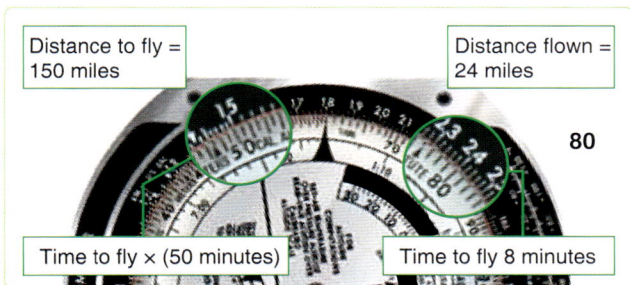

Distance to fly = 150 miles

Distance flown = 24 miles

80

Time to fly x (50 minutes)

Time to fly 8 minutes

**Figure 4-15.** *Solving for time when speed and distance are known.*

A problem that often occurs is to find the GS of the aircraft when a given distance is traveled in a given time. This is solved in the same manner, except the computer is marked with a speed index to aid in finding the correct proportion. In the problem just stated, if 24 is set over 8 as in the original problem, the GS of the aircraft, 180 knots, is read above the speed index, as shown.

Problem: GS is 204 knots. Find the distance traveled in 1 hour 15 minutes.

Solution: Set the speed index on the minutes scale to 204 on the miles scale. Opposite 75 on the minutes scale, read 255 NM on the miles scale. The computer solution is shown in *Figure 4-16*. The solution for time and speed when the

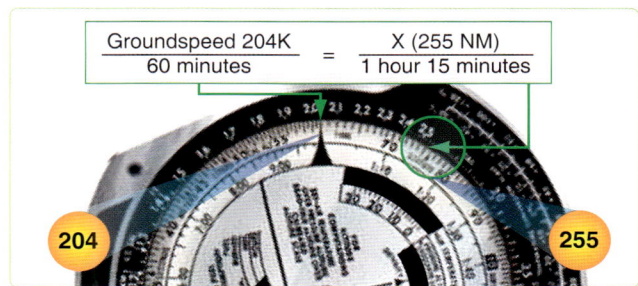

$$\frac{\text{Groundspeed 204K}}{\text{60 minutes}} = \frac{\text{X (255 NM)}}{\text{1 hour 15 minutes}}$$

204

255

**Figure 4-16.** *Solving for distance when speed and time are known.*

other variables are known follows the same basic format. [*Figures 4-17 and 4-18*]

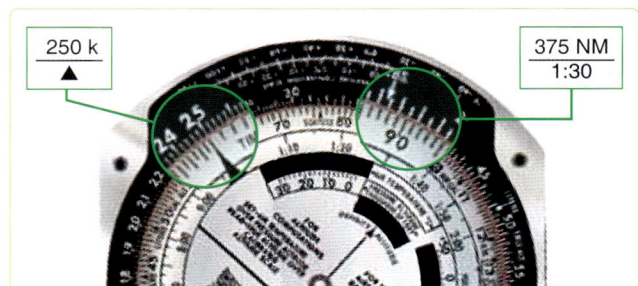

250 k

375 NM
1:30

**Figure 4-17.** *Solving for speed when time and distance are known.*

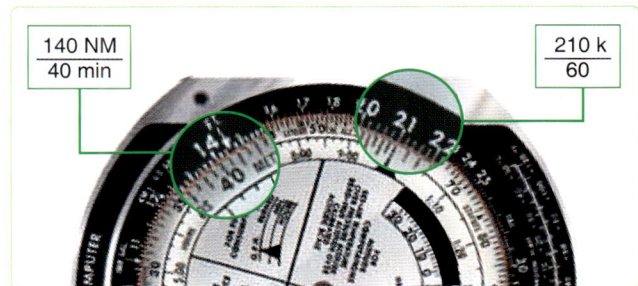

140 NM
40 min

210 k
60

**Figure 4-18.** *Solving for speed when time and distance are known.*

## Seconds Index

Since 1 hour is equivalent to 3,600 seconds, a subsidiary index mark, called seconds index, is marked at 36 on the minutes scale of some computers. When placed opposite a speed on the miles scale, the index relates the scales for converting distance to time in seconds. Thus, if 36 is placed opposite a GS of 144 knots, 50 seconds is required to go 2 NM; and in 150 seconds (2 minutes 30 seconds), 6.0 NM are covered. Similarly, if 4 NM are covered in 100 seconds, GS is 144 knots. *[Figure 4-19]*

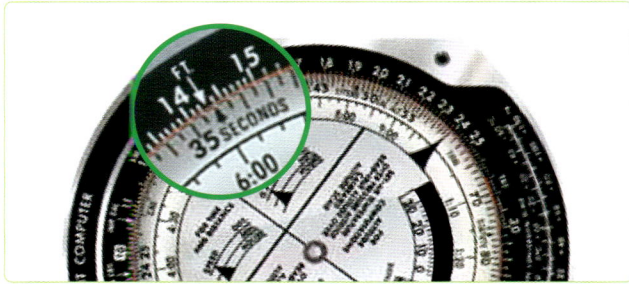

**Figure 4-19.** *Seconds index.*

## Conversion of Distance

Subsidiary indexes are placed on some computers to aid in the conversion of distances from one unit of measure to another. The most common interconversions are those involving SM, NM, and kilometer (km).

### Statute Mile-Nautical Mile Interconversion

The miles scale of the computer is marked with a SM index at 76 and a NM index at 66. The units are interconverted by setting the known distance under the appropriate index and reading the desired unit under the other.

Example: To convert 136 SM to NM, set 136 on the minutes scale under the STAT index on the miles scale. Under the NAUT index on the miles scale, read the number of NM (118) on the minutes scale. *[Figure 4-20]*

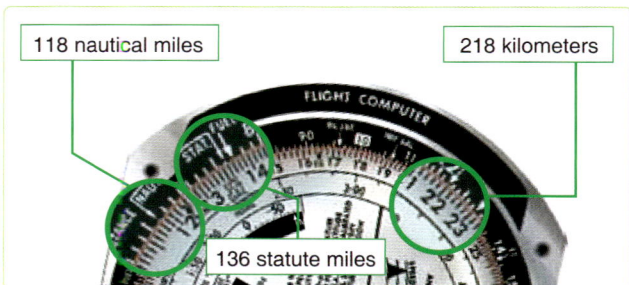

**Figure 4-20.** *Statute mile, nautical mile, and kilometer interconversion.*

### Conversion of NM or SM to km

A km index is indicated on the miles scale of the computer at 122. When NM or SM are placed under their appropriate index on the miles scale, kms may be read, on the minutes scale, under the km index.

Example: To convert 118 NM to kms, place 118 on the minutes scale under the NAUT index on the miles scale. Under the km index on the miles scale, read km (218) on the minutes scale.

### Multiplication and Division

To multiply two numbers, for example 12 × 2, the index (printed as 10 on the minutes scale) is placed opposite one of the numbers to be multiplied (12), and the product (24) is read on the miles scale above the other number (2) on the minutes scale. *[Figure 4-21]*

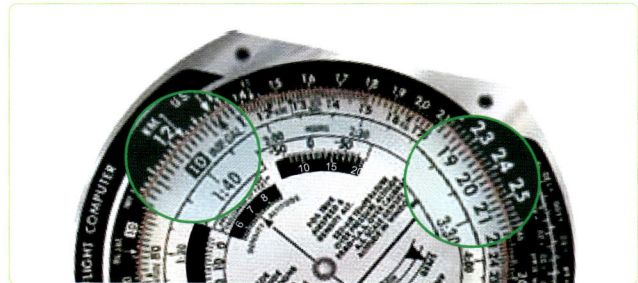

**Figure 4-21.** *Multiplying two numbers.*

To divide one number by another, for example 24/8, set the divisor (8) on the minutes scale opposite the dividend (24) on the miles scale, and read the quotient (3) on the miles scale opposite the index on the minutes scale. *[Figure 4-22]* In the computations encountered in air navigation, as in the above examples, a mental estimate aids in placing the decimal point.

**Figure 4-22.** *Dividing one number by another.*

## Effect of Wind on Aircraft

Any vehicle traveling on the ground, such as an automobile, moves in the direction in which it is steered or headed and is affected very little by wind. However, an aircraft seldom travels in exactly the direction in which it is headed because of the wind effect.

Any free object in the air moves downwind with the speed of the wind. This is just as true for an aircraft as it is for a balloon. If an aircraft is flying in a 20-knot wind, the body of air in which it is flying moves 20 NM in 1 hour. Therefore, the aircraft also moves 20 NM downwind in 1 hour. This movement is in addition to the forward movement of the aircraft through the body of air.

The path of an aircraft over the earth is determined by the two unrelated factors shown in *Figure 4-23:*

1. The motion of the aircraft through the airmass, and

2. The motion of the airmass across the earth's surface.

The motion of the aircraft through the airmass is directly forward in response to the pull of the propellers or thrust of the jet engines, and its rate of movement through the airmass is TAS. This motion takes place in the direction of true heading (TH). This motion of the airmass across the earth's surface may be from any direction and at any speed. The measurement of its movement is called wind and is expressed in direction and speed (wind vector (W/V)).

## Drift Caused by Wind

The effect of wind on the aircraft is to cause it to follow a different path over the ground than it does through the airmass. The path over the ground is its track. The terms true course (TC) and track are often considered synonymous. TC represents the intended path of the aircraft over the earth's surface. Track is the actual path that the aircraft has flown over the earth's surface. TC is considered to be future, while track is considered to be past.

The lateral displacement of the aircraft caused by the wind is called drift. Drift is the angle between the TH and the track. As shown in *Figure 4-24,* the aircraft has drifted to the right; this is known as right drift.

With a given wind, the drift changes on each heading. A change of heading also affects the distance flown over the earth's surface in a given time. This rate traveled relative to the earth's surface is known as GS. Therefore, with a given wind, the GS varies on different headings.

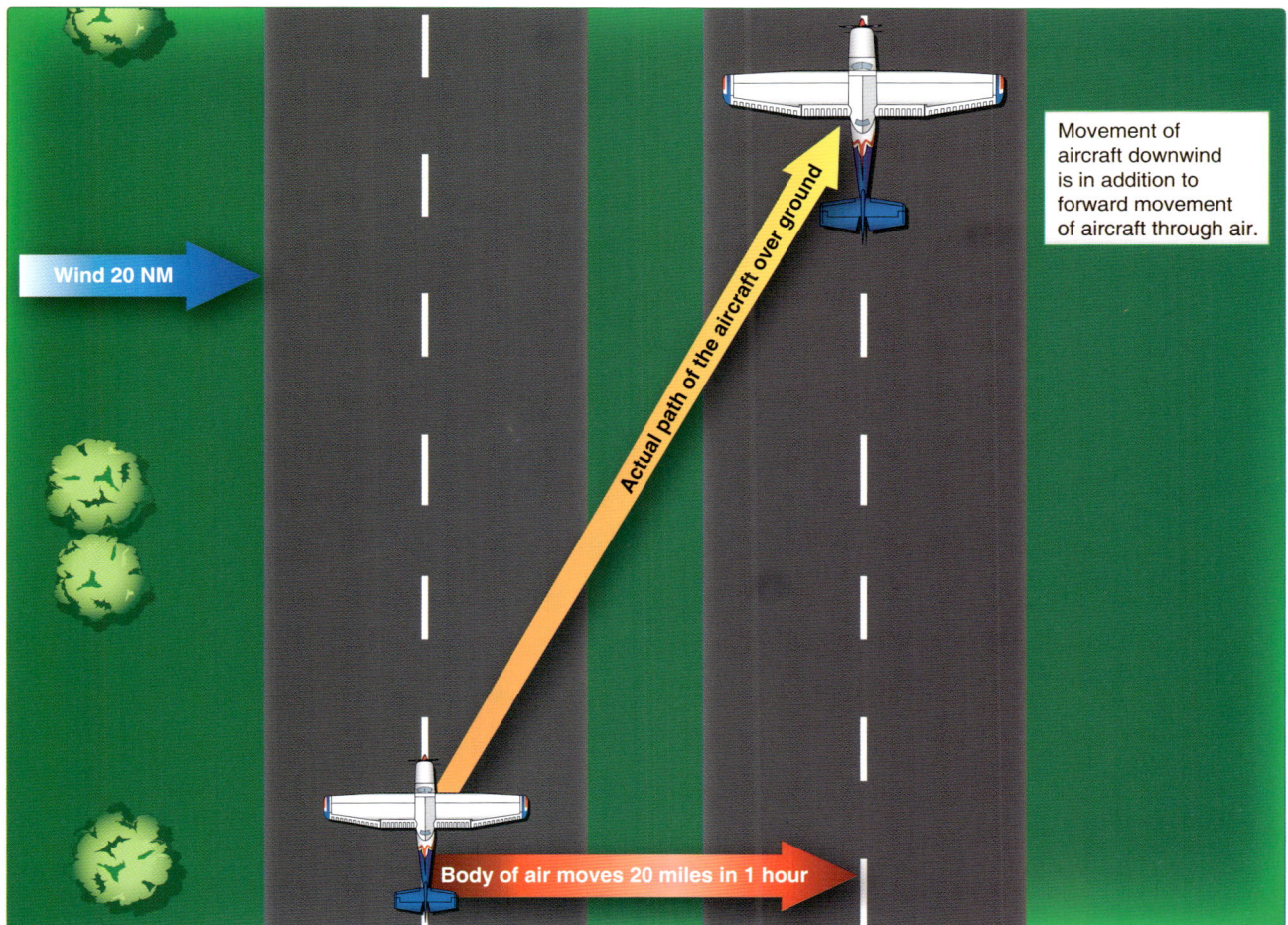

**Figure 4-23.** *Two factors determine the aircraft's path.*

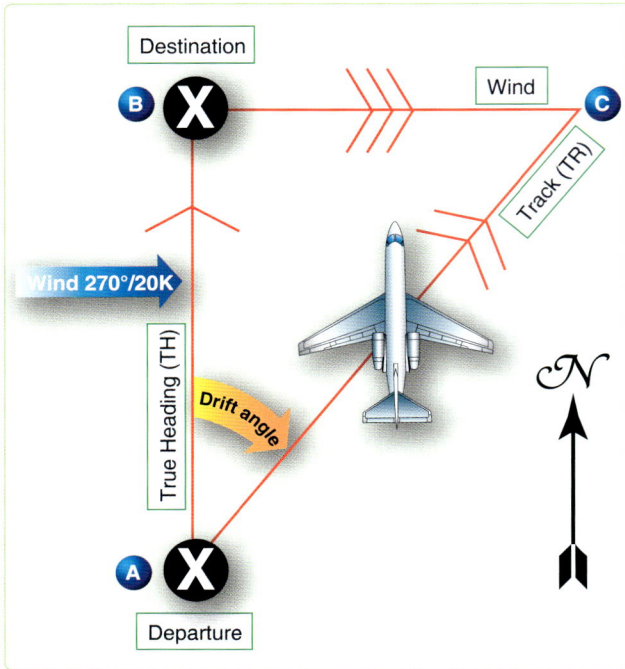

**Figure 4-24.** *In 1 hour, aircraft drifts downwind an amount equal to windspeed.*

*Figure 4-25* shows the effect of a 270°/20 knots wind on the GS and track of an aircraft flying on headings of 000°, 090°, 180°, and 270°. The aircraft flies on each heading from point X for 1 hour at a constant TAS.

Note that on a TH of 000°, the wind causes right drift; whereas on a TH of 180°, the same wind causes left drift. On the headings of 090° and 270°, there is no drift at all. Note further that on a heading of 090°, the aircraft is aided by a tailwind and travels farther in 1 hour than it would without a wind; thus, its GS is increased by the wind. On the heading of 270°, the headwind cuts down on the GS and also cuts down the distance traveled. On the headings of 000° and 180°, the GS is unchanged.

### Drift Correction Compensates for Wind

In *Figure 4-26*, suppose the navigator wants to fly from point A to point B, on a TC of 000°, when the wind is 270°/20 knots. If the navigator flew a TH of 000°, the aircraft would not end up at point B but at some point downwind from B.

By heading the aircraft upwind to maintain the TC, drift is compensated for. The angle BAC is called the drift correction angle or, more simply, the drift correction. Drift correction is the correction that is applied to a TC to find the TH.

*Figure 4-27* shows the drift correction necessary in a 270°/20-knot wind if the aircraft is to make a good TC of 000°, 090°, 180°, or 270°. When drift is right, correct to the left, and the sign of the correction is minus. When the drift is left, correct to the right, and the sign of the correction is plus.

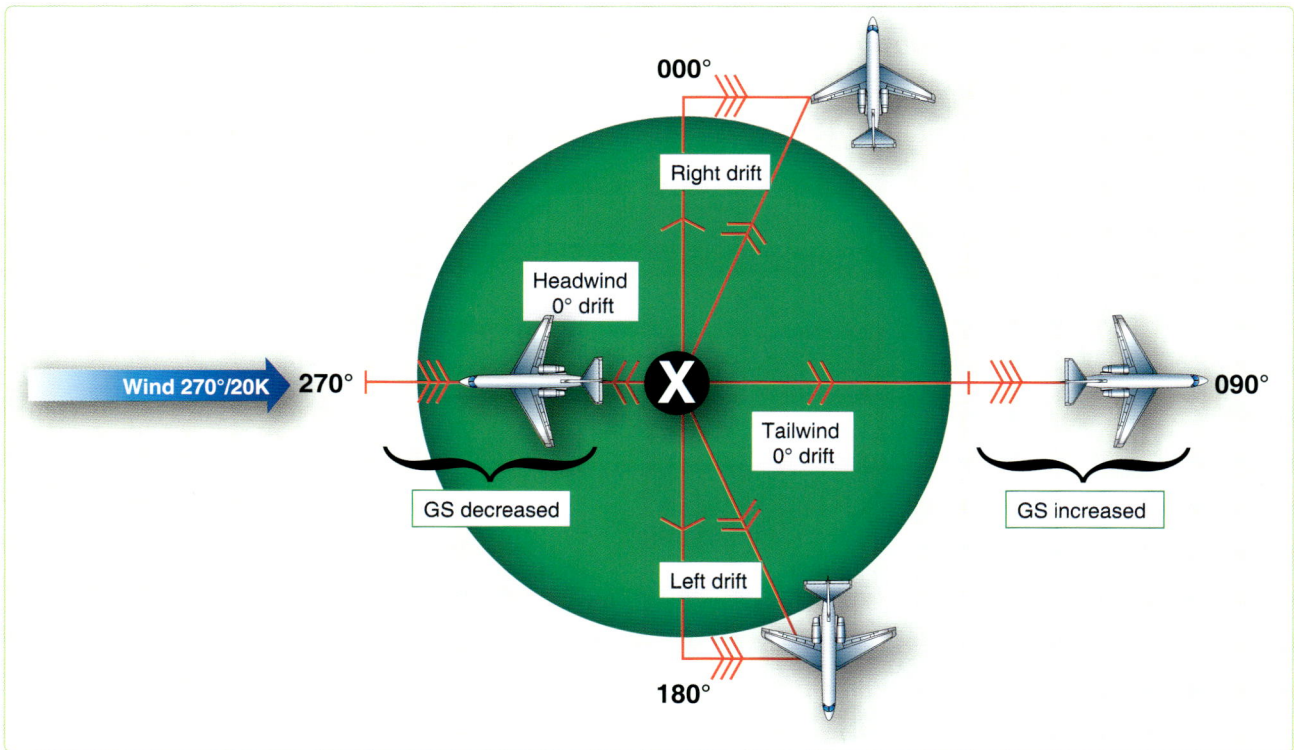

**Figure 4-25.** *Effects of wind on aircraft flying in opposite directions.*

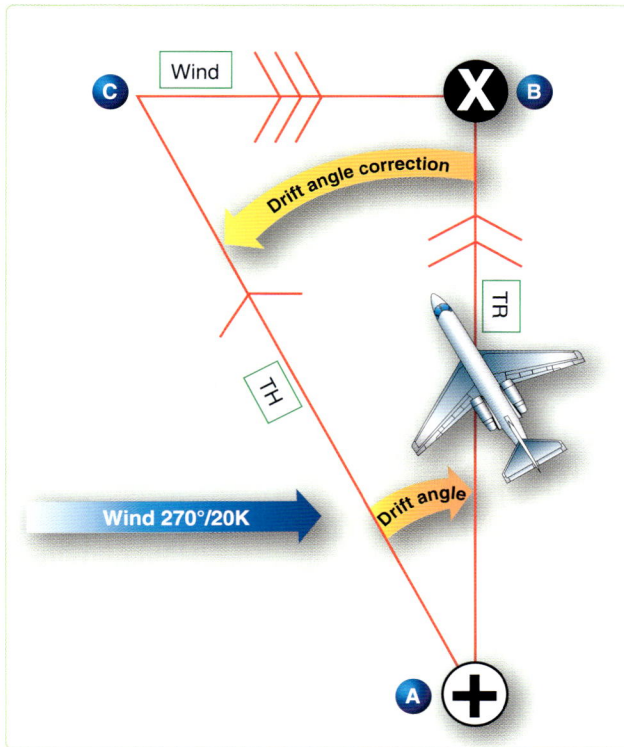

**Figure 4-26.** *Aircraft heads upwind to correct for drift.*

## Vectors and Vector Diagrams

In aerial navigation, there are many problems to solve involving speeds and directions. These speeds and directions fit together in pairs: one speed with one direction.

By using vector solution methods, unknown quantities can be found. For example, TH, TAS, and W/V may be known, and track and GS unknown. To solve such problems, the relationships of these quantities must be understood.

The vector can be represented on paper by a straight line. The direction of this line would be its angle measured clockwise from true north (TN), while the magnitude or speed is the length of the line compared to some arbitrary scale. An arrowhead is drawn on the line representing a vector to avoid any misunderstanding of its direction. This line drawn on paper to represent a vector is known as a vector diagram, or often it is referred to simply as a vector. *[Figure 4-28]* Future references to the word vector means its graphic representation. Two or more vectors can be added together simply by placing the tail of each succeeding vector at the head of the previous vector. These vectors added together are known as component vectors.

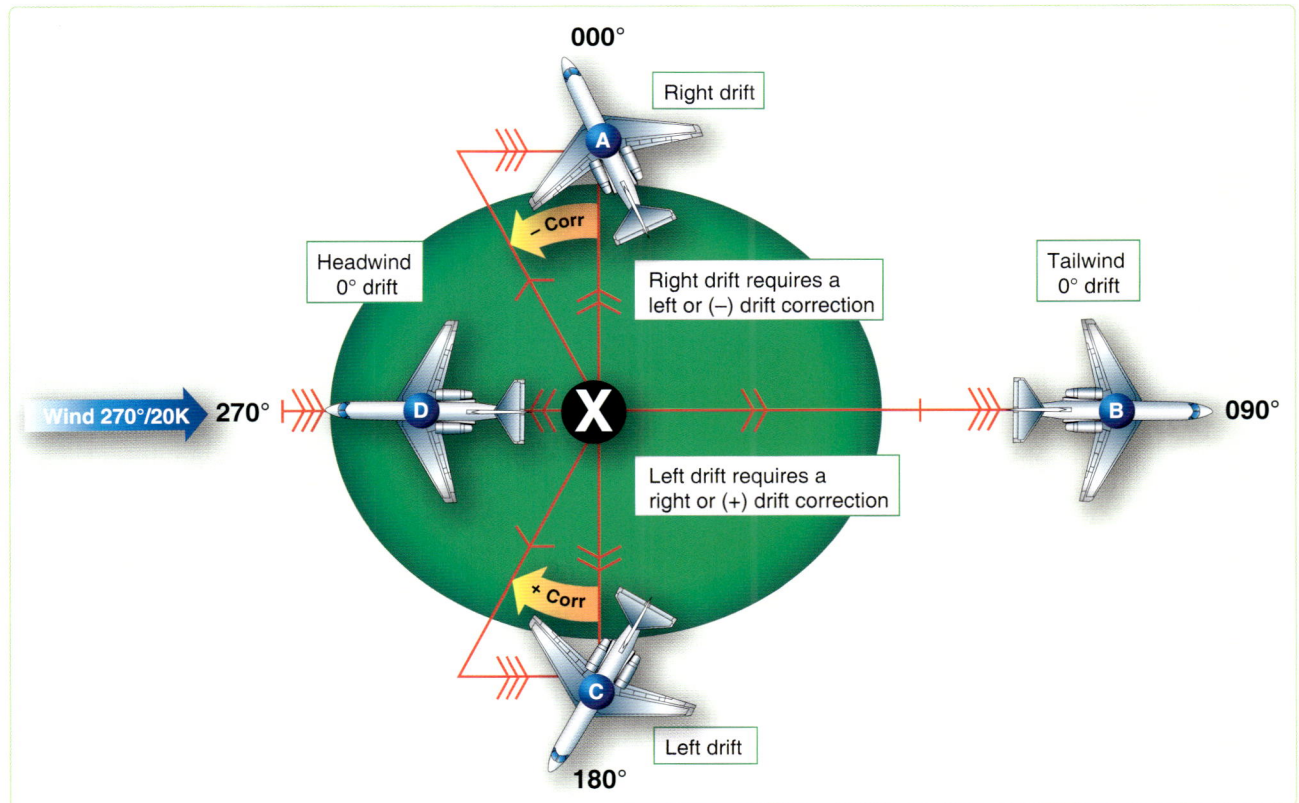

**Figure 4-27.** *Maintaining course in wind.*

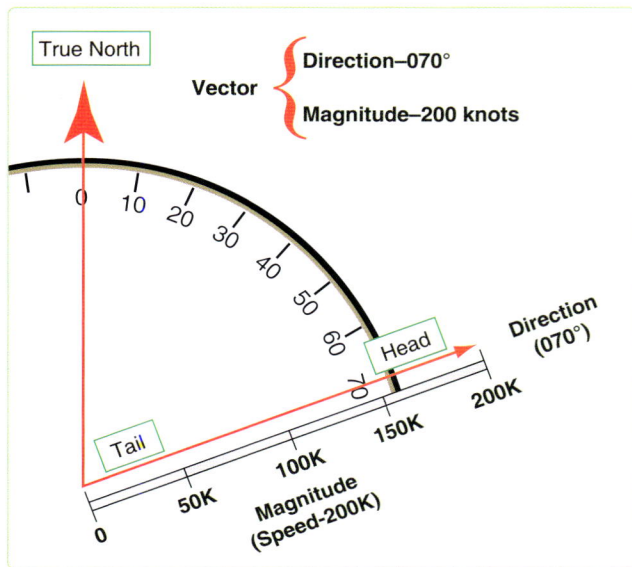

**Figure 4-28.** *Vector has both magnitude and direction.*

The sum of the component vectors can be determined by connecting, with a straight line, the tail of one vector to the head of the other. This sum is known as the resultant vector. By its construction, the resultant vector forms a closed figure. [*Figure 4-29*] Notice the resultant is the same, regardless of the order, as long as the tail of one vector is connected to the head of another.

The points to remember about vectors are as follows:

1. A vector possesses both direction and magnitude. In aerial navigation, these are referred to as direction and speed.

2. When the components are represented tail to head in any order, a line connecting the tail of the first and the head of the last represents the resultant.

3. All component vectors must be drawn to the same scale.

## Wind Triangle and Its Solution

### Vector Diagrams and Wind Triangles

A vector illustration showing the effect of the wind on the flight of an aircraft is called a wind triangle. Draw a line to show the direction and speed of the aircraft through airmass (TH and TAS); this vector is called the air vector. Using the same scale, connect the tail of the wind vector to the head of the air vector. Draw a line to show the direction and speed of the wind (W/V); this is the wind vector. A line connecting the tail of the air vector with the head of the wind vector is the resultant of these two component vectors; it shows the direction and speed of the aircraft over the earth (track and GS). It is called the ground vector.

To distinguish one from another, it is necessary to mark each vector. Accomplish this by placing one arrowhead at midpoint on the air vector pointing in the direction of TH. The ground vector has two arrowheads at midpoint in the direction of track. The wind vector is labeled with three arrowheads in the direction the wind is blowing. The completed wind triangle is shown in *Figure 4-30*.

Remember that WD and WS compose the wind vector. TAS and TH form the air vector and GS and track compose the ground vector. The ground vector is the resultant of the other two; hence, the air vector and the wind vector are always drawn head to tail. An easy way to remember this is that the wind always blows the aircraft from TH to track.

Consider just what the wind triangle shows. In *Figure 4-31*, the aircraft departs from point A on the TH of 360° at a TAS of 150 knots. In 1 hour, if there is no wind, it reaches point B at a distance of 150 NM.

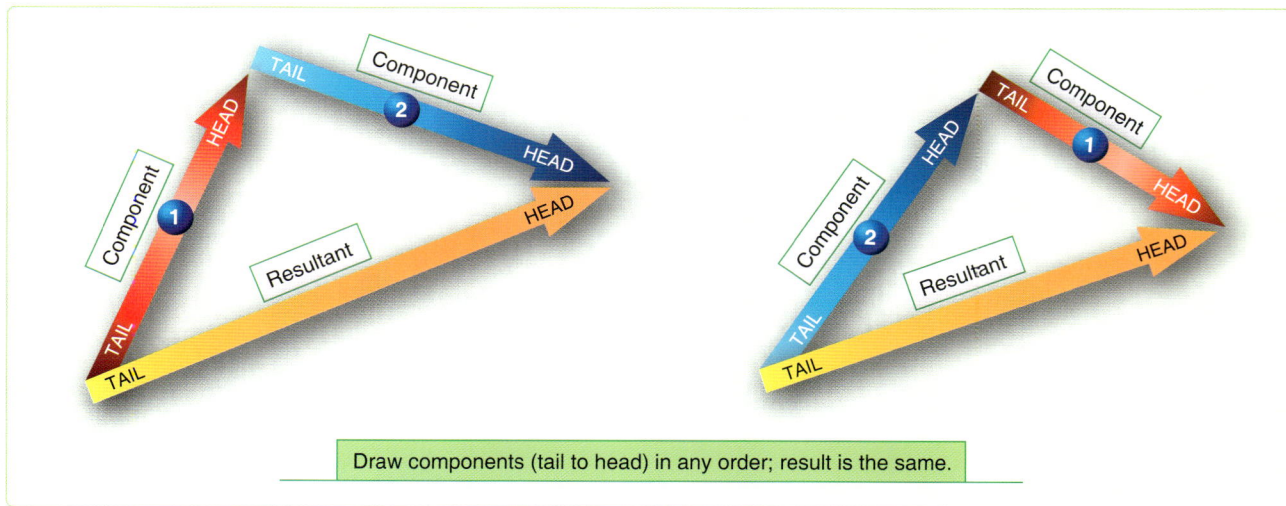

Draw components (tail to head) in any order; result is the same.

**Figure 4-29.** *Resultant vector is sum of component vectors.*

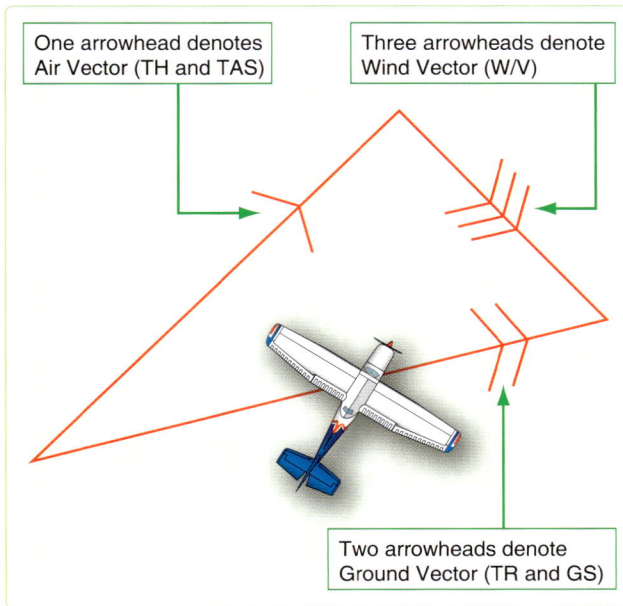

**Figure 4-30.** *Mark each vector of wind triangle.*

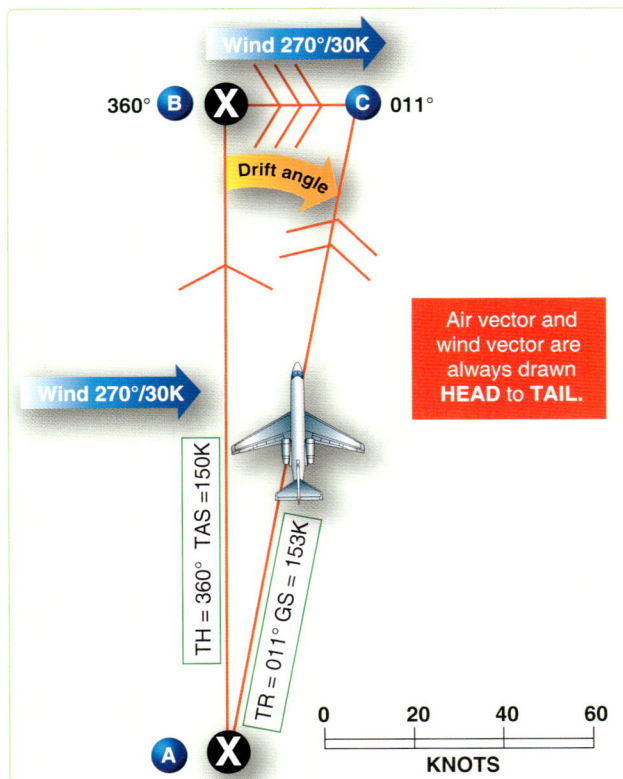

**Figure 4-31.** *Wind triangle.*

In actuality, the wind is blowing from 270° at 30 knots. At the end of 1 hour, the aircraft is at point C 30 NM downwind. Therefore, the length BC represents the speed of the wind drawn to the same scale as the TAS. The length of BC represents the wind and is the wind vector.

Line AC shows the distance and direction the aircraft travels over the ground in 1 hour. The length of AC represents the GS drawn to the same scale as the TAS and windspeed. The line AC, which is the resultant of AB and BC, represents the motion of the aircraft over the ground and is the ground vector.

Measuring the length of AC determines that the GS is 153 knots. Measuring the drift angle, BAC, and applying it to the TH of 360°, result in the track of 011°.

If two vectors in a wind triangle are known, the third one can be found by drawing a diagram and measuring the parts. Actually, the wind triangle includes six quantities; three speeds and three directions. Problems involving these six quantities make up a large part of DR navigation. If four of these quantities are known, the other two can be found. This is called solving the wind triangle and is an important part of navigation.

The wind triangle may be solved by trigonometric tables; however, this is unnecessary since the accuracy of this method far exceeds the accuracy of the data available and the results needed. In flight, the wind triangle is solved graphically, either on the chart or on the vector or wind face of the computer.

The two graphic solutions of the wind triangle (chart solution and computer solution) perhaps appear dissimilar at first glance. However, they work on exactly the same principles. Plotting the wind triangle on paper has been discussed; now, the same triangle is plotted on the wind face of the computer.

### Wind Triangles on DR Computer

The wind face of the computer has three parts: a frame, a transparent circular plate that rotates in the frame, and a slide or card that can be moved up and down in the frame under the circular plate. *[Figure 4-32]*

The frame has a reference mark called the true index. A drift scale is graduated 45° to the left and 45° to the right of the true index; to the left this is marked drift left, and to the right, drift right.

The circular plate has around its edge a compass rose graduated in units of 1 degree. The position of the plate may be read on the compass rose opposite the true index. Except for the edge, the circular plate is transparent so that the slide can be seen through it. Pencil marks can be made on the transparent surface. The centerline is cut at intervals of two units by arcs of concentric circles called speed circles; these are numbered at intervals of 10 units.

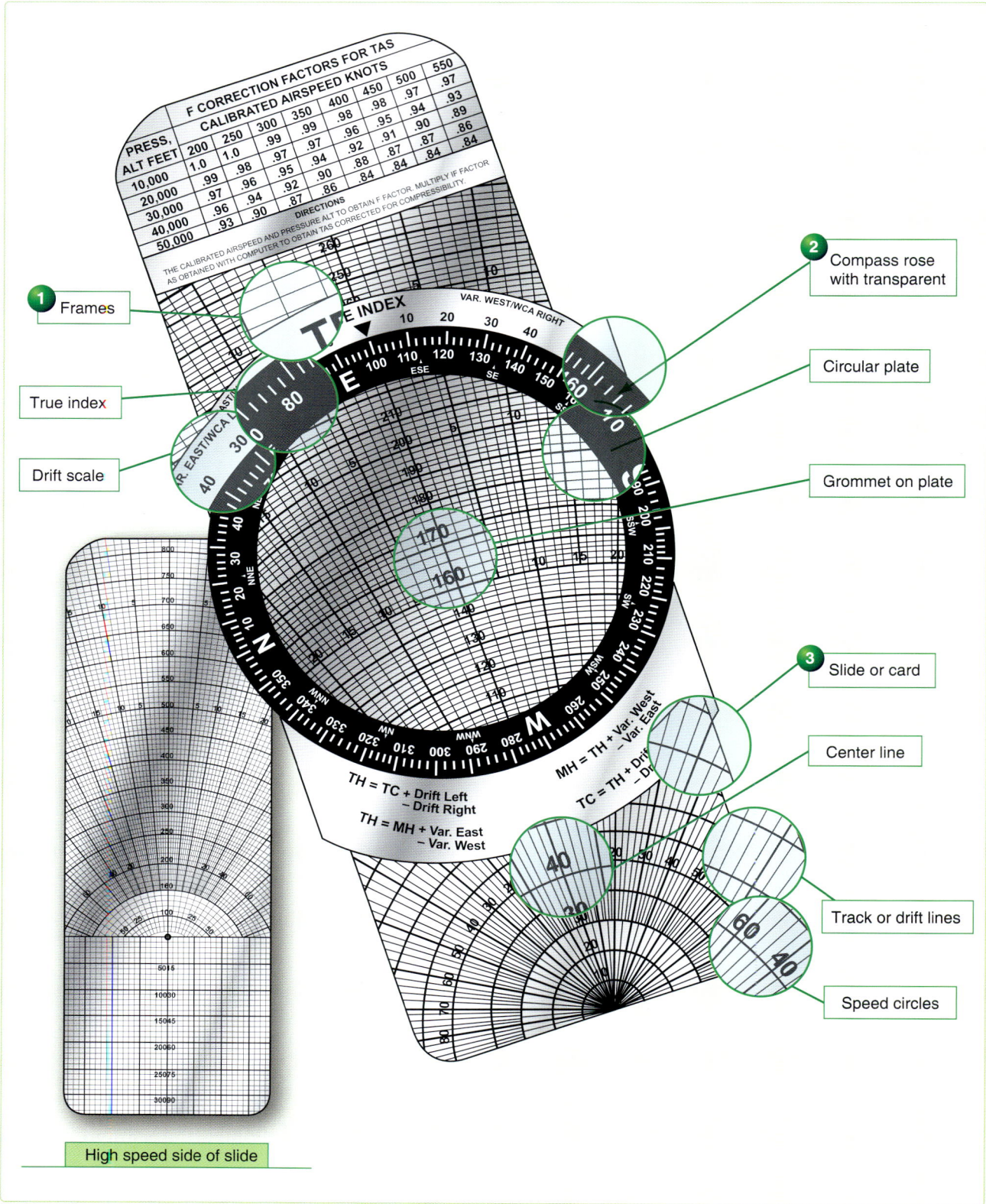

**Figure 4-32.** *Wind face of DR computer.*

F CORRECTION FACTORS FOR TAS

| PRESS, ALT FEET | CALIBRATED AIRSPEED KNOTS | | | | | | | |
|---|---|---|---|---|---|---|---|---|
| | 200 | 250 | 300 | 350 | 400 | 450 | 500 | 550 |
| 10,000 | 1.0 | 1.0 | .99 | .99 | .98 | .98 | .97 | .97 |
| 20,000 | 1.0 | .98 | .97 | .97 | .96 | .95 | .94 | .93 |
| 30,000 | .99 | .96 | .95 | .94 | .92 | .91 | .90 | .89 |
| 40,000 | .97 | .96 | .92 | .90 | .88 | .87 | .87 | .86 |
| 50,000 | .96 | .94 | .87 | .86 | .84 | .84 | .84 | .84 |
| | .93 | .90 | | | | | | |

**DIRECTIONS**

THE CALIBRATED AIRSPEED AND PRESSURE ALT TO OBTAIN F FACTOR. MULTIPLY IF FACTOR AS OBTAINED WITH COMPUTER TO OBTAIN TAS CORRECTED FOR COMPRESSIBILITY.

TRUE INDEX

VAR. WEST/WCA RIGHT

VAR. EAST/WCA LEFT

TH = TC + Drift Left − Drift Right

TH = MH + Var. East − Var. West

MH = TH + Var. West − Var. East

TC = TH + Drift − Drift

1 Frames

True index

Drift scale

2 Compass rose with transparent

Circular plate

Grommet on plate

3 Slide or card

Center line

Track or drift lines

Speed circles

High speed side of slide

4-18

On each side of the centerline are track lines that radiate from a point of origin off the slide. *[Figure 4-33]* Thus, the 14° track line on each side of the centerline makes an angle of 14° with the centerline at the origin.

In solving a wind triangle on the computer, plot part of the triangle on the transparent surface of the circular plate. For the other parts of the triangle, use the lines that are already drawn on the slide. Actually, there is not room for the whole triangle on the computer, for the origin of the centerline is one vertex of the triangle. When learning to use the wind face of the computer, it may help to draw in as much as possible of each triangle.

The centerline from its origin to the grommet always represents the air vector. If the TAS is 150k, move the slide so that 150 is under the grommet; then the length of the vector from the origin to the grommet is 150 units. *[Figure 4-34A]*

The ground vector is represented by one of the track lines, with its tail at the origin and its head at the appropriate speed circle. If the track is 15° to the right of the TH, and the GS is 180 knots, use the track line 15° to the right of the centerline and consider the intersection of this line with the 180 speed circle as the head of the vector. *[Figure 4-34B]*

The tail of the wind vector is at the grommet and its head is at the head of the ground vector. *[Figure 4-34C]*

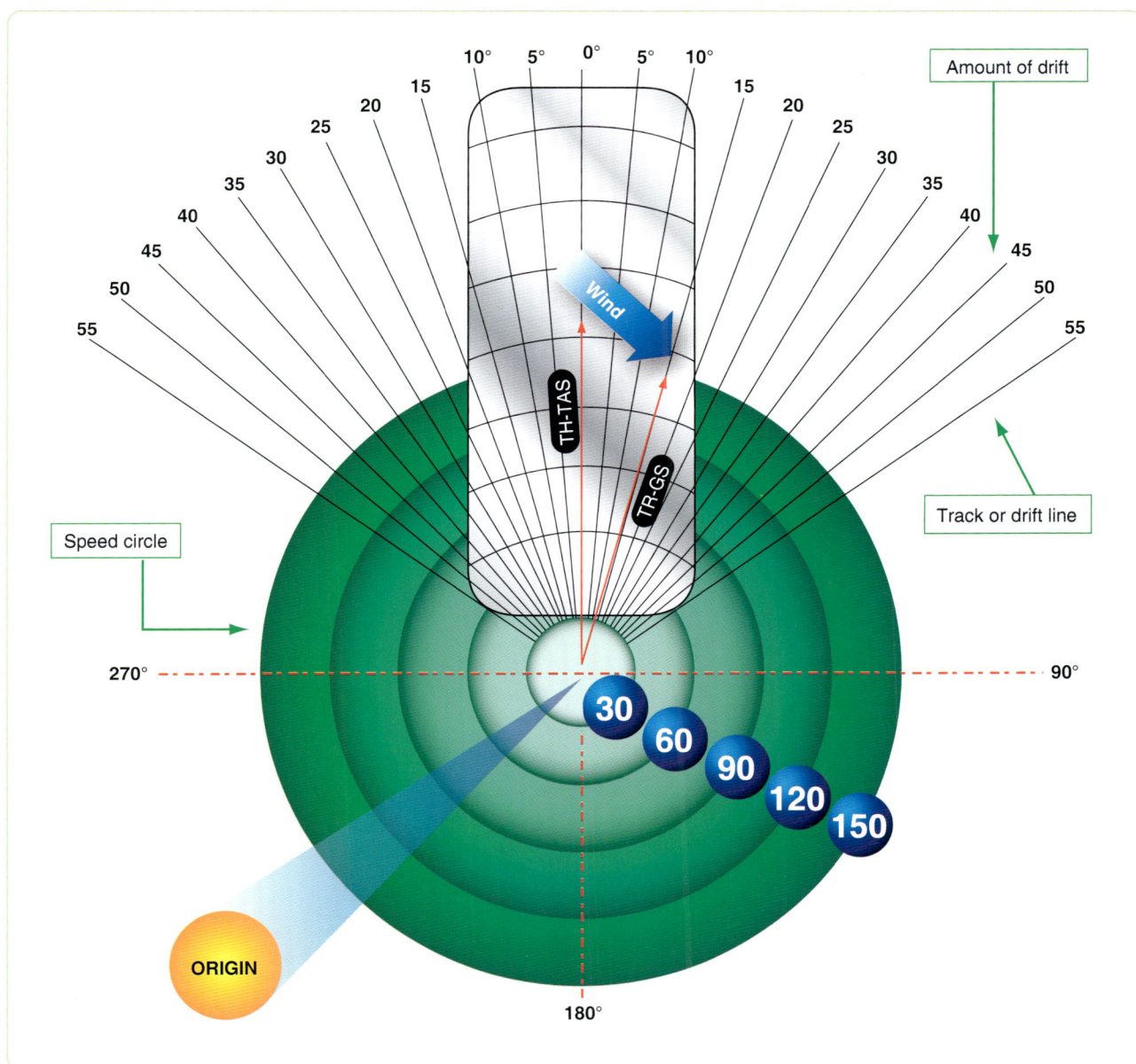

**Figure 4-33.** *Speed circles and track lines.*

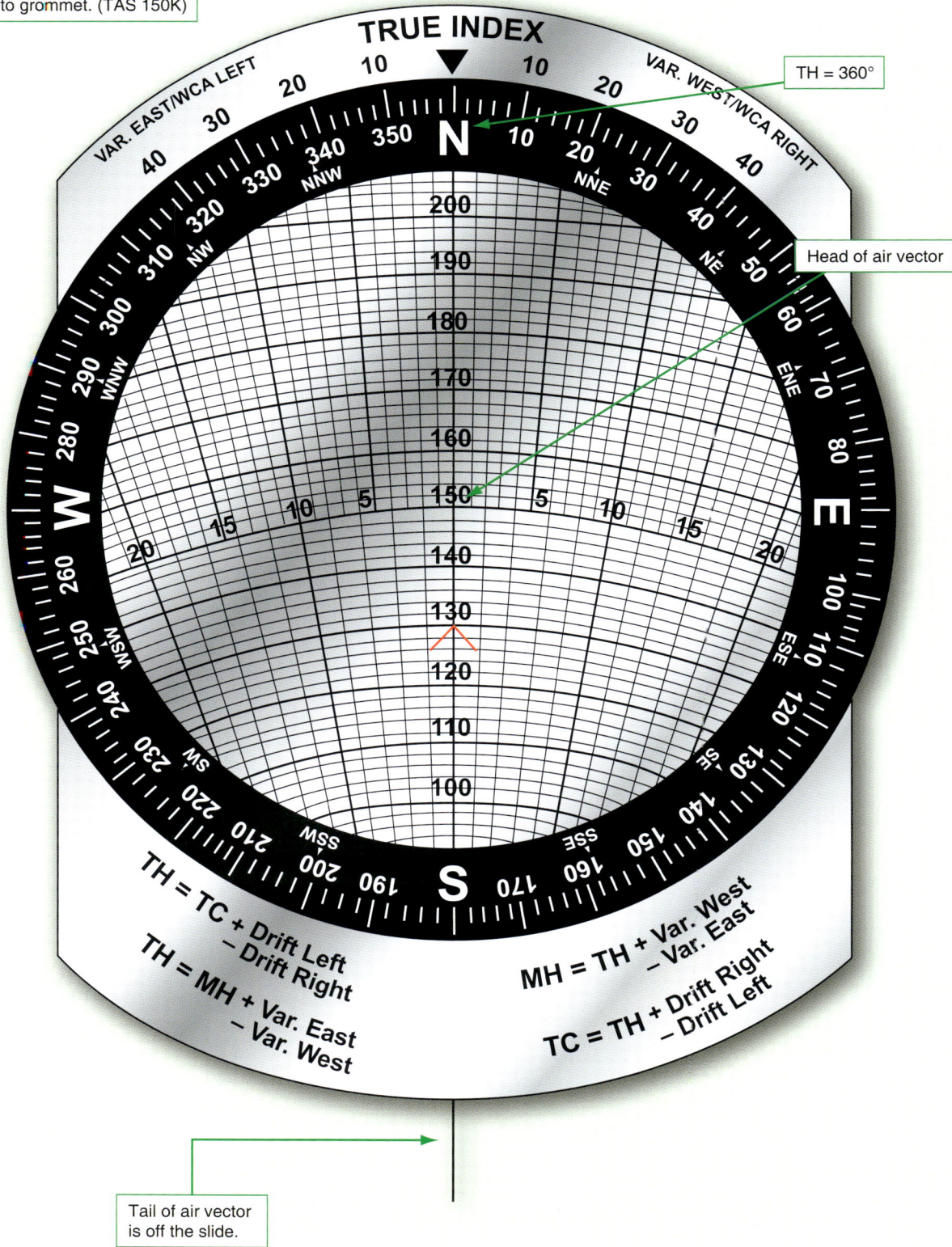

**A** Draw air vector up to grommet. (TAS 150K)

TRUE INDEX

TH = 360°

Head of air vector

Tail of air vector is off the slide.

VAR. EAST/WCA LEFT

VAR. WEST/WCA RIGHT

TH = TC + Drift Left − Drift Right

TH = MH + Var. East − Var. West

MH = TH + Var. West − Var. East

TC = TH + Drift Right − Drift Left

**Figure 4-34.** *Plotting a wind triangle on computer.*

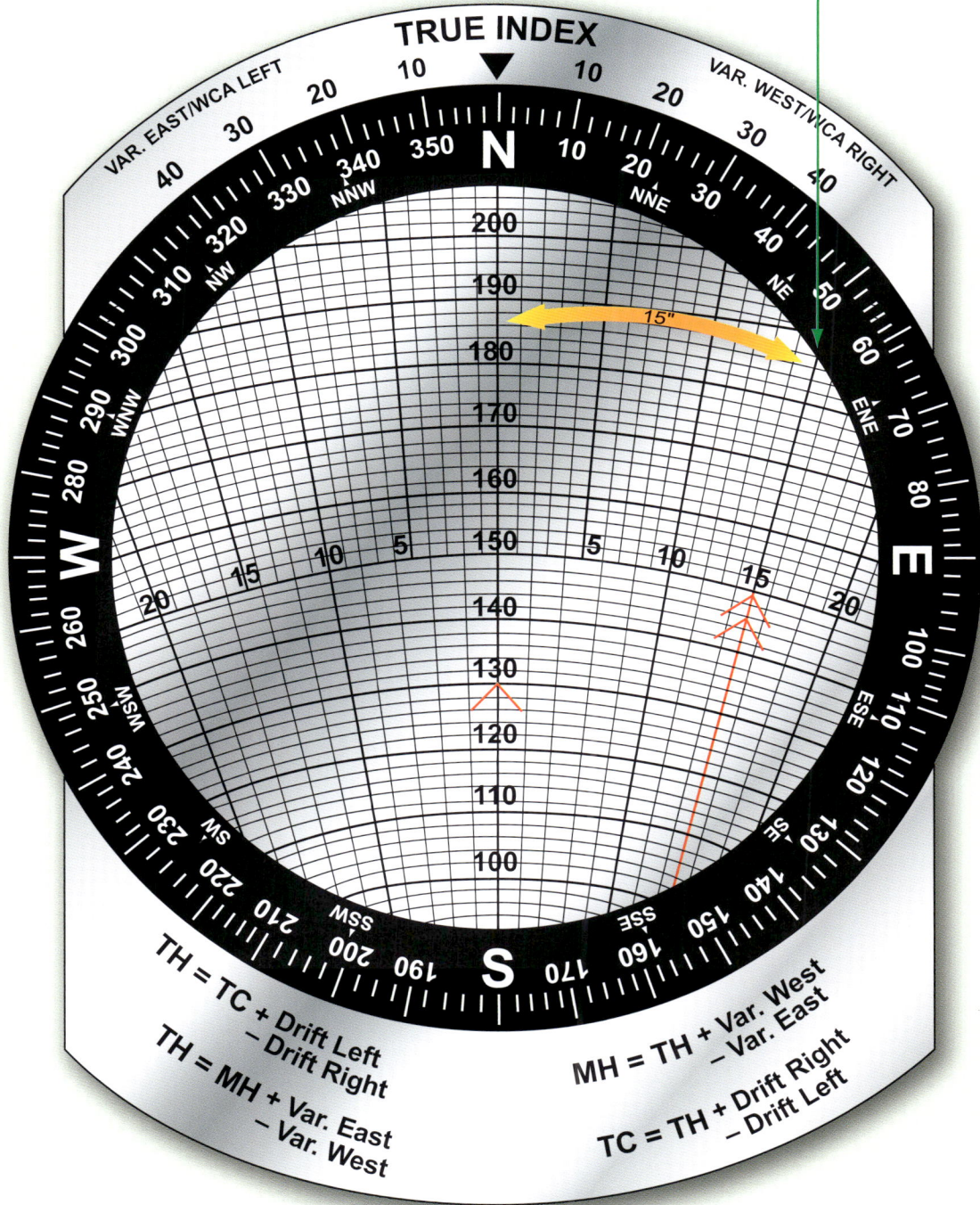

**Figure 4-34.** *Plotting a wind triangle on computer (continued).*

4-21

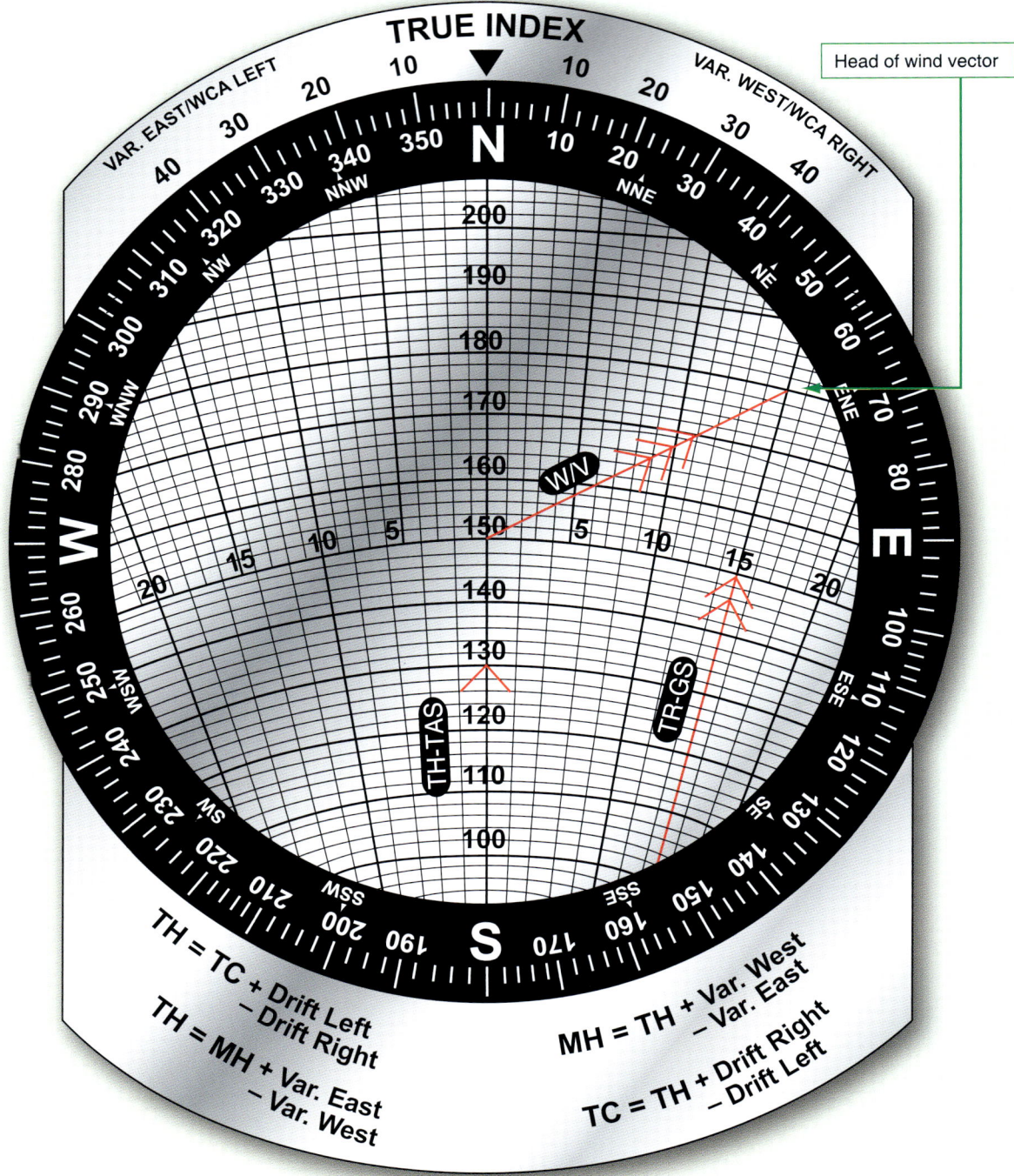

**C** Locate head of wind vector at head of ground vector.

Head of wind vector

TRUE INDEX

**Figure 4-34.** *Plotting a wind triangle on computer (continued).*

Thus far, nothing has been said about the direction of the vectors. Since the true index is over the centerline beyond the head of the air vector, this vector always points toward the index. Therefore, TH is read on the compass rose opposite the true index.

Since track is TH with the drift angle applied, the value of track can be found on the scale of the circular plate opposite the drift correction on the drift scale. The wind vector is drawn with its tail at the grommet. *[Figure 4-35]* Since WD is the direction from which the wind blows, it is indicated on the compass rose by the rearward extension of the wind vector. Therefore, the most convenient way to draw the wind vector is to set WD under the true index and draw the vector down the centerline from the grommet; the scale on the centerline can then be used to determine the length of the vector.

Conversely, to read a wind already determined, place the head of the wind vector on the centerline below the grommet and read WD below the true index.

### Wind Triangle Problems

Depending on which of the six quantities of the wind triangle are known and which are unknown, there are three principal types of problems to solve: ground vector, wind vector, and TH and GS. The following discussion gives the steps for the computer solution for each type. Work each sample problem and notice that the same wind triangle is shown on the computer that is shown on the chart, even though it is not completely drawn on the computer.

To find ground vector when air vector and wind vector are known:
Given:
TH = 100°
TAS = 210 knots
W/V = 020°/25 knots
Find: Track and GS

This type of problem arises when TH and TAS are known by reading the flight instruments and when the WD and velocity are known from either the metro forecast or from calculations in flight.

Study *Figure 4-36* and determine what has happened. By flying a TH of 100° at a TAS of 210 knots in a wind of 020°/25 knots, the aircraft has actually moved over the ground along a track of 107° at a GS of 208 knots.

*Computer Solution*
First, set the data:

1. Draw the wind vector from the grommet down the centerline, making its length (25 units) along the speed scale to conform with the windspeed (25 knots).

2. Set the TH (100°) under the true index by rotating the compass rose.

3. Slide the card up or down until the TAS (210 knots) is under the grommet. The graphic solution is now displayed on the computer. *[Figure 4-37]* The ground vector lies along one of the radiating track lines with its head at the head of the wind vector.

4. Read GS (208 knots) on the speed circle that passes through the head of the ground vector.

5. Read the drift angle (7° right) by counting the number of degrees from the centerline to the ground vector; that is, to the head of the wind vector.

6. Determine track (107°) by applying the drift angle to the TH. If the track is right of the center line, it is greater than the TH; so the drift angle must be added to the TH. An alternate method of determining track on the computer is to read the drift angle at the head of the ground vector, then transform this value to the drift scale on the same side of the true index and read the track on the compass rose of the circular disk.

To find wind vector when air vector and ground vector are known.

Given:

TH = 270°

Track = 280°

TAS = 230 knots

GS = 215 knots

Find: wind vector

This type of problem arises when determination of TH and TAS can be done by reading the flight instruments and finding track and GS either by measuring the direction and distance between two established positions of the aircraft or by determining the drift angle and GS by reference to the ground. Refer to *Figure 4-38* for a graphic solution.

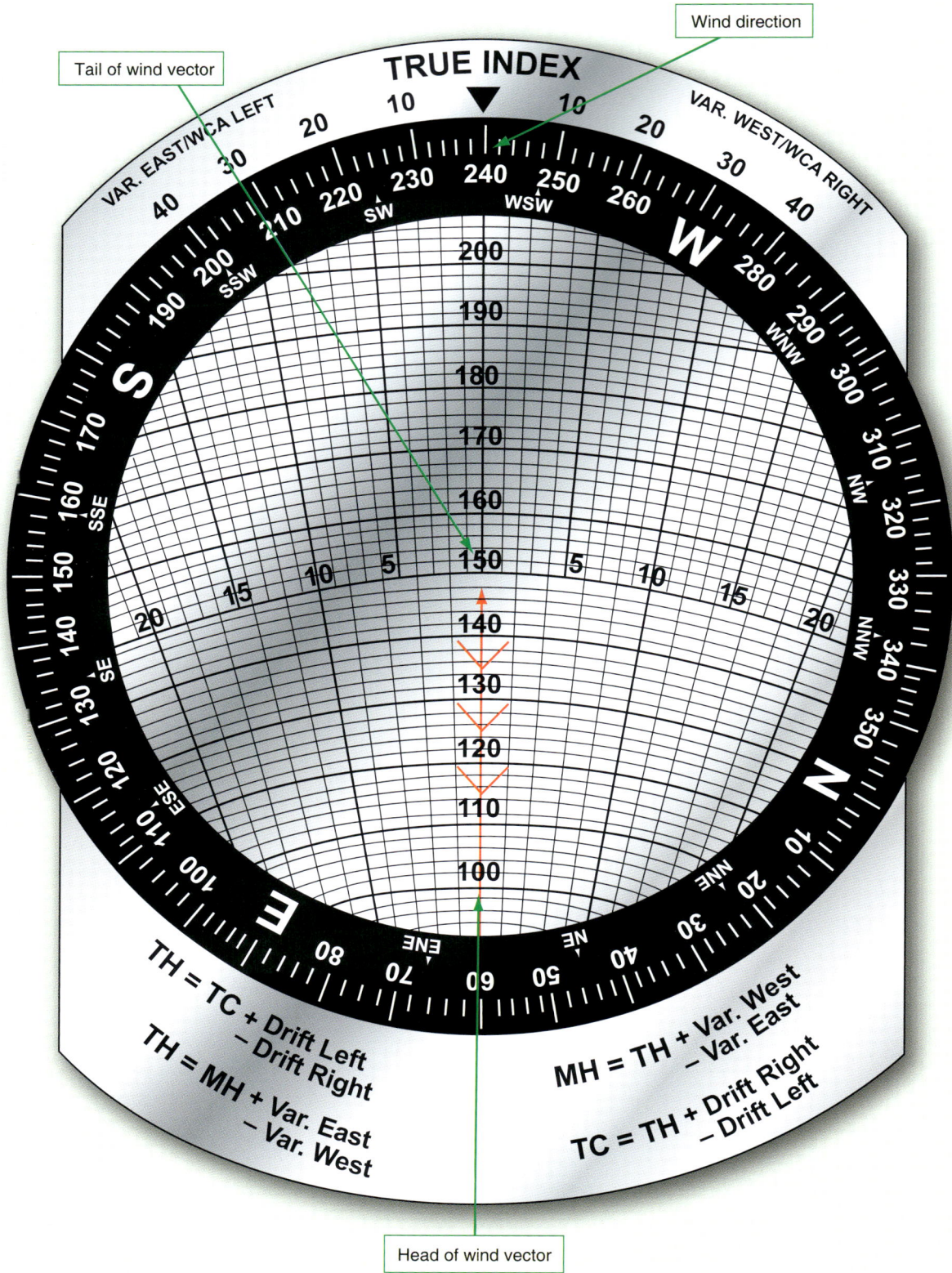

**Figure 4-35.** *Draw wind vector down from grommet.*

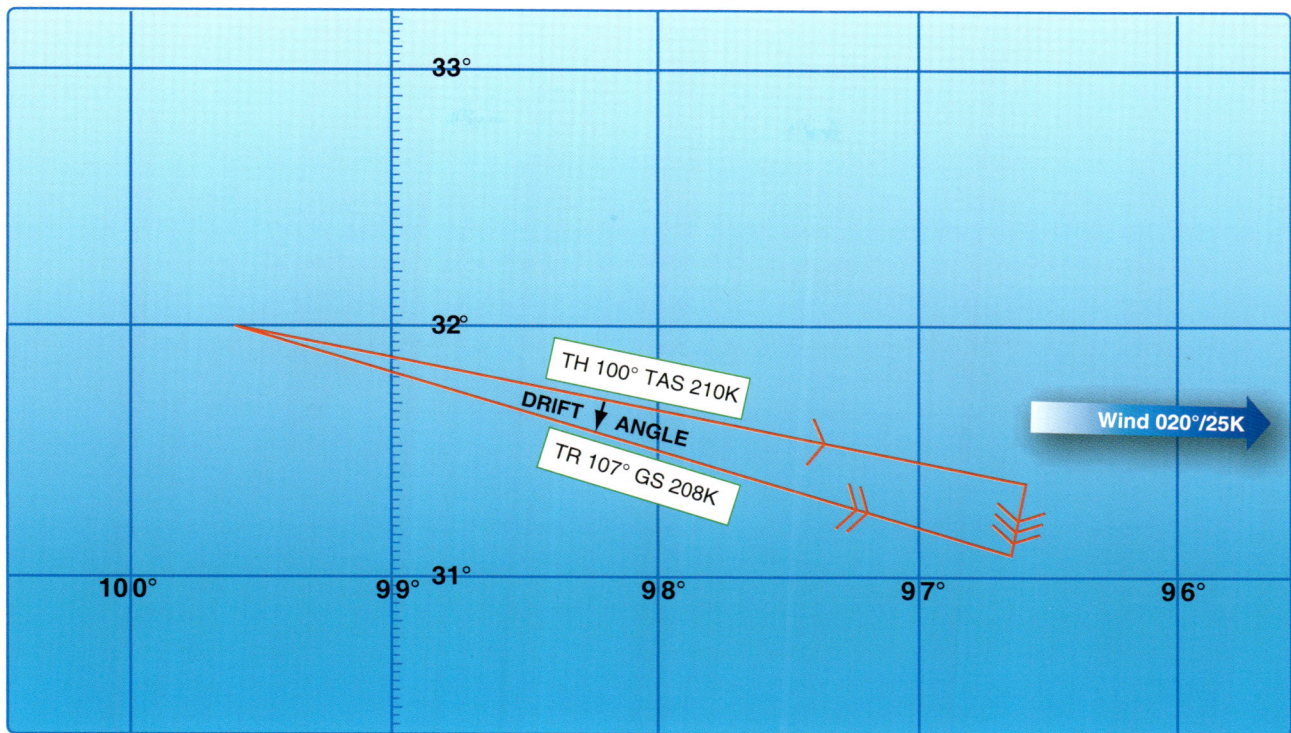

**Figure 4-36.** *Solving for track and groundspeed using chart.*

*Computer Solution*

First, set in the data *[Figure 4-39]:*

1. Set the TH (270°) under the true index.

2. Set the TAS (230 knots) under the grommet.

3. Find the drift angle (10° right) by comparing the TH (270°) with the track (280°). If the track is greater than the TH, drift is right; if it is less, drift is left. Find the appropriate track line on the computer (10° right of centerline).

4. Find the speed circle (215 knots) corresponding to the GS circle. The wind triangle is now constructed. The mark made is the head of the wind vector and the head of the ground vector.

5. Rotate the compass rose until the head of the wind vector is on the centerline below the grommet. Read the WD (207°) under the true index.

6. Read the windspeed (42 knots) on the speed scale between the grommet and the head of the wind vector.

To find true heading (TH) and groundspeed (GS) when true course (TC), true airspeed (TAS) and wind vector are known:
Given:
TC   = 230°
TAS = 220 knots
W/V = 270°/50 knots
Find: TH and GS

This type of problem arises before a flight or during a flight, when a TH is needed to be determined in order to fly and a GS with which to compute an estimated time of arrival (ETA).

*Chart Solution*

First, construct the triangle. *[Figure 4-40]*

1. Draw the wind vector in any convenient scale in the direction toward which the wind is blowing (090°) and to the length representing the windspeed (50 knots) (from any origin).

2. Draw a line in the direction of the TC (230°) and of indefinite length, since the GS is not known (from same origin).

3. Open the dividers an amount equal to TAS (220 knots); then, from the head of the wind arrow, swing an arc with a radius of 220 NM to intersect the TC line (using the same scale as in step 1).

4. Draw a line from the point of intersection of the arc and the TC line to the head of the wind arrow.

To determine the TH (238°), measure the direction of the air vector.

To determine the GS (179 knots), measure the length of the ground vector, using the same scale as before.

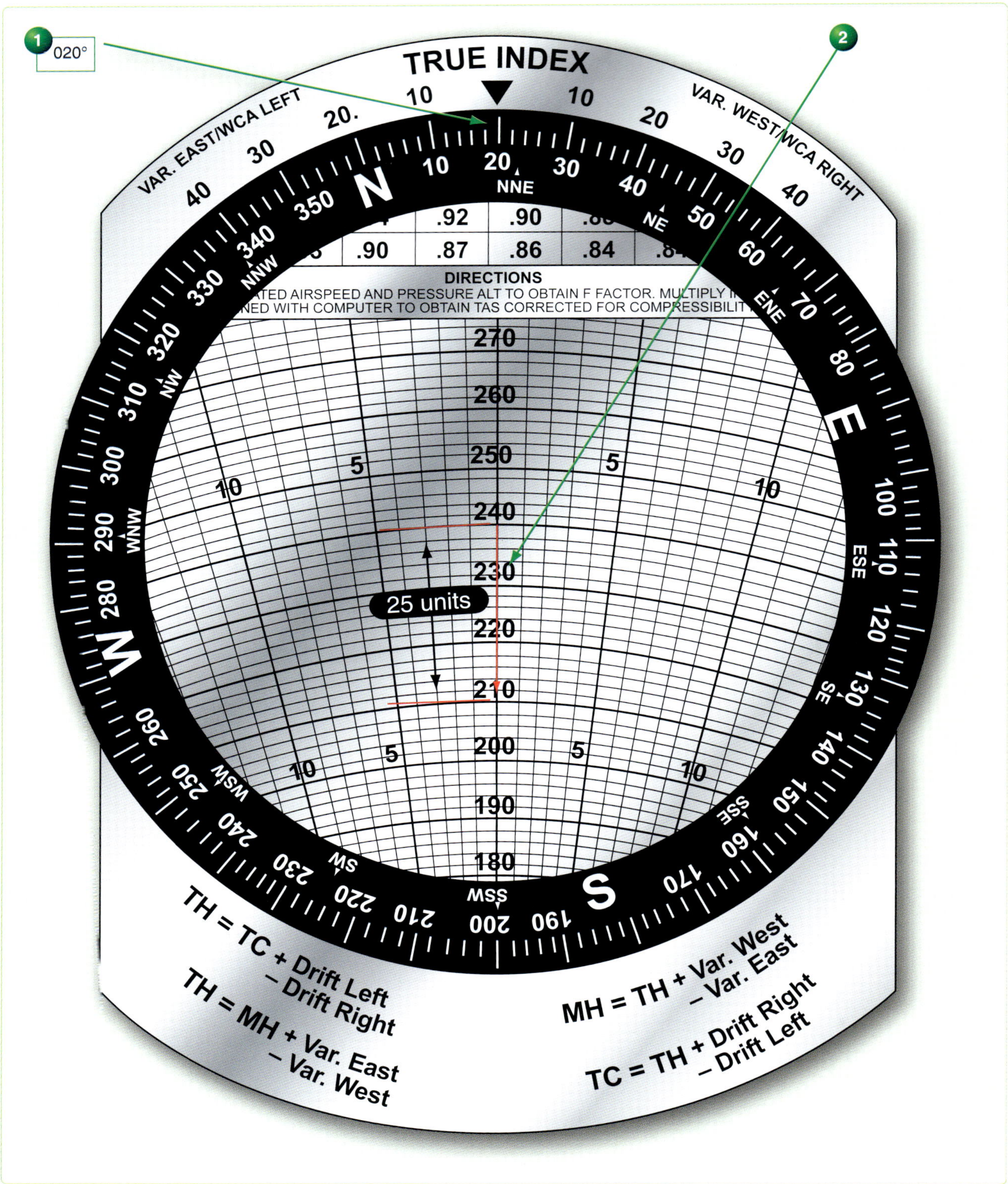

**Figure 4-37.** *Solving for track and groundspeed using computer.*

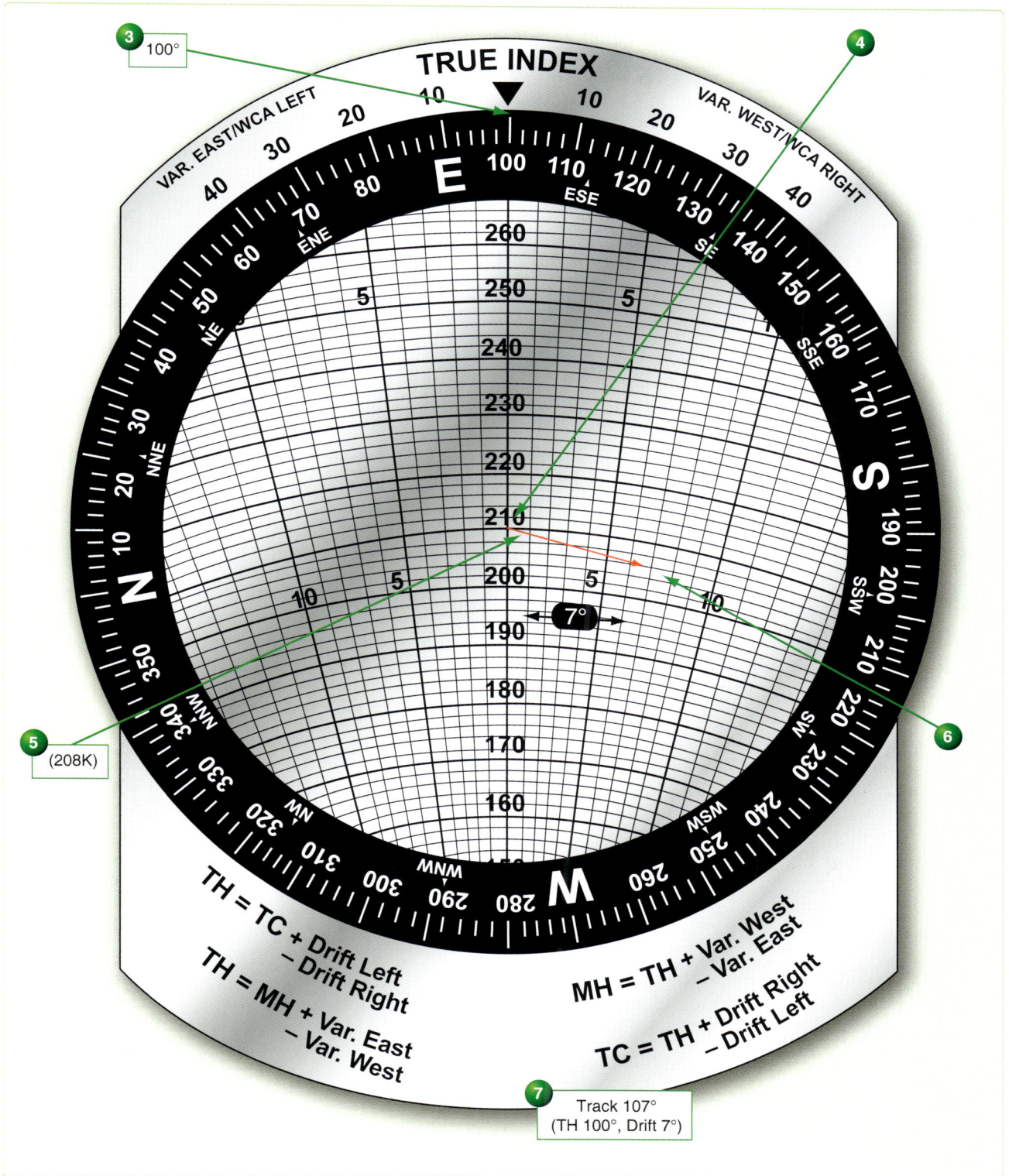

**Figure 4-37.** *Solving for track and groundspeed using computer (continued).*

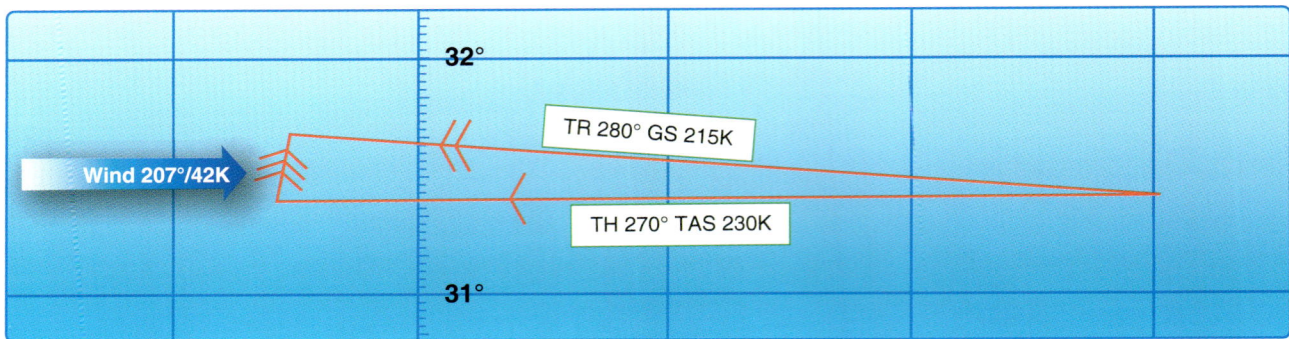

**Figure 4-38.** *Solving for wind using chart.*

## Computer Solution

There are two methods to solve for TH and GS: slip-and-slide method and the juggle method. Both are discussed; however, the slip-and-slide method is normally preferred.

### Slip-and-Slide Method

1. Set WD (270°) under the true index.

2. Draw the wind vector down the center from the grommet, making its length along the speed scale correspond to the windspeed (50 knots).

3. Set the TC (230°) under the true index.

4. Set end of wind vector on the TAS (220 knots) by moving the slide.

5. Read drift left or right (8° left).

6. Apply drift correction mathematically to TC and set this computed TH under the true index (238°).

7. Move the slide up until the grommet is on TAS (220 knots). The wind triangle is now set up correctly.

8. Read GS at the end of the wind vector (179 knots).

### Juggle Method *[Figure 4-42]*

1. Set WD (270°) under the true index.

2. Draw the wind vector down the center from the grommet, making its length along the speed scale correspond to the windspeed (50 knots).

3. Set the TAS (220 knots) under the grommet.

4. Set the TC (230°) under the true index. *[Figure 4-42]* The wind triangle is set up incorrectly, for TC rather than TH is set under the true index. However, since the TH is not known, the TC is used as a first approximation of the TH. This gives a first approximation of the drift angle, which can be applied to the TC to get a more accurate idea of the TH.

5. Determine the drift angle (10° left) on the approximate heading (230°) to obtain a second approximation of the TH (240°). If the drift angle is right, the drift correction is minus; if it is left, the drift correction is plus.

6. Set the second approximate heading (240°) under the true index. Read the drift angle for this heading (8° left). To correct the wind triangle, the drift angle which is read at the head of the wind vector must equal the difference between the TC and the TH, which is set under the true index. As it stands, the drift angle is 8° left, while the difference between TC and the indicated TH is 10° left.

7. Juggle the compass rose until the drift angle equals the difference between TC and TH. In this example, the correct drift angle is 8° left. Now the wind triangle is set up correctly.

8. Read the TH (238°) under the true index.

9. Read the GS (179 knots) on the speed circle passing through the head of the wind vector.

## Average Wind Affecting Aircraft

An average wind is an imaginary wind that would produce the same wind effect during a given period as two or more actual winds that affect the aircraft during that period. Sometimes an average wind can be applied once instead of applying each individual wind separately.

## Authentically Averaging WDs

If the WDs are fairly close together, a satisfactory average wind can be determined by arithmetically averaging the WDs and windspeeds. However, the greater the variation in WD, the less accurate the result. It is generally accepted that winds should not be averaged arithmetically if the difference in directions exceeds 090° and/or the speed of less than 15 knots. In this case, there are other methods that may be used to obtain a more accurate average wind. A chart solution is shown in *Figure 4-43.*

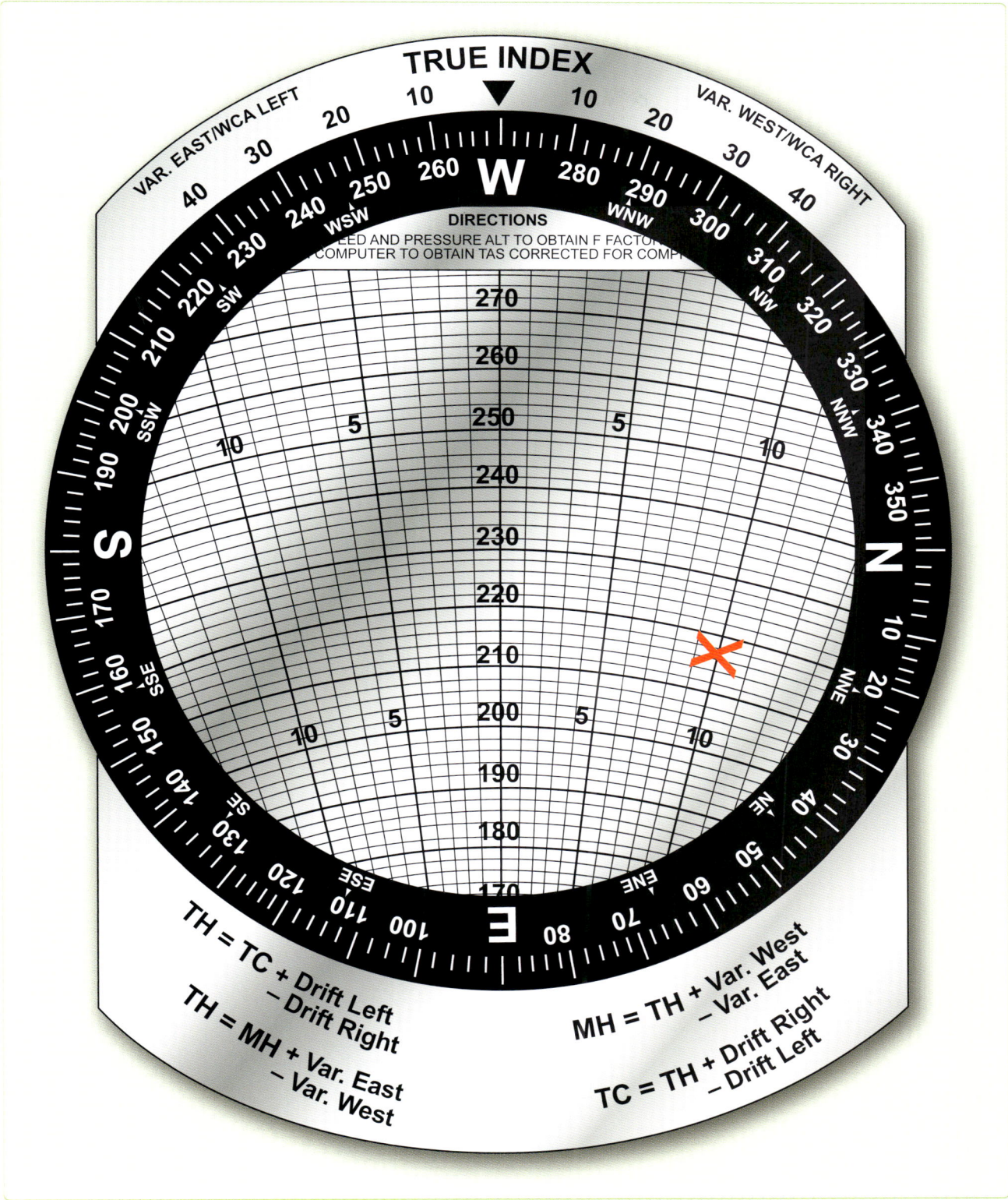

**Figure 4-39.** *Solving for wind using computer.*

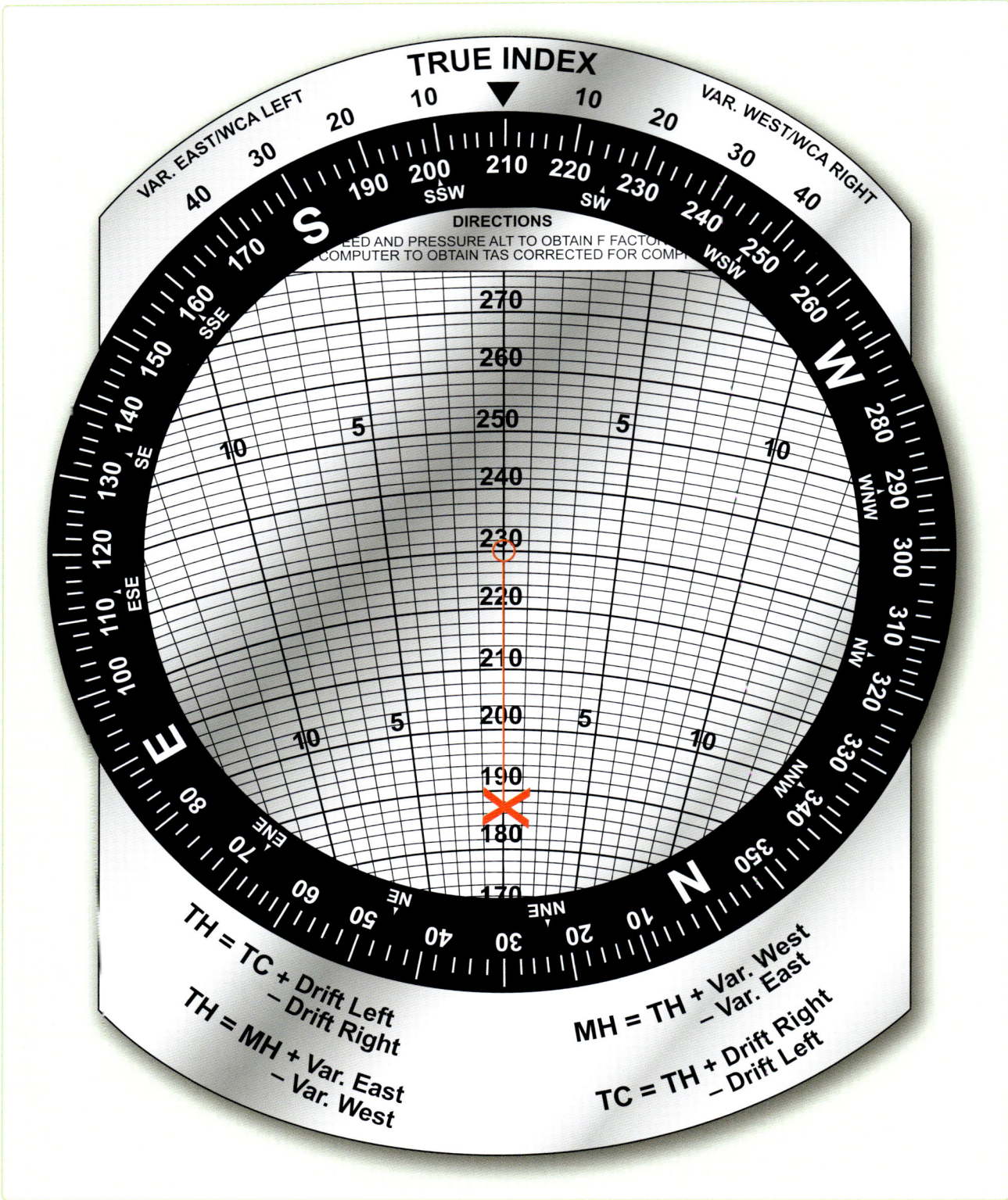

**Figure 4-39.** *Solving for wind using computer (continued).*

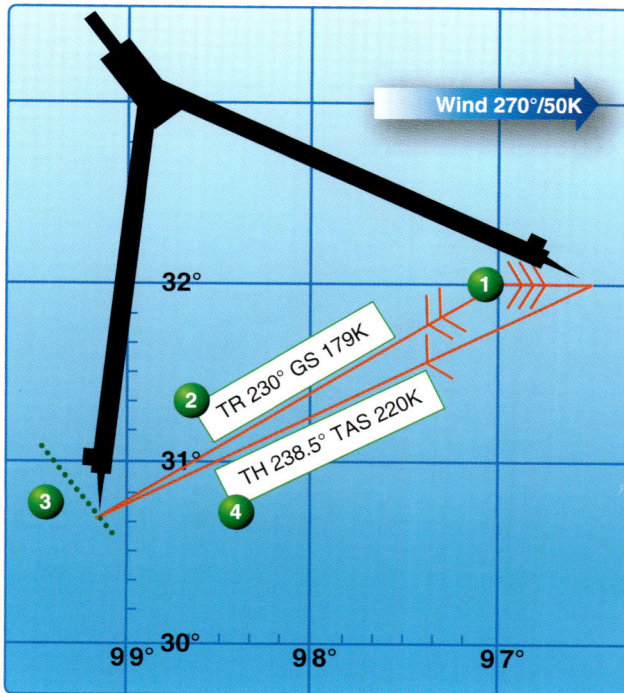

**Figure 4-40.** *Solving for the true heading and groundspeed using chart.*

### Computer Solution

Winds can be averaged by vectoring them on the wind face of the DR computer, using the square grid portion of the slide and the rotatable compass rose. Average the following three winds by this method: 030°/l5 knots, 080°/20 knots, and 150°/35 knots:

Place the slide in the computer so that the top line of the square grid portion is directly under the grommet and the compass rose is oriented so that the direction of the first wind (030°) is under the true index. The speed of the wind (15 knots) is drawn down from the grommet. *[Figure 4-44A]*

Turn the compass rose until the direction of the second wind (080°) is under the true index, and then reposition the slide so that the head of the first wind vector is resting on the top line of the square grid section of the slide. Draw the speed of the second wind (20 knots) straight down (parallel to the vertical grid lines) from the head of the first wind arrow. *[Figure 4-44B]*

Turn the compass rose so that the direction of the third wind (150°) is under the true index, and reposition the slide so that the head of the second wind vector is resting on the top line of the square grid section of this slide. Draw the speed of the third wind (35 knots) straight down from the head of the second wind arrow. *[Figure 4-44C]*

Turn the compass rose so the head of the third wind arrow is on centerline below the grommet. Reposition the slide to place the grommet on the top line of the square grid section. The resultant or average wind direction is read directly beneath the true index (108°). Measuring the length of the resultant wind vector (46) on the square grid section and divide it by the number of winds used (3) to determine the windspeed. This gives a WS of about 151/2 knots. The average wind then is 108°/15 1/2 knots. *[Figure 4-44D]*

With a large number of winds to be averaged or high windspeeds, it may not possible to draw all the wind vectors on the computer unless the windspeeds are cut by 1/2 or 1/3. If one windspeed is cut, all windspeeds must be cut. In determining the resultant windspeed, the length of the total vector must be multiplied by 2 or 3, depending on how the windspeed was cut, and then divided by the total number of winds used. In cutting the speeds, the direction is not affected and the WD is read under the true index.

Wind effect is proportional to time. *[Figure 4-45]* To sum up two or more winds that have affected the aircraft for different lengths of time, weigh them in proportion to the times. If one wind has acted twice as long as another, its vector should be drawn in twice as shown. In determining the average windspeed, this wind must be counted twice.

### Resolution of Rectangular Coordinates

Data for radar equipment is often given in terms of rectangular coordinates; therefore, it is important that the navigator be familiar with the handling of these coordinates. The DR computer provides a ready and easy method of interconversion.

Given:
A wind of 340°/25 knots to be converted to rectangular coordinates. *[Figure 4-46]*

1. Plot the wind on the computer in the normal manner. Use the square grid side of the computer slide for the distance.

2. Rotate the compass rose until north, the nearest cardinal heading, is under the true index.

3. Read down the vertical scale to the line upon which the head of the wind vector is now located. The component value (23) is from the north under the true index.

4. Read across the horizontal scale from the center line to the head of the wind vector. The component value (9) is from the west. The wind is stated rectangularly as N-23, W-9.

**B**

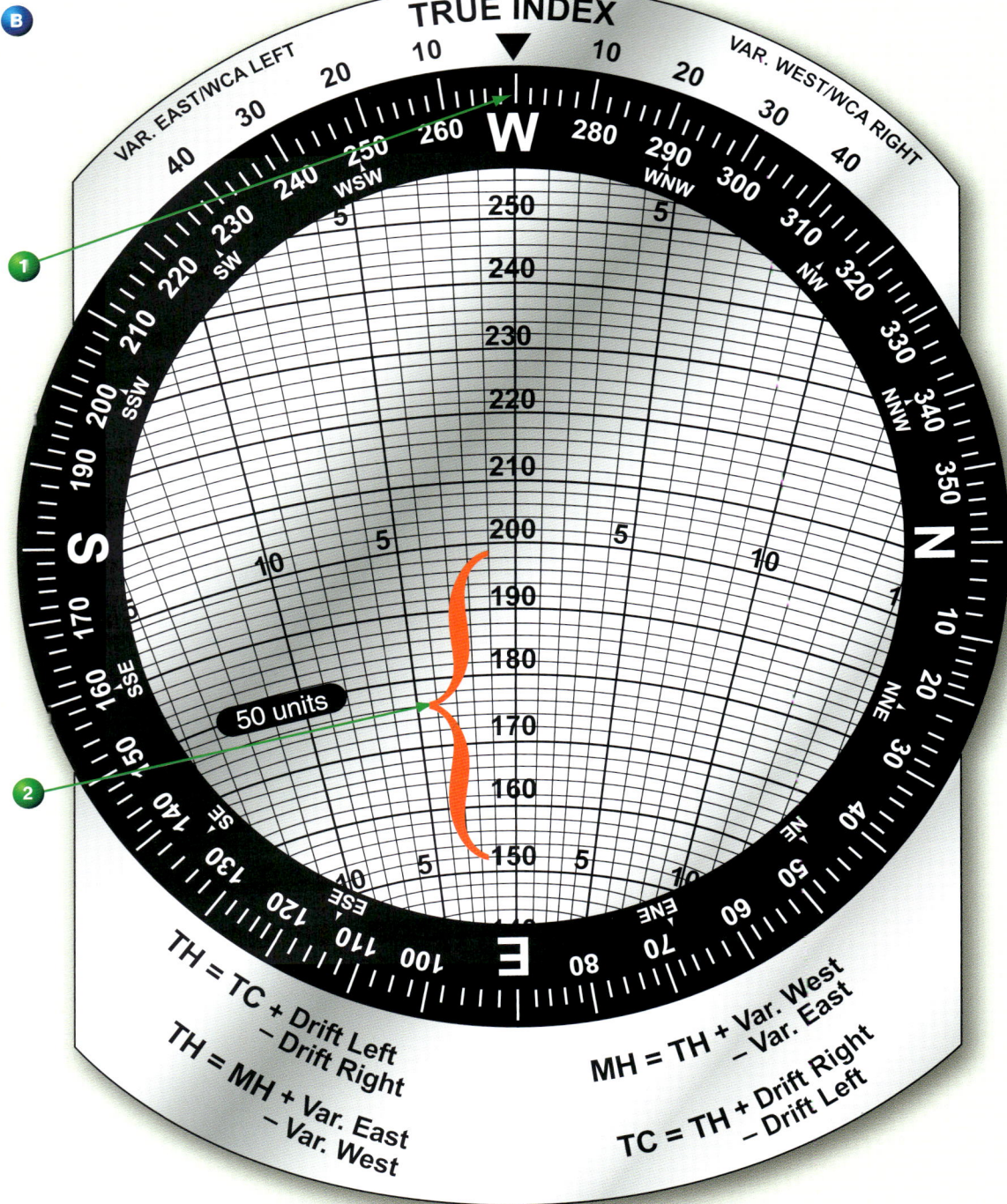

**Figure 4-41.** *Solving for TH and GS using slip-and-slide method.*

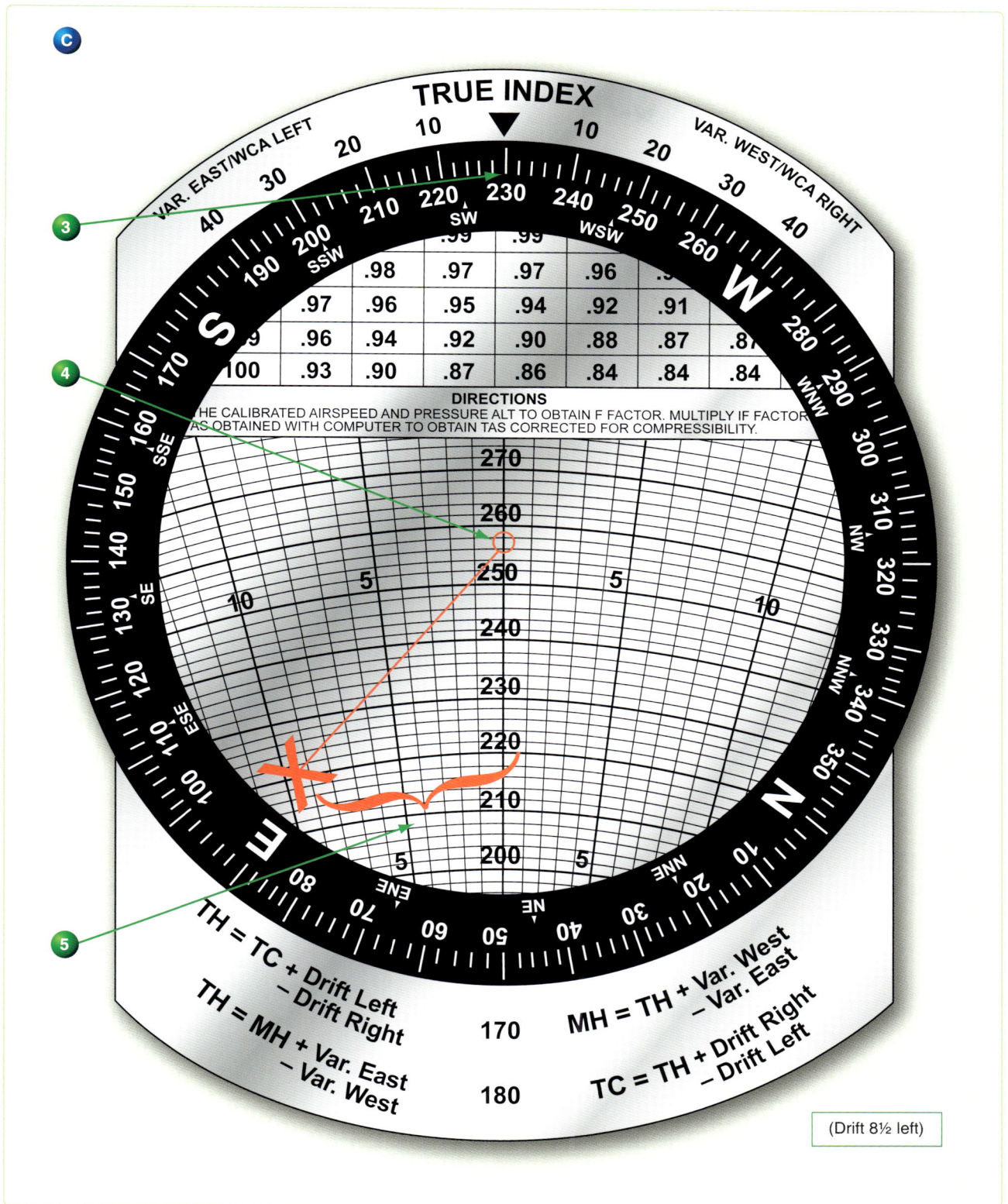

**Figure 4-41.** *Solving for TH and GS using slip-and-slide method (continued).*

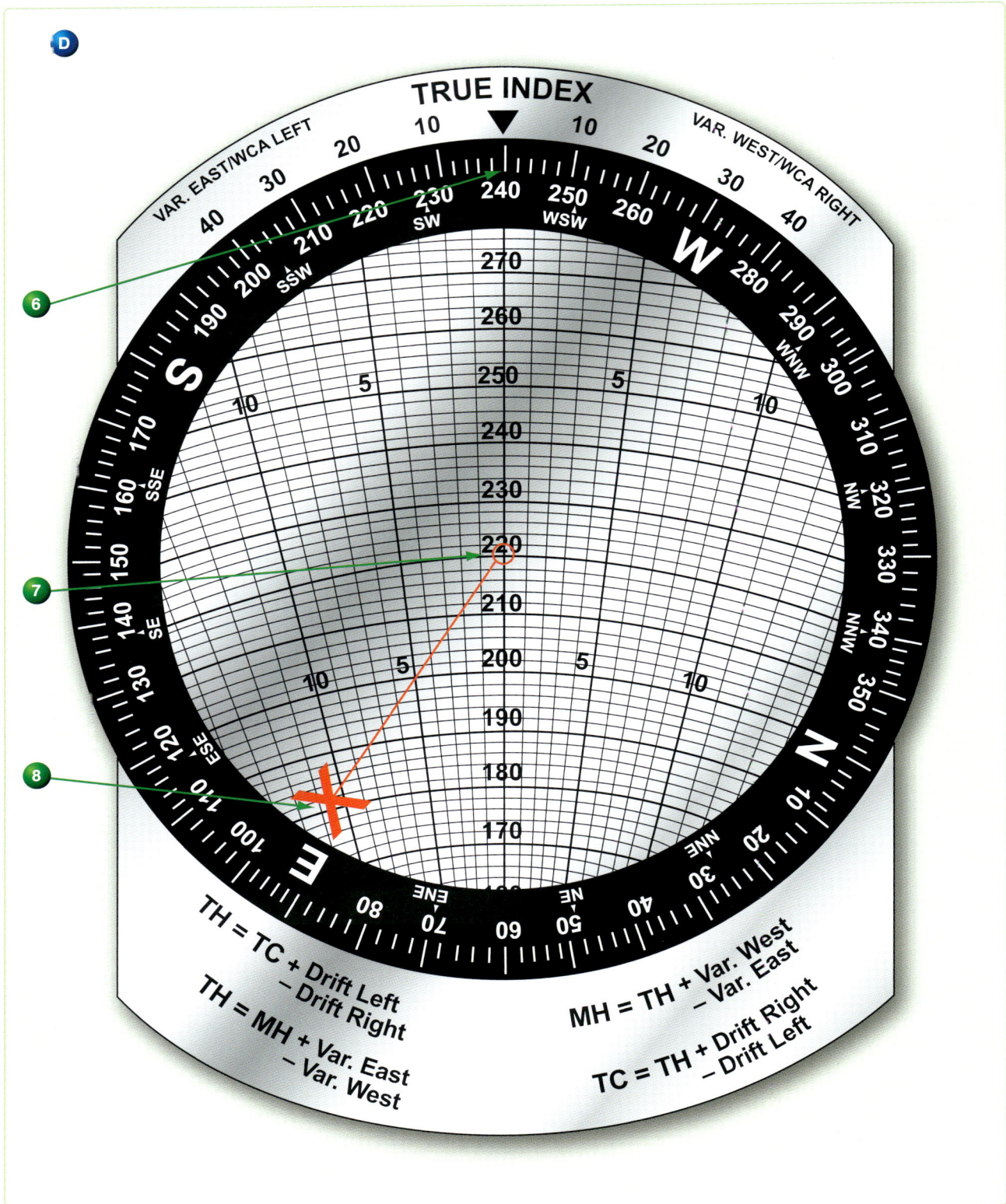

**Figure 4-41.** *Solving for TH and GS using slip-and-slide method (continued).*

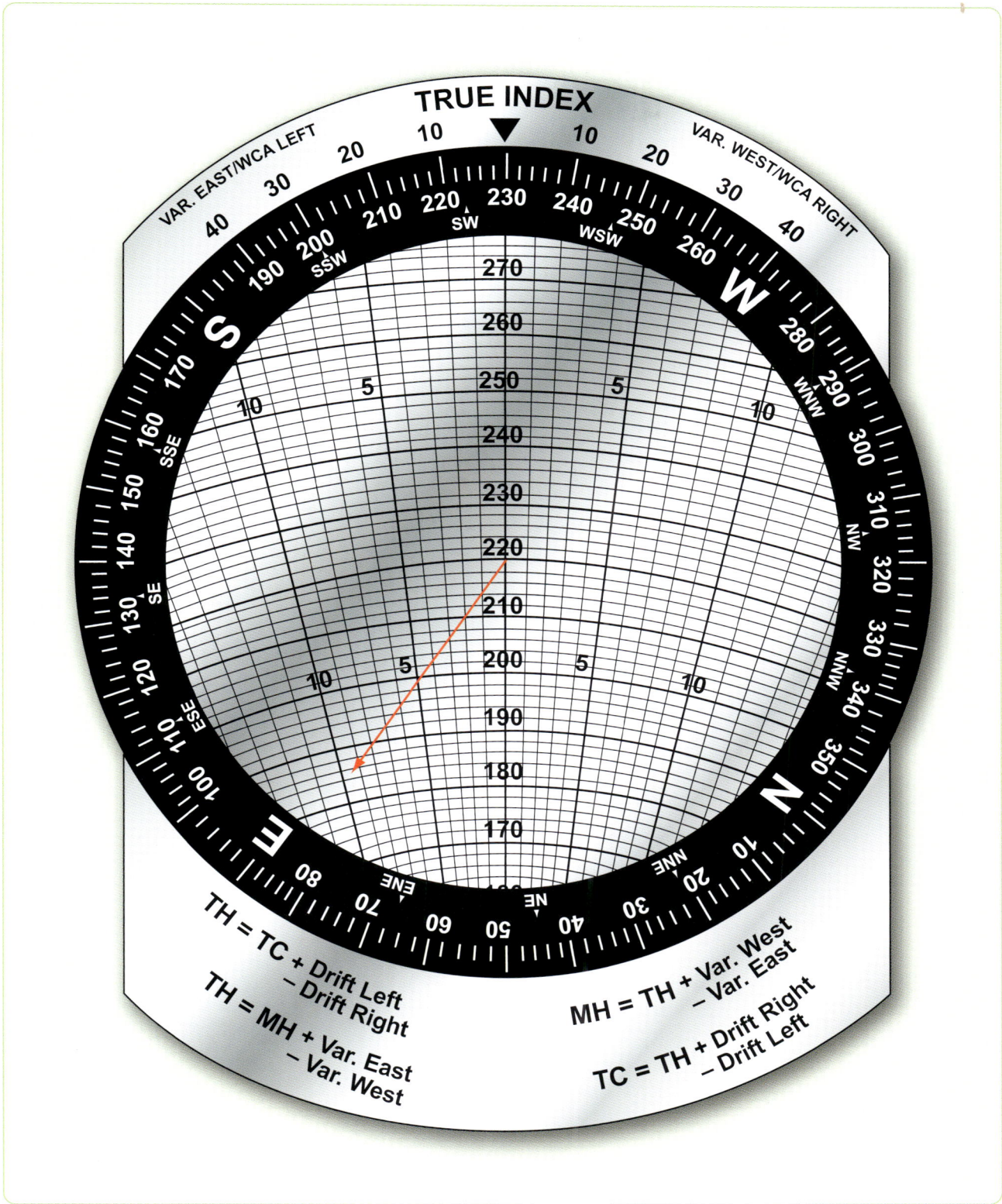

**Figure 4-42.** *Solving for TH and GS using the juggle method.*

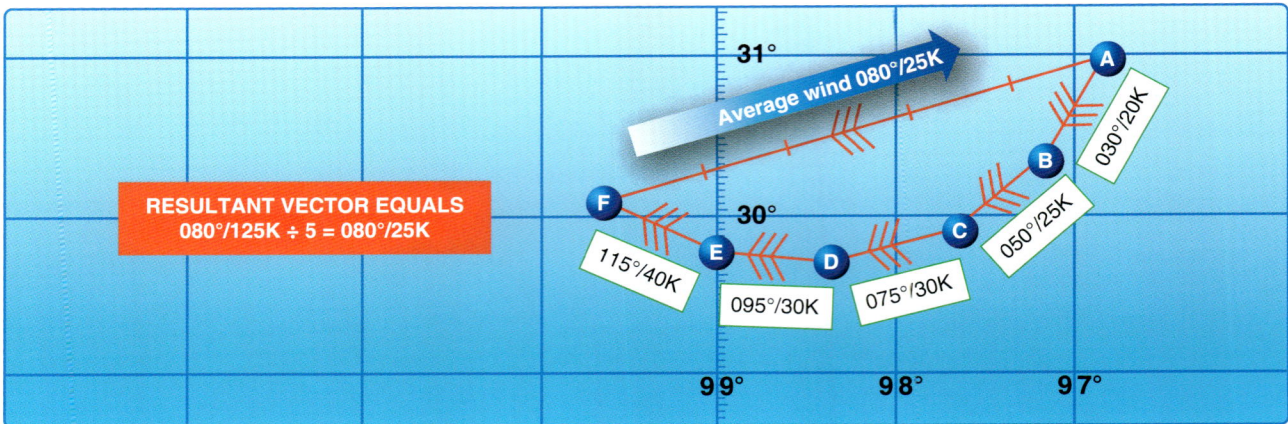

**Figure 4-43.** *Solving for average wind using chart.*

Given:
Coordinates, S-30, E-36, to convert to a wind.

1. Use the square grid side of the computer.

2. Place south cardinal heading under the true index and the grommet on zero of the square grid.

3. Read down from the grommet along the centerline for the value (30) of the cardinal direction under the true index.

4. Place east cardinal heading, read horizontally along the value located in Step 3 from the centerline of the value of the second cardinal direction and mark the point.

5. Rotate the compass rose until the marked point is over the centerline of the computer.

6. Read the WD (130) under the true index and velocity (47 knots) from the grommet to the point marked.

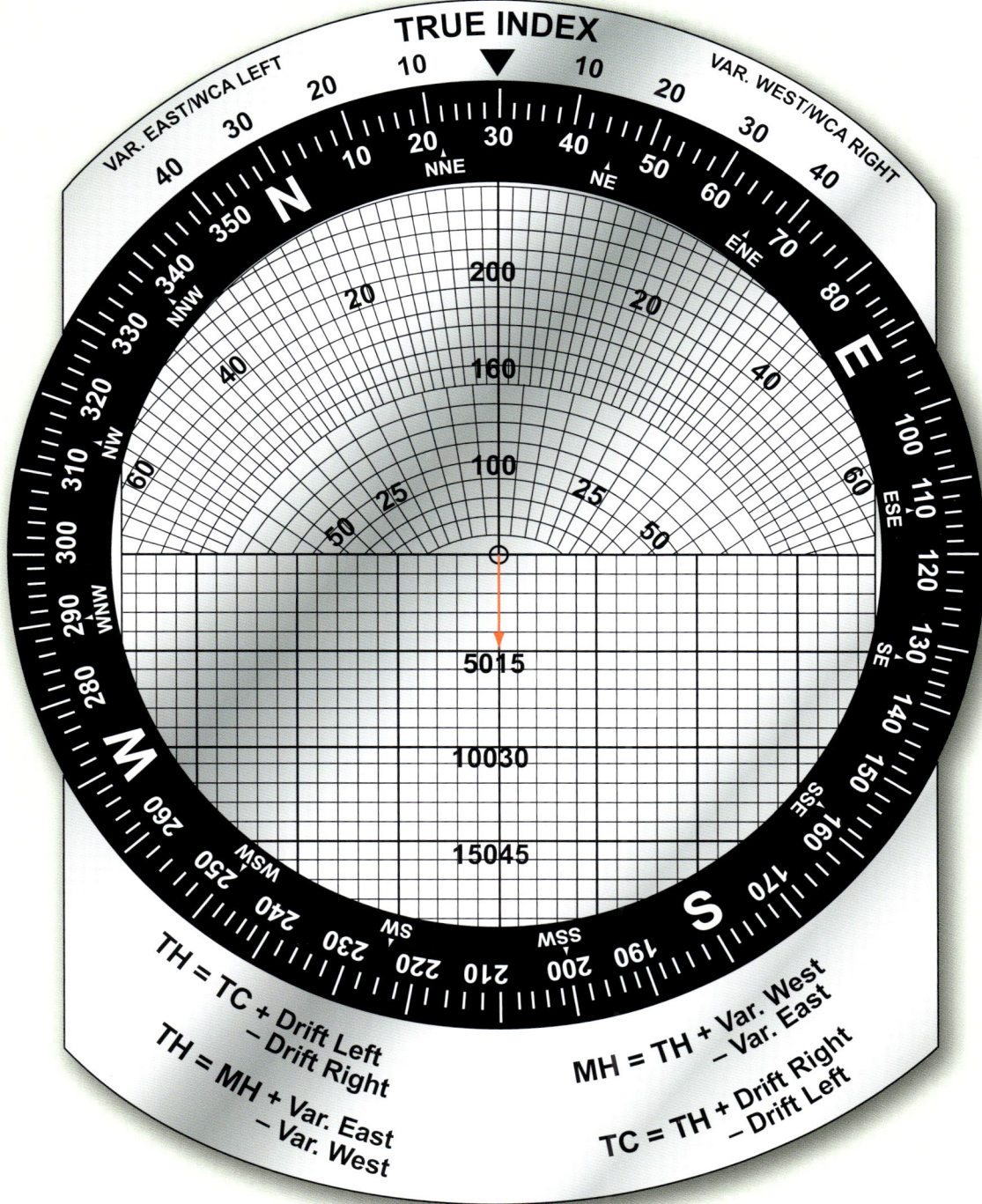

**Figure 4-44.** *Solving for average wind using computer.*

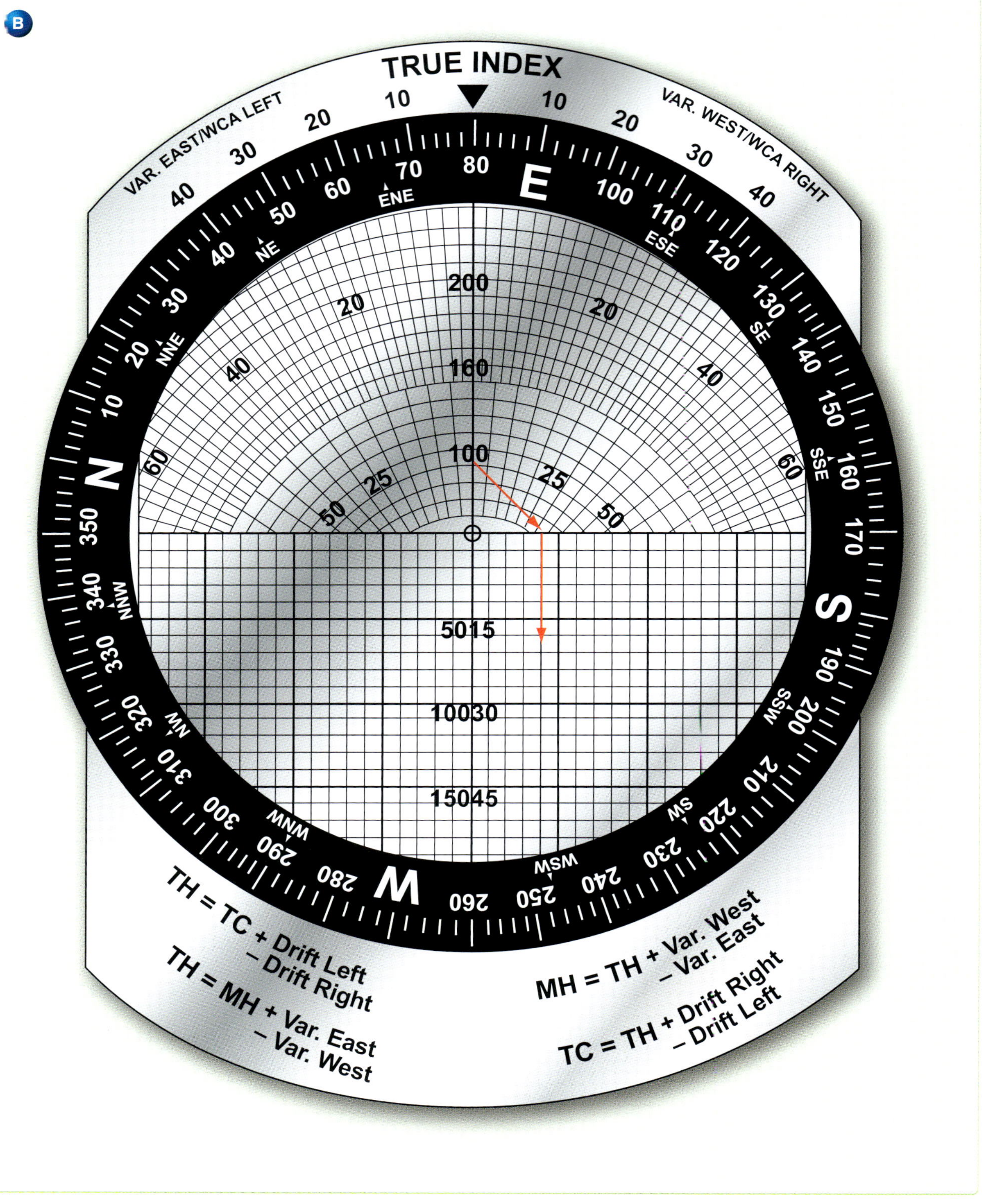

**Figure 4-44.** *Solving for average wind using computer (continued).*

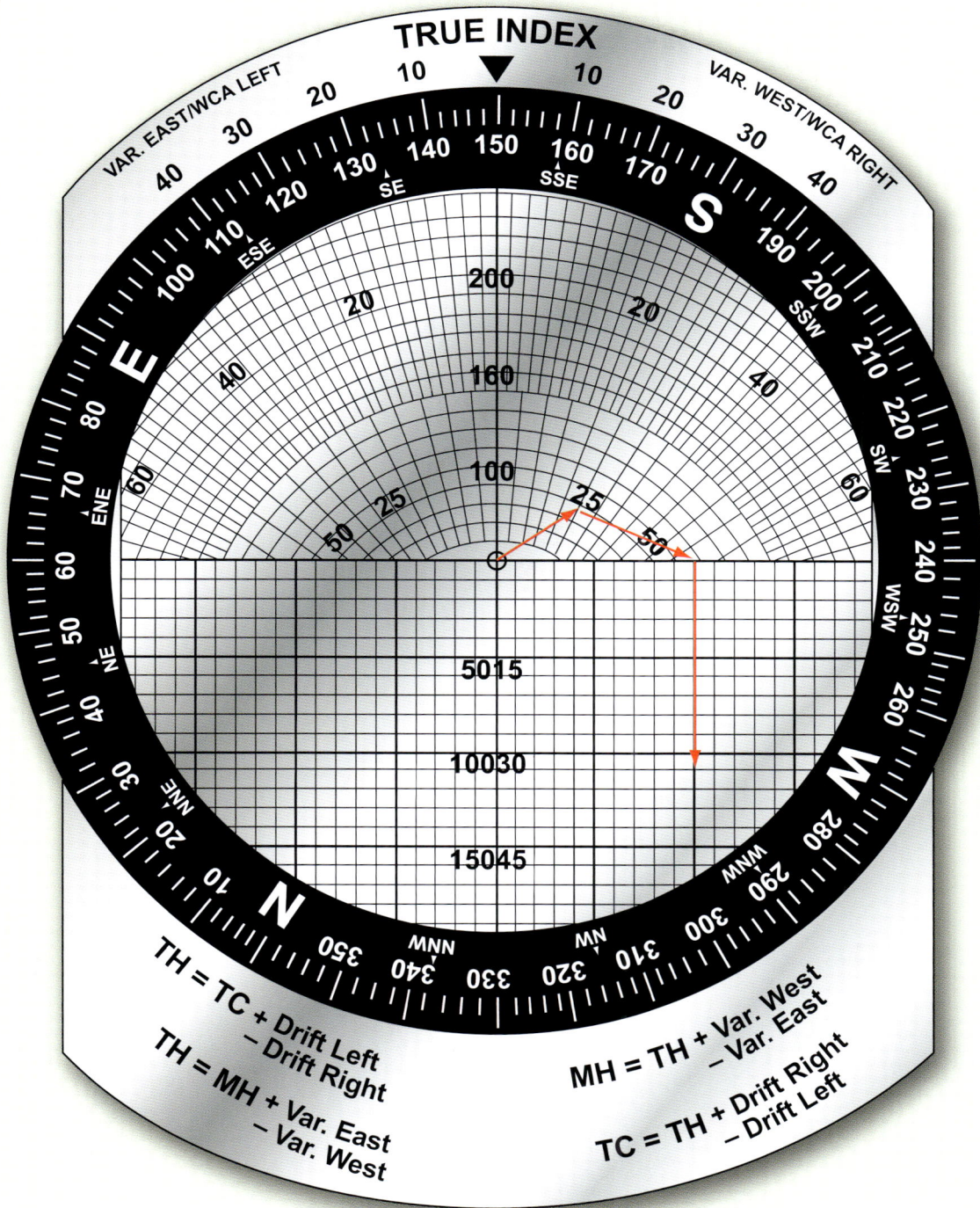

**Figure 4-44.** *Solving for average wind using computer (continued).*

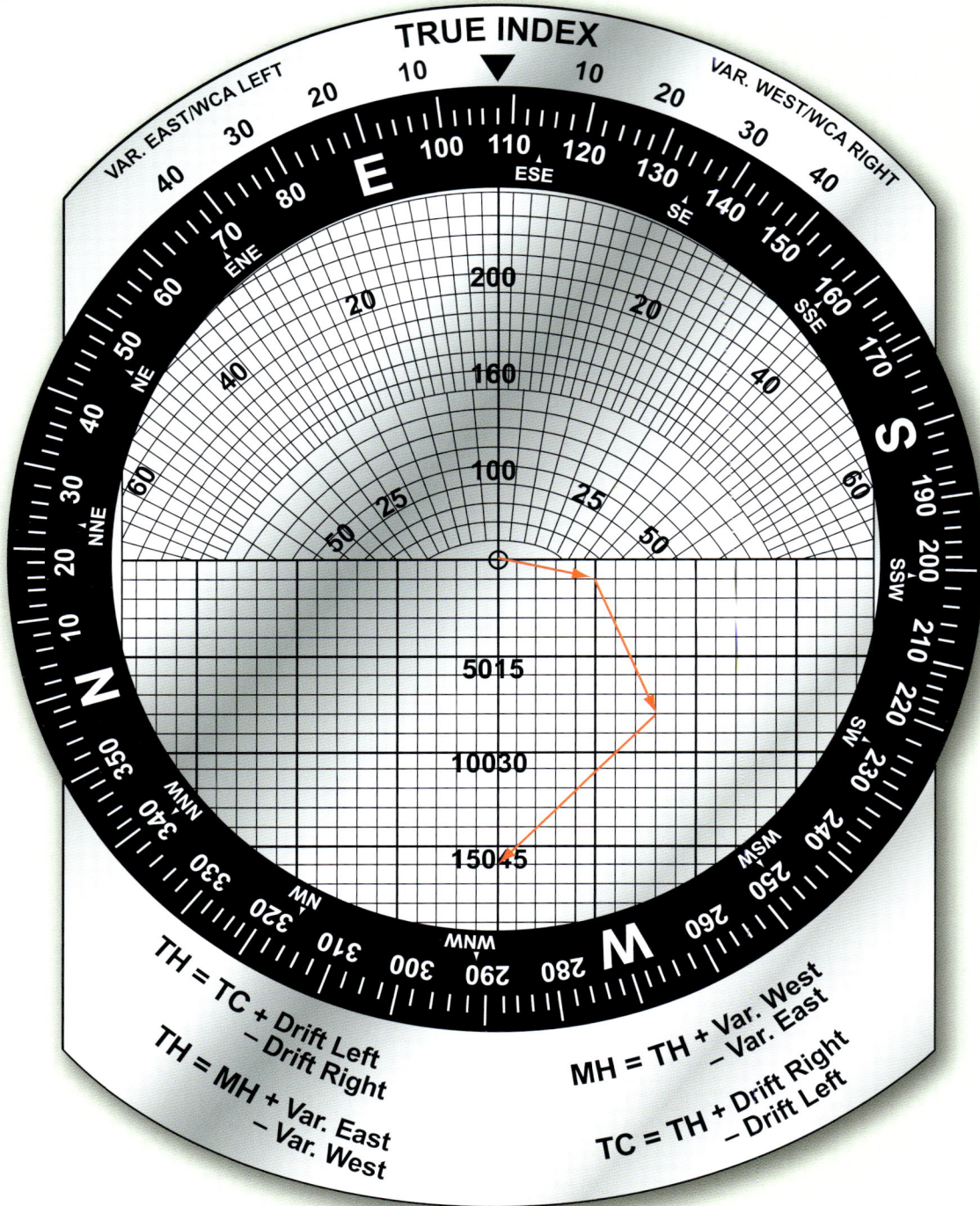

**Figure 4-44.** *Solving for average wind using computer (continued).*

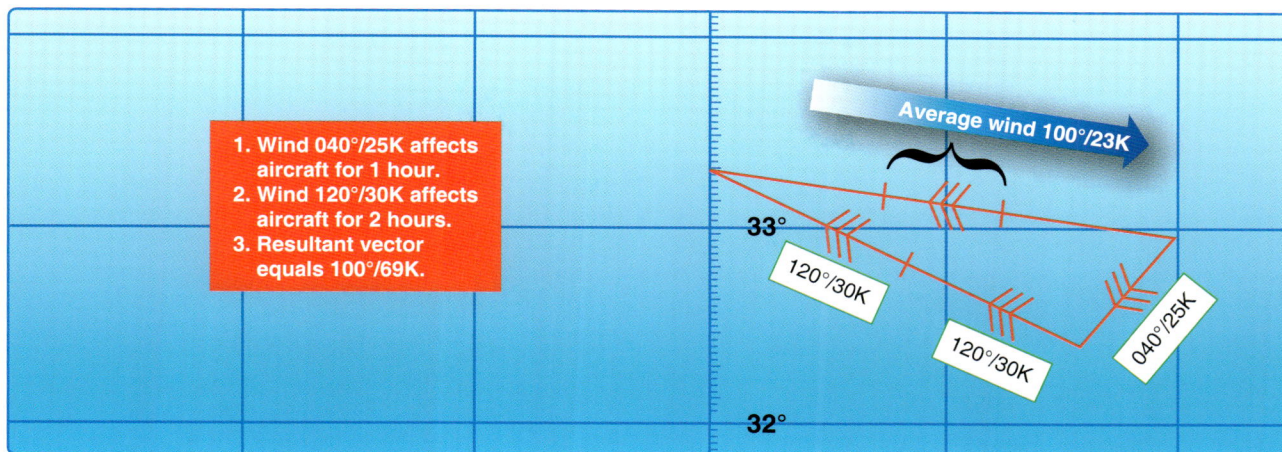

**Figure 4-45.** *Weigh winds in proportion to time.*

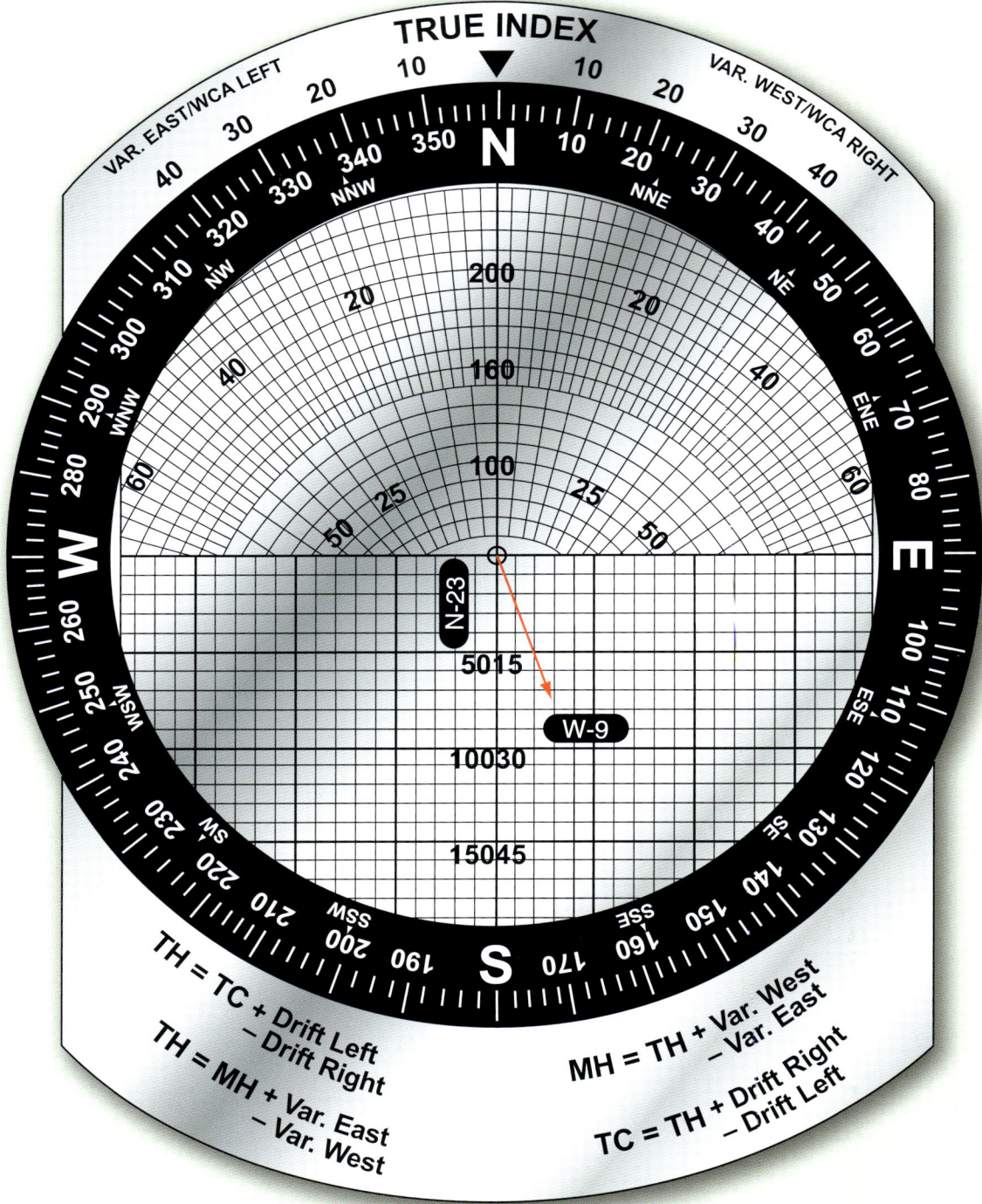

**Figure 4-46.** *Convert wind to rectangular coordinates.*

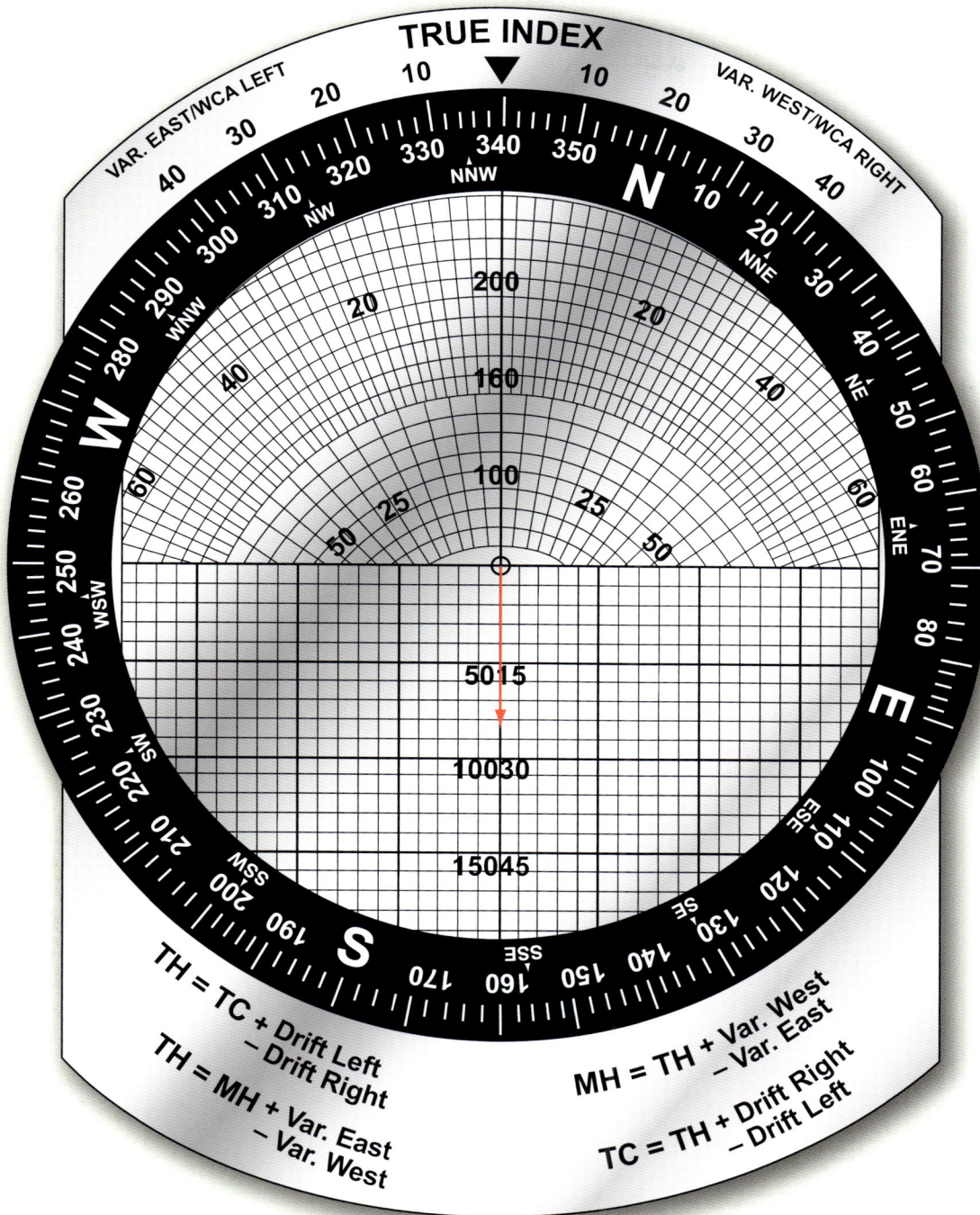

**Figure 4-46.** *Convert wind to rectangular coordinates (continued).*

# Chapter 5

# Radio Aid Fixing

## Bearings and Lines of Position

Good dead reckoning (DR) techniques can result in fairly accurate positions. But, even when employing the very best techniques, the DR position becomes less accurate as time increases beyond the last known position. Small errors tend to accumulate into one total error, which is unacceptable. To minimize this error, the navigator must be able to establish an accurate position from which to restart DR. This accurate position is free of any DR errors and is called a fix. A fix is simply a point from which the navigator can restart DR, just as if it were the takeoff point. We begin our discussion of fixing with an explanation of lines of position (LOP).

## Lines of Position (LOP)

It is possible to solve part of the fix problem without knowing an exact location. For example, assume you are in a strange town and you call a friend to meet you downtown. If you tell this person that you are somewhere on Park Street, your friend can limit any search for you to that particular street. In this case, Park Street is an LOP. An LOP is a series of possible positions or fixes. It can be a straight line, such as a city street, or a curved line, such as a river, but it gives a definite clue to position.

If you tell your friend that you are at Park Street where it crosses the Karuzas River, it would then establish your exact location. You have used two LOPs to determine your exact position. Thus, two intersecting LOPs identify a point that establishes a fix.

You can use the same procedure as a navigator. You may be flying along a railroad that you identify as the Jedicke Railroad on your chart. As you continue on this course, you notice the railroad crosses a river that is labeled the King River on your chart. When you fly over the point where these two visual LOPs cross, you know your exact location over the ground and on your chart. You now have a fix from which you can continue to DR.

## Types of LOPs

A fix gives definite information as to both track and groundspeed (GS) of an aircraft since the last fix, but a single LOP can only define either the track or the GS—not both. And it may not clearly define either. The evidence obtained from an LOP depends upon the angle at which it intersects the track. LOPs are sometimes classified according to this angle.

## Course Line

An LOP that is parallel or nearly parallel to the course is called a course line. *[Figure 5-1]* It gives information as to possible locations of the aircraft laterally in relation to the course; that is, whether it is to the right or left of course. Because it does not indicate how far the aircraft is along the track, no speed information is provided.

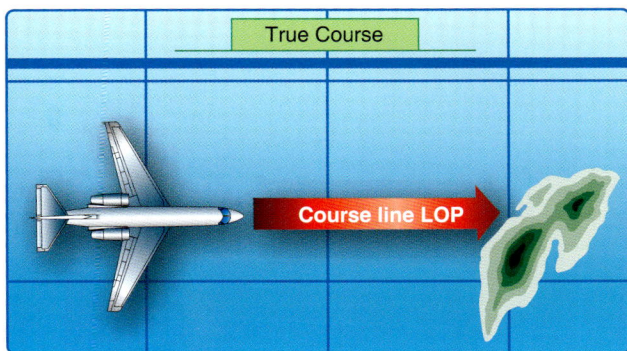

**Figure 5-1.** *Line of position parallel to track is the course line.*

## Speed Line

An LOP that is perpendicular, or nearly so, to the track is called a speed line because it indicates how far the aircraft has traveled along the track and, thus, is a measure of GS. *[Figure 5-2]* It does not indicate whether the aircraft is to the right or left of the course.

**Figure 5-2.** *Line of position perpendicular to track is the speed line.*

## LOPs by Bearings

One method of determining an LOP is to establish the direction of the line of sight (LOS) to a known fixed object. The direction of the LOS is the bearing of the object from the aircraft. A line plotted in the direction of the bearing is an LOP. At the time of the observation, the aircraft was on the LOP.

## Relative Bearings (RB)

An RB is the angle between the fore-and-aft axis of the aircraft and the LOS to the object, always measured clockwise from 000° at the nose of the aircraft through 360°. In *Figure 5-3*, the RB of the object is shown as 070°. Convert this to a true bearing (TB) before it can be plotted. To do this, simply add the RB to the true heading (TH) the aircraft was flying when the bearing was obtained. (Subtract 360° if the total exceeds this amount.) Thus:

$$RB + TH = TB \text{ (RuB THe TuB)}$$

Where:
RB is the relative bearing, TH is the true heading, and TB is the true bearing.

Assuming the aircraft was on a TH of 210° when the bearing was taken, the corresponding TB of the object is 280°. (070° RB + 210° TH = 280° TB)

## Plotting the LOP

As previously stated, two intersecting LOPs determine the position of the aircraft. The only other possible point from

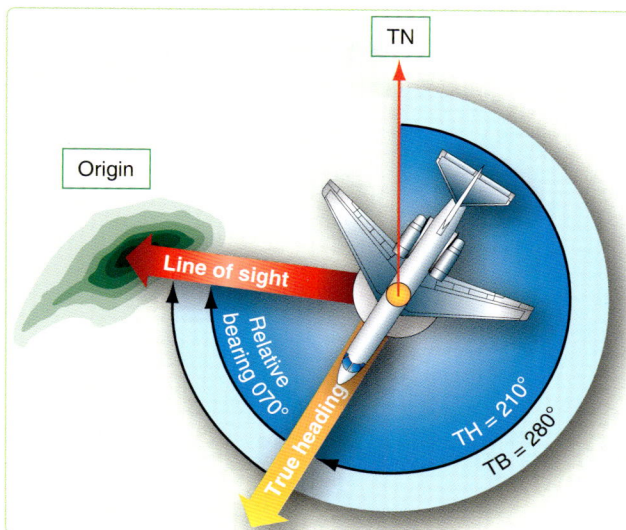

**Figure 5-3.** *True bearing equals relative bearing plus true heading.*

which to begin plotting the LOP is the object on which you took the bearing. The procedure is to use the reciprocal of the TB of the object, thus drawing an LOP toward the aircraft. In actual practice, it is not necessary to compute the reciprocal of the bearing; the TB is measured with the plotter, and the LOP is drawn toward the opposite end of the plotter. To establish an LOP by RB, the navigator must know:

1.  Position of the source (object) of the bearing,

2.  TH of the aircraft,

3.  RB of the object, and

4.  Exact time at which the TH and RBs were taken.

## Fixes

### Adjusting LOPs for a Fix

Sometimes it is impossible for a navigator to obtain more than one LOP at a given time. If two LOPs are for two different times, their intersection does not constitute a fix because the aircraft moved between the time it was on the first LOP and the second LOP. *Figure 5-4* shows a bearing taken at 1055Z and another at 1100Z. At 1055Z when the navigator took the first bearing, the aircraft was somewhere along the 1055Z LOP (single-barbed LOP) and, at 1100Z, it was somewhere along the 1100Z LOP. The intersection of these two lines, as plotted, does not constitute a fix. For an intersection to become a fix, the navigator must either obtain the LOPs at the same time or adjust them to a common time by using the motion of the aircraft between the observations. The usual method of adjusting an LOP for the motion of the aircraft is to advance one line to the time of the other. *Figure 5-4* shows how this is done. The desired time of the fix is 1100Z.

Determine the time to advance the 1055Z LOP (5 minutes). Multiply this time by the aircraft GS (300 knots).

*   Measure the distance computed in the first step in the direction of the track of the aircraft (045°).

*   Draw a line through this point parallel to the 1055Z LOP (double-barbed LOP). This represents the advanced LOP. The intersection of the advanced LOP and the 1100Z LOP is the fix. The advanced LOP is usually plotted on the chart with two arrowheads, while the unadvanced LOP is marked with a single arrowhead.

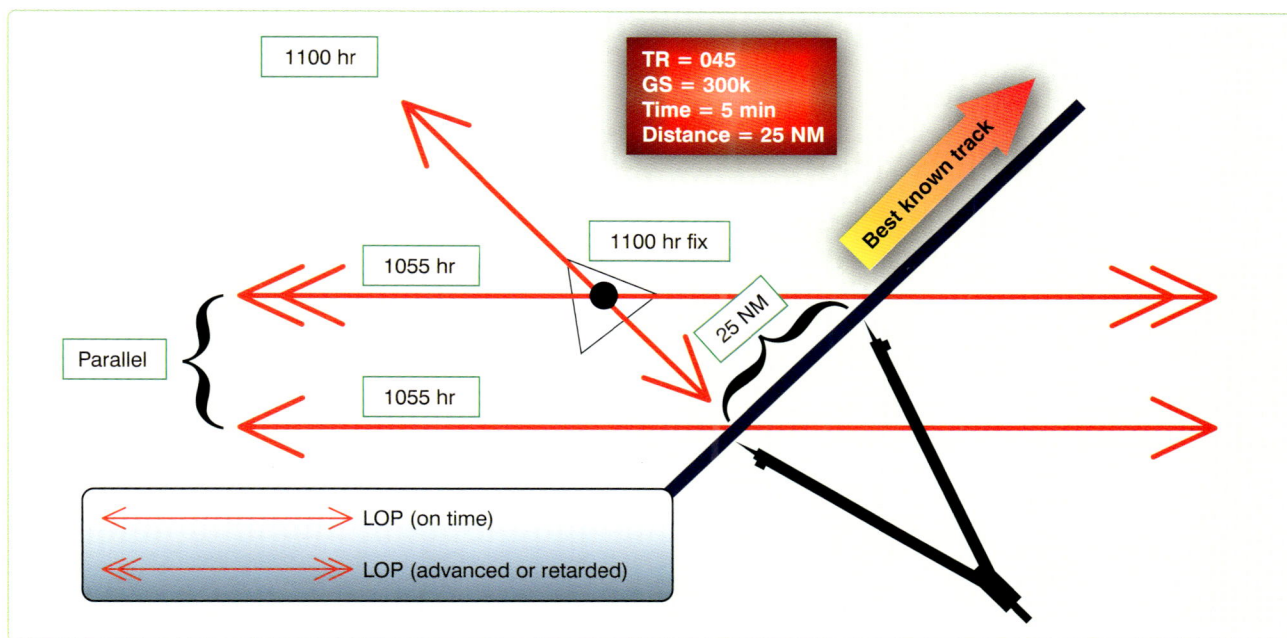

**Figure 5-4.** *Adjusting lines of position for fix.*

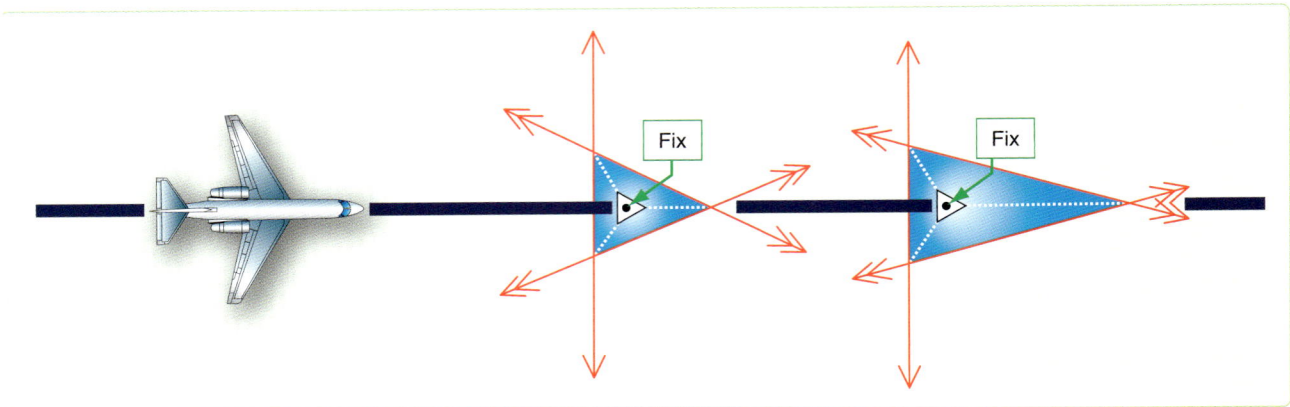

**Figure 5-5.** *Bisector method.*

- When three LOPs are involved, the procedure is exactly the same as for two. The resolution of three LOPs, however, may result in a triangle instead of a point, and the triangle may be large enough to vary the position of the fix. The technique many navigators use is to place the fix at the center of the triangle. *Figure 5-5* shows a technique for finding the center of the triangle by bisecting the angles of the triangle. The point of intersection of the bisectors is equal distance from all three LOPs and is the fix position.

## The Running Fix

It is possible to establish an aircraft position by a series of bearings on the same object. For best accuracy, these RBs are taken when the object is approximately 45°, 90°, and 135°

from the aircraft. The navigator then advances or retards the LOPs to a common time. The result is a running fix. The accuracy is based on the aircrafts distance from object and the amount of time it takes to go from the first bearing to the last bearing since you must move two of the LOPs for the aircrafts track and GS. The running fix is illustrated in *Figure 5-6*.

## Accuracy of a Fix

The accuracy of a fix can sometimes be improved by the use of a little foresight. If the track of the aircraft is known more accurately than the GS, the course line should be adjusted since any error in the GS has little effect on it. If, however, you desire to adjust a speed line under these conditions, the accuracy of the fix is in doubt. Similarly, if the GS is known more accurately than the track, the speed line should

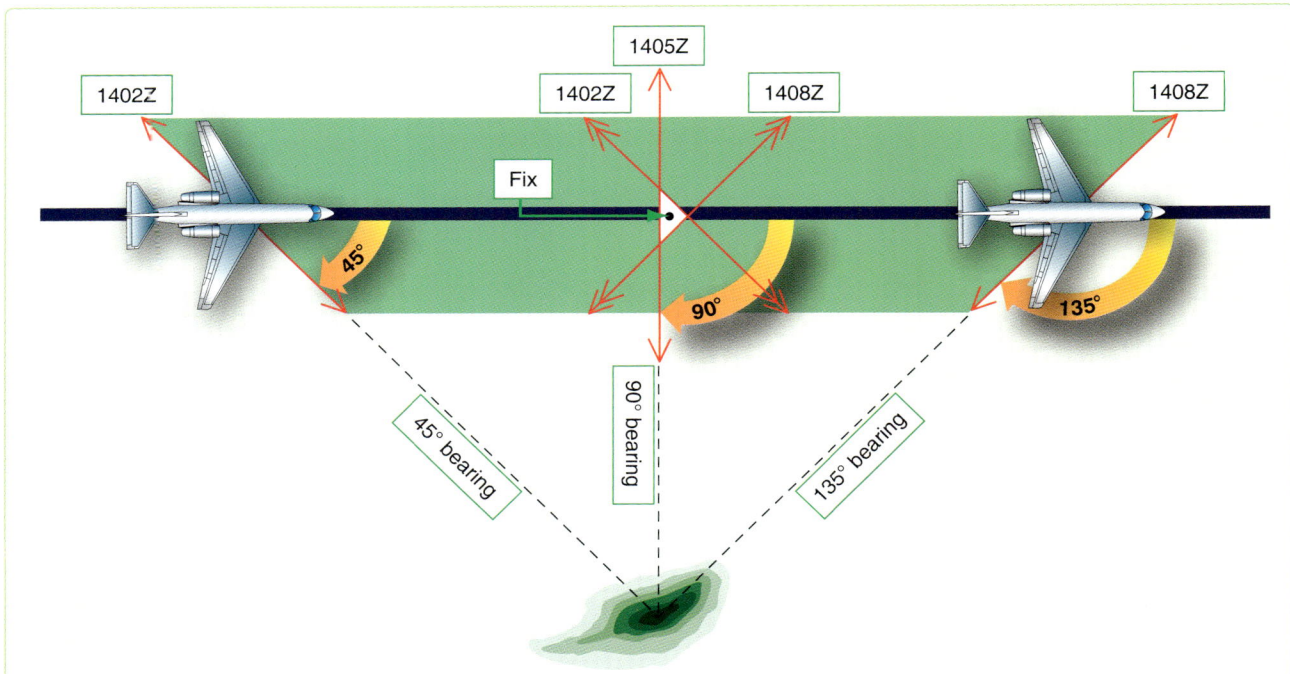

**Figure 5-6.** *The running fix.*

be adjusted to the time of the course line. The line that is affected least by the information in doubt is the line that should be adjusted.

# Radio Aids

Radio aids are ground based navigation facilities that transmit electronic signals received by airborne units. Radio aids can be used for departure, en route navigation or arrivals, and procedures for obtaining and plotting a fix vary by category. Explanations of different types of radio aids and how to fix with them follow.

## Nondirectional Radio Beacon (NDB)

This is a low, medium, or ultra high frequency (UHF) radio beacon that transmits nondirectional signals whereby the user can determine a bearing and home to the station. nondirectional radio beacons (NDBs) normally operate in the frequency band of 190 to 535 kilocycles (kHz) and transmit a continuous carrier with either 400 or 1020 cycles per second (Hz) modulation. All radio beacons, except the compass locators, transmit a continuous three-letter identification code except during voice transmissions.

## Disturbance

Radio aids are subject to disturbances that may result in erroneous bearing information. Such disturbances result from intermittent or unpredictable signal propagation due to such factors as lightning, precipitation, static, etc. At night, radio beacons are vulnerable to interference from distant stations. Nearly all disturbances that affect the automatic direction finder (ADF) bearing also affect the facility's identification. Noisy identification usually occurs when the ADF needle is erratic. Voice, music, or erroneous identification is usually heard when a steady false bearing is being displayed. Since ADF receivers do not have a flag to warn the user when erroneous bearing information is being displayed, the NDB's identification should be continuously monitored.

## Control Panels

There are several different types of control panels installed in specific aircraft. Refer to the aircraft technical manuals for specific guidance pertaining to equipment operation.

## Plotting on a Chart

Before an ADF bearing can be plotted on a navigation chart, two things must be done. First, the bearing obtained must be converted to a TB. If a nonrotatable compass card is used, the resultant RB may be converted to TB by adding the aircraft true heading (TH) (TH + RB = TB). If a rotatable compass card is used, the TB can be found by applying the magnetic variation at the vicinity of the aircraft.

## Ultra High Frequency (UHF) Direction Finders (DF)

Some aircraft are equipped with ADFs in the UHF frequency range (225.0–399.9 mHz), which utilize loop and sensing (antennas) to give bearing information. Operation of the direction finders (DF) is controlled from the UHF radio panel. It is used to obtain bearing to other aircraft and to emergency locator beacons.

## VHF Omnidirectional Range (VOR)

Very high frequency (VHF) omnidirectional range (VOR) stations operate between 108.00 and 117.95 megacycles per second (mHz). VHF communications operate between 118.00 and 135.90 mHz. Station identifiers for VOR NAVAIDs are given in code or voice, or by alternating code and voice transmissions. VOR transmissions are limited by LOS and a combination of aircraft altitude and distance to the station. Accurate information may be obtained from 40 to 100 NM around the facility, although the usable range may be much greater (300 NM). VOR may be used by flying courses from one station to another as part of the high or low jet navigation airways system. It may be used as a fixing aid by taking a bearing and applying magnetic variation at the station (converting magnetic bearing (MB) to TB) and plotting an LOP. In aircraft equipped with two VORs, the bearings to two different stations may be taken simultaneously and plotted, and a fix position obtained. The aircraft is directly overflying a VOR when the bearing pointer drops rapidly below the 3 or 9 o'clock position.

## Control Panel

A VOR control panel contains a power switch, frequency window, volume control, equipment self-test capability, and frequency selector controls. [Figure 5-7] To tune a VOR, turn power switch to PWR, select desired frequency and identify the station. For positive test indications, consult applicable aircraft flight or operator's manual.

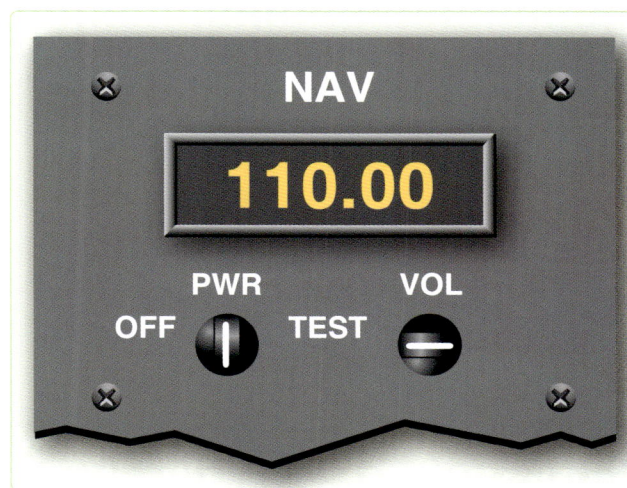

Figure 5-7. VOR navigation control panel.

**Figure 5-8.** *Course indicator.*

## Indicator Panel

Several types of indicators exist that display VOR information. Examples shown here are the course indicator [*Figure 5-8*], the radio magnetic indicator (RMI) [*Figure 5-9*], and the bearing direction heading indicator (BDHI). [*Figure 5-10*] The course indicator has eight significant features: TO-FROM indicator, glideslope and course warning flags, course selector window, marker beacon light, glideslope indicator, heading pointer, course deviation indicator (CDI), and course set knob.

## Indicators

The TO-FROM indicator shows whether the radial set in the course selector window is to or from the station, and the CDI represents this radial. If the aircraft is to the right of the radial, the CDI is displaced to the left of center on the course indicator. The glideslope indicator is similar to the CDI but represents the glideslope transmitted by an instrument landing system (ILS). If the glideslope indicator is below the center of the course indicator, the aircraft is above the glideslope. The glideslope and course warning flags inform the user that either the glideslope indicator or CDI is inoperative, or that signals received are too weak to be used. The heading pointer

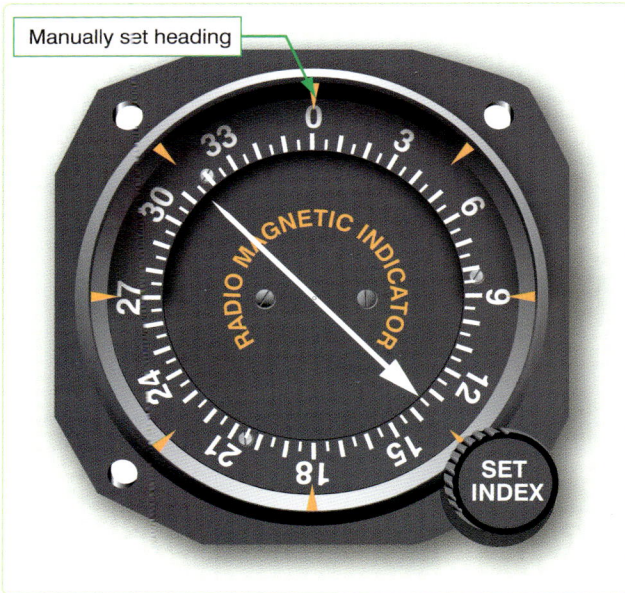

**Figure 5-9.** *Radio magnetic indicator.*

**Figure 5-10.** *Bearing distance heading indicator.*

indicates the difference, left or right, between the aircraft MH and the radial set in the course selection window.

The marker beacon light flashes when passing over a marker beacon, such as outer marker of the ILS. The RMI is a bearing indicator, usually with two pointers and a movable compass rose. The compass rose rotates as the aircraft turns, indicating the compass heading of the aircraft under the top of the index at all times. Therefore, all bearings taken from an RMI are magnetic. Consult the specific flight or operator's manual as to which pointer is the VOR.

### Bearing Direction Heading Indicator (BDHI)

The bearing direction heading indicator (BDHI) is similar to the RMI because the needles provides MB information. Additional information concerning the BDHI is in the tactical air navigation (TACAN) section.

### Tactical Air Navigation (TACAN)

The TACAN system was developed to provide information needed for precise positioning within 200 NM. As with VOR, TACAN provides an infinite number of radials radiating outwardly from the station. In addition, distance measuring equipment (DME), an integral part of TACAN, provides continuous slant-range distance information. TACAN operates in the UHF band and has 126 channels available in the X-band pulse code. Development of pulse coding has given ground equipment the capability of an additional 126 channels in the Y-band. The station identifier is transmitted at 35-second intervals in international Morse code. Airborne DME transmits on 1025–1150 mHz; associated ground-to-air frequencies are in the 962–1024 mHz and 1151–1213 mHz ranges. Channels are separated at 1 mHz intervals in these bands.

### Ground Equipment

The ground equipment consists of a rotating-type antenna for transmitting bearing information and a receiver-transmitter (transponder) for transmitting distance information. Permanent ground stations are dual transmitter-equipped (one operating and one in standby) installations that automatically switch to the standby transmitter when a malfunction occurs. Each station has a ground monitor that is set to alarm at a radial shift of 1° from the alignment to magnetic north (MN). This alarm is usually located in the control tower or approach control, and sets off a light and buzzer to warn when an out-of-tolerance condition exists. It is possible to select a TACAN station and get erroneous DME and azimuth lock-on when the station is undergoing maintenance. This can be detected by an absence of signal identifier. Checks of en route or radio NAVAIDs may be made by consulting NOTAMs prior to flight or by contacting air traffic control (ATC) for advisories when airborne.

### Airborne Equipment

The airborne equipment also contains a multichannel transmitter-receiver (transceiver). Bearing information is automatically obtained with the correct channel selected. Distance is determined by measuring the elapsed time between transmission of interrogating pulses of the airborne set and reception of corresponding reply pulses from the ground stations. This sequence is initiated by the aircraft transmitter and requires about 12 microseconds per NM round trip. Since the DME gives a readout of slant range rather than ground range, a correction has to be applied to the reading.

DME is designed to provide reliable information to a maximum distance of 199 NM, dependent on aircraft equipment and LOS. Accuracy is plus or minus one-half NM or 3 percent of the distance, whichever is greater.

Since a large number of aircraft could be interrogating the same station, the aircraft TACAN must sort out the pulses that are replies to its own signal. Interrogation pulses are transmitted on an irregular and random basis by the airborne set which then searches for replies synchronized to its own interrogations. If the signals are interrupted, a memory circuit maintains the last distance indications on the range indicator for approximately 10 seconds to prevent the search operation from recurring. This process starts automatically when a new station is tuned or when there is a major interruption of signals. Depending upon the actual distance from the station, the searching process may require up to 22 seconds. The maximum number of aircraft that can be accommodated by one station at any one time is 100. With the development of the X and Y bands, this number can be doubled.

### TACAN Characteristics

#### Bearing and/or Distance Unlock

Since TACAN bearing and DME are subject to LOS restrictions, this information could be lost any time signals are blocked. Temporary obstructions can occur in flight any time any part of the aircraft gets between the ground and aircraft antenna. Other aircraft, terrain, and buildings are external causes for unlock. Any time the signal is obstructed for more than 10 seconds for DME and 2 seconds for azimuth, the unlock conditions are indicated by a rotating bearing needle and a tumbling DME readout.

#### Azimuth Cone of Confusion

TACAN antennas transmit radio energy in circular patterns out from the transmitter. However, waves are not transmitted directly above the station. Therefore, as the aircraft approaches a TACAN station, signals are lost. This is indicated by a rotating TACAN bearing needle in the RMI. The azimuth cone can be up to 100° or more in width or approximately 15 NM wide at 40,000 feet. Thus, one may enter the cone of confusion at approximately 7.5 DME at this altitude. Approaching the station, usable TACAN information is lost before the cone is reached as aircraft memory circuits maintain the last information.

### Range Indicator Fluctuations

Slight oscillations up to approximately one forth NM are normal for range indicator operation due to the pulses generated by the transmit or receive function. When a usable signal is lost, the memory circuit maintains the indicated range for about 10 seconds, after which unlock occurs unless usable signals are regained.

### Forty Degree Azimuth Error Lock-on

The construction of the TACAN ground antenna is such that it transmits a series of nine signal lobes (eight auxiliary and one main reference pulse) 40° apart. With the airborne receiver working correctly, these pulses lock on the airborne equipment with the main reference at 090°. With a weak signal, the main reference pulse may slide over or miss the 090° slot and lock on at one of the auxiliary positions. When this occurs, azimuth indications are 40° or a multiple of 40° in error. Forty degrees azimuth lock-on error does not cause a course warning flag to appear on the indicator. Rechanneling the airborne receiver may give the set another chance to lock on properly.

## Co-Channel Interference

Co-channel interference occurs when an aircraft is in a position to receive TACAN signals from more than one ground station on the same frequency. This normally occurs only at high altitudes when distance separation between like frequencies is inadequate. DME, azimuth, or identification from either station may be received. This is not a malfunction of either airborne or ground equipment, but a result of position.

## Tuning and Controls

The basic controls of most TACAN systems are shown in *Figure 5-11*. The proper channel is tuned by rotating channel selector knobs (1) to any of 126 channels. Knob (2), the internal test mode, validates working condition of TACAN. The channel mode selector (4) allows X or Y band to be selected. These controls are presented in the channel indicator (3). A volume control (5) adjusts the audio level of the station identifier signal. The TACAN test button permits the user to perform a system self-test. The function selector (6) has four settings:

1. OFF—Removes power to the set.

2. REC—Energizes the receiver to obtain bearing information.

| 1 | Channel selecting knobs | 4 | X or Y band selection |
| 2 | Test button | 5 | Audio level |
| 3 | Channel selected | 6 | Function selector |

**Figure 5-11.** *TACAN control panel.*

3. T/R—Energizes both receiver and transmitter to obtain both bearing and distance information.

4. A/A—Transmits and receives interrogations and replies to measure range to another A/A TACAN-equipped aircraft. Bearing information is not provided on this set.

## VORTAC

In order to provide both military and civilian pilots the capability of positioning from the same radio NAVAIDs, a combination of VOR and TACAN station was developed. Each facility offers three services. VOR azimuth signals are transmitted on the published VOR frequency. TACAN azimuth and DME signals are broadcast on the published UHF channel.

## Fix-to-Fix Navigation (Using RMI and BDHI)

Flying from one radial and DME to another is basic to many departures and approaches. A heading to the desired point may be derived quickly through the use of an RMI, providing a radial and a separate readout of DME. The same procedures apply for a BDHI. The following technique and example are provided in order to demonstrate how to compute a heading. *[Figure 5-12]*

Example:
Present position = 180°/60
Desired position = 090°/30
Present heading = 000°

1. Tune, identify, and monitor correct VOR and TACAN.

2. Turn the aircraft in the general direction of the desired fix by turning to a heading approximately halfway between the head of the bearing pointer (000°) and the radial on which the desired fix is located (090°). In this case, turn to 045°. *[Figure 5-12A]*

**Figure 5-12.** *Fix-to-fix solution.*

3. Visualize the aircraft position and the desired fix on the RMI as follows:

- The center of the RMI is considered to be the VOR or TACAN, and the compass rose simulates the radials around the station.

- The fix with the greater range (180°/60) is established at the outer edge of the compass card.

- The fix with the lesser range (090°/30) is established at a point that is proportional to the distance represented by the outer edge of the compass card.

- Determine the heading to the desired fix by connecting the present position to the desired fix with an imaginary line on the RMI. *[Figure 5-12D to C]* Establish another imaginary line parallel to the line labeled B to C through the center of the RMI. This line indicates the no-wind heading to the desired fix (030°).

- Turn to 030° and apply any drift correction. With 5° right drift, turn to 025°. Cross-check position continually and correct as necessary.

### Fix-to-Fix Navigation (Using the MB-4 Computer)

A fix-to-fix can also be computed on the wind face side of an MB-4 computer. First, give the pilot a general heading

toward the fix. (NOTE: Work in bearings; however, all work must be done in either bearings or radials to compute the solution.) For the following example, radials are used. The fix to navigate to is the 280° radial at 30 DME. Set up a graphic depiction on the wind face side of the computer with present position (350° radial at 050 DME) and the desired fix (280°/030). Use the following steps:

1. Place present position (350°/050) on the wind face side using the square grid at the bottom of the MB-4. Align 350° on the compass rose under the true index. Mark the point by counting down 50 NM from the TAS grommet and mark with a ×. Use the scale set up on the square grid or set up an applicable scale. The scale used must remain constant throughout the problem. *[Figure 5-13]*

2. Place the fix radial and DME (280°/030) on the computer the same as in step one. *[Figure 5-14]* Mark as a fix symbol (Δ).

3. Determine the no wind heading by rotating the compass rose so that the present position (×) is directly above the fix (Δ). Use the square grid at the bottom to help with alignment. *[Figure 5-15]* Turn the aircraft to MC under the true index (206° for this example) and kill the drift. (NOTE: Place present position (×) on the 0 NM horizontal baseline then, using a NM increment scale, count down to the fix position (Δ) to determine how far you are from the fix (48 NM in this example).

4. Repeat the procedure as necessary to keep all progress updated.

## Communication

### Long and Short Range Communication

Air-to-ground communications can be achieved through the use of many types of radio equipment. High frequency bands (HF, VHF, UHF) are relatively static-free and are less susceptible to outside interference than lower frequencies. It must be remembered, however, that the higher the frequency, the more nearly the transmission follows an LOS path. As frequency increases, therefore, communication range decreases.

### Long Range

Systems used for long-range radio communications between aircraft and ground stations may be either amplitude modulation (AM) or single sideband (SSB) transmissions. Single sideband transmitters concentrate all available power into one sideband; therefore, SSB is much more efficient and has greater range than an AM transmitter of the same power. Although HF ground waves attenuate rapidly, sky waves at these frequencies are capable of transmitting at distances

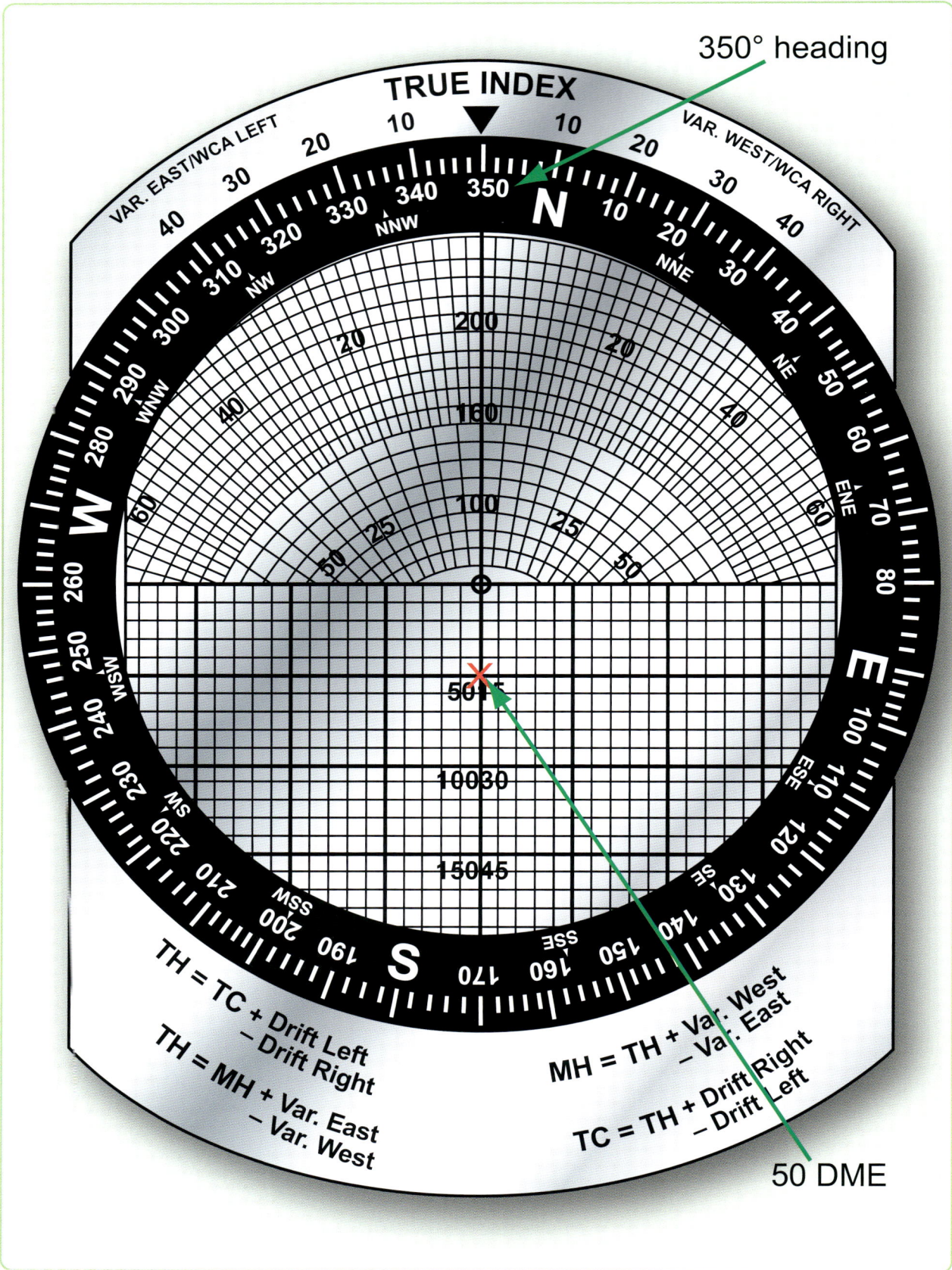

**Figure 5-13.** *Setting present position on MB-4 computer.*

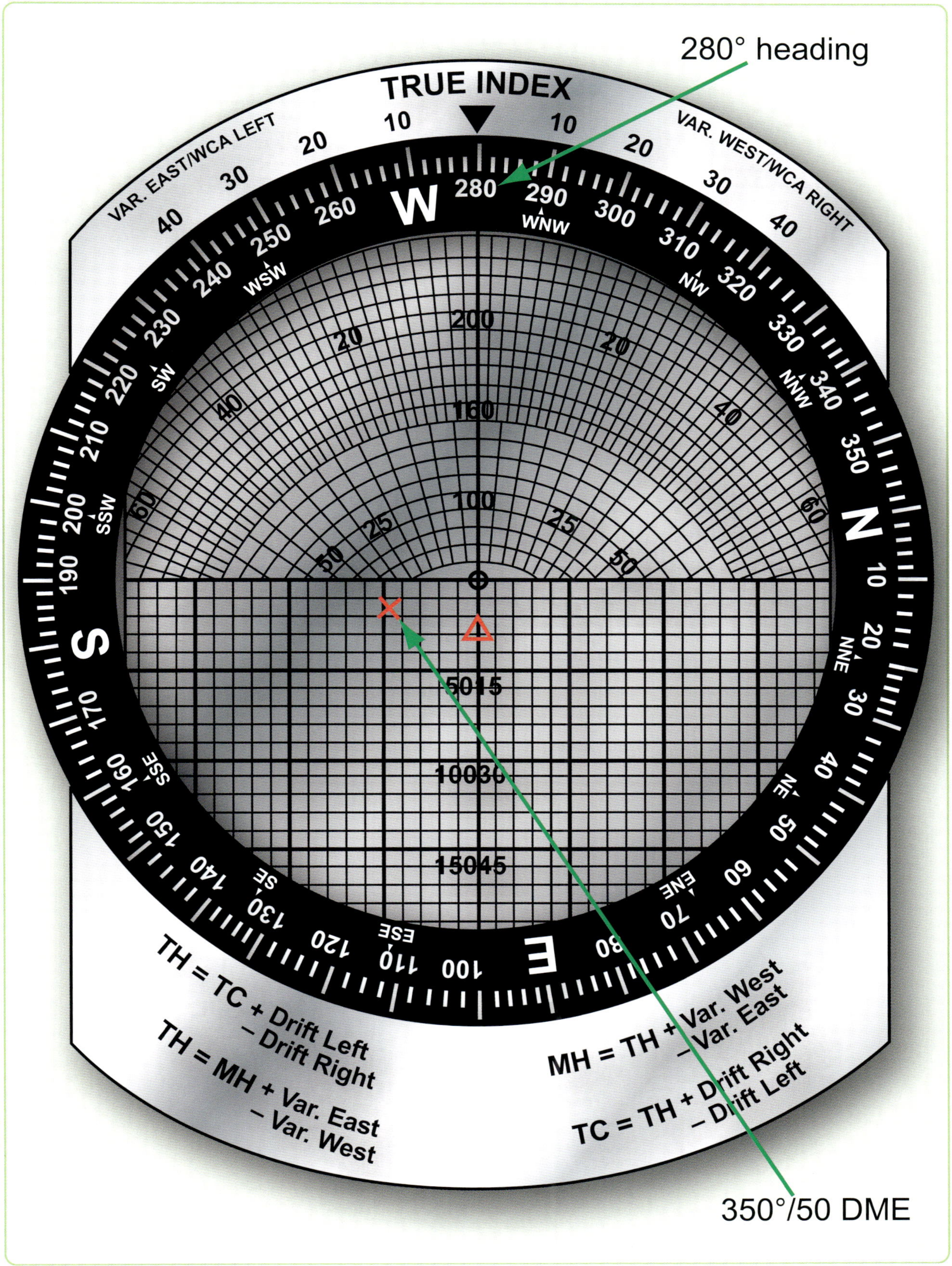

**Figure 5-14.** *Setting fix on MB-4 computer.*

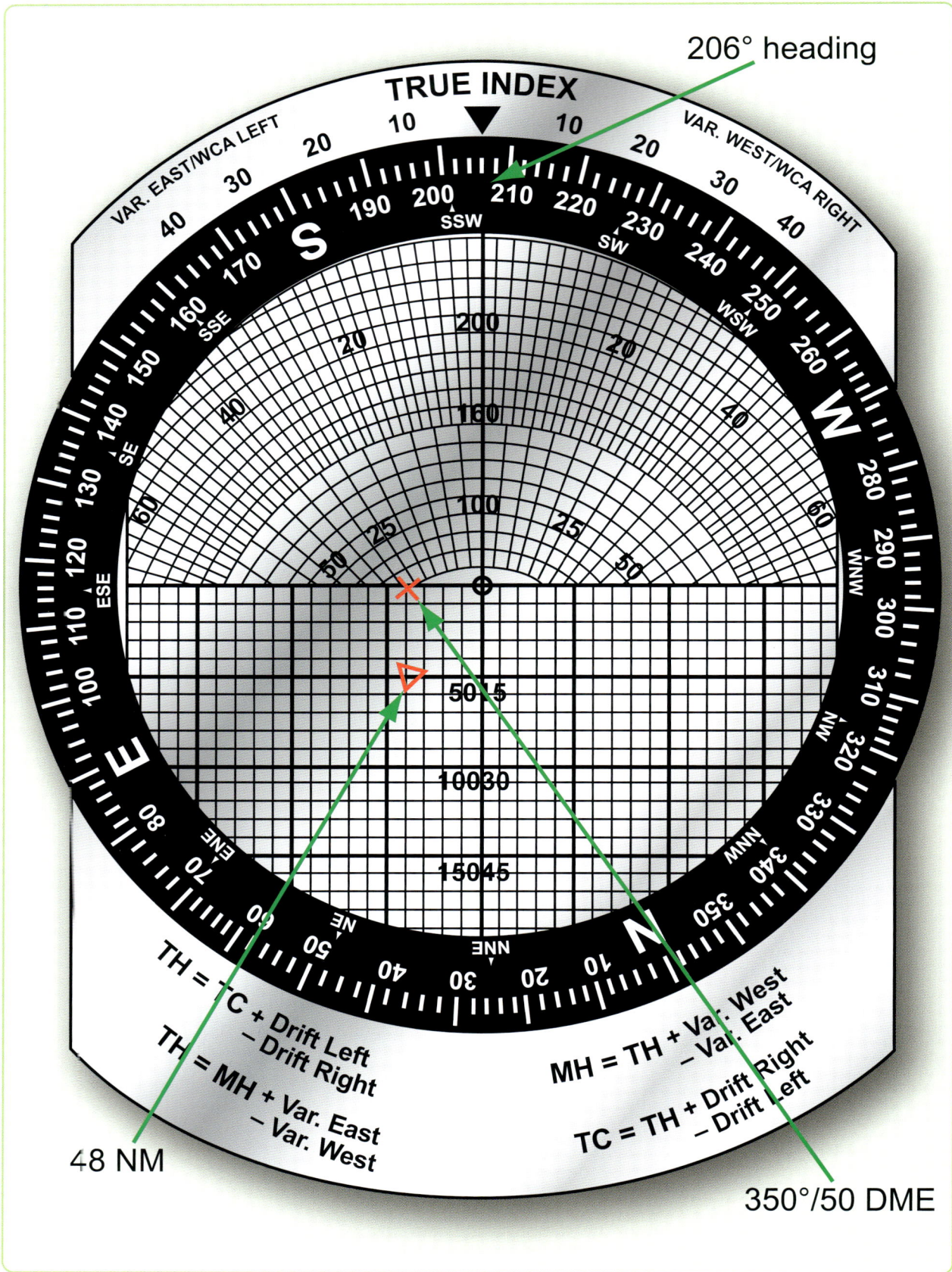

**Figure 5-15.** *Solution to fix-to-fix and distance to fix.*

up to 12,000 miles or more, depending on ionospheric conditions. HF equipment is used mostly in remote areas where VHF or UHF communication is not possible because of the great distance that must be spanned.

### Short Range Air-to-Air and Air-to-Ground

Short range air-to-air and air-to-ground communications are confined to the VHF and UHF bands. VHF channels are spaced at 25 kHz intervals from 116 to 151.975 mHz, and UHF channels are spaced at 50 kHz intervals from 225.0 to 399.9 mHz. Most UHF and HF transceivers have a manual frequency selection capability in addition to a number of preset channels. Transmission and reception are accomplished with a single antenna.

## Chapter Summary

This chapter explained how to navigate from one point to another using a fix. LOPs are discussed in relation to plotting a fix and how to develop multiple lines of position accurately. Fixes are discussed in detail, as well as the various aids that are used to plot them, such as radio aids, aircraft instruments, airborne equipment, ground equipment, flight computers, and transponders. The chapter explained the use of a NDB, VOR, TACAN, and how they are used in direct relation with the aircraft navigational instruments for plotting a fix. Fix-to-fix navigation is discussed using the MB-4 flight computer, as well as a detailed discussion on the different features of a transponder.

## Chapter 6

# Map Reading

## Introduction

Map reading is the determination of aircraft position by matching natural or built-up features with their corresponding symbols on a chart. It is one of the more basic aids to dead reckoning (DR) and certainly the earliest used form of aerospace navigation. The degree of success in map reading depends upon a navigator's proficiency in chart interpretation, ability to estimate distance, and the availability of landmarks.

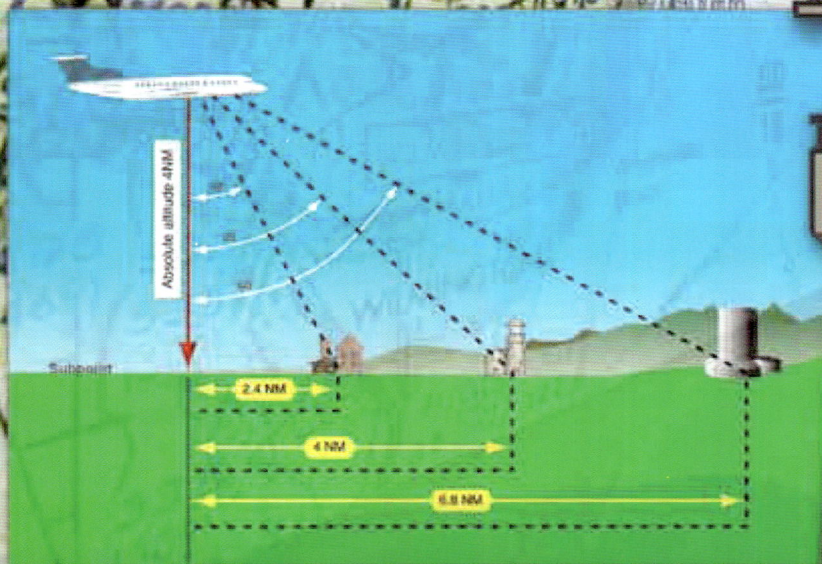

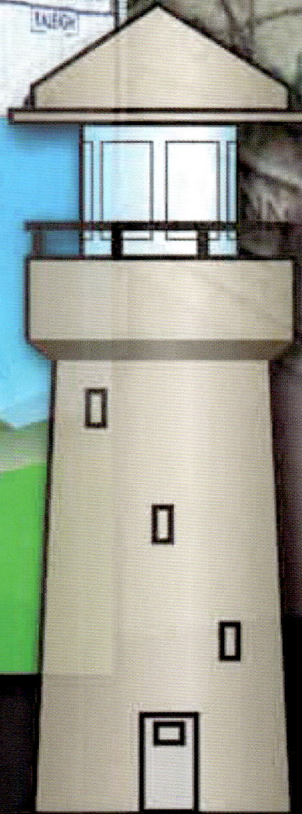

## Checkpoints

Checkpoints are landmarks or geographic coordinates used to fix the position of the aircraft. By comparing the aircraft position to that of the checkpoint, the navigator fixes the aircraft's location. Arrival over checkpoints at planned times is a confirmation of the wind predication and indicates reliability of the predicted track and groundspeed. If the aircraft passes near but not over a checkpoint, the anticipated track was not made good. If checkpoints are crossed but not at the predicted time, the anticipated ground speed (GS) was in error. Prudent navigators are quick to observe and evaluate the difference between an anticipated position and an actual position. They must make corrections to maintain their intended course as soon as possible because small errors can be cumulative and may eventually result in becoming lost.

Before fixing each position, navigators should look for several related details around each checkpoint to make sure it has been positively identified. For example, if the checkpoint is a small town, there may be a lake to the north, a road intersection to the south, and a bridge to the east.

Generally, it is better to select a feature on the chart and then seek it on the ground rather than to work from the ground to the chart. The chart does not show all the detail that is on the ground, and one could easily become confused. Checkpoints should be features, or groups of features, that stand out from the background and are easily identifiable. In open areas, any town or road intersection can be used; however, these same features in densely populated areas are difficult to distinguish. *Figures 6-1* and *6-2* compare various chart and corresponding photo areas and list the features to look for when identifying landmarks as checkpoints.

## Chart Selection

Use a chart for map reading that provides sufficient natural and built-up features to accurately position the aircraft. The Operational Navigation Chart (ONC), with a scale of 1:1,000,000, has excellent cultural and relief portrayal. For increased detail, a Sectional Aeronautical, with a scale of 1:500,000, or suitable scaled USGS topographic charts with a scale of 1:10,000, may be used.

## Map Reading Procedures

When in flight, orient the chart so that north on the chart is toward true north (TN). The course line on the chart is then aligned with the intended course of the aircraft so that landmarks on the ground appear in the same relative position

**Figure 6-1.** *Landmarks as checkpoints, populated areas.*

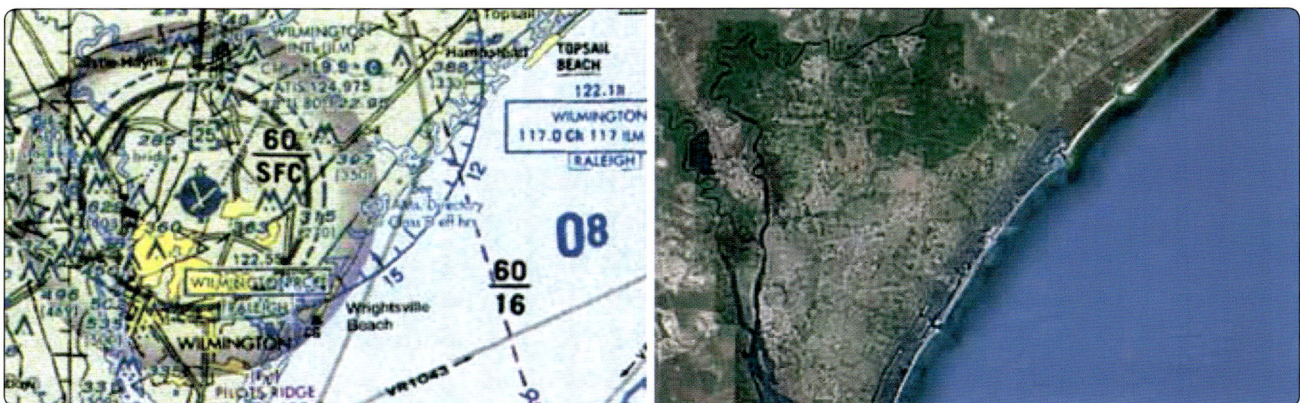

**Figure 6-2.** *Landmarks as checkpoints, coastal areas.*

as the features on the chart. Obtain the approximate position of the aircraft by DR. Select an identifiable landmark on the chart at or near the DR position. It is important to work from the chart to the ground. Identify the landmark selected and fix the position of the aircraft. The importance of a good DR cannot be over emphasized. When there is any uncertainty of position, every possible detail should be checked before identifying a checkpoint. The relative positions of roads, railroads, airfields, and bridges make good checkpoints. Intersections and bends in roads, railroads, and rivers are equally good. When a landmark is a large feature, such as a major metropolitan area, select a small prominent checkpoint within the large landmark to fix the position of the aircraft. When a landmark is not available as a reference at a scheduled turning point, make the turn on the estimated time of arrival (ETA). Extend the DR position to the next landmark and fix the position of the aircraft to make sure the desired course and GS are being maintained. Remember, the desired magnetic course on any given leg corrected for drift is the magnetic heading which parallels course. This helps to keep from getting any farther off course.

## Map Reading While Flying at Low Altitudes

When flying at lower altitudes, additional difficulties may be encountered. Turbulence increases the difficulty of reading instruments. Depending on the aircraft's altitude above ground level (AGL), the circle of visibility can be greatly reduced, and those objects that are visible pass by so rapidly only the largest landmarks can be easily identified.

In low altitude navigation, flight planning is especially important as there is little time for inflight computations. An important part of good flight planning is proper chart preparation. Normally radius-of-turn procedures are used when drawing the chart, but depending on your tactics, point to point is also an option. Time elapsed marks and distance remaining marks along the course line of each leg gives navigators a running DR with the aid of a stopwatch.

In low altitude flight, one should be particularly alert to possible danger from obstructions. Hills and mountains are easily avoided if the visibility is good. Radio and television towers, which may extend as much as 1,000 feet or more into the air, often from elevated ground, are less conspicuous. All such obstructions may or may not be shown on the aeronautical charts. Flights need to be above the highest elevation listed for that grid square on the sectional chart to ensure obstruction clearance. Maximum elevation figures (MEF) are explained on the inside panel (left side) of the sectional chart.

## Map Reading at Night

At night, unlighted landmarks may be difficult or impossible to see. Lights can be confusing because they appear closer than they really are. Fixing on points other than those directly beneath the aircraft is very difficult. Objects may be more easily seen by scanning or looking at them indirectly to eliminate the eye's visual blindspot. Preserve night vision by working with red or green light, being aware that red light can detract from the chart color. Moonlight makes it possible to see prominent landmarks like land-water contrast. Reflected moonlight often causes a river or lake to stand out brightly for a moment, but this condition is usually too brief for accurate fixing. Roads and railroads may be seen after the eyes are accustomed to the darkness. Lighted landmarks, such as cities and towns, stand out more clearly at night than in daytime. Large cities can often be recognized by their distinctive shapes. Many small towns are dark at night and are not visible to the unaided eye. Some airfields have distinctive light patterns and may be used as checkpoints. Military fields use a double white and single green rotating beacon, while civilian fields use a single white and single green rotating beacon. Busy highways are discernible because of automobile headlights.

## Estimating Distance

A landmark often falls right or left of course and the navigator must estimate the distance to it. While the ability to estimate distance from a landmark rests largely in skill and experience, the following methods may be of assistance. One method is to compare the distance to a landmark with the distance between two other points as measured on the chart. Another method is to estimate the angle between the aircraft subpoint and the line of sight. *[Figure 6-3]* The distance in NM from the landmark to the subpoint of the aircraft depends on the sighting angle:

(60°) horizontal distance = absolute altitude of aircraft × 1.7
(45°) horizontal distance = absolute altitude of aircraft
(30°) horizontal distance = absolute altitude of aircraft × .6

## Seasonal Changes

Seasonal changes can conceal landmarks or change their appearance. Small lakes and rivers may dry up during the summer. Their outlines may change considerably during the wet season. Snow can cover up almost all of the normally used landmarks. When flying in the winter, it is often necessary to rely on more prominent checkpoints, such as river bends, hills, or larger towns. However, due to the size of these checkpoints, course control can be somewhat degraded.

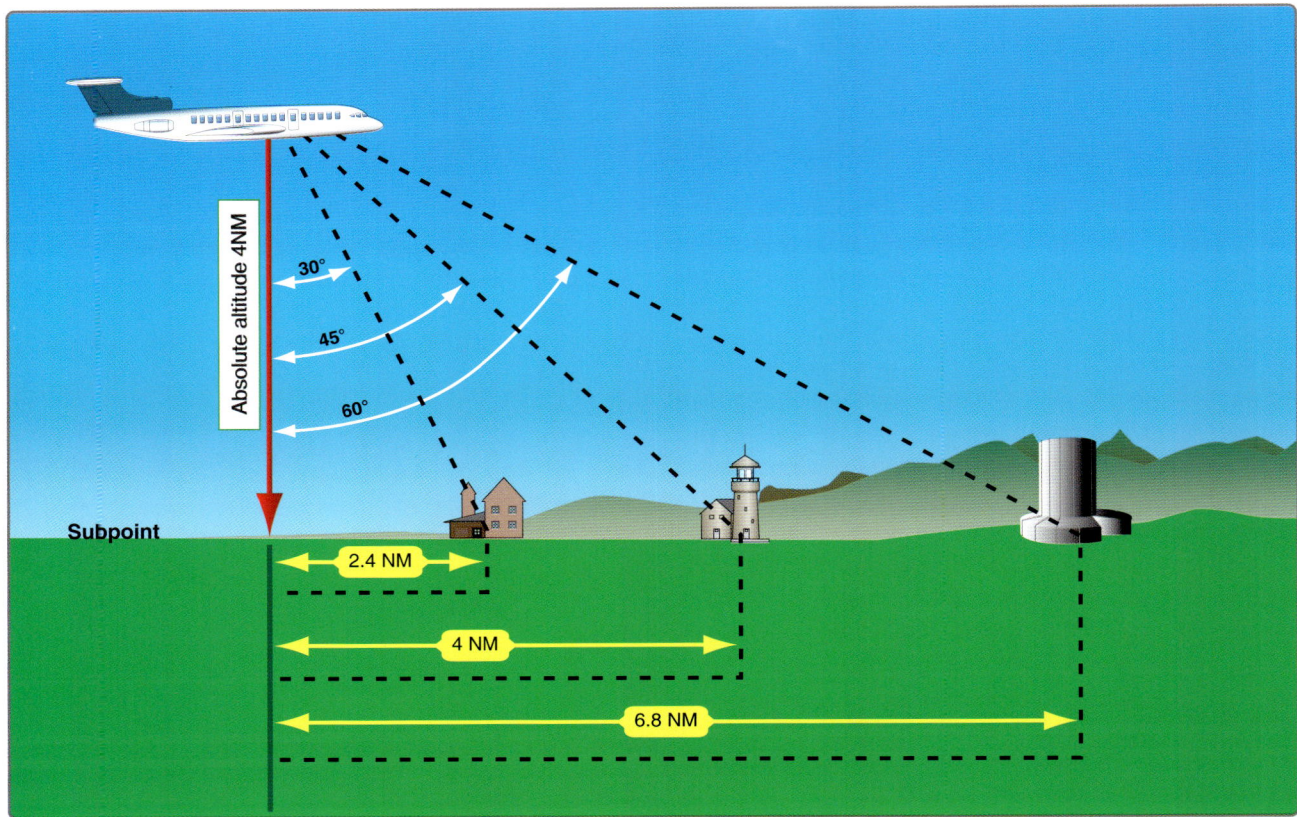

**Figure 6-3.** *Estimating distances.*

## Map Reading in High Latitudes

Map reading in high latitudes is considerably more difficult than map reading in the lower latitudes. The nature of the terrain is drastically different, charts are less detailed and less precise, seasonal changes may alter the terrain appearance or hide it completely from view, and there are fewer cultural features.

In high latitudes, navigators find few distinguishable features from which to determine a position. Built-up features are practically nonexistent. The few that do exist are closely grouped, offering little help to the navigator flying long navigation legs. Natural features that do exist are in limited variety and are difficult to distinguish from each other. Lakes seem endless in number and identical in appearance. The countless inlets are extremely difficult to identify, particularly in winter. What appears to be land may in reality be floating ice, the shape of which can change from day to day. Recognizable, reliable checkpoints are few and far between.

Map reading in high latitudes is further complicated by inadequate charting. Some polar areas are yet to be thoroughly surveyed. The charts portray the appearance of general locales, but many individual terrain features are merely approximated or omitted entirely. In place of detailed outlines of lakes, for example, charts often carry the brief annotation—many lakes. Fixing is possible, but requires extended effort and keen judgment.

When snow blankets the terrain from horizon to horizon, navigation by map reading becomes acutely difficult. Coastal ice becomes indistinguishable from the land; coastal contours appear radically changed; and many inlets, streams, and lakes disappear. Blowing snow may extend to heights of 200 to 300 feet and may continue for several days, but visibility is usually excellent in the absence of interfering clouds or ice crystal haze. However, when snow obliterates surface features and the sky is covered with a uniform layer of clouds so that no shadows are cast, the horizon disappears, causing earth and sky to blend together. This forms an unbroken expanse of white called whiteout. In this complete lack of contrast, distance and height above ground are virtually impossible to estimate. Whiteout is particularly prevalent in northern Alaska during late winter and spring. The continuous darkness of night presents another hazard; nevertheless, surface features are often visible because the snow is an excellent reflector of light from the moon, the stars, and the aurora.

### Contour Map Reading

Use of contours is the most common method of showing relief features on a chart. Contours are lines that, at certain intervals, connect points of equal elevation. To understand contours better, think of the zero contour line to be sea level. If the sea were to rise 10 feet, the new shoreline would be the 10-foot contour line. Similarly, successive 10-foot contour lines could be easily determined. Contour lines are closer together where the slope is steep and farther apart where the slope is gentle. Within the limits of the contour intervals, the height of points and the angle of slope can also be determined from the chart.

Contour intervals are determined by the scale of the chart, the amount of relief, and the accuracy of the survey. These intervals may range from 1 foot on a large-scale chart through 2,000 feet or greater on a smaller scale chart. Contours may be annotated in feet or meters. Contours may be shown on charts in varying colors and are frequently labeled with figures of elevation. To further accentuate the terrain, a gradient system of coloring is also employed. The lighter colors are used to show lower areas while a gradual increase in density (darkness) is used to portray the higher terrain.

## Chapter Summary

Map reading is a critical skill for navigators in many aircraft, but it takes time to become proficient. Keep a good DR, work from chart to ground, and remember the effect varying conditions have on what is seen outside a window.

# Chapter 7
# Radar Navigation

## Radar Principles

In the hands of the skilled operator, radar provides precise updates to dead reckoning (DR) for navigation. At cruising altitudes, it provides information on land and water characteristics, as well as hazardous weather conditions over hundreds of miles around the aircraft. At low-level, it provides detailed terrain information used to navigate at high speed over changing courses. It is adapted to terrain-avoidance and terrain-following equipment. Radar is a source of track and drift angle (DA) information for wind computations.

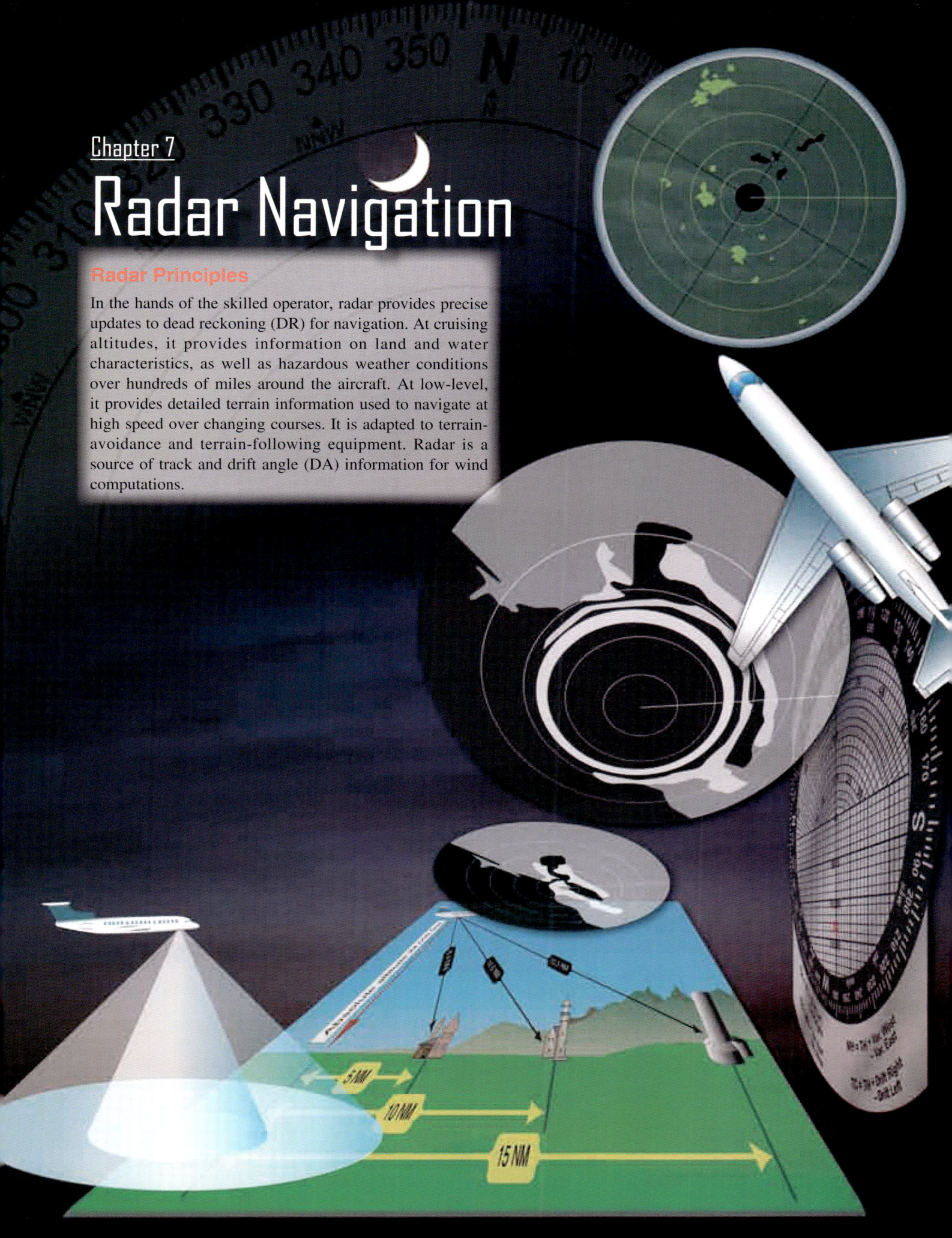

The basis of the system has been known theoretically since 1888, when Heinrich Hertz successfully demonstrated the transfer of electromagnetic energy in space and showed that such energy is capable of reflection. The transmission of electromagnetic energy between two points was developed as radio, but it was not until 1922 that practical use of the reflection properties of such energy was conceived. The idea of measuring the elapsed time between the transmission of a radio signal and receipt of its reflected echo from a surface originated nearly simultaneously in the United States and England. In the United States, two scientists working with air-to-ground signals noticed that ships moving in the nearby Potomac River distorted the pattern of these signals. In 1925, the same scientists were able to measure the time required for a short burst, or pulse, of radio energy to travel to the ionosphere and return. Following this success, it was realized the radar principle could be applied to the detection of other objects, including ships and aircraft.

By the beginning of World War II, the Army and Navy had developed equipment appropriate to their respective fields. During and following the war, the rapid advance in theory and technological skill brought improvements and additional applications of the early equipment. It is now possible to measure accurately the distance and direction of a reflecting surface in space, whether it is an aircraft, ship, hurricane, or prominent feature of the terrain, even under conditions of darkness or restricted visibility. For these reasons, radar has become a valuable navigational tool.

As noted previously, the fundamental principle of radar may be likened to that of relating sound to its echo. Thus, a ship sometimes determines its distance from a cliff at the water's edge by blowing its whistle and timing the interval until the echo is received. The same principle applies to radar, which uses the reflected echo of electromagnetic radiation traveling at the speed of light. This speed is approximately 162,000 nautical miles (NM) per second; it may also be expressed as 985 feet per microsecond. If the interval between the transmission of the signal and return of the echo is 200 microseconds, the distance to the target is:

$$\frac{985 \times 200}{2} = 98,500 \text{ ft} = 16.2 \text{ NM}$$

## Radar Set Components

The principle of radar is accomplished by developing a pulse of microwave energy that is transmitted from the aircraft and is reflected by objects in its path. The reflected pulse is amplified and converted by the receiver for display on the display. The timing unit, or synchronizer, synchronizes all the actions in the set. To this basic unit, improvements are added for special purposes, such as weather avoidance, filtering, and terrain following.

### Components

The receiver and transmitter are usually one unit (the R/T) with separate functions that, for this description, are dealt with separately. *[Figure 7-1]*

### *Radar System Components*

The transmitter produces the radio frequency (RF) energy using magnetrons. A magnetron generates radar pulses by bunching electrons using alternately charged grids that the electrons travel past. The spurts of energy are of high power and short duration. The energy is released at intervals (the pulse recurrence rate) determined by the selected operating range.

The generated pulse travels through either coaxial cable or, more frequently, a hollow tube called the wave guide. The wave guide requires pressurization to ensure the maintenance of conditions for proper microwave conduction. The energy passes an electronic switching device that directs outgoing

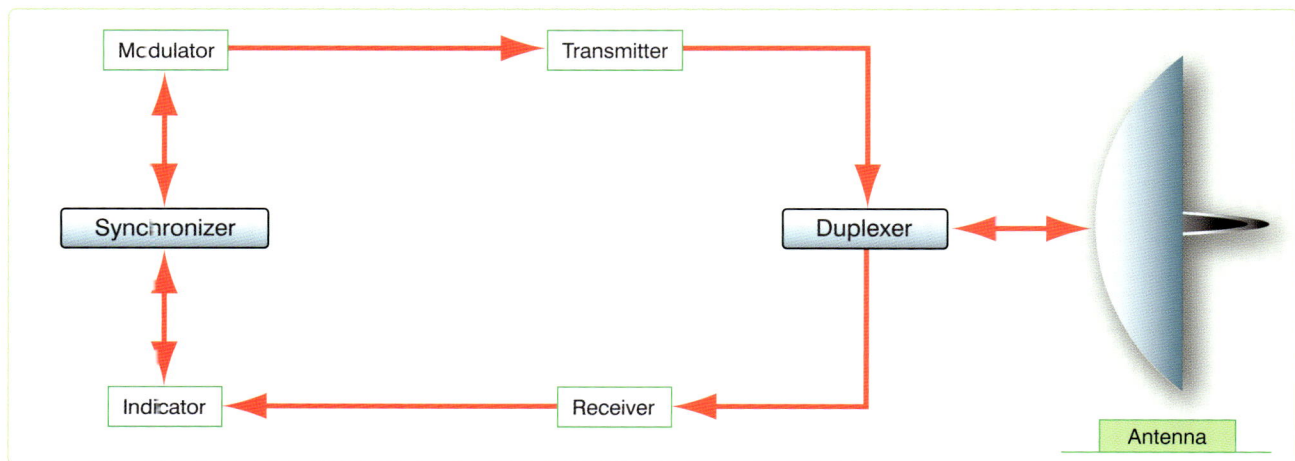

**Figure 7-1.** *Radar system components.*

pulses to the antenna and incoming pulses from the antenna to the receiver.

The antenna is a parabolic dish with a protruding wave guide. It is gimbal-mounted to allow rotation of the dish and, in most cases, to allow stabilization of the dish relative to the earth's surface when the aircraft turns. Rotation of the antenna could be through 360° or in a sector (either variable or preset). The 360° rotation, or scan, is usually for mapping, whereas a sector is used in aircraft with limited space for the antenna or where the intent is to concentrate energy in a small area. The antenna assembly is either permanently locked to the longitudinal axis of the aircraft (boresighted) or only so aligned when stabilization units are inactive. When not caged, the antenna stabilization is accomplished by using gyroservo mechanisms. A sensor system that provides information to a computer keeps the antenna radiation plane parallel to the earth even when the aircraft is in a climb or a bank.

There are two radiation patterns popular in airborne radar design: fan and pencil beams. The fan beam is a wide pattern that distributes the RF energy across the beam in proportion to the distance it must travel. [Figure 7-2] The fan beam is best for general mapping. To concentrate the energy emitted, the pencil beam antenna is used. The pencil beam dish allows scanning for weather or aircraft while eliminating ground clutter. It can be used to put more energy on a section of ground to increase returns.

The antenna can be manipulated to aim the emissions through a control that tilts the dish from the horizontal plane. At cruising altitudes, in the mapping mode, it is sufficient to slightly tilt the dish down, but tilt should be constantly adjusted for optimum returns.

After transmission, the reflected energy is directed back to the wave guide where it travels past the switching device that directs the returns to the receiver. The receiver converts the microwave returns to electrical signals that are amplified and sent to a display called the planned position indicator (PPI). The amplification of the returns is controllable through a gain circuit. Depending on the type of return desired on the display, the operator adjusts the receiver gain. Other booster circuits, such as sweep intensity or video gain, are available but operation of the receiver is most important. If adequate receiver amplification of weak returns is not applied, no amount of later stage adjustments put the target on the scope.

The display or scope, offers both range and azimuth information about targets to the operator. This information is relative to the aircraft's position, which can be referenced at either the center of the scope or offset to the side of the screen. [Figure 7-3] The controls manipulates the display so that returns can be presented on the scope in their correct position relative to the observer. [Figure 7-4]

Applying a polarization to the signals going to a display produces the actual presentation of the return. The null return has a predominantly positive charge; therefore, the trace is suppressed. A polarization shift is produced in the current to produce a blooming of the trace corresponding to the strength and position of the received signal.

Range is determined by the travel time of a pulse from and back to the R/T unit. Knowing that RF energy travels at the constant speed of light, range determination is simple. The synchronizer coordinates its display on the display.

At the same instant that the timer triggers the transmitter, it also sends a trigger signal to the indicator. Here, a circuit is actuated that causes the current in the deflection coils to rise

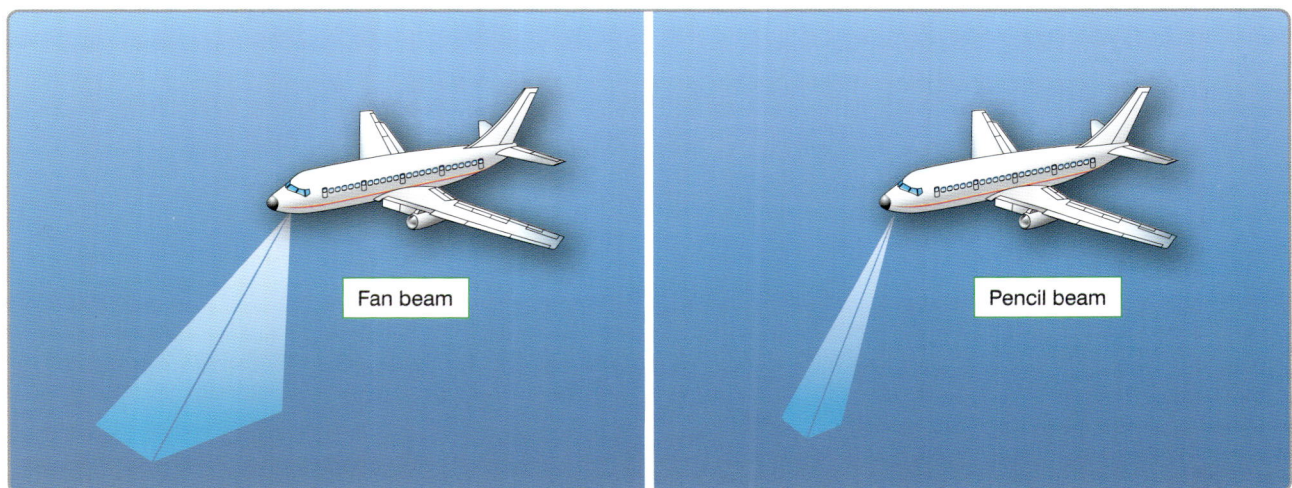

**Figure 7-2.** *Antenna radiation patterns.*

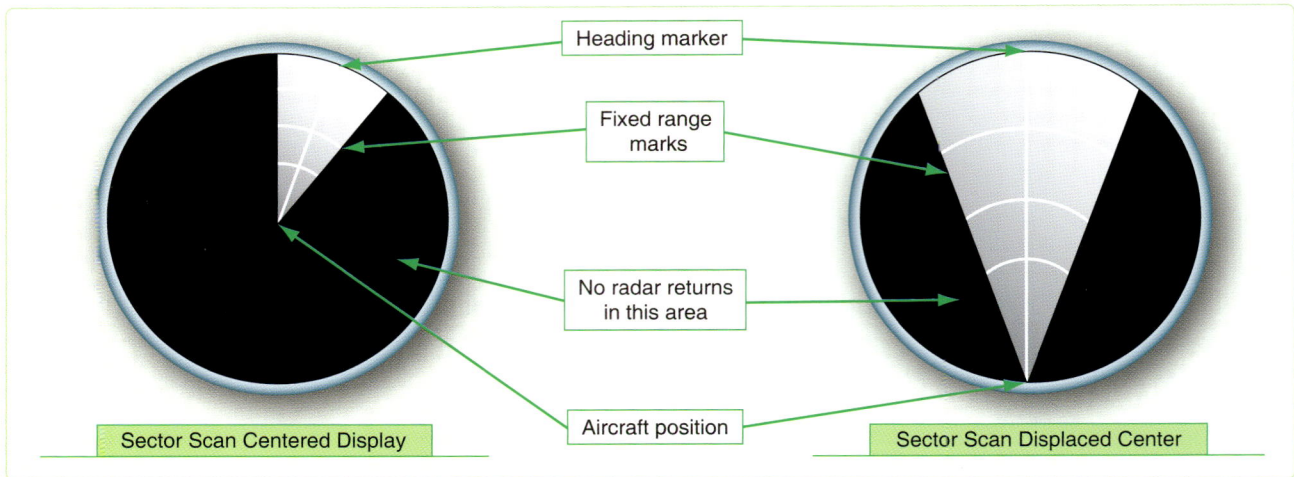

Heading marker

Fixed range marks

No radar returns in this area

Aircraft position

Sector Scan Centered Display

Sector Scan Displaced Center

**Figure 7-3.** *Sector scan displays.*

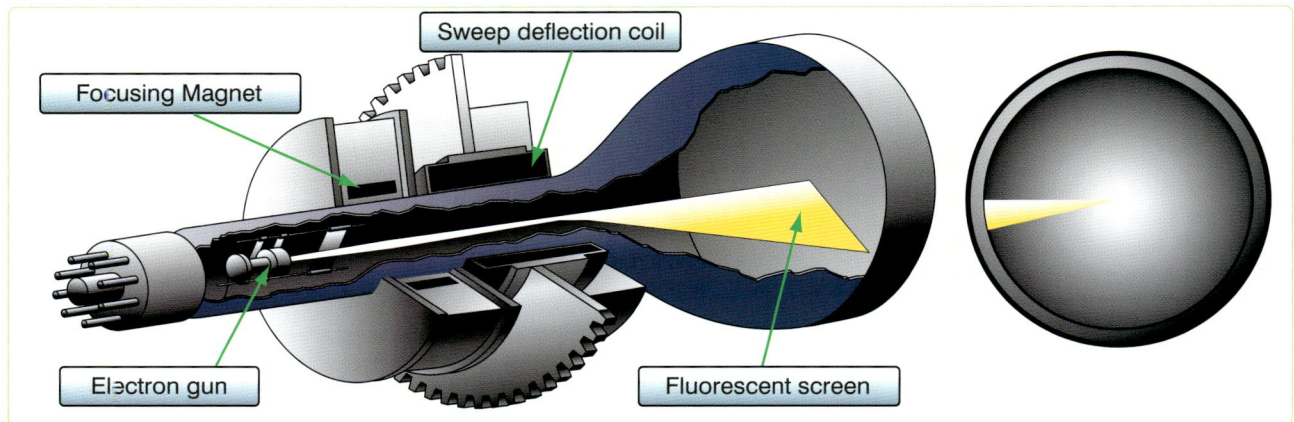

Sweep deflection coil

Focusing Magnet

Electron gun

Fluorescent screen

**Figure 7-4.** *Electromagnetic cathode ray tube.*

at a linear (uniform) rate. The rising current, in turn, causes the spot to be deflected radially outward from the center of the scope. The spot traces a faint line on the scope; this line is called the sweep. If no echo is received, the intensity of the sweep remains uniform throughout its entire length. However, if an echo is returned, it is so applied to the display that it intensifies the spot and momentarily brightens a segment of the sweep relative to the size of the target. Since the sweep is linear and begins with the emission of the transmitted pulse, the point at which the echo brightens the sweep is an indication of the range to the object causing the echo.

The progressive positions of the pulse in space also indicate the corresponding positions of the electron beam as it sweeps across the face of the display. If the radius of the scope represents 40 miles and the return appears at three-quarters of the distance from the center of the scope to its periphery, the target is represented as being about 30 miles away.

In the preceding example, the radar is set for a 40-mile range operation. The sweep circuits operates only for an equivalent time interval so that targets beyond 40 miles do not appear on the scope. The time equivalent to 40 miles of radar range is only 496 microseconds ($496 \times 10^{-6}$ seconds). Thus, 496 microseconds after a pulse is transmitted (plus an additional period of perhaps 100 microseconds to allow the sweep circuits to recover), the radar is ready to transmit the next pulse. The actual pulse repetition rate in this example is about 800 pulses per second. The return, therefore, appears in virtually the same position along the sweep as each successive pulse is transmitted, even though the aircraft and the target are moving at appreciable speeds.

At times, the display does not display targets across the entire range selected on the scope. In these cases, atmospheric refraction and the line of sight (LOS) characteristics of radar energy have affected the effective range of the set. The following formula can determine the radar's range in these situations where D is distance and h is the aircraft altitude:

$$D = 1.23 \sqrt{h}$$

Azimuth measurement is achieved by synchronizing the deflection coil with the antenna. In the basic radar unit, when the antenna is pointed directly off the nose of the aircraft, the deflection coils are aligned to fire the trace at the 12 o'clock position on the scope. As the antenna rotates, the deflection coil moves at the same rate. Relative target presentations are displayed as the sweep rotation is combined with the range display.

## Scope Interpretation

The display presents a map-like picture of the terrain below and around the aircraft. Just as map reading skill is largely dependent upon the ability to correlate what is seen on the ground with the symbols on the chart, so the art of scope presentation analysis is largely dependent upon the ability to correlate what is seen on the scope with the chart symbols. Application of the concept of radar reflection and an understanding of how received signals are displayed on the display are prerequisites to scope interpretation. Furthermore, knowledge of these factors applied in reverse enables the navigator to predict the probable radarscope appearance of any area.

### Factors Affecting Reflection

A target's ability to reflect energy is based on the target's composition, size, and the radar beam's angle of reflection. [*Figure 7-5*] The range of the target from the aircraft is definitive in the quantity of returned energy. The range of a target produces an inverse effect on the target's radar cross-section, and there is some atmospheric attenuation of the pulse proportional to the distance that the energy must travel. Generally, all four factors contribute to the displayed return. A single factor can, in some cases, either prevent a target from reflecting sufficient energy for detection or cause a disproportionate excess of reflected energy to be received and displayed. The following are general rules of radarscope interpretation:

1. The greatest return potential exists when the radar beam forms a horizontal right angle with the frontal portion of the reflector.

2. Radar return potential is roughly proportional to the target size and the reflective properties (density) of the target.

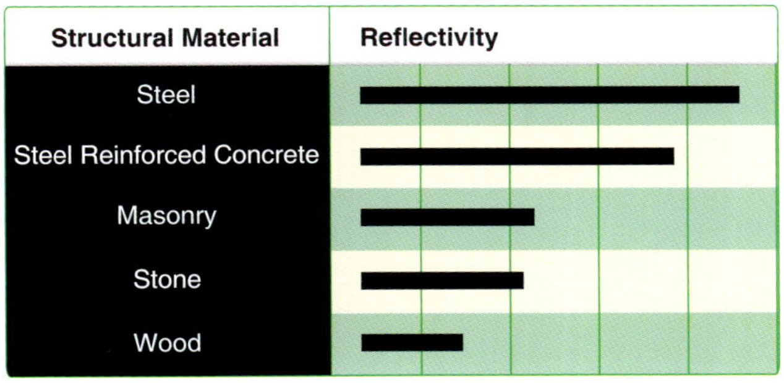

**Figure 7-5.** *Relative reflectivity of structural materials.*

3. Radar return potential is greatest within the zone of the greatest radiation pattern of the antenna.

4. Radar return potential decreases as altitude increases, because the vertical reflection angle becomes more and more removed from the optimum. (There are many exceptions to this general rule since there are many structures that may present better reflection from roof surfaces than from frontal surfaces, or in the case of weather.)

5. Radar return potential decreases as range increases because of the greater beam width at long ranges and because of atmospheric attenuation.

NOTE: All of the factors affecting reflection must be considered to determine the radar return potential.

## Typical Radar Returns
### Returns from Land

All land surfaces present minute irregular parts of the total surface for reflection of the radar beam; thus, there is usually a certain amount of radar return from all land areas. The amount of return varies considerably according to the nature of the land surface scanned. This variance is caused by the difference in reflecting materials of which the land area is composed and the texture of the land surface. These are the primary factors governing the total radar return from specific land areas.

### *Flat Land*

A certain amount of any surface, however flat in the overall view, is irregular enough to reflect the radar beam. Surfaces that are apparently flat are actually textured and may cause returns on the scope. Ordinary soil absorbs some of the radar energy, the return that emanates from this type of surface is not strong. Irregularly textured land areas present more surface to the radar beam than flat land and cause more return. The returns from irregularly textured land areas are most intense when the radar beam scans the ridges or similar features at a right angle. This effect is particularly helpful in detecting riverbeds, gullies, or other sharp breaks in the surface height. At times, in desolate areas that are flat, these occasional surface changes are apparent where it would not have appeared in more irregular topography. Such returns provide recognizable targets in otherwise sparse circumstances. In other cases, especially at low-level over broken terrain, this effect could complicate scope interpretation.

### *Hills and Mountains*

Hills and mountains normally give more radar returns than flat land, because the radar beam is more nearly perpendicular to the sides of these features. The typical return is a bright return from the near side of the feature and an area of no return on the far side. The area of no return, called a mountain shadow, exists because the radar beam cannot penetrate the mountain and its LOS transmission does not allow it to intercept targets behind the mountain. *[Figure 7-6]* The shadow area varies in size, depending upon the height of the aircraft with respect to the mountain. As an aircraft approaches a mountain, the shadow area becomes smaller at higher altitudes. Furthermore, the shape of the shadow area and the brightness of the return from the peak varies as the aircraft's position changes. As the aircraft closes on the mountainous area, shadows may disappear completely as the beam covers the entire surface area. At this point, a great deal of energy is reflected back at the antenna and recognizable features in that area are rare.

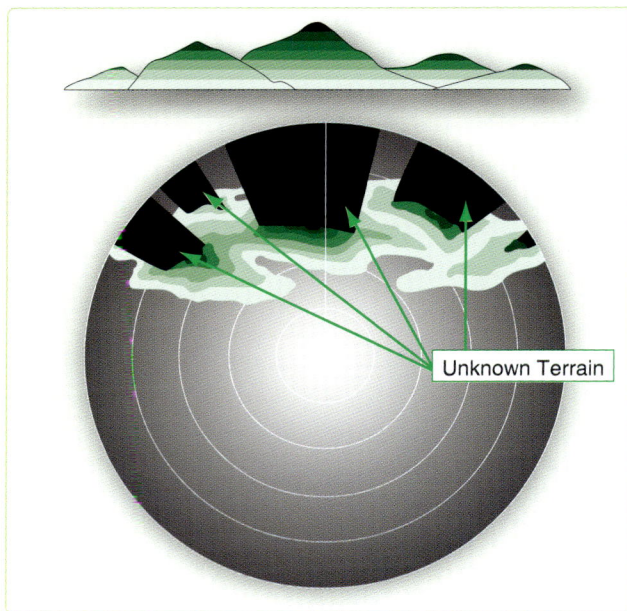

**Figure 7-6.** *Mountain shadows.*

Recognition of mountain shadow is important because any target in the area behind the mountain cannot be seen on the scope. In areas with isolated high peaks or mountain ridges, contour navigation may be possible, because the returns from such features assume an almost three-dimensional appearance. This allows specific peaks to be identified.

In more rugged mountainous areas, however, there may be so many mountains with resulting return and shadow areas that contour navigation is almost impossible. But these mountainous areas are composed of patches of mountains or hills, each having different relative sizes and shapes and relative positions from other patches. By observing these relationships on a chart, general aircraft positioning is feasible.

## *Coastlines and Riverbanks*

The contrast between water and land is very sharp, so that the configuration of coasts and lakes are seen with map-like clarity in most cases. *[Figure 7-7]* When the radar beam scans

**Figure 7-7.** *Radar returns.*

the banks of a river, lake, or larger body of water, there is little or no return from the water surface itself, but there is usually a return from the adjoining land. The more rugged the bank or coastline, the more returns are experienced. In cases where there are wide, smooth mud flats or sandy beaches, the exact definition of the coastline requires careful tuning. Since both mountains and lakes present a dark area on the scope, it is sometimes easy to mistake a mountain shadow for a lake. This is particularly true when navigating in mountainous areas that also contain lakes.

One difference between returns from mountain areas and lakes is that returns from mountains are bright on the near side and dark on the far side, while returns from lakes are of more uniform brightness all around the edges. Another characteristic of mountain returns is that the no-show area changes its shape and position quite rapidly as the aircraft moves; returns from lakes change inconsequentially.

## Cultural Returns

The overall size and shape of the radar return from any given city can usually be determined with a fair degree of accuracy by referring to a current map of the area. *[Figure 7-7]* However, the brightness of one cultural area as compared to another may vary greatly, and this variance can hardly be forecasted by reference to the navigation chart. In general, due to the collection of dense materials therein, urban and suburban areas generate strong returns, although the industrial and commercial centers of the cities produce a much greater brightness than the outlying residential areas. Many isolated or small groups of structures create radar returns. The size and brightness of the radar returns these features produce are dependent on their construction. If these structures are not plotted on the navigation charts, they are of no navigational value. However, some of them give very strong returns, such as large concrete dams and steel bridges. If any are plotted on the chart and can be properly identified, they can provide valuable navigational assistance.

## Weather Returns

Cloud returns that appear on the scope are of interest for two reasons. First, since the brightness of a given cloud return is an indication of the intensity of the weather within the cloud, intense weather areas can be avoided by directing the pilot through the areas of least intensity or by circumnavigating the entire cloud return. *[Figure 7-8]* Second, cloud returns obscure useful natural and cultural features on the ground.

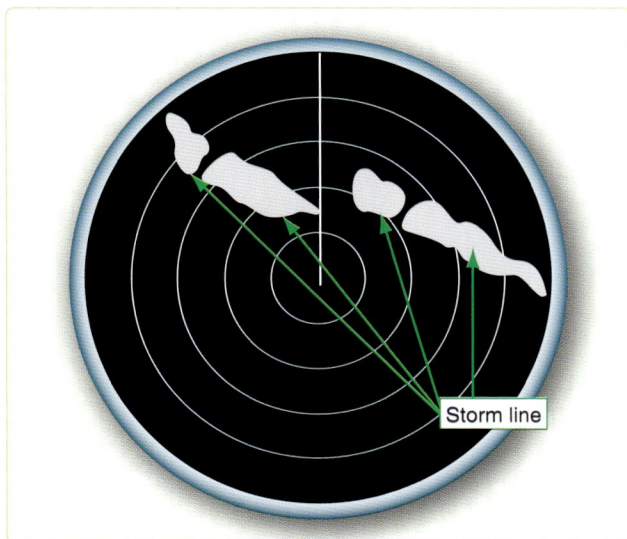

**Figure 7-8.** *Weather returns.*

They may also be falsely identified as a ground feature that can lead to gross errors in radar fixing. Clouds must be reasonably large to create a return on the scope. However, size alone is not the sole determining factor. The one really important characteristic that causes clouds to create radar returns is the size of the water droplets forming them. Radar waves are reflected from large rain droplets and hail that fall through the atmosphere or are suspended in the clouds by strong vertical air currents. Thunderstorms are characterized by strong vertical air currents; therefore, they give very strong radar returns. Cloud returns may be identified as follows:

- Brightness varies considerably, but the average brightness is greater than a normal ground return.

- Returns generally present a hazy, fuzzy appearance around their edges.

- Returns often produce shadow areas similar to mountain shadows, because the radar beam does not penetrate clouds completely.

- Returns do not fade away as the antenna tilt is raised, but ground returns do tend to decrease in intensity with an increase in antenna tilt.

- Returns can appear in the altitude hole when altitude delay is not used and the distance to the cloud is less than the altitude.

## Effects of Snow and Ice

The effects of snow and ice are similar to the effect of water. If a land area is covered to any great depth with snow:

1. Some of the radar beam reflects from the snow, and

2. Some of the energy is absorbed by the snow.

The overall effect is to reduce the return that would normally come from the snow-blanketed area.

Ice reacts in a slightly different manner, depending upon its roughness. If an ice coating on a body of water remains smooth, the return appears approximately the same as a water return. However, if the ice is formed in irregular patterns, the returns created are comparable to terrain features of commensurate size. For example, ice ridges or ice mountains would create returns comparable to ground embankments or mountains, respectively. Also, offshore ice floes tend to disguise the true shape of a coastline so that the coastline may appear vastly different in winter as compared to summer. This phenomenon is termed arctic reversal, because the resultant display is often the opposite of the anticipated display.

## Inherent Scope Errors

Another factor that must be considered in radarscope interpretation is the inherent distortion of the radar display. This distortion is present to a greater or lesser degree in every radar set, depending upon its design. Inherent scope errors may be attributed to three causes: width of beam, the length (time duration) of the transmitted pulse, and the diameter of the electron spot.

## Beam-Width Error

Beam-width error is not overly significant in radar navigation. Since the distortion is essentially symmetrical, it may be nullified by bisecting the return with the bearing cursor when a bearing is measured. Reducing the receiver gain control also lessens beam-width distortion.

## Pulse-Length Error

Pulse-length error is caused by the fact that the radar transmission is not instantaneous but lasts for a brief period of time. There is a distortion in the range depiction on the far side of the reflector, and this pulse-length error is equal to the range equivalent of one-half of the pulse time. Since pulse-length error occurs on the far side of the return, it may be nullified by reading the range to, and plotting from, the near side of a reflecting target when taking radar ranges.

## Spot-Size Error

Spot-size error is caused by the fact that the electron beam that displays the returns on the scope has a definite physical diameter. No return that appears on the scope can be smaller than the diameter of the beam. Furthermore, a part of the glow produced when the electron beam strikes the phosphorescent coating of the display radiates laterally across the scope. As a result of these two factors, all returns displayed on the scope appear to be slightly larger in size than they actually are. Spot-size distortion may be reduced by using the lowest practicable receiver gain, video gain, and bias settings and by keeping the operating range at a minimum so that the area represented by each spot is kept at a minimum. Further, the operator should check the focus control for optimum setting.

## Total Distortion

For navigational purposes, these errors are often negligible. However, the radar navigator should realize that they do exist and that optimum radar accuracy demands that they be taken into account. They are usually most significant when the target is a thin, no-show (river), when it is very reflective but small, or when it is in close proximity to another show target. Thin no-shows are erased except for their wider points. With tiny, but very reflective targets, the crosssection of the return would normally be negligible on the display. Their extremely strong reflectance, coupled with the inherent errors, causes them to appear larger and of seemingly more significance on the indicator. When show targets are close to each other, these errors cause them to blend together, diminishing the scope resolution. Generally, the combined effects of the inherent errors cause reflecting targets to appear larger and nonreflecting targets to dwindle. *[Figure 7-9]*

## Radar Enhancements

### Variable Range Marker and Crosshairs

Most radar sets provide a range marker that may be moved within certain limits by the radar operator. This variable range marker permits more accurate measurement of range, because the marker can be positioned more accurately on the scope. Furthermore, visual interpolation of range is simplified when using the variable range marker. On many radar sets, an electronic azimuth marker has been added to the variable range marker to facilitate fixing. The intersection of the azimuth marker and the variable range marker is defined as radar crosshairs.

### Altitude Delay

It is obvious that the ground directly beneath the aircraft is the closest reflecting object. Therefore, the first return that can appear on the scope is from this ground point. Since it takes some finite period of time for the radar pulses to travel to the ground and back, it follows that the sweep must travel some finite distance radially from the center of the scope before it displays the first return. Consequently, a hole appears in the center of the scope within which no ground returns can appear. Since the size of this hole is proportional to altitude, its radius can be used to estimate altitude. If the radius of the altitude hole is 12,000 feet, the absolute altitude of the aircraft is about 12,000 feet.

Although the altitude hole may be used to estimate altitude, it occupies a large portion of the scope face, especially when the aircraft is flying at a high altitude and using a short range. *[Figure 7-10]* In this particular case, the range selector

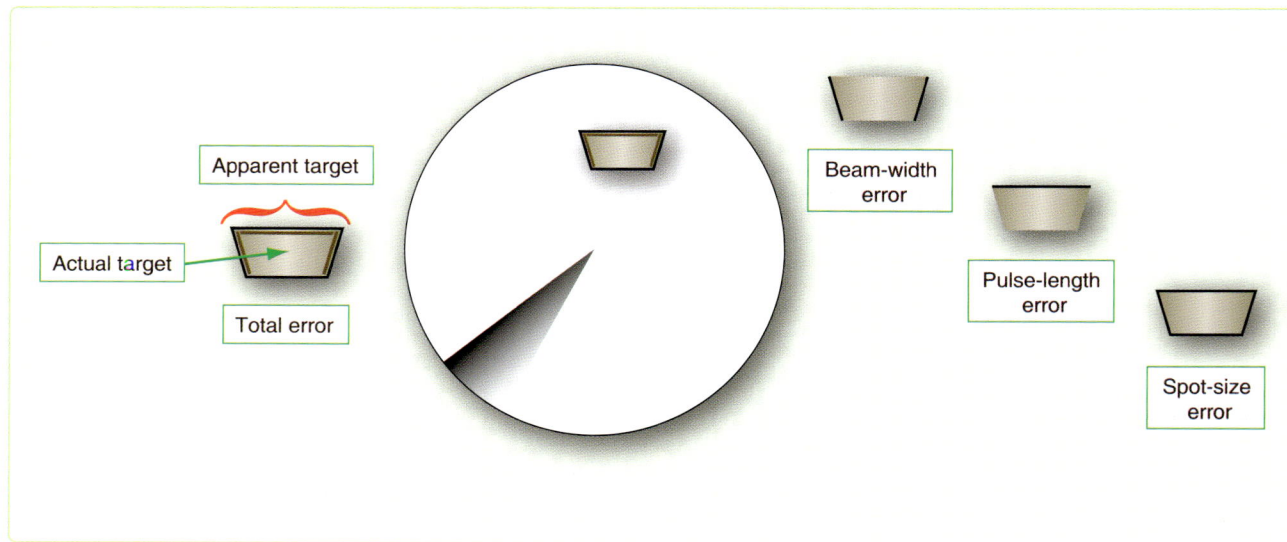

**Figure 7-9.** *Combined effects of inherent errors.*

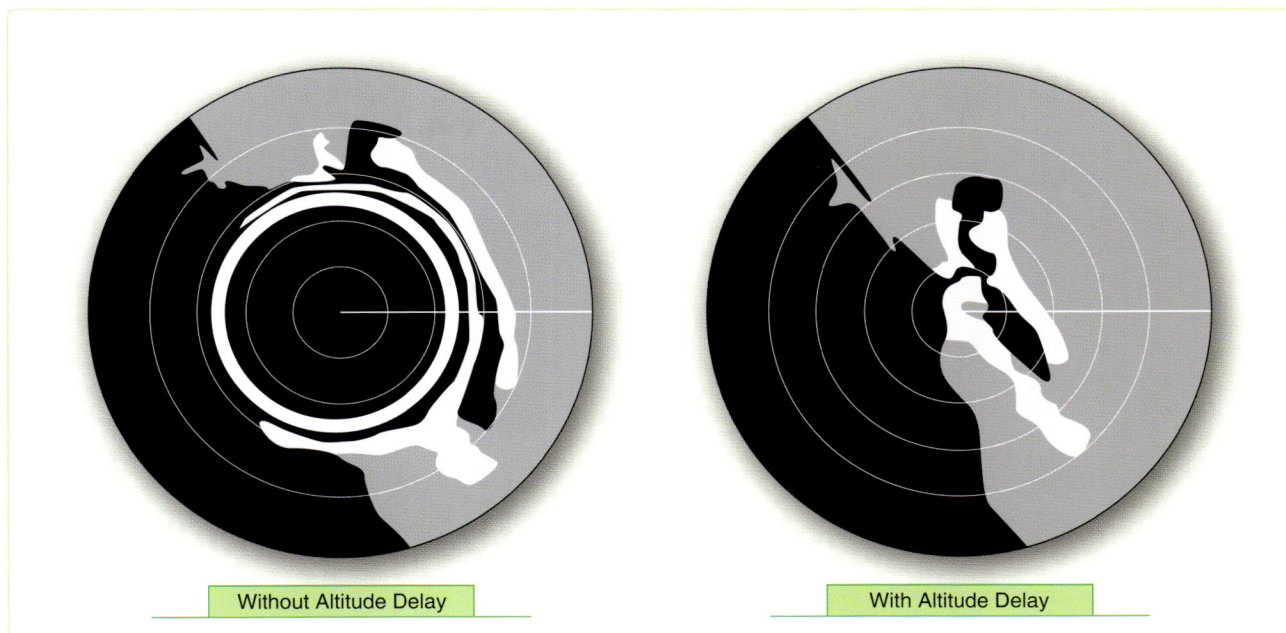

| Without Altitude Delay | With Altitude Delay |

**Figure 7-10.** *Altitude delay eliminates the hole.*

switch is set for a 50/10-mile range presentation. Without altitude delay, the return shown on the inside part of the scope consists of the altitude hole, and the return shown on the remaining part is a badly distorted presentation of all of the terrain below the aircraft.

Many radar sets incorporate an altitude delay circuit that permits the removal of the altitude hole. This is accomplished by delaying the start of the sweep until the radar pulse has had time to travel to the ground point directly below the aircraft and back. Hence, the name altitude delay circuit. The altitude delay circuit also minimizes distortion and makes it possible for the radarscope to present a ground picture that preserves the actual relationships between the various ground objects.

### Sweep Delay

Sweep delay is a feature that delays the start of the sweep until after the radar pulse has had time to travel some distance into space. In this respect, it is very similar to altitude delay. The use of sweep delay enables the radar operator to obtain an enlarged view of areas at extended ranges. For example, two targets that are 75 miles from the aircraft can only be displayed on the scope if a range scale greater than 75 miles is being used. On the 100-mile range scale, the two targets might appear very small and close together. By introducing 50 miles of sweep delay, the display of the two targets is enlarged. *[Figure 7-11]* The more this range is reduced, the greater the enlarging effect. On some sets, the range displayed during sweep delay operation is fixed by the design of the set and cannot be adjusted by the operator.

### Iso-Echo

Detecting hazardous weather is not difficult in the normal mapping mode with most radar units. The weather mode offers increased sensitivity to weather phenomenon. But to discriminate between areas of varying hazards presents a dilemma. Reflected energy from weather is dependent on the density of the rain and hail it contains. The limitations of display capabilities to display these dynamic characteristics make detection of the more intense areas difficult. Also, computer circuitry is more effective at judging slight variations in shading than the human eye.

The iso-echo control compensates for this deficiency by presenting a void area on the display corresponding to a hazardous area in the weather environment. This void area, the black hole, is dependent on a control that the operator sets to define the intensity of the area that is to be avoided. For instance, say only the largest cells of weather are desired to be displayed. The operator would set the appropriate control and, on the display, the weather depiction would be present. The areas within the weather where the most hazardous cells were located would be no-show areas or black holes.

The iso-echo circuits are capable of sensing the variation in the received signals and act like a radio squelch control to block presentation of selected intensities. *[Figure 7-12]* A word of caution, the iso-echo is not selective in the targets it blocks. If ground returns are received by the radar and a portion of their intensity falls into the range selected to be blocked, they too are blocked from the scope.

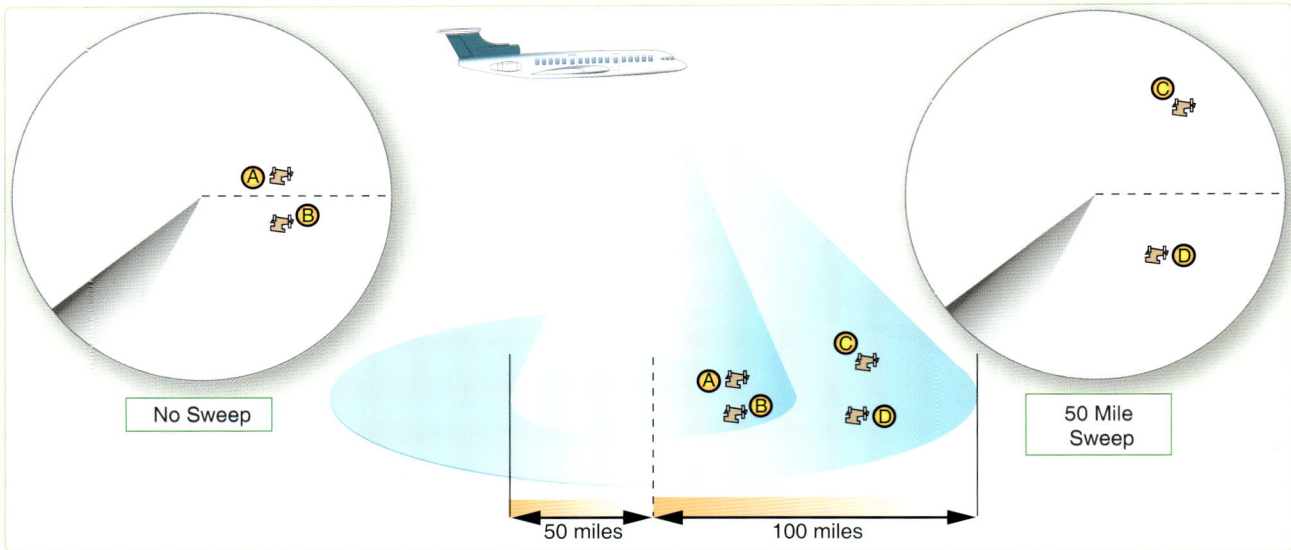

**Figure 7-11.** *Sweep delay provides telescopic view.*

## Radar Beacon

Radar beacons have been used for many years in aviation. In the past, airfields had beacons visible on radar much like a nondirectional beacon (NDB), but most are now decommissioned.

Radar beacons consist of interrogator and responder units operating from different locations. The interrogator transmits a pulse that causes the responder to transmit a corresponding pulse. The interrogator receives the coded return and uses time lapse and azimuth, or sweep relationships, to display the returns on the display. The time needed for generation of the return pulse causes a range error amounting to one-half mile, generally.

Beacons are sometimes coded with a mixture of aircraft identification and flight parameters for air route traffic control centers (ARTCC). Aircraft equipped with beacons, like the APN-69, can interrogate and respond to like-equipped aircraft. Beacons, like the APN-69, use a pulsed code of up to six pulses. The pulse codes are set by the responder aircraft and appear on the interrogators display. The first pulse is in the relative position of the responder with successive pulses trailing. The range between aircraft is equal to the range of the first pulse (minus one half NM) and the azimuth is measured through the middle of the pulse length.

Two blocking circuits are included in the units to prevent interference from radar on other frequencies or a return of the interrogating pulse. This sometimes prevents a ring around where false azimuth inputs are presented on the display. In such cases, excessive gain causes returns to be picked up by side lobes of the antenna. *Figure 7-13* is an example of a beacon return on the scope.

## Sensitivity Time Constant (STC)

Most radar sets produce a hot spot in the center of the radarscope, because the high-gain setting required to amplify the weak echoes of distant targets overamplifies the strong

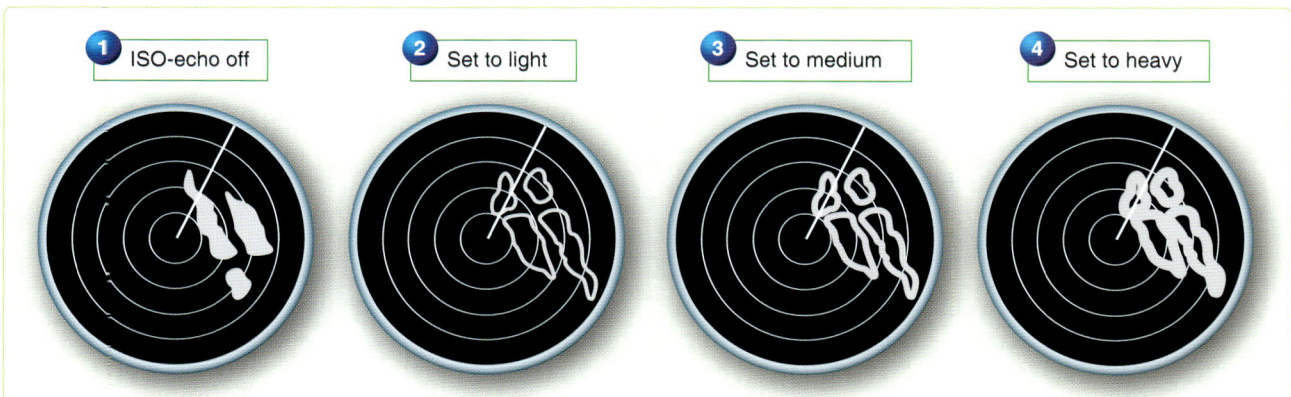

**Figure 7-12.** *Iso-echo.*

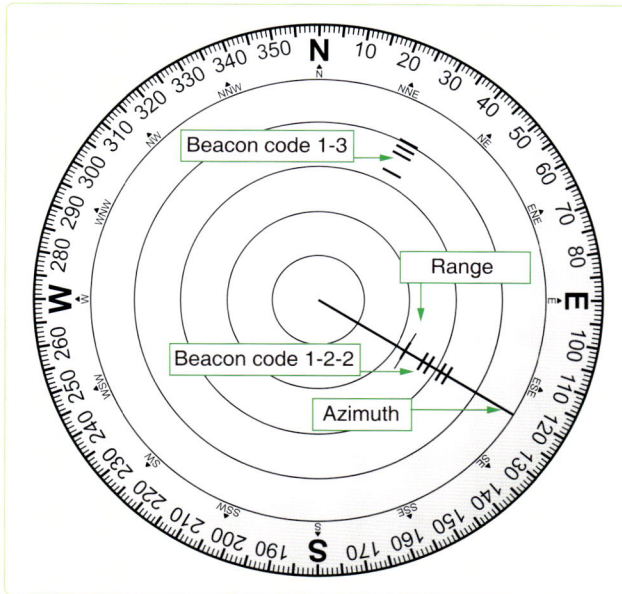

**Figure 7-13.** *Radar beacon returns.*

echoes of nearby targets. If the receiver gain setting is reduced sufficiently to eliminate the hot spot, distant returns are weakened or eliminated entirely. The difficulty is most pronounced when radar is used during low-level navigation; to make best use of the radar, the navigator is forced to adjust the receiver gain setting constantly. Sensitivity time constant (STC) solves the problem by increasing the gain as the electron beam is deflected from the center to the edge of the radarscope, automatically providing an optimum gain setting for each range displayed. In this manner, the hot spot is removed while distant targets are amplified sufficiently. STC controls vary from one model radar set to another. Refer to the appropriate technical order for operating instructions.

## Plan Display

The plan display is a sector scan presentation that indicates the range and direction of obstructions projecting above a selected clearance plane. The clearance plane can be manually set at any level from 3,000 feet below the aircraft up to the level of the aircraft. Only those peaks projecting above the clearance plane are displayed; all other returns are inconsequential and are eliminated. The sector scan presentation limits the returns to those ahead of the aircraft. The vertical line represents the ground track of the aircraft, and ranges are determined by range marks.

## Profile Display

The profile display, normally received only by the pilot, provides an outline of the terrain 1,500 feet above and below the clearance plane. Elevations of returns are represented vertically; azimuth is represented horizontally. This display gives the operator a look up the valley. The returns seen

represent the highest terrain within the selected range. The position of the aircraft is represented by an aircraft symbol on the indicator overlay.

## Techniques on Radar Usage

Radars currently in use offer variations of special equipment and capabilities. The following are techniques to use with radar in common situations and with special equipment designed to enhance radar usage. These are basic suggestions that can and should be adapted to specific aircraft.

### Radar Fixing

Techniques in radar fixing change from operator to operator and most provide accurate results. The following are reminders that affect the fix accuracy if not considered.

Radar is an aid to DR. Before any radar return can be accurately identified, the operator should be familiar with a chart of the target area. This chart study relies on knowing the approximate location of the aircraft and, therefore, it is essential to radar fixing that the best possible DR position is ascertained.

In examining the area surrounding the DR on the chart, attention should be given to details like roadways and waterways, as well as the more prominent urban returns. Cultural returns build up along such byways; therefore, discrepancies between the chart (which could be years old) and the display can be more successfully analyzed.

Prior to fixing, take care to adjust gain, antenna tilt, and heading marker. If you use a mechanical cursor, ensure its center is aligned with the sweep origin, or risk parallax error. Do not accept a return on the scope as the chosen target unless you have verified it using surrounding returns. Work from chart to scope. If your desired target does not show, but you see a return you think you recognize, go back to the chart and verify it before fixing from it.

When obtaining fix readings, remember to compensate for inherent scope errors. If the fix is a multirange or multibearing type, choose the targets to provide the optimum cut. When using multiple targets, read the returns that are changing their values the fastest closest to fix time. (With multirange, a target off the nose changes range faster than the one off the wing.)

### Slant Range

Once you identify a return, use it to fix the position of the aircraft by measuring its bearing and distance from a known geographical point. Of particular significance in any discussion of radar ranging is the subject of slant range versus ground range. *[Figure 7-14]* Slant range is the straight-line distance between the aircraft and the target, while ground

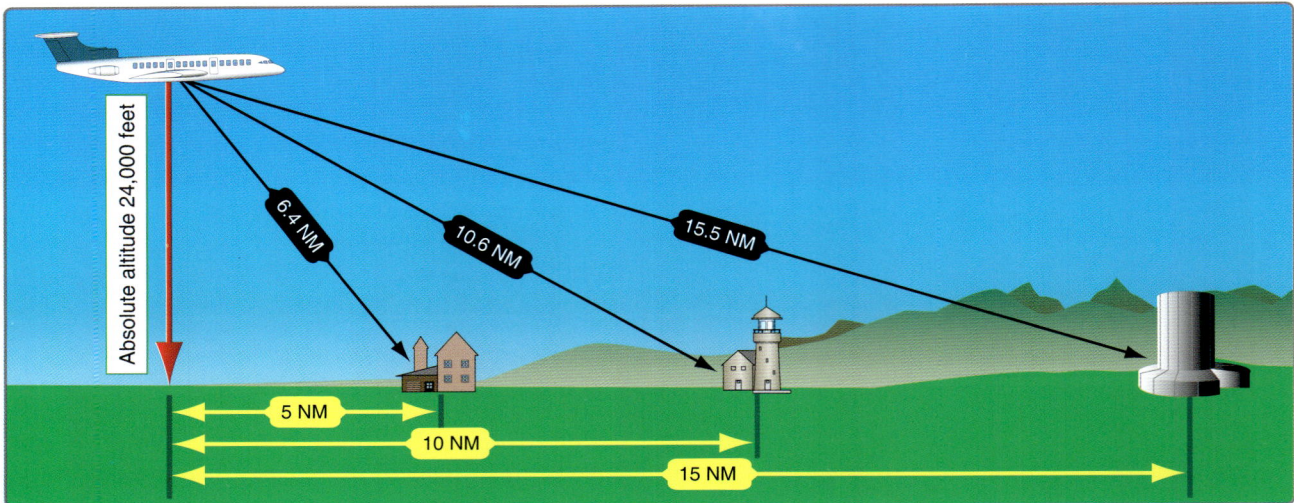

**Figure 7-14.** *Slant range (black arrows) compared to ground range (yellow arrows).*

range is the range between the point directly below the aircraft and the target.

To fix the position of the aircraft, the navigator is interested in the ground range from the fixing point, yet the fixed range markers give slant range. The trick is to determine the critical range below which the navigator must convert slant range to ground range to keep fixes accurate. This range may be determined by a simple formula:

Critical slant range = Absolute altitude (in K)–5

Slant range can be converted to ground range, using the latitude and longitude lines of a chart if the slant range table is not available. Set dividers at the slant range distance to the target. Place one point of the dividers at the equivalent (in NM) of the aircraft's altitude on the longitude line. Set the other point where it meets a nearby latitude line. Without moving point, reset the first point along the latitude line at the intersection of the latitude and longitude lines. The distance is the ground range in NM. *[Figure 7-15]* Slant range correction charts are provided in *Figures 7-16.*

## Side Lobe Interference

Side lobes are small extra fields of energy separate from the main beam and are an inherent flaw in any radar. These side lobes are rarely strong enough to generate a return. However, when a large or very reflective target comes into this field, or when the transmitter power increases the size of the lobes, multiple shadow returns may appear on the display. Curved strobes originating at the center of the radarscope are also caused by the side lobes of the radar receiving energy from your radar or others in the same frequency range. Solutions to this problem include reducing the gain or changing transmitter frequencies.

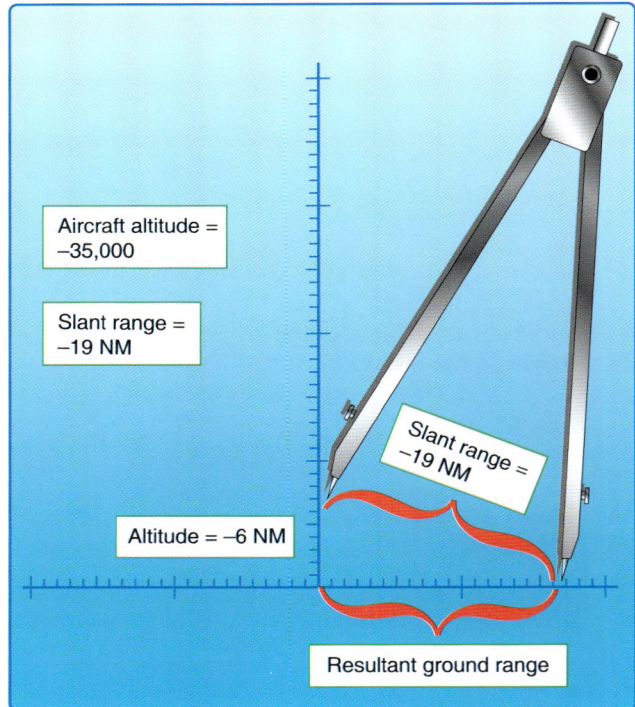

**Figure 7-15.** *Slant range from chart.*

## Target-Timing Wind

This is a technique for obtaining a wind by using radar targets to provide track and groundspeed (GS) of an aircraft. *[Figure 7-17]* The MB-4 computer solution for wind requires true heading (TH), true airspeed (TAS), drift angle (DA), and GS. The first two can be derived from basic aircraft instruments (indicated airspeed (IAS) and compass). The other two require a target that can be tracked for about 4 minutes and which is preferably within 20 degrees of the radar heading marker. The identity of the target is irrelevant, but it

7-12

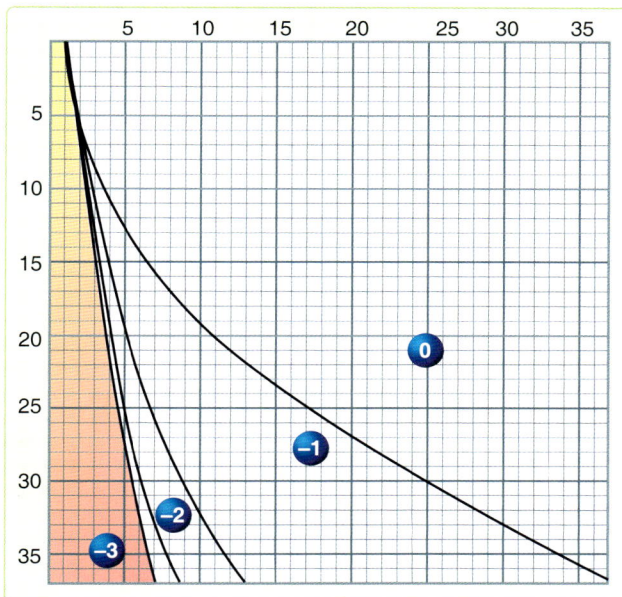

**Figure 7-16.** *Slant range correction chart.*

| EXAMPLE | | | |
|---|---|---|---|
| TH = 320° | Fix 1 | 310° | 40 NM |
| TAS = 400 knots | Fix 2 | 308° | 30 NM |
| Time = 3 + 30 min | Fix 3 | 305° | 20 NM |
| **SOLUTION** | | | |
| TC = 316° | | | |
| GS = 352 knots | | | |
| Wind = 349°/53 knots | | | |

**Figure 7-17.** *Tartget-timing wind example.*

should not be too big to make range and bearing determination vague, or so small that it disappears. Choose a target that has just appeared on the scope and read its range and bearing. Also, start a stopwatch or note the minute and seconds on a clock so elapsed time can be measured. At least two ranges and bearings should be taken over a distance of 20 to 25 NM. One technique is to fix at the 40, 30, and 20 NM range marks to space the fixes evenly. At the last observation, stop the watch and determine the elapsed time. On the windface grid of the MB-4, place the grommet over the center mark of the top reference line. Turn the compass rose to the azimuth of the first fix. Using your own values for each of the horizontal grid lines, plot a point representing the range of the first fix (going down). Then, turn the compass rose to the azimuth of the second fix and plot a point (measuring from the top line again) representing the range of the second fix. Repeat for the successive fixes.*[Figure 7-18]* To solve for the wind, rotate

the compass rose so that the three plotted lines are parallel to the vertical grid lines and read the track under the true index of the compass rose. Then, determine the GS by measuring the distance between the first and last plotted points using the grid lines. Using track and TH, find the DA and use the standard MB-4 wind solution.

## Weather Avoidance

Severe turbulence, hail, and icing associated with thunderstorms constitute severe hazards to flight. You must avoid these thunderstorms whenever possible. Airborne weather radar, if operated and interpreted properly, can be an invaluable aid in avoiding thunderstorm areas.

You must be aware of factors and limitations affecting thunderstorm radar returns to get the most out of the radar. Some of these factors are not meteorological and depend on the characteristics of the radar and the way it is used. The same weather target can vary considerably in its appearance from ground mapping mode to weather mode. Navigators must ensure they use the radar as intended for weather avoidance. Primary meteorological factors that affect radar returns are the amount of moisture in the weather target and atmospheric absorption characteristics between the radar antenna and the target.

The predominant weather-induced returns on most radarscopes are caused by precipitation-size water droplets, not by clouds. Intense returns indicate the presence of very large droplets. These large droplets are generally associated with the most hazardous phenomena; those with strong vertical currents that are necessary to maintain these droplets in the cloud. It is possible, however, to encounter such strong turbulence in an echo-free area or even in an adjacent cloud-free area, so avoiding areas giving intense returns does not necessarily guarantee safe flight in the vicinity of thunderstorms. Make careful note of all areas forecast to have the potential for hazardous weather.

Generally, the map mode of the radar with a moderate amount of gain applied is adequate for obtaining a return from hazardous cells. Sometimes, ground returns hamper detection in the area by hiding the storm. This can occur in mountainous areas where ground returns are similar and airmass lifting action breeds the cells. For these reasons, raising the radar tilt or switching to pencil beam, or both, are techniques that aid weather detection.

There are two types of weather avoidance with radar: avoidance of isolated thunderstorms and penetration of a line of thunderstorms. Avoid an isolated return by first identifying it and then circumnavigating it at a safe distance.

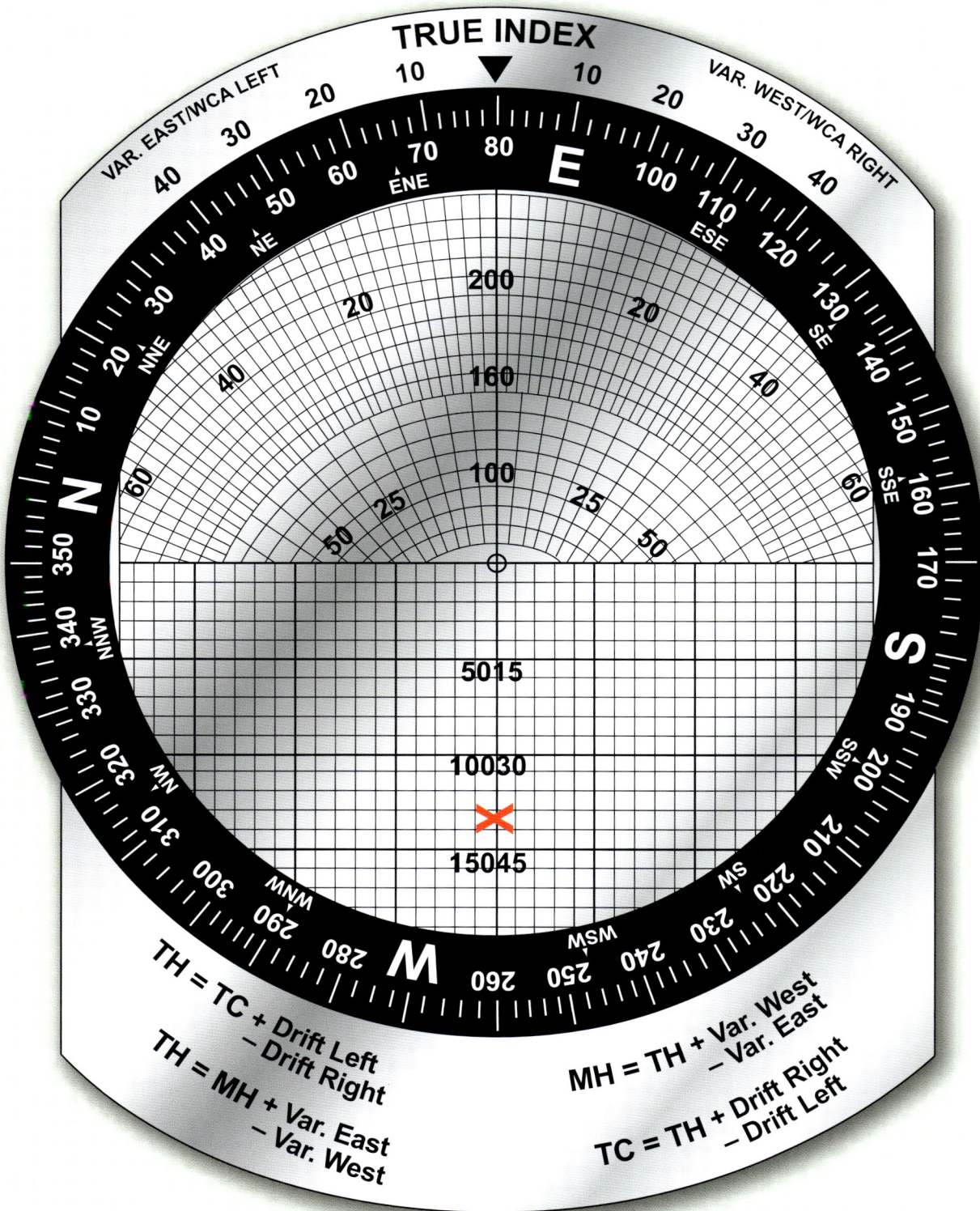

**Figure 7-18.** *Target timing wind solution.*

## STEPS 2 & 3
Repeat Step 1 for second and third observations.

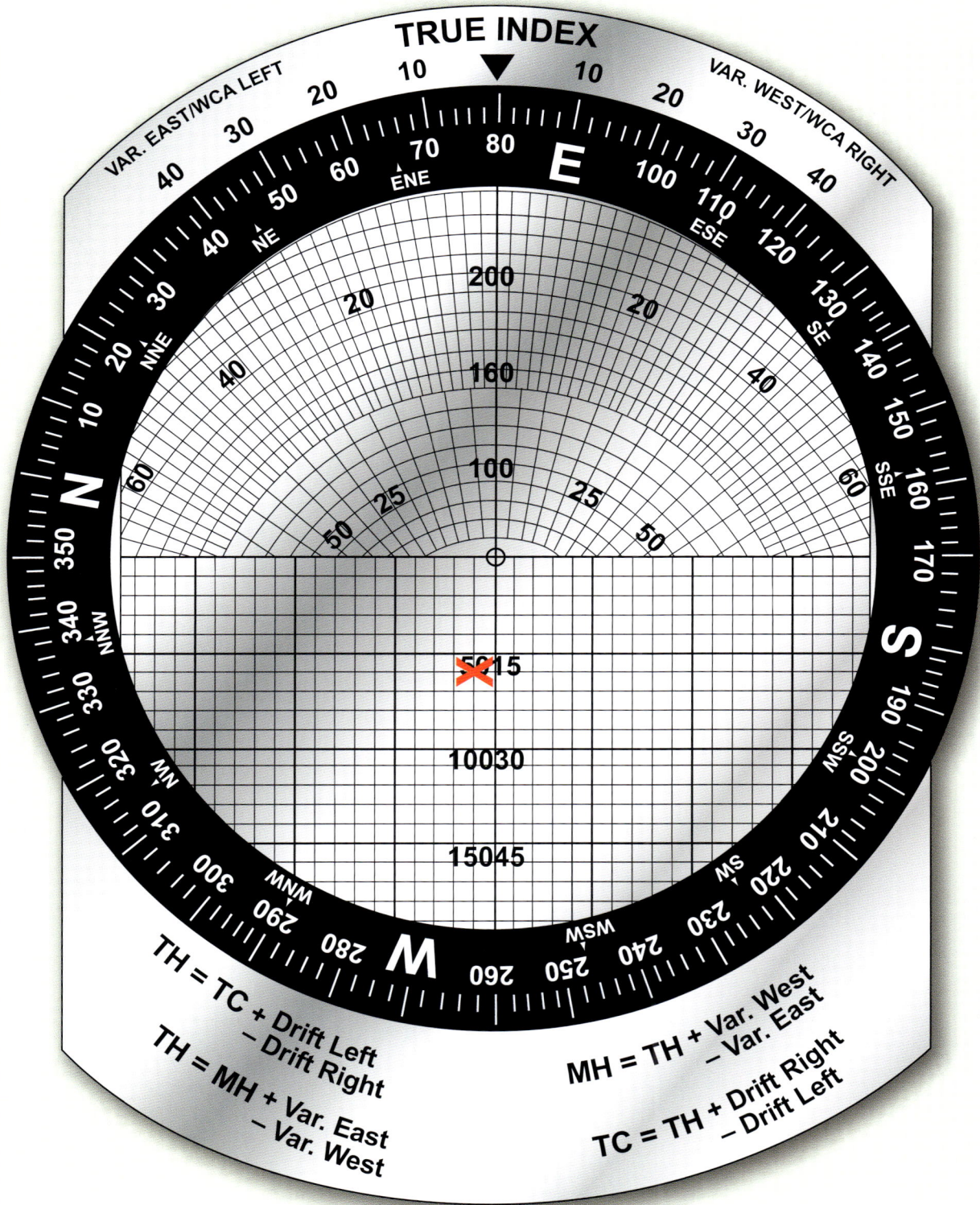

**Figure 7-18.** *Target timing wind solution (continued).*

## STEP 4
Rotate compass rose to align X with grid lines. Read track under index. Determine distance from first to last X. Use track, distance, and instrument readings to compute wind.

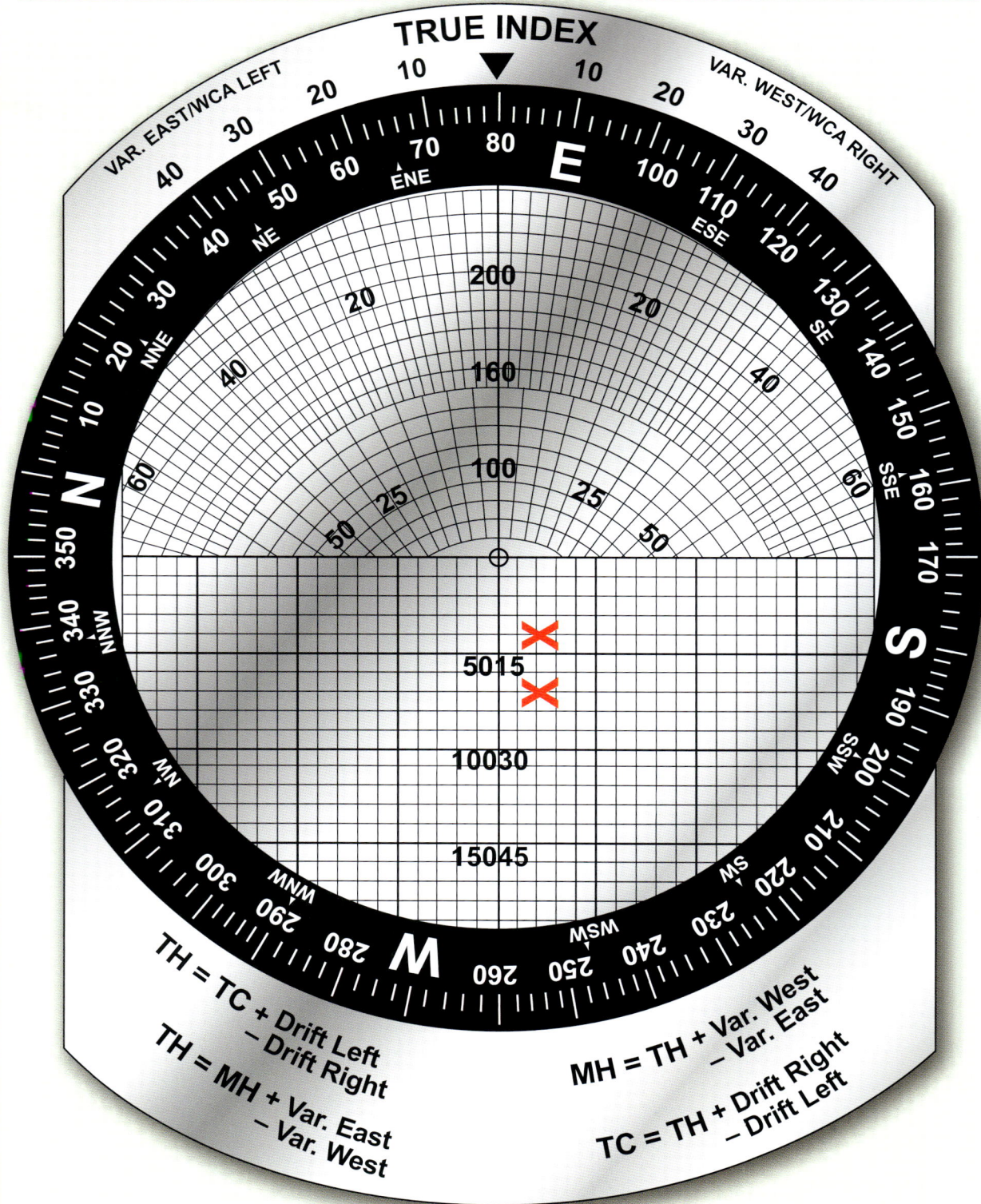

**Figure 7-18.** *Target timing wind solution.*

After detecting a weather system, determine its extent. Analyze the weather's layout relative to planned track and decide either to deviate around it or penetrate the line. If the system is complex, remember your deviation could worsen the situation by flying into a sucker hole, where a solid system could surround the aircraft. Sometimes, what seems to be a good heading at short range seems foolish when viewed at long range. Remember that turning around is always an option, and ARTCC can sometimes assist in weather analysis. A simple technique for flying around weather at a preferred distance (say 20 NM) is the flying disc technique. Imagine the aircraft is a disc defined by the 20 NM range mark on the display. The heading marker is the nose of the disc. Draw an imaginary tangent from the disc to the edge of the weather, or use a pencil or plotter. Turn the aircraft the same number of degrees that it would take to get the heading marker to fire parallel to the tangent. After the turn, recheck the heading in the same manner. This technique works best with a scan of more than 180°. [Figure 7-19]

Penetration of a line of thunderstorms is a last resort and presents a different problem. Since the line may extend for hundreds of miles, circumnavigation is not practical or even possible. If no other course of action exists, the main objective is to avoid the more dangerous areas in the line.

Figure 7-20 shows an example of frontal penetration using radar. An iso-echo equipped radar can discriminate between the safe and violent areas. Without it, decreasing the gain works to highlight the worst areas by leaving the densest water cells as the last returns on the display. Upon approaching the line, the navigator determines an area that has weak or no returns and that is large enough to allow avoidance of all intense returns by the recommended distances throughout penetration. The navigator directs the aircraft to that point, making the penetration at right angles to the line so as to remain in the bad weather areas for the shortest possible time. Avoid the dangerous echoes by a safe distance. Penetration of a line of severe thunderstorms is always a potentially dangerous procedure. Attempt it only when you must continue the flight and cannot circumnavigate the line. Always advise ARTCC of your intentions when deviating from your flight-planned route.

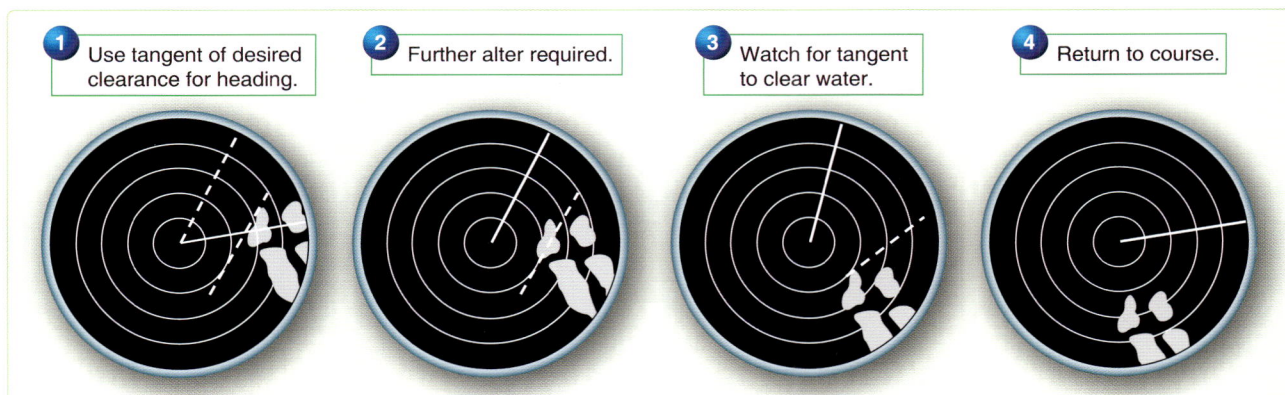

Figure 7-19. *Weather avoidance.*

1. Use tangent of desired clearance for heading.
2. Further alter required.
3. Watch for tangent to clear water.
4. Return to course.

Figure 7-20. *Penetration of thunderstorm area.*

1. Select area of least radar return.
2. Alter through safe penetration area.
3. Maintain heading through thunderstorms.
4. All clear—resume normal navigation.

### Heading Marker Correction

For optimum accuracy, it may sometimes become necessary to correct the bearings taken on the various targets. This necessity arises when the heading marker reading does not agree with the TH of the aircraft when azimuth stabilization is used, or the heading marker reading does not agree with 360° when azimuth stabilization is not used.

For example, if the TH is 125° and the heading marker reads 120°, all of the returns on the scope indicate a bearing that is 5° less than it should be. Therefore, if a target indicates a bearing of 50°, add 5° to the bearing before plotting it. Conversely, if the heading marker reads 45° when the TH is 040°, all of the scope returns indicate a bearing that is 5° more than it should be. Therefore, if a target indicates a bearing of 275°, subtract 5° from the bearing before plotting it. The greater the distance to the target from the aircraft, the more important this heading marker correction becomes.

## Chapter Summary

This chapter discussed the history of radar navigation and how radar technology has improved throughout the years. Basic radar principals are explained including the different types and how the components work. Scope interpretation is introduced in this chapter along with radar reflection. The many variables affecting radar return are discussed, such as varying terrain and weather. Inherent scope errors are explained along with the different radar enhancements that help the navigator. The many different uses for the radar are discussed and how each can help the navigator navigate through different terrain, weather, and altitude changes.

## Chapter 8

# Celestial Concepts

## Introduction

Celestial navigation is a universal aid to dead reckoning (DR). Because it is available worldwide and is independent of electronic equipment, it is a very reliable method of fixing the position of the aircraft. It cannot be jammed and emanates no signals. Each celestial observation yields one line of position (LOP). In the daytime, when the sun may be the only visible celestial body, a single LOP may be all you can get. At night, when numerous bodies are available, LOPs obtained observing two or more bodies may be crossed to determine a fix.

It is impossible to predict, in so many miles, the accuracy of a celestial fix. Celestial accuracy depends on the navigator's skill, the type and condition of the equipment, and the weather. With the increase in aircraft speed and range, celestial navigation is very demanding. Fixes must be plotted and used as quickly as possible.

You do not have to be an astronomer or mathematician to establish a celestial LOP. Your ability to use a sextant is a matter of practice, and specially designed celestial tables have reduced the computations to simple arithmetic.

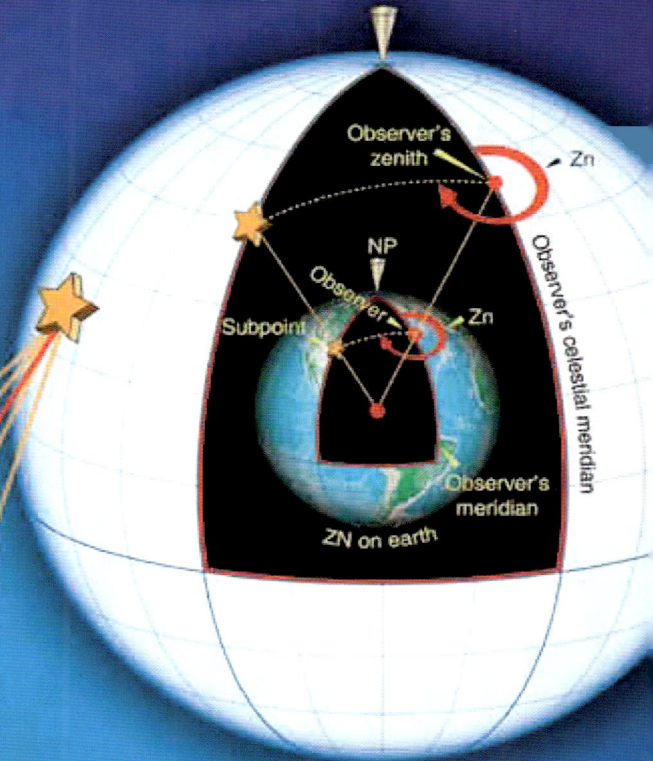

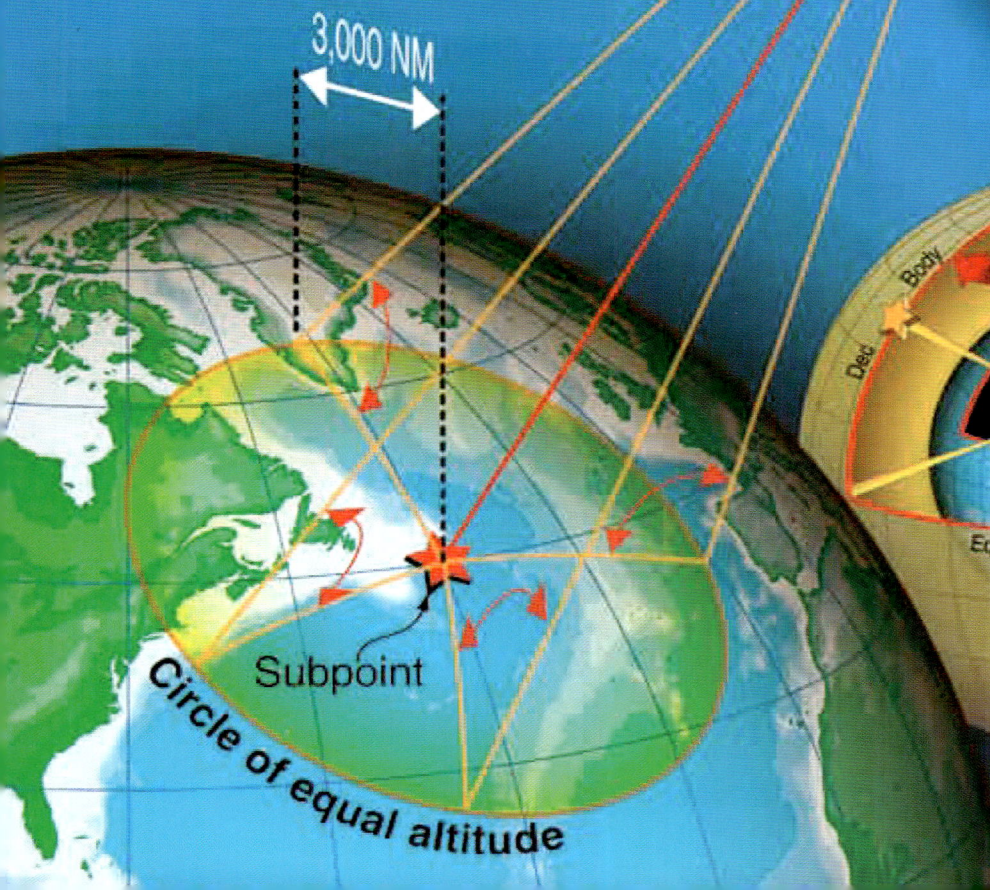

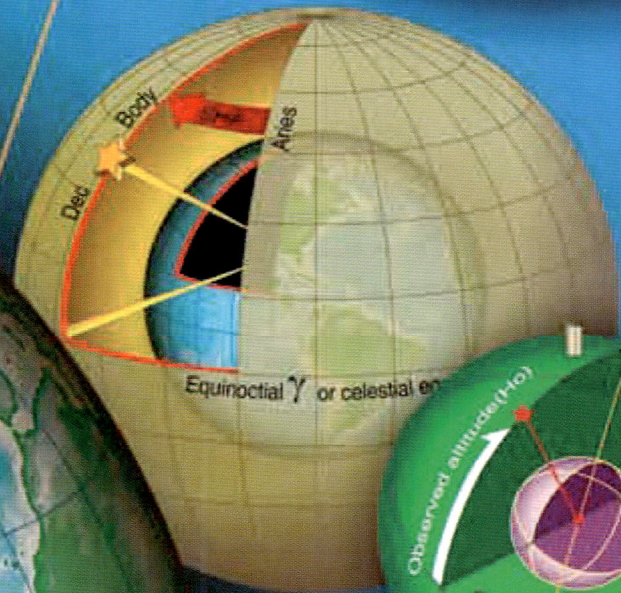

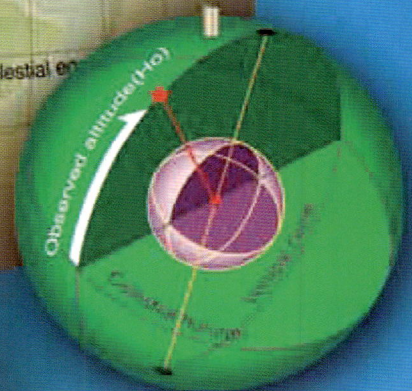

Although you do not need to understand astronomy in detail to establish an accurate celestial position, celestial work and celestial LOPs mean more if you understand the basics of celestial astronomy. Celestial astronomy includes the navigational bodies in the universe and their relative motions. Although there are an infinite number of heavenly bodies, celestial navigation utilizes only 63 of them: 57 stars, the moon, the sun, Venus, Jupiter, Mars, and Saturn.

## Assumptions

We make certain assumptions to simplify celestial navigation. These assumptions help you obtain accurate LOPs without a detailed knowledge of celestial astronomy. However, celestial positioning is more than extracting numbers from various books. A working knowledge of celestial concepts helps you crosscheck your computations.

First, assume the earth is a perfect sphere. That puts every point on the earth's surface equidistant from the center, forming the terrestrial sphere. Next, assume the terrestrial sphere is the center of an infinite universe. Finally, assume all other bodies, except the moon, are an infinite distance from the terrestrial sphere. Imagine them on the inside surface of an enormous concentric sphere, the celestial sphere. If the stars, planets, and sun are infinitely distant from the earth's center, then the earth's surface (or aircraft's altitude) is approximately the center of the universe.

Ptolemy proposed the celestial concept of the universe in AD 127. He said the earth is the center of the universe, and all bodies rotate about the earth from east to west. In the relatively short periods of time involved with celestial positioning, you can assume that all bodies on the celestial sphere rotate at the same rate. In actuality, over months or years, the planets move among the stars at varying rates.

Establishing an artificial celestial sphere with an infinite radius simplifies computations for three celestial spheres has a corresponding point on the terrestrial sphere and; conversely, every point on the terrestrial sphere has a corresponding point on the celestial sphere.

Second, the celestial sphere's infinite radius dwarfs variations in the observer's location. An infinite radius means all light rays from the celestial body arrive parallel, so the angle is the same whether viewed at the earth's center, on the surface, or at the aircraft's altitude.

Third, the relationships are valid for all bodies on the celestial sphere. Because the moon is relatively close to the earth, it must be treated differently. With certain corrections, the moon still provides an accurate LOP. This is addressed in Chapter 10, Celestial Precomputation.

Because the celestial sphere and terrestrial sphere are concentric, each sphere contains an equator, two poles, meridians, and parallels of latitude or declination. The observer on earth has a corresponding point directly overhead on the celestial sphere called the zenith. A celestial body has a corresponding point on the terrestrial sphere directly below it called the subpoint or geographic position. At the subpoint, the light rays from the body are perpendicular to the earth's surface. [*Figures 8-1* and *8-2*]

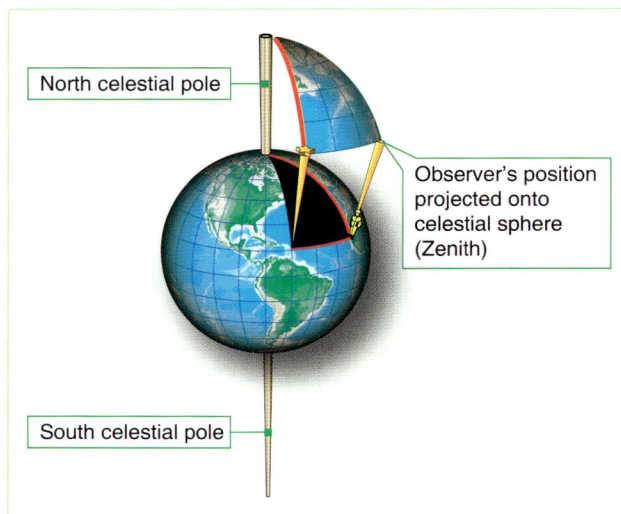

**Figure 8-1.** *Celestial points and subpoints on earth have the same relationship.*

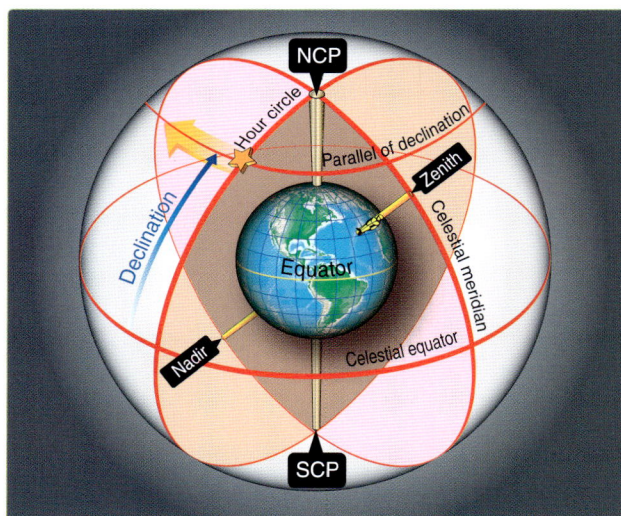

**Figure 8-2.** *Elements of the celestial sphere.*

Consistent with the celestial assumptions, the earth and the celestial meridians do not rotate. All bodies on the celestial sphere rotate 15° per hour past the celestial meridians. The moon moves at approximately 14.5° per hour.

## Motion of Celestial Bodies

All the celestial bodies have two types of motion: absolute and apparent. Apparent motion is important to navigators. Apparent motion is the motion of one celestial body as perceived by an observer on another moving celestial body. Since apparent motion is relative, it is essential to establish the reference point for that motion. For example, the apparent motion of Venus would be different if observed from the earth or the sun.

### Apparent Motion

The earth's rotation and revolution causes the apparent motion of the celestial bodies. Rotation causes celestial bodies to appear to rise in the east, climb to a maximum height, then set in the west. All bodies appear to move along a diurnal circle, approximately parallel to the plane of the equator.

The apparent effect of rotation varies with the observer's latitude. At the equator, the bodies appear to rise and set perpendicular to the horizon. Each body is above the horizon for approximately 12 hours each day. At the North and South Poles, a different phenomenon occurs. The same group of stars is continually above the horizon; they neither rise nor set, but move on a plane parallel to the equator. This characteristic explains the periods of extended daylight, twilight, and darkness at higher latitudes. The remainder of the earth is a combination of these two extremes; some bodies rise and set, while others continually remain above the horizon.

The greater the northerly declination (Dec) of a body, the higher it appears in the sky to an observer at the North Pole. Polaris, with a Dec of almost 90°, appears overhead. Bodies with southern Dec are not visible from the North Pole.

A circumpolar body appears to revolve about the pole and never set. If the angular distance of the body from the elevated pole is less than the observer's latitude, the body is circumpolar. For example, the Dec of Dubhe is 62° N. Therefore, it is located at an angle of 90°– 62° from the North Pole, or 28°. So, an observer located above 28° N views Dubhe as circumpolar. Although *Figure 8-3* uses the North Pole, the same characteristics can be observed from the South Pole.

If the earth stopped rotating, the effect of the earth's revolution on the apparent motion of celestial bodies would be obvious. The sun would appear to circle around the earth once each year. It would cover 360° in 365 days or move eastward at slightly less than 1 degree per day. The stars would move at the same rate. That is why different constellations are visible at different times of the year. Every evening, the same star appears to rise 4 minutes earlier.

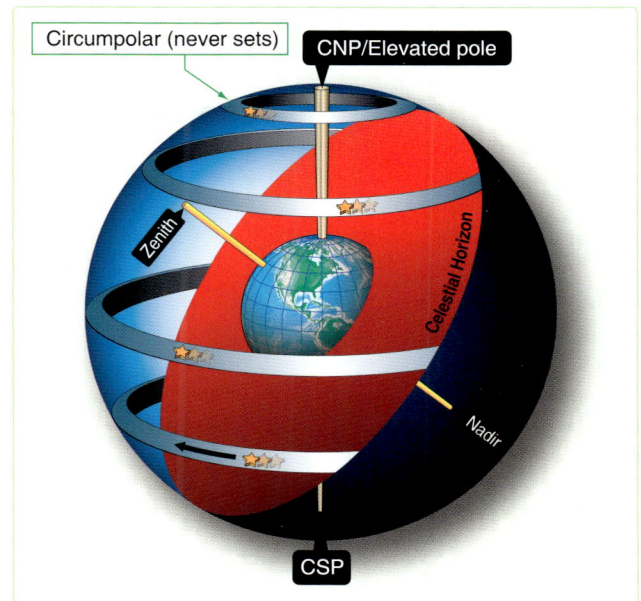

**Figure 8-3.** *Some bodies are circumpolar.*

After half a year, when the earth reached the opposite extreme of its orbit, its dark side would be turned in the opposite direction in space, facing a new field of stars. Hence, an observer at the equator would see an entirely different sky at midnight in June, than the one that appeared at midnight in December. In fact, the stars seen at midnight in June are those that were above the horizon at midday in December.

### Seasons

The annual variation of the sun's declination and the consequent change of the seasons are caused by the revolution of the earth. *[Figure 8-4]* If the celestial equator coincided with the ecliptic, the sun would always be overhead at the equator, and its Dec would always be zero. However, the earth's axis is inclined about 66.5° to the plane of the earth's orbit, and the plane of the equator is inclined about 23.5°. Throughout the year, the axis points in the same direction. That is, the axis of the earth in one part of the orbit is parallel to the axis of the earth in any other part of the orbit. *[Figure 8-5]*

In June, the North Pole is inclined toward the sun so that the sun is at a maximum distance from the plane of the equator. About June 22, at the solstice, the sun has its greatest northern Dec.

The solstice brings the long days of summer, while in the Southern Hemisphere, the days are shortest. This is the beginning of summer for the Northern Hemisphere and of winter for the Southern Hemisphere. Six months later, the axis is still pointing in the same direction; but, since the earth is at the opposite side of its orbit and the sun, the North Pole is inclined away from the Sun. At the winter solstice, about

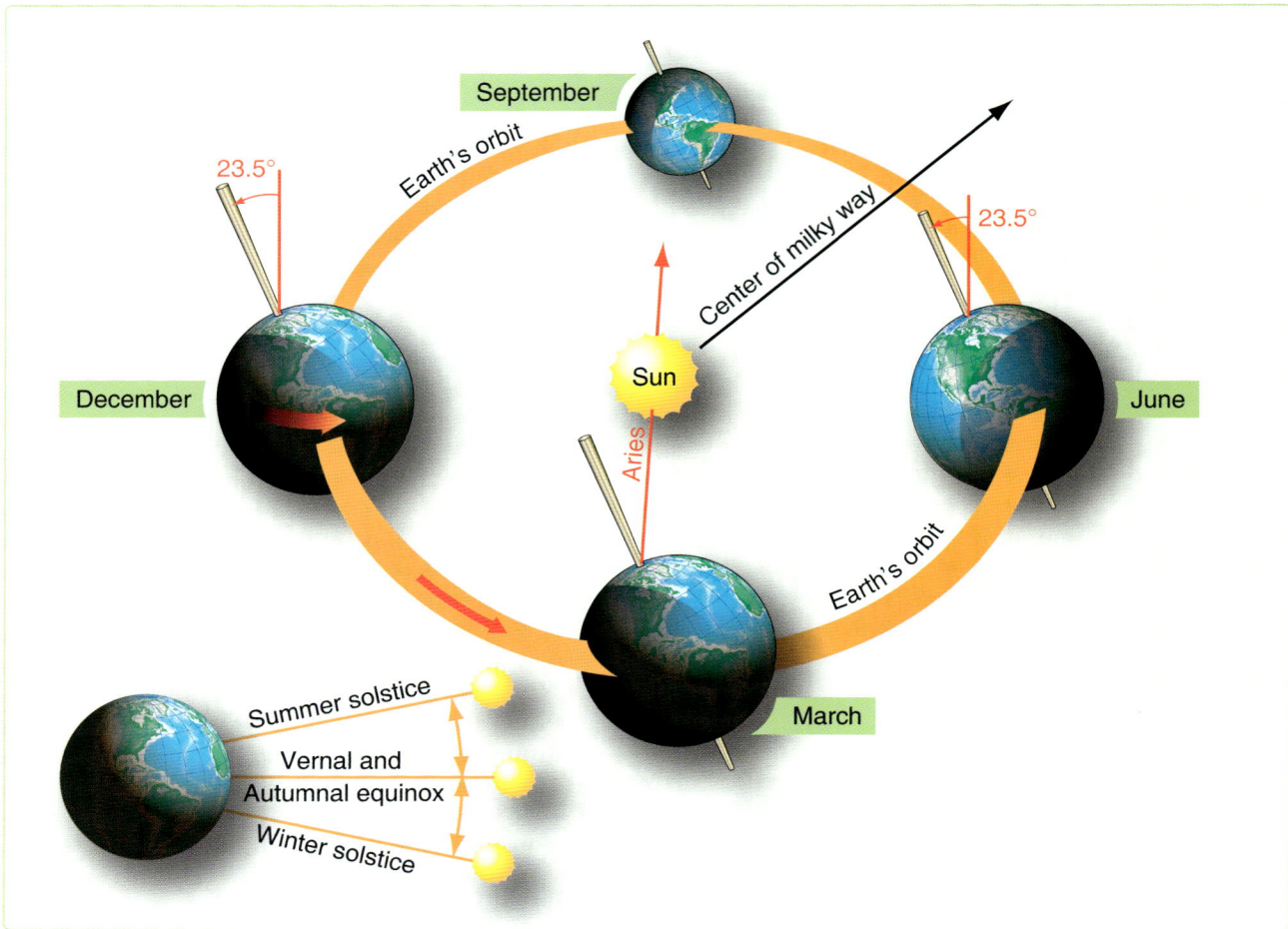

**Figure 8-4.** *Seasonal changes of earth's position.*

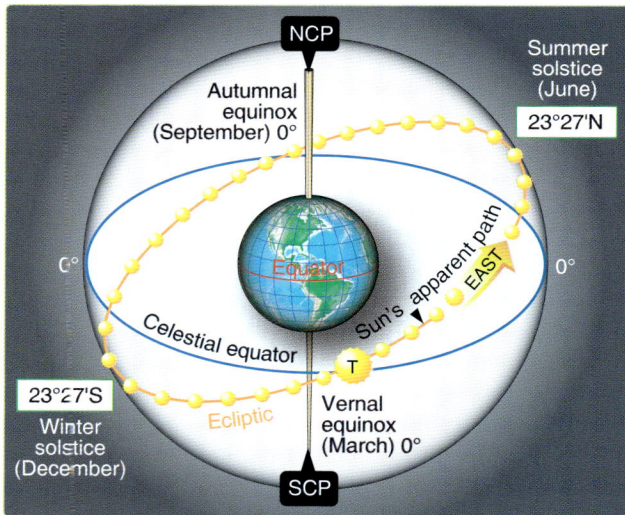

**Figure 3-5.** *Ecliptic with solstices and equinoxes.*

December 21, the sun has its greatest southern Dec. Days are shortest in the Northern Hemisphere, and winter is beginning.

Halfway between the two solstices, the axis of the earth is inclined neither toward nor away from the sun, and the sun is on the plane of the equator. These positions correspond to the beginning spring and fall.

## Celestial Coordinates

Celestial bodies and the observer's zenith may be positioned on the celestial sphere using a coordinate system similar to that of the earth. Terrestrial lines of latitude correspond to celestial parallels of Dec. Lines of longitude establish the celestial meridians.

The observer's celestial meridian is a great circle containing the zenith, the nadir, and the celestial poles. *[Figure 8-2]* A line extended from the observer's zenith, through the center of the earth, intersects the celestial sphere at the observer's nadir, the point on the celestial sphere directly beneath the observer's position. The poles divide the celestial meridians into upper and lower branches. The upper branch contains the observer's zenith. The lower branch contains the nadir.

A second great circle on the celestial sphere is the hour circle. The hour circle contains the celestial body and the celestial poles. Unlike celestial meridians, which remain stationary, hour circles rotate 15° per hour. Hour circles also contain upper and lower branches. The upper branch contains the body. Again, the moon's hour circle moves at a different rate. The subpoint is the point on the earth's surface directly beneath the celestial body.

You can locate any body on the celestial sphere relative to the celestial equator and the Greenwich meridian using Dec and Greenwich hour angle.

### Declination (Dec)

Dec is the angular distance a celestial body is north or south of the celestial equator measured along the hour circle. It ranges from 0° to 90° and corresponds to latitude.

### Greenwich Hour Angle (GHA)

GHA is the angular distance measured westward from the Greenwich celestial meridian to the upper branch of the hour circle. It has a range of 0° to 360°. The Air Almanac lists the GHA and the Dec of the sun, moon, four planets, and Aries. The subpoint's latitude matches its Dec, and its longitude correlates to its GHA, but not exactly. GHA is always measured westward from the Greenwich celestial meridian, and longitude is measured in the shortest direction from the Greenwich meridian to the observer's meridian.

The following are examples of converting a body's celestial coordinates to its subpoint's terrestrial coordinates. If the GHA is less than 180°, then the subpoint is in the Western Hemisphere and GHA equals longitude. When the GHA is greater than 180°, the subpoint is in the Eastern Hemisphere and longitude equals 360° GHA. Again, Dec and latitude are equal. [Figure 8-6]

You will use two other hour angles in celestial navigation in addition to GHA, local hour angle (LHA), and sidereal hour angle (SHA). [Figure 8-7] LHA is the angular distance from the observer's celestial meridian clockwise to the hour circle. LHA is computed by applying the local longitude to the GHA of the body. In the Western Hemisphere, LHA equals GHA – W Long, and in the Eastern Hemisphere, LHA equals GHA + E Long. [Figure 8-8] When the LHA is 0, the body's hour circle and the upper branch of the observer's celestial meridian are collocated, and the body is in transit. If the LHA is 180, the hour circle is coincident with the lower branch of the observer's celestial meridian. SHA is used with the first point of Aries. The first point of Aries is the point where the sun appears to cross the celestial equator from south to north on the vernal equinox or first day of spring. Though not absolutely stationary relative to the

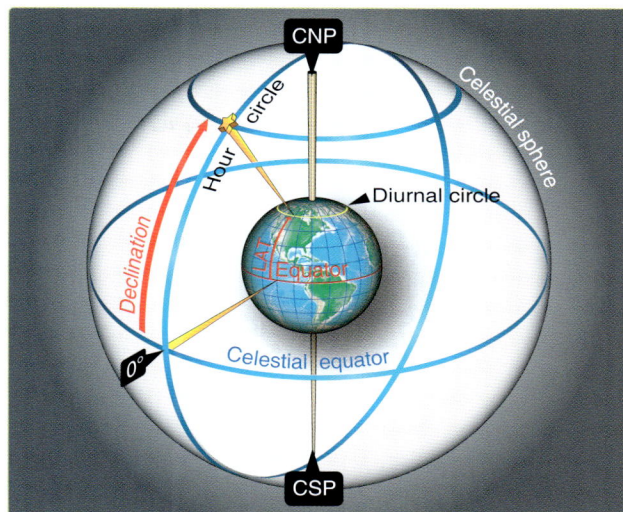

**Figure 8-6.** *Declination of a body corresponds to a parallel of latitude.*

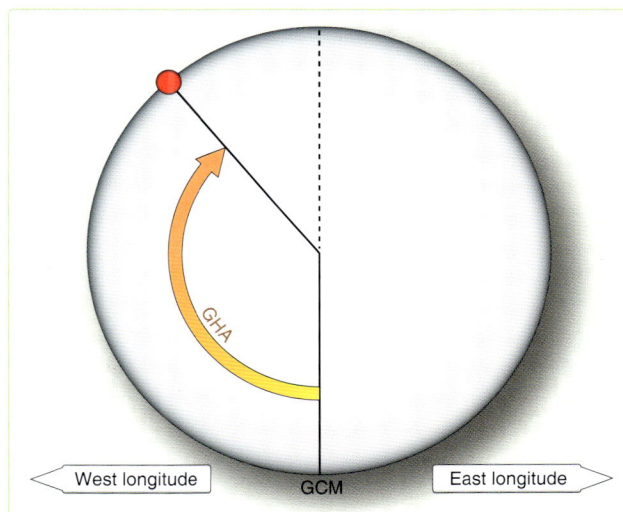

**Figure 8-7.** *Greenwich hour angle.*

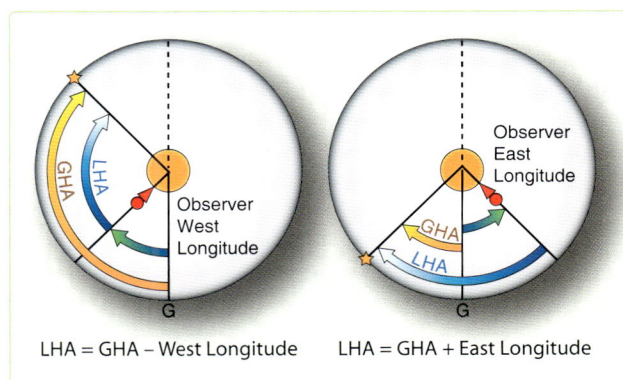

LHA = GHA – West Longitude        LHA = GHA + East Longitude

**Figure 8-8.** *Local hour angle.*

stars, Aries moves so slowly that we consider it fixed on the celestial equator for as long as a year. The SHA is the angular measurement from the hour circle of Aries to the star's hour circle. *[Figure 8-9]* Aries and the stars move together so the SHA remains constant for a year.

## Use of the Air Almanac

Although the Air Almanac contains astronomical amounts of data, most of it is devoted to tabulating the GHA of Aries and the GHA and Dec of the sun, moon, and the three navigational planets most favorably located for observation. Enter the daily pages with Greenwich date and GMT to extract the GHA and Dec of a celestial body.

### Finding GHA and Dec

The GHA is listed for 10-minute intervals on each daily sheet. If the observation time is listed, read the GHA and Dec directly under the proper column opposite the time.

For example, find the sun's GHA and Dec at GMT 0540 on 11 August 1995. *[Figure 8-10]* The GHA is 263°–41' and Dec is N 15°–24'. (Extractions of GHA and Dec are to the nearest whole minute.) To convert these values to the subpoint's geographical coordinates, latitude is North 15°–24'. When GHA is greater than 180°, subtract it from 360° to get east longitude. The subpoint's longitude in this example is (360°–00' minus 263°–41') East 96°–19'.

When you do not observe at a 10-minute interval, use the time immediately before the observation time. Then, use the Interpolation of GHA table on the inside front cover of the Air Almanac or the back of the star chart and add the increment to the GHA. *[Figure 8-11]*

For example, on 11 August 1995, you observe the sun at 1012 GMT. Enter *Figure 8-10* to find the GHA listed for 1010 (331°–1 1'). Since the observation was 2 minutes after

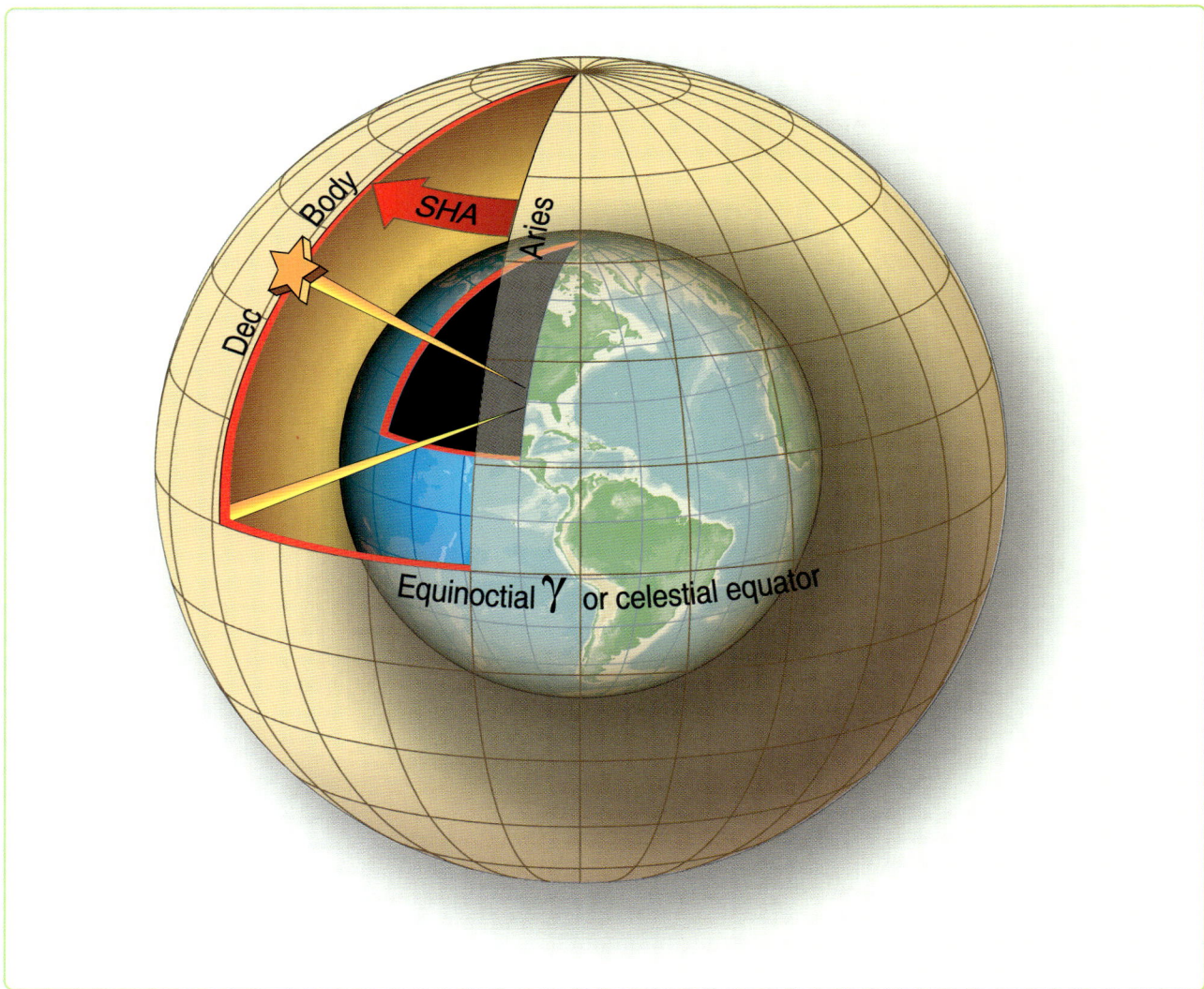

**Figure 8-9.** *Sidereal hour angle.*

### (DAY 223) Greenwich A.M. 1995 August 11 (Friday)

| UT (GMT) | SUN GHA | SUN Dec. | ARIES GHA | MARS 1.4 GHA | MARS Dec. | JUPITER 2.3 GHA | JUPITER Dec. | SATURN 0.9 GHA | SATURN Dec. | MOON GHA | MOON Dec. | Lat. N | Moon rise h m | Diff. m |
|---|---|---|---|---|---|---|---|---|---|---|---|---|---|---|
| **h m** | | | | | | | | | | | | | | |
| 00 00 | 178 40.2 | N15 28.3 | 318 59.5 | 127 27 | S 4 49 | 75 09 | S20 40 | 323 50 | S 4 35 | 356 59 | S10 01 | | | |
| 00 10 | 181 10.3 | 28.2 | 321 29.9 | 129 57 | | 77 40 | | 326 21 | | 359 24 | 9 59 | 72 | 20 18 | −04 |
| 00 20 | 183 40.3 | 28.1 | 324 00.3 | 132 27 | | 80 10 | | 328 51 | | 1 48 | 57 | 70 | 20 09 | −04 |
| 00 30 | 186 10.3 | 27.9 | 326 30.8 | 134 57 | | 82 40 | | 331 22 | | 4 13 | 56 | 68 | 20 02 | +02 |
| 00 40 | 188 40.3 | 27.8 | 329 01.2 | 137 28 | | 85 11 | | 333 52 | | 6 37 | 54 | 66 | 19 55 | 04 |
| 00 50 | 191 10.3 | 27.7 | 331 31.6 | 139 58 | | 87 41 | | 336 22 | | 9 02 | 52 | 64 | 19 50 | 06 |
| | | | | | | | | | | | | 62 | 19 45 | 08 |
| 01 00 | 193 40.3 | N15 27.6 | 334 02.0 | 142 28 | S 4 50 | 90 12 | S20 40 | 338 53 | S 4 35 | 11 26 | S 9 50 | 60 | 19 41 | 10 |
| 01 10 | 196 10.3 | 27.4 | 336 32.4 | 144 58 | | 92 42 | | 341 23 | | 13 51 | 49 | 58 | 19 37 | 11 |
| 01 20 | 198 40.4 | 27.3 | 339 02.8 | 147 28 | | 95 12 | | 343 54 | | 16 15 | 47 | 56 | 19 34 | 12 |
| 01 30 | 201 10.4 | 27.2 | 341 33.2 | 149 59 | | 97 43 | | 346 24 | | 18 40 | 45 | 54 | 19 31 | 13 |
| 01 40 | 203 40.4 | 27.1 | 344 03.6 | 152 29 | | 100 13 | | 348 55 | | 21 04 | 43 | 52 | 19 29 | 14 |
| 01 50 | 206 10.4 | 27.0 | 346 34.0 | 154 59 | | 102 44 | | 351 25 | | 23 29 | 42 | | | |
| 02 00 | 208 40.4 | N15 26.8 | 349 04.5 | 157 29 | S 4 50 | 105 14 | S20 40 | 353 55 | S 4 35 | 25 54 | S 9 40 | 50 | 19 26 | 15 |
| 02 10 | 211 10.4 | 26.7 | 351 34.9 | 159 59 | | 107 44 | | 356 26 | | 28 18 | 38 | 45 | 19 21 | 17 |
| 02 20 | 213 40.5 | 26.6 | 354 05.3 | 162 29 | | 110 15 | | 358 56 | | 30 43 | 37 | 40 | 19 16 | 18 |
| 02 30 | 216 10.5 | 26.5 | 356 35.7 | 165 00 | | 112 45 | | 1 27 | | 33 07 | 35 | 35 | 19 13 | 20 |
| 02 40 | 218 40.5 | 26.3 | 359 06.1 | 167 30 | | 115 16 | | 3 57 | | 35 32 | 33 | 30 | 19 09 | 21 |
| 02 50 | 221 10.5 | 26.2 | 1 36.5 | 170 00 | | 117 46 | | 6 28 | | 37 56 | 31 | | | |
| | | | | | | | | | | | | 20 | 19 03 | 23 |
| 03 00 | 223 40.5 | N15 26.1 | 4 06.9 | 172 30 | S 4 51 | 120 16 | S20 40 | 8 58 | S 4 35 | 40 21 | S 9 30 | 10 | 18 58 | 24 |
| 03 10 | 226 10.5 | 26.0 | 6 37.3 | 175 00 | | 122 47 | | 11 29 | | 42 46 | 28 | 0 | 18 53 | 26 |
| 03 20 | 228 40.6 | 25.9 | 9 07.7 | 177 30 | | 125 17 | | 13 59 | | 45 10 | 26 | 10 | 18 48 | 28 |
| 03 30 | 231 10.6 | 25.7 | 11 38.1 | 180 01 | | 127 48 | | 16 29 | | 47 35 | 24 | 20 | 18 43 | 30 |
| 03 40 | 233 40.6 | 25.6 | 14 08.6 | 182 31 | | 130 18 | | 19 00 | | 49 59 | 23 | | | |
| 03 50 | 236 10.6 | 25.5 | 16 39.0 | 185 01 | | 132 48 | | 21 30 | | 52 24 | 21 | 30 | 18 37 | 32 |
| | | | | | | | | | | | | 35 | 18 34 | 33 |
| 04 00 | 238 40.6 | N15 25.4 | 19 09.4 | 187 31 | S 4 51 | 135 19 | S20 40 | 24 01 | S 4 35 | 54 48 | S 9 19 | 40 | 18 30 | 34 |
| 04 10 | 241 10.6 | 25.2 | 21 39.8 | 190 01 | | 137 49 | | 26 31 | | 57 13 | 17 | 45 | 18 25 | 36 |
| 04 20 | 243 40.7 | 25.1 | 24 10.2 | 192 32 | | 140 20 | | 29 02 | | 59 38 | 16 | 50 | 18 20 | 38 |
| 04 30 | 246 10.7 | 25.0 | 26 40.6 | 195 02 | | 142 50 | | 31 32 | | 62 02 | 14 | | | |
| 04 40 | 248 40.7 | 24.9 | 29 11.0 | 197 32 | | 145 20 | | 34 02 | | 64 27 | 12 | 52 | 18 17 | 39 |
| 04 50 | 251 10.7 | 24.8 | 31 41.4 | 200 02 | | 147 51 | | 36 33 | | 66 51 | 10 | 54 | 18 15 | 40 |
| | | | | | | | | | | | | 56 | 18 11 | 41 |
| 05 00 | 253 40.7 | N15 24.6 | 34 11.8 | 202 32 | S 4 52 | 150 21 | S20 40 | 39 03 | S 4 35 | 69 16 | S 9 09 | 58 | 18 08 | 42 |
| 05 10 | 256 10.7 | 24.5 | 36 42.3 | 205 02 | | 152 51 | | 41 34 | | 71 41 | 07 | 60 | 18 04 | +44 |
| 05 20 | 258 40.8 | 24.4 | 39 12.7 | 207 33 | | 155 22 | | 44 04 | | 74 05 | 05 | S | | |
| 05 30 | 261 10.8 | 24.3 | 41 43.1 | 210 03 | | 157 52 | | 46 34 | | 76 30 | 03 | | | |
| 05 40 | 263 40.8 | 24.1 | 44 13.5 | 212 33 | | 160 23 | | 49 05 | | 78 54 | 02 | | | |
| 05 50 | 265 10.8 | 24.0 | 46 43.9 | 215 03 | | 162 53 | | 51 35 | | 81 19 | 9 00 | | | |
| 06 00 | 268 40.8 | N15 23.9 | 49 14.3 | 217 33 | S 4 53 | 165 23 | S20 40 | 54 06 | S 4 35 | 83 44 | S 8 56 | | | |
| 06 10 | 271 10.8 | 23.8 | 51 44.7 | 220 03 | | 167 54 | | 56 36 | | 86 08 | 56 | | | |
| 06 20 | 273 40.9 | 23.6 | 54 15.1 | 222 34 | | 170 24 | | 59 07 | | 88 33 | 55 | | | |
| 06 30 | 276 10.9 | 23.5 | 56 45.5 | 225 04 | | 172 55 | | 61 37 | | 90 57 | 53 | | | |
| 06 40 | 278 40.9 | 23.4 | 59 16.0 | 227 34 | | 175 25 | | 64 07 | | 93 22 | 51 | | | |
| 06 50 | 281 10.9 | 23.3 | 61 46.4 | 230 04 | | 177 55 | | 66 38 | | 95 47 | 49 | | | |
| 07 00 | 283 40.9 | N15 23.2 | 64 16.8 | 232 34 | S 4 53 | 180 26 | S20 40 | 69 08 | S 4 35 | 98 11 | S 8 48 | | | |
| 07 10 | 286 10.9 | 23.0 | 66 47.2 | 235 04 | | 182 56 | | 71 39 | | 100 36 | 46 | | | |
| 07 20 | 288 40.9 | 22.9 | 69 17.6 | 237 35 | | 185 27 | | 74 09 | | 103 00 | 44 | | | |
| 07 30 | 291 11.1 | 22.8 | 71 48.0 | 240 05 | | 187 57 | | 76 40 | | 105 25 | 42 | | | |
| 07 40 | 293 41.1 | 22.7 | 74 18.4 | 242 35 | | 190 27 | | 79 10 | | 107 50 | 41 | | | |
| 07 50 | 296 11.1 | 22.5 | 76 48.8 | 245 05 | | 192 58 | | 81 41 | | 110 14 | 39 | | | |
| 08 00 | 298 41.1 | N15 22.4 | 79 19.2 | 247 35 | S 4 54 | 195 28 | S20 40 | 84 11 | S 4 35 | 112 39 | S 8 37 | | | |
| 08 10 | 301 11.1 | 22.3 | 81 19.6 | 250 05 | | 195 59 | | 86 41 | | 115 04 | 35 | | | |
| 08 20 | 303 41.1 | 22.2 | 84 20.1 | 252 36 | | 200 29 | | 89 12 | | 117 28 | 33 | | | |
| 08 30 | 306 11.2 | 22.1 | 86 50.5 | 255 06 | | 202 59 | | 91 42 | | 119 53 | 32 | | | |
| 08 40 | 308 41.2 | 21.9 | 89 20.9 | 257 36 | | 205 30 | | 94 13 | | 122 17 | 30 | | | |
| 08 50 | 311 11.2 | 21.8 | 91 51.3 | 260 06 | | 208 00 | | 96 43 | | 124 42 | 28 | | | |
| **h m** | | | | | | | | | | | | | | |
| 09 00 | 313 41.1 | N15 21.7 | 94 21.7 | 262 36 | S 4 55 | 210 31 | S20 40 | 99 14 | S 4 35 | 127 07 | S 8 26 | | | |
| 09 10 | 316 11.1 | 21.6 | 96 52.1 | 265 07 | | 213 01 | | 101 44 | | 129 31 | 25 | | | |
| 09 20 | 318 41.1 | 21.4 | 99 22.5 | 267 37 | | 215 31 | | 104 14 | | 131 56 | 23 | | | |
| 09 30 | 321 11.2 | 21.3 | 101 52.9 | 270 07 | | 218 02 | | 106 45 | | 134 21 | 21 | | | |
| 09 40 | 323 41.2 | 21.2 | 104 23.3 | 272 37 | | 220 32 | | 109 15 | | 136 45 | 19 | | | |
| 09 50 | 326 11.2 | 21.1 | 106 53.8 | 275 07 | | 223 03 | | 111 46 | | 139 10 | 17 | | | |
| 10 00 | 328 41.2 | N15 20.9 | 109 24.2 | 277 37 | S 4 55 | 225 33 | S20 41 | 114 16 | S 4 35 | 141 35 | S 8 16 | | | |
| 10 10 | 331 11.2 | 20.8 | 111 54.6 | 280 08 | | 228 03 | | 116 47 | | 143 59 | 14 | | | |
| 10 20 | 333 41.2 | 20.7 | 114 25.0 | 282 38 | | 230 34 | | 119 17 | | 146 24 | 12 | | | |
| 10 30 | 336 11.3 | 20.6 | 116 55.4 | 285 08 | | 233 04 | | 121 47 | | 148 49 | 10 | | | |
| 10 40 | 338 41.3 | 20.5 | 119 25.8 | 287 38 | | 235 35 | | 124 18 | | 151 13 | 09 | | | |
| 10 50 | 341 11.3 | 20.3 | 121 56.2 | 290 08 | | 238 05 | | 126 48 | | 153 38 | 07 | | | |
| 11 00 | 343 41.3 | N15 20.2 | 124 26.6 | 292 38 | S 4 56 | 240 35 | S20 40 | 129 19 | S 4 35 | 156 03 | S 8 05 | | | |
| 11 10 | 346 11.3 | 20.1 | 126 57.0 | 295 09 | | 243 06 | | 131 49 | | 158 27 | 03 | | | |
| 11 20 | 348 41.3 | 20.0 | 129 27.4 | 297 39 | | 245 36 | | 134 20 | | 160 52 | 01 | | | |
| 11 30 | 351 11.4 | 19.8 | 131 57.9 | 300 09 | | 248 07 | | 136 50 | | 163 16 | 8 00 | | | |
| 11 40 | 353 41.4 | 19.7 | 134 28.3 | 302 39 | | 250 37 | | 139 20 | | 165 41 | 7 58 | | | |
| 11 50 | 356 11.4 | 19.6 | 136 58.7 | 305 09 | | 253 07 | | 141 51 | | 168 06 | 56 | | | |
| Rate | 15 00.1 | S0 00.7 | | 15 01.0 | S0 00.6 | 15 02.4 | 0 00.0 | 15 02.6 | S0 00.1 | 14 27.6 | N0 10.5 | | | |

**Moon's P. in A.**

| Alt ° | Corr +, | Alt ° | Corr +, |
|---|---|---|---|
| 0 | 60 | 53 | 35 |
| 3 | 59 | 54 | 34 |
| 11 | 58 | 55 | 33 |
| 15 | 57 | 56 | 32 |
| 21 | 56 | 58 | 31 |
| 23 | 55 | 59 | 30 |
| 26 | 54 | 60 | 29 |
| 28 | 53 | 61 | 28 |
| | | 62 | |
| 30 | 52 | 63 | 27 |
| 32 | 51 | 64 | 26 |
| 33 | 50 | 65 | 25 |
| 35 | 49 | 66 | 24 |
| | 48 | | 23 |

| Alt ° | Corr +, | Alt ° | Corr +, |
|---|---|---|---|
| 37 | 47 | 67 | 22 |
| 38 | 46 | 68 | 21 |
| 40 | 45 | 69 | 20 |
| 41 | 44 | 70 | 19 |
| 43 | 43 | 71 | 18 |
| 44 | 42 | 72 | 17 |
| 45 | 41 | 73 | 16 |
| 47 | 40 | 74 | 15 |
| 48 | 39 | 75 | 14 |
| 49 | 38 | 76 | 13 |
| 51 | 37 | 77 | 12 |
| 52 | 36 | 78 | 11 |
| 53 | 35 | 79 | 10 |
| 54 | | 80 | |

Sun SD 15'8
Moon SD 16'
Age 15d

**Figure 8-10.** *Daily page from Air Almanac—11 August 1995.*

**Interpolation of G.H.A.**
Increment to be added for intervals of G.M.T. to G.H.A. of:
Sun, Aries (♈) and planets; Moon.

Block 1

| Sun, etc. (m s) | Incr. | Moon (m s) |
|---|---|---|
| 00 00 | 0 00 | 00 00 |
| 01 | 0 01 | 00 02 |
| 05 | 0 02 | 00 06 |
| 09 | 0 03 | 00 10 |
| 13 | 0 04 | 00 14 |
| 17 | 0 05 | 00 18 |
| 21 | 0 06 | 00 22 |
| 25 | 0 07 | 00 26 |
| 29 | 0 08 | 00 31 |
| 33 | 0 09 | 00 35 |
| 37 | 0 10 | 00 39 |
| 41 | 0 11 | 00 43 |
| 45 | 0 12 | 00 47 |
| 49 | 0 13 | 00 51 |
| 53 | 0 14 | 00 55 |
| 00 57 | 0 15 | 01 00 |
| 01 | 0 16 | 01 04 |
| 05 | 0 17 | 01 08 |
| 09 | 0 18 | 01 12 |
| 13 | 0 19 | 01 16 |
| 17 | 0 20 | 01 20 |
| 21 | 0 21 | 01 24 |
| 25 | 0 22 | 01 29 |
| 29 | 0 23 | 01 33 |
| 33 | 0 24 | 01 37 |
| 37 | 0 25 | 01 41 |
| 41 | 0 26 | 01 45 |
| 45 | 0 27 | 01 49 |
| 49 | 0 28 | 01 53 |
| 53 | 0 29 | 01 58 |
| 01 57 | 0 30 | 02 02 |
| 02 01 | 0 31 | 02 06 |
| 05 | 0 32 | 02 10 |
| 09 | 0 33 | 02 14 |
| 13 | 0 34 | 02 18 |
| 17 | 0 35 | 02 22 |
| 21 | 0 36 | 02 27 |
| 25 | 0 37 | 02 31 |
| 29 | 0 38 | 02 35 |
| 33 | 0 39 | 02 39 |
| 37 | 0 40 | 02 43 |
| 41 | 0 41 | 02 47 |
| 45 | 0 42 | 02 51 |
| 49 | 0 43 | 02 56 |
| 53 | 0 44 | 03 00 |
| 02 57 | 0 45 | 03 04 |
| 03 01 | 0 46 | 03 08 |
| 05 | 0 47 | 03 12 |
| 09 | 0 48 | 03 16 |
| 13 | 0 49 | 03 20 |
| 17 | 0 50 | 03 25 |
| 03 21 | | 03 29 |

Block 2

| Sun, etc. (m s) | Incr. | Moon (m s) |
|---|---|---|
| 03 17 | 0 50 | 03 25 |
| 21 | 0 51 | 03 29 |
| 25 | 0 52 | 03 33 |
| 29 | 0 53 | 03 37 |
| 33 | 0 54 | 03 41 |
| 37 | 0 55 | 03 45 |
| 41 | 0 56 | 03 49 |
| 45 | 0 57 | 03 54 |
| 49 | 0 58 | 03 58 |
| 53 | 0 59 | 04 02 |
| 03 57 | 1 00 | 04 06 |
| 04 01 | 1 01 | 04 10 |
| 05 | 1 02 | 04 14 |
| 09 | 1 03 | 04 19 |
| 13 | 1 04 | 04 23 |
| 17 | 1 05 | 04 27 |
| 21 | 1 06 | 04 31 |
| 25 | 1 07 | 04 35 |
| 29 | 1 08 | 04 39 |
| 33 | 1 09 | 04 43 |
| 37 | 1 10 | 04 48 |
| 41 | 1 11 | 04 52 |
| 45 | 1 12 | 04 56 |
| 49 | 1 13 | 05 00 |
| 53 | 1 14 | 05 04 |
| 04 57 | 1 15 | 05 08 |
| 05 01 | 1 16 | 05 12 |
| 05 | 1 17 | 05 17 |
| 09 | 1 18 | 05 21 |
| 13 | 1 19 | 05 25 |
| 17 | 1 20 | 05 29 |
| 21 | 1 21 | 05 33 |
| 25 | 1 22 | 05 37 |
| 29 | 1 23 | 05 41 |
| 33 | 1 24 | 05 46 |
| 37 | 1 25 | 05 50 |
| 41 | 1 26 | 05 54 |
| 45 | 1 27 | 05 58 |
| 49 | 1 28 | 06 02 |
| 53 | 1 29 | 06 06 |
| 05 57 | 1 30 | 06 10 |
| 06 01 | 1 31 | 06 15 |
| 05 | 1 32 | 06 19 |
| 09 | 1 33 | 06 23 |
| 13 | 1 34 | 06 27 |
| 17 | 1 35 | 06 31 |
| 21 | 1 36 | 06 35 |
| 25 | 1 37 | 06 39 |
| 29 | 1 38 | 06 44 |
| 33 | 1 39 | 06 48 |
| 37 | 1 40 | 06 52 |
| 06 41 | | 06 56 |

Block 3

| Sun, etc. (m s) | Incr. | Moon (m s) |
|---|---|---|
| 06 37 | 1 40 | 06 52 |
| 41 | 1 41 | 06 56 |
| 45 | 1 42 | 07 00 |
| 49 | 1 43 | 07 04 |
| 53 | 1 44 | 07 08 |
| 06 57 | 1 45 | 07 13 |
| 07 01 | 1 46 | 07 17 |
| 05 | 1 47 | 07 21 |
| 09 | 1 48 | 07 25 |
| 13 | 1 49 | 07 29 |
| 17 | 1 50 | 07 33 |
| 21 | 1 51 | 07 37 |
| 25 | 1 52 | 07 42 |
| 29 | 1 53 | 07 46 |
| 33 | 1 54 | 07 50 |
| 37 | 1 55 | 07 54 |
| 41 | 1 56 | 07 58 |
| 45 | 1 57 | 08 02 |
| 49 | 1 58 | 08 06 |
| 53 | 1 59 | 08 11 |
| 07 57 | 2 00 | 08 15 |
| 08 01 | 2 01 | 08 19 |
| 05 | 2 02 | 08 23 |
| 09 | 2 03 | 08 27 |
| 13 | 2 04 | 08 31 |
| 17 | 2 05 | 08 35 |
| 21 | 2 06 | 08 40 |
| 25 | 2 07 | 08 44 |
| 29 | 2 08 | 08 48 |
| 33 | 2 09 | 08 52 |
| 37 | 2 10 | 08 56 |
| 41 | 2 11 | 09 00 |
| 45 | 2 12 | 09 04 |
| 49 | 2 13 | 09 09 |
| 53 | 2 14 | 09 13 |
| 08 57 | 2 15 | 09 17 |
| 09 01 | 2 16 | 09 21 |
| 05 | 2 17 | 09 25 |
| 09 | 2 18 | 09 29 |
| 13 | 2 19 | 09 33 |
| 17 | 2 20 | 09 38 |
| 21 | 2 21 | 09 42 |
| 25 | 2 22 | 09 46 |
| 29 | 2 23 | 09 50 |
| 33 | 2 24 | 09 54 |
| 37 | 2 25 | 09 58 |
| 41 | 2 26 | 10 00 |
| 45 | 2 27 | |
| 49 | 2 28 | |
| 53 | 2 29 | |
| 09 57 | 2 30 | |
| 10 00 | | |

**Figure 8-11.** *Interpolation of Greenwich hour angle, Air Almanac.*

the listed time, enter the Interpolation of GHA table and find the correction listed for 2 minutes of time (30'). *[Figure 8-11]* Add this correction to the listed GHA to determine the sun's exact GHA at 1012 (331°–41'). The Dec for the same time is N 15°–21'. Thus, at the time of the observation, the subpoint of the sun is at latitude 15°–21' N, longitude (360°–00' minus 331°–41') 028°–19' E.

For example, at 0124 GMT on 11 August 1995, you observe Altair. To find the GHA and Dec, look at the extracts from the tables in *Figures 8-11* and *8-12*.

You can find the GHA and Dec of a planet in almost the same way as the sun. Because the planet's Dec change slowly, they are recorded only at hourly intervals. Use the Dec listed for the entire hour. For example, to find the GHA and Dec of Jupiter at 1109 GMT, 11 August 1995, enter the correct daily page for the time of 1100 GMT. *[Figure 8-10]* The GHA is 240°–35' and the Dec is S20°–40'. Enter the Interpolation of GHA table under sun, etc., to get the adjustment for 9 minutes of time, 2°–15'. *[Figure 8-11]* Therefore, GHA is 242°–50'. Jupiter's subpoint is at latitude 20°–40' S, longitude (360°–00' minus 242° –50') 1 17°–10' E.

If you need to find an accurate GHA and Dec without the Air Almanac, you can find the procedures and applicable tables in Publication No. 249, Volume 1 for Aries or Volume 2 or 3 for the sun.

## Finding GHA and Dec of Moon

The moon moves across the sky at a different rate than other celestial bodies. In the Interpolation of GHA table, the intervals for the moon are listed in the right column where the values for the sun, Aries, and the planets are in the left column.

The interpolation of GHA table is a critical table and the increment is opposite the interval in which the difference of GMT occurs. If the difference (for example, 06'-3 1" for the moon) is an exact tabular value, take the upper, or right, of the two possible increments (that is, 1°–34'). The up, or right, rule applies to all critical tables.

For example, at 1136 GMT on 11 August 1995, you observe the moon. The following information is from the Air Almanac *[Figures 8-10 and 8-11]*:

GHA of moon at 1130 GMT 163° 16'

GHA correction for 6 minutes 1° 27'

GHA 164° 43' Dec S8° 00'

Thus, at 1136Z, the moon's subpoint is located at S 8°–00', longitude 164°–43'W.

## Finding GHA and Dec of a Star

The stars and the first point of Aries remain fixed in their relative positions in space, so the gas of the stars and Aries change at the same rate. Rather than list the GHA and Dec of every star throughout the day, the Air Almanac lists the GHA of Aries at 10-minute intervals and gives the sidereal hour angle (SHA) of the stars. The GHA of a star for any time can be found by adding the GHA of Aries and the SHA of the star. The GHA of a star is used to precomp any star that falls within 29° (Dec) of the equator using Volume 2 or Volume 3.

| STARS, SEPT.-DEC 1981 | | | | |
|---|---|---|---|---|
| No. | Name | Mag. | S.H.A. | Dec. |
| 7* | Acamar | 3.1 | 315 45 | S.40 26 |
| 5* | Achernar | 0.6 | 335 52 | S.57 24 |
| 30* | Acrux | 1.1 | 173 50 | S.62 55 |
| 19 | Adhara † | 1.6 | 255 40 | S.28 55 |
| 10* | Aldebaran † | 1.1 | 291 30 | N.16 27 |
| 32* | Alioth | 1.7 | 166 52 | N.56 08 |
| 34* | Alkaid | 1.9 | 153 27 | N.49 29 |
| 55 | Al Na'ir | 2.2 | 28 28 | S.47 08 |
| 15 | Alnilam † | 1.8 | 276 22 | S. 1 13 |
| 25* | Alphard † | 2.2 | 218 31 | S. 8 31 |
| 41* | Alphecca † | 2.3 | 126 41 | N.26 50 |
| 1* | Alpheralz | 2.2 | 358 20 | N.28 55 |
| 51* | Altair † | 0.9 | 62 43 | N. 8 47 |
| 2* | Ankaa | 2.4 | 353 50 | S.42 29 |
| 42* | Anlares † | 1.2 | 113 10 | S.26 22 |
| 3* | Schedar | 2.5 | 350 21 | N.56 21 |
| 45* | Shaula | 1.7 | 97 10 | S.37 05 |
| 18* | Sirius † | 1.6 | 259 05 | S.16 40 |
| 33* | Spica † | 1.2 | 159 09 | S.10 59 |
| 23* | Suhail | 2.2 | 223 19 | S.43 18 |
| 49* | Vega | 0.1 | 81 03 | N.38 45 |
| 39* | Zuben'ubi † | 2.9 | 137 45 | S.15 54 |

**INTER POLATION OF G.H.A.**
Increment to be added for intervals of G.M.T. to G.H.A. of:
Sun, Aries (♈) and planets ; Moon

| Sun, etc. m s | | Moon m s | Sun, etc. m s | | Moon m s | Sun, etc. m s | | Moon m s |
|---|---|---|---|---|---|---|---|---|
| 00 00 | | 00 00 | 03 17 | | 03 25 | 06 37 | | 06 52 |
| 01 | 0 00 | 00 02 | 21 | 0 50 | 03 29 | 41 | 1 40 | 06 56 |
| 05 | 0 01 | 00 06 | 25 | 0 51 | 03 33 | 45 | 1 41 | 07 00 |
| 09 | 0 02 | 00 10 | 29 | 0 52 | 03 37 | 49 | 1 42 | 07 04 |
| 13 | 0 03 | 00 14 | 33 | 0 53 | 03 47 | 53 | 1 43 | 07 08 |
| 17 | 0 04 | 00 18 | 37 | 0 54 | 03 45 | 06 57 | 1 44 | 07 13 |
| 21 | 0 05 | 00 22 | 41 | 0 55 | 03 49 | 07 01 | 1 45 | 07 17 |
| 25 | 0 06 | 00 26 | 45 | 0 56 | 03 54 | 05 | 1 46 | 07 21 |
| 29 | 0 07 | 00 31 | 49 | 0 57 | 03 58 | 09 | 1 47 | 07 25 |
| 33 | 0 08 | 00 35 | 53 | 0 58 | 04 02 | 13 | 1 48 | 07 29 |
| 37 | 0 09 | 00 39 | 03 57 | 0 59 | 04 06 | 17 | 1 49 | 07 33 |
| 41 | 0 10 | 00 43 | 04 01 | 1 00 | 04 10 | 21 | 1 50 | 07 37 |
| 45 | 0 11 | 00 47 | 05 | 1 21 | 04 14 | 25 | 1 51 | 07 42 |
| 05 | 0 46 | 03 12 | 25 | 1 36 | 06 39 | 45 | 2 26 | |
| 09 | 0 47 | 03 16 | 29 | 1 37 | 06 44 | 49 | 2 27 | |
| 13 | 0 48 | 03 20 | 33 | 1 38 | 06 48 | 53 | 2 28 | |
| 17 | 0 49 | 03 25 | 37 | 1 39 | 06 52 | 09 57 | 2 29 | |
| 03 21 | 0 50 | 03 29 | 06 41 | 1 40 | 06 56 | 10 00 | 2 30 | |

\* Stars used in H.O. 249 (A.P. 3270) Vol. 1
† Stars that may be used with Vols. 2 and 3.

**Figure 8-12.** *Sidereal hour angle obtained from table.*

The table, STARS, is inside the front cover of the almanac and on the back of the star chart. This table lists navigational stars and the following information for each star: the number corresponding to the sky diagram in the back, the name, the magnitude or relative brightness, the SHA, the Dec, whether used in Publication No. 249, and stars that can be used with Dec tables. NOTE: If you need a higher degree of accuracy, the SHA and Dec of the stars are listed to tenths of a degree in the Air Almanac's appendix. Thus, the subpoint of Altair is 08°–52' N 042°–24' W.

All the celestial concepts and assumptions you have learned may help you obtain a cestial LOP. A celestial LOP is simply a circle plotted with the center at the subpoint and a radius equal to the distance from the observer to the subpoint. To accurately compute this distance and the direction to the subpoint of the body, you must initially position the subpoint and then measure the angular displacement of the body above the horizon. GHA and Dec position the body, and the sextant measures the height above the horizon. A basic knowledge of celestial theory and LOPs will help you appreciate celestial navigation and detect errors. The next section explains how angular displacement is measured.

## Celestial Horizon

You use a sextant to measure a body's angular displacement above the horizon. The celestial horizon is a plane passing through the earth's center perpendicular to the zenith-nadir axis. The visual horizon approximates this plane at the earth's surface. *Figure 8-13* depicts the zenith-nadir axis and the celestial horizon. The angular displacement you see through

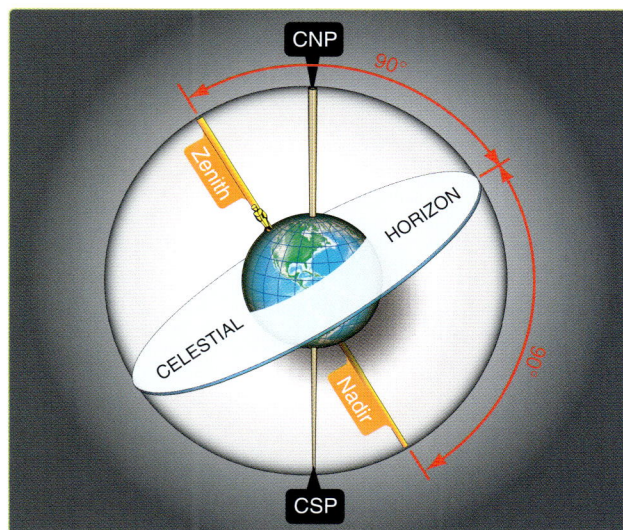

**Figure 8-13.** *Celestial horizon is 90° from observer Zenith and Nadir.*

a sextant is the height observed (Ho). Ho is measured along the vertical circle above the horizon. The vertical circle is a great circle containing the zenith, nadir, and celestial body. The body's altitude is the same whether measured at the earth's surface from an artificial horizon or at the center of the earth from the celestial horizon, because these horizons are parallel and the light rays from the body are essentially parallel. *Figure 8-14* shows that the infinite celestial sphere makes the difference in angle for light rays arriving at different points on the earth infinitesimal.

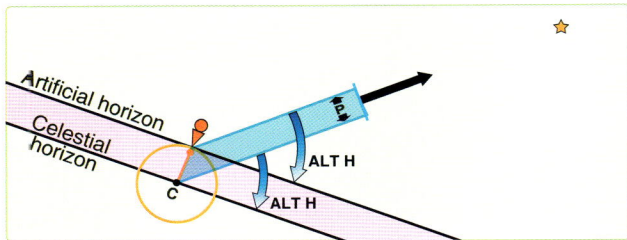

**Figure 8-14.** *Parallel lines make equal angles with parallel planes.*

The angle between light rays is called parallax. In *Figure 8-15,* parallax is shown at its maximum; that is, when the observer and the subpoint are separated by 90°. Since the earth's radius is tiny compared to the infinite distance to the stars, the angle p is very small. For the sun, angle p is a negligible 9 seconds of arc or 0.15 nautical miles (NM). Observed altitudes from either the artificial or celestial horizon are practically the same.

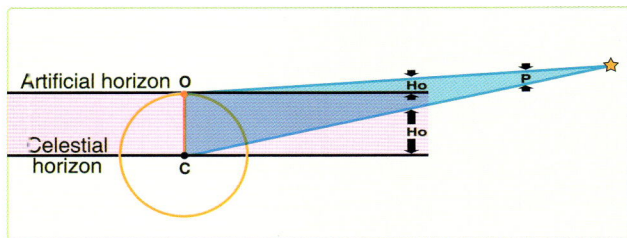

**Figure 8-15.** *Parallax.*

The bubble in a sextant or artificial horizon is most used by navigators. As in a carpenter's level, a bubble indicates the apparent vertical and horizontal. With the bubble, the navigator can level the sextant and establish an artificial horizon parallel to the plane of the celestial horizon. *Figure 8-16* shows that the plane of the artificial (bubble) horizon and the plane of the celestial horizon are parallel and separated by the earth's radius. Compared to the vast distances of space, the radius of the earth is inconsequential. Thus, the artificial horizon and the celestial horizon are nearly identical.

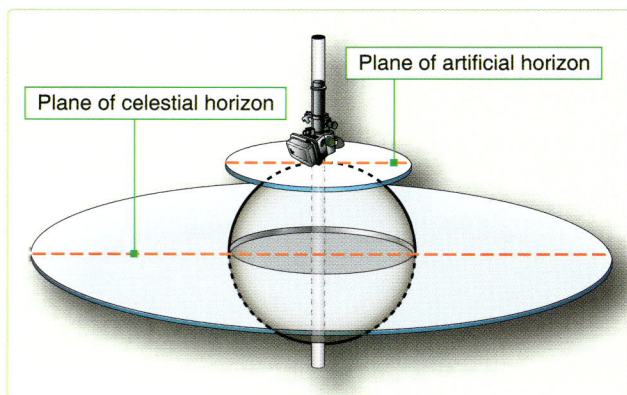

**Figure 8-16.** *The two planes are parallel.*

## Observed Altitude

The distance of the observer from the subpoint and a body's Ho are related. *[Figure 8-17]* When the body is directly overhead, the Ho is 90°, and the subpoint and the observer's position are collocated. When the Ho is 0°, the body is on the horizon and the subpoint is 90° (5,400 NM) from the observer's position. (See *Figure 8-18,* where C is the center of the earth, AB is the observer's horizon, and S is the subpoint of the body.) Since the sum of the angles in a triangle equals 180°, the angle OX is equal to 180° − (Ho + P). The sum of the angles on a straight line equals 180°, so angle OXC is equal to Ho + P. The horizon AB being tangent to the earth at O is perpendicular to OC, a radius of the earth. Thus, angle OCX equals 90° (Ho + P). The preceding discussion showed that

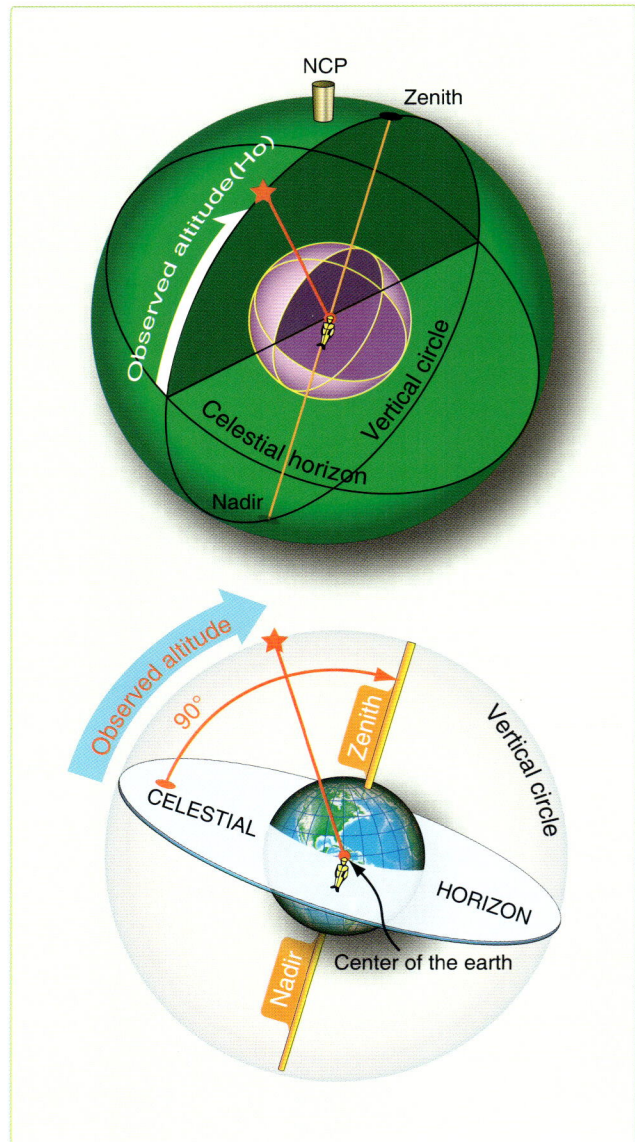

**Figure 8-17.** *Measure altitude from celestial horizon along vertical circle.*

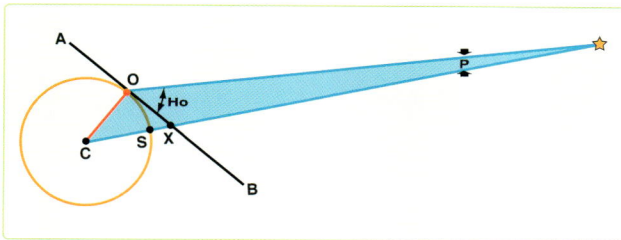

**Figure 8-18.** *Finding observed altitude.*

angle P is negligible, so this angle becomes 90° – Ho. The arc on the surface subtended by the angle OCX at the center of the earth is arc OS. This arc then is equal to 90° – Ho.

The distance from the subpoint to the observer is the zenith distance or co-alt and is computed using the astronomical triangle described in Chapter 9. Basically, the zenith distance equals 90° minus the Ho. *[Figure 8-19]* The figures are then converted to NM by multiplying the number of degrees by 60 and adding in the odd minutes of arc. Zenith distance is the radius of the circle which becomes the celestial LOP. This circle is called the circle of equal altitude *[Figure 8-20]*, as anyone located on it views an identical Ho. Now that you can determine the distance to the subpoint, you must next find the direction.

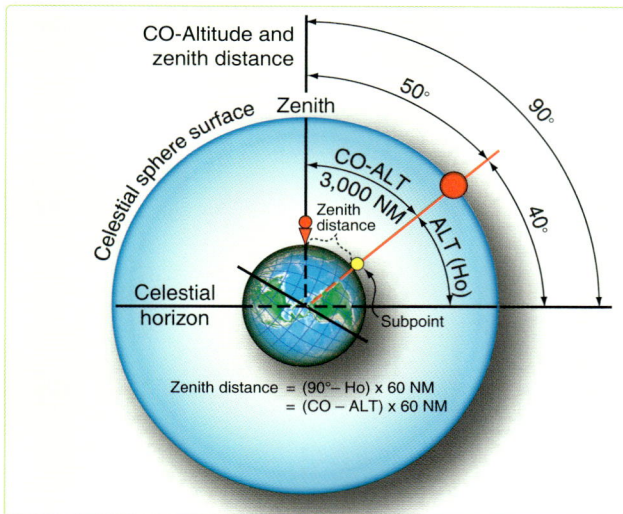

**Figure 8-20.** *Constructing a circle of equal altitude.*

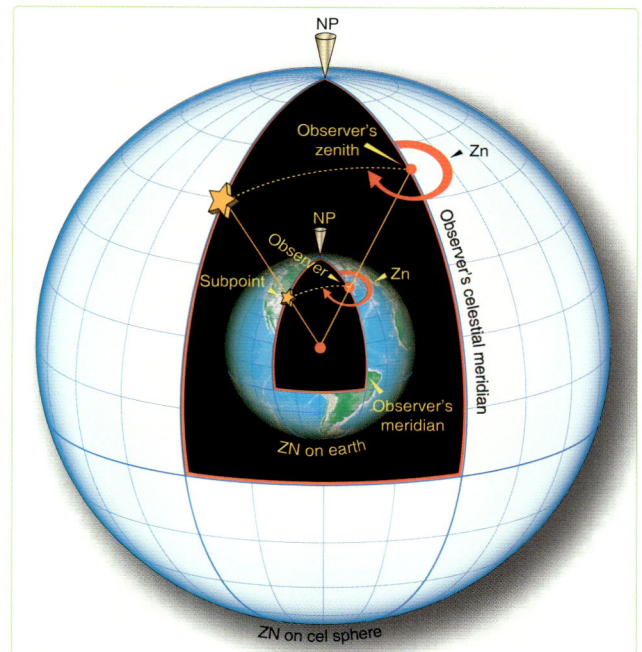

**Figure 8-19.** *Co-altitude and Zenith distance.*

## True Azimuth (Zn)

The direction to a body from an observer is called Zn. A celestial body's Zn is the true bearing (TB) to its subpoint. The Zn is the angle measured at the observer's position from true north (TN) clockwise through 360° to the great circle arc joining the observer's position with subpoint. *[Figure 8-21]* If you could measure the Zn when you measure its altitude, you could have a fix. Unfortunately, there is no instrument in the aircraft that measures Zn accurately enough. Except in the case of a very high body (85–90°), if you observe a

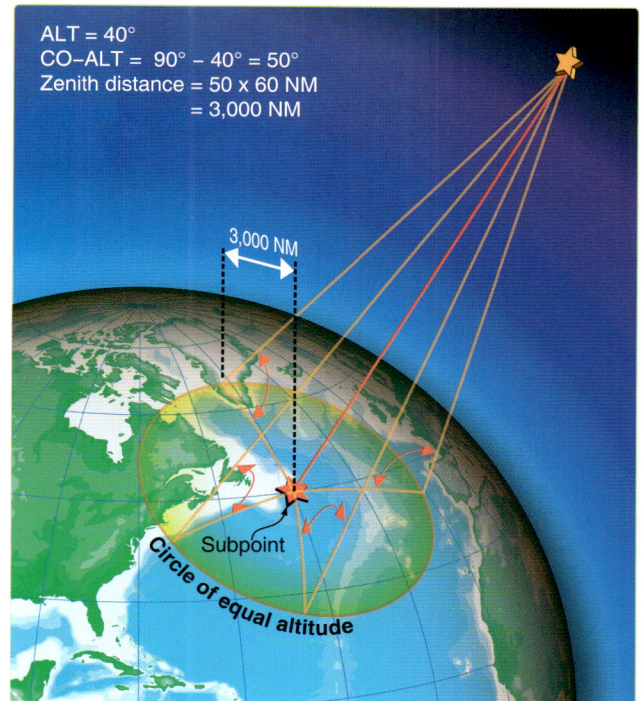

**Figure 8-21.** *Relationship of true azimuth to an observer.*

body with a Ho of 40° and you mismeasure the Zn by 1°, the fix will be 50 NM off.

## Celestial Fix

Since you cannot normally fix off a single body, you usually need to cross two or more LOPs. The fix position is the intersection of the LOPs. A celestial LOP is a circle as shown in *Figure 8-22*. When two celestial LOPs are plotted, they intersect at two points, only one of which can be your position. In practice, these two intersections usually are so far apart that DR removes all doubt as to which is correct.

## Chapter Summary

This chapter introduced celestial concepts and how they relate to celestial navigation. The motion of celestial bodies and celestial coordinates are discussed, as well as the use of the Air Almanac. The celestial horizon is explained and detail is given on how to determine observed altitude, true azimuth, and how to find a celestial fix.

# Chapter 9
# Computing Altitude and True Azimuth

## Introduction

This chapter discusses the procedures and some of the tables used to compute a celestial line of position (LOP). Some of the tables used to resolve the LOP, including the Air Almanac, were previously mentioned. First, we will discuss the astronomical triangle upon which the tables are based. Then, we will cover how to determine the local hour angle (LHA) of Aries and the LHA of a star.

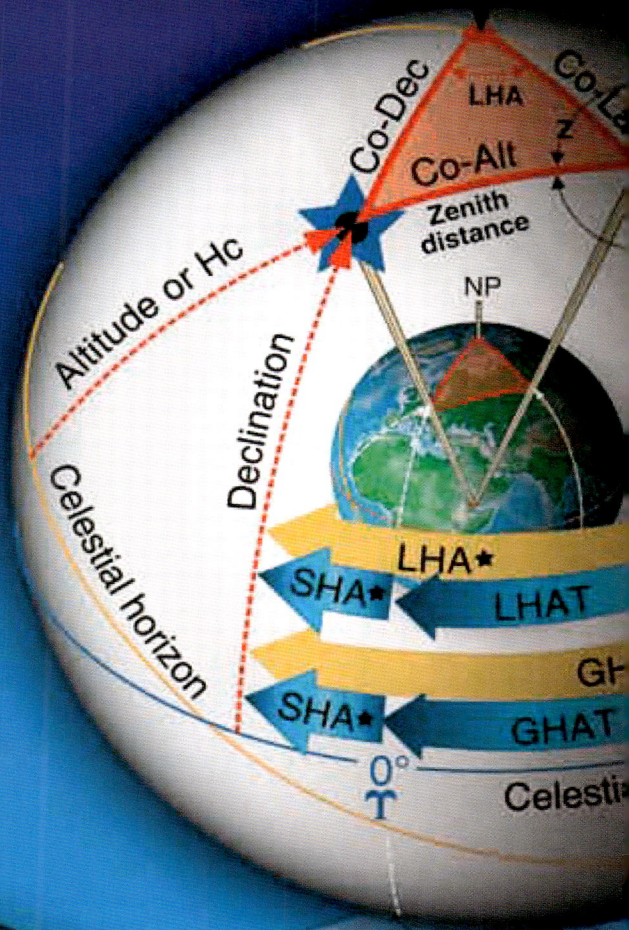

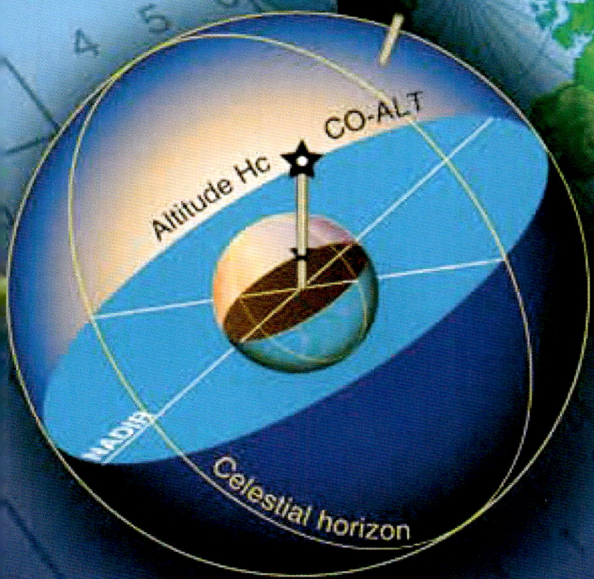

# LHA and the Astronomical Triangle

The basic principle of celestial navigation is to consider yourself to be at a certain assumed position at a given time; then, by means of the sextant, determining how much your basic assumption is in error. At any given time, an observer has a certain relationship to a particular star. The observer is a certain number of nautical miles (NM) away from the subpoint, and the body is at a certain true bearing (TB) called true azimuth (Zn), measured from the observer's position. [Figure 9-1]

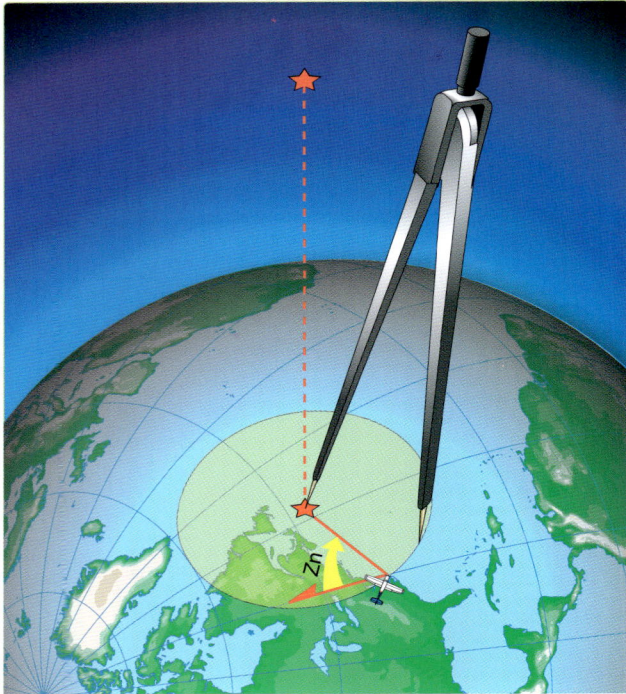

**Figure 9-1.** *Subpoint of a star.*

## Intercept

Assume yourself to be at a given point called the assumed position. At a given time, there exists at that instant a specific relationship between your assumed position and the subpoint. The various navigational tables provide you with this relationship by solving the astronomical triangle for you. From the navigational tables, you can determine how far away your assumed position is from the subpoint and the Zn of the subpoint from the assumed position. This means, in effect, that the tables give you a value called computed altitude (Hc) which would be the correct observed altitude (Ho) if you were anywhere on the circle of equal altitude through the assumed position. Any difference between the Hc determined for the assumed position and the Ho as determined by the sextant for the actual position is called intercept. Intercept is the number of NM between your actual circle of equal altitude and the circle of equal altitude through the assumed position. It is by means of the astronomical triangle that you can solve for

Hc and Zn in the Sight Reduction Tables for Air Navigation found in Pub. No. 249.

## Construction of the Astronomical Triangle

Consider the solution of a star as it appears on the celestial sphere. Start with the Greenwich meridian and the equator. Projected on the celestial sphere, these become the celestial meridian and the celestial equator (called equinoctial) as shown in *Figure 9-2*. Notice also in the same illustration how other known information is derived, namely the LHA of the star Aries—equal to the Greenwich hour angle (GHA) of Aries minus longitude west. You can also see that if the LHA of Aries and sidereal hour angle (SHA) of the star are known, the LHA of the star is their sum. It should also be evident that the GHA of Aries plus SHA of the star equals GHA of the star. Also, the GHA of the body minus west longitude (or plus east longitude) of the observer's zenith equals LHA of the body. These are important relationships used in the derivation of the Hc and Zn.

*Figure 9-3* shows part of the celestial sphere and the astronomical triangle. Notice that the known information of the astronomical triangle is the two sides and the included angle; that is, Co-Dec, Co-Lat, and LHA of the star. Co-Dec, or polar distance, is the angular distance measured along the hour circle of the body from the elevated pole to the body. The side, Co-Lat, is 90° minus the latitude of the assumed position. The included angle in this example is the LHA. With two sides and the included angle of the spherical triangle known, the third side and the interior angle at the observer are easily solved. The third side is the zenith distance, and the interior angle at the observer is the azimuth angle (Z). Instead of listing the zenith distance, the astronomical tables list the remaining portion of the 90° from the zenith, or the Hc. Hc equals 90° minus zenith distance of the assumed position, just as zenith distance of the assumed position equals 90° – Hc. Note that when measured with reference to the celestial horizon, zenith distance is synonymous with co-altitude. *Figure 9-4* is a side view of this solution.

So far, the astronomical triangle has been defined only on the celestial sphere. Refer again to *Figure 9-3* and notice the same triangle on the terrestrial sphere (earth). The same triangle with its corresponding vertices may be defined on the earth as follows: (1) celestial pole—terrestrial pole; (2) zenith of assumed position—assumed position; and (3) star—-subpoint of the star. The three interior angles of this triangle are exactly equal to the angles on the celestial sphere. The angular distance of each of the three sides is exactly equal to the corresponding side on the astronomical triangle. Celestial and terrestrial terms are used interchangeably. For example, refer to *Figure 9-3* and notice that Co-Lat on the terrestrial triangle is also called Co-Lat on the celestial

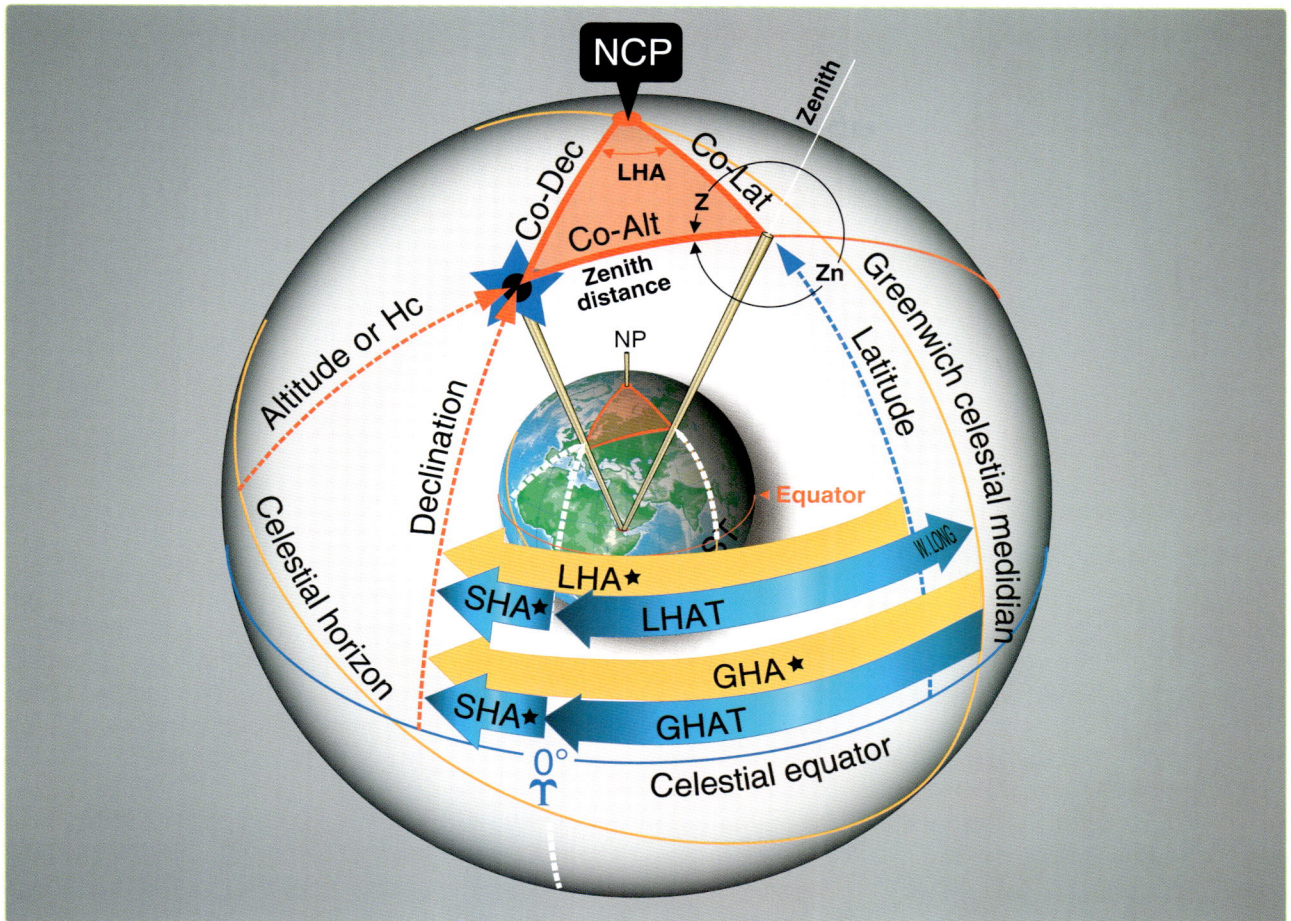

**Figure 9-2.** *Astronomical triangle.*

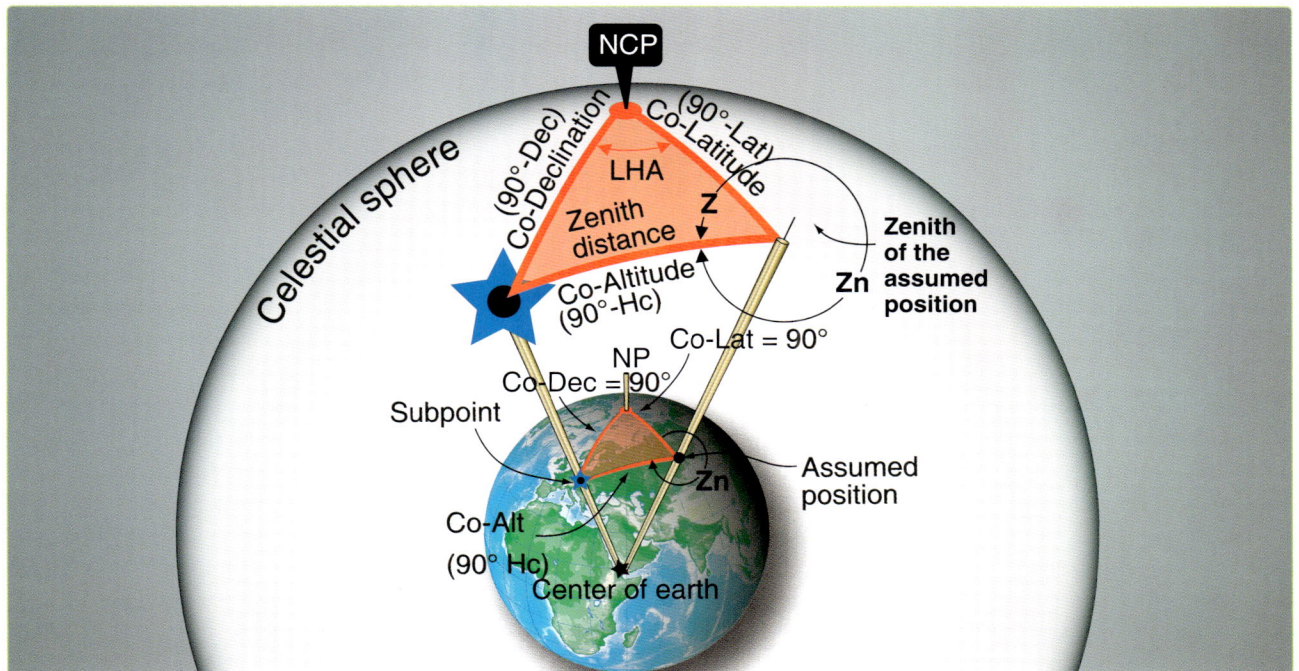

**Figure 9-3.** *Celestial-terrestrial relationship.*

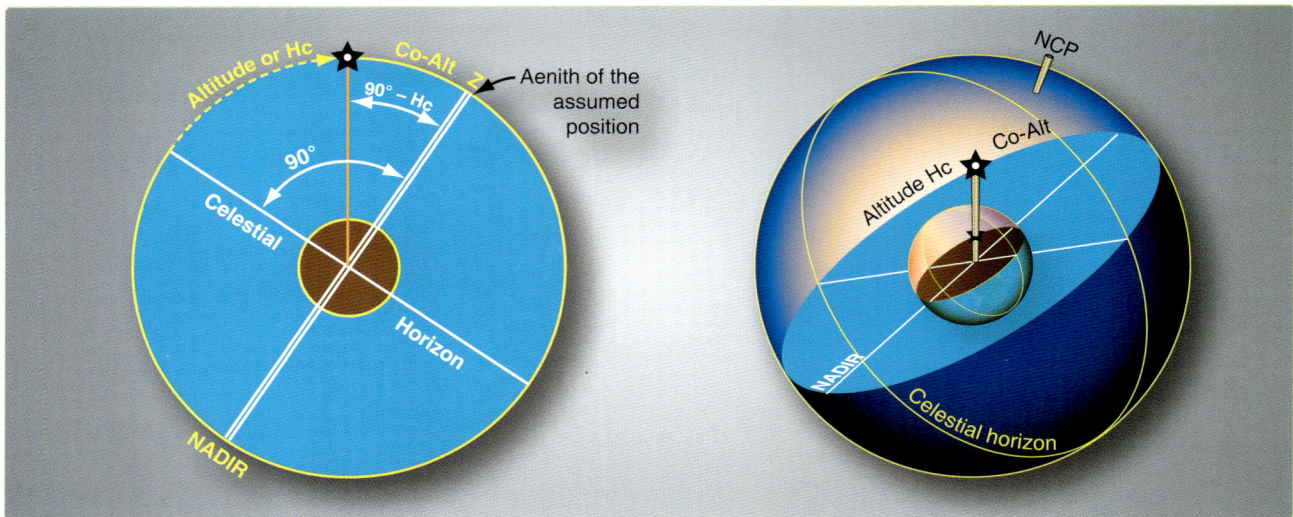

**Figure 9-4.** *Co-altitude equals 90 minus Hc.*

triangle. To be perfectly correct, the term on the celestial sphere corresponding to latitude on the earth is declination (Dec); therefore, the celestial side could well be called co-declination of the zenith of the assumed position.

Rather than have this confusion, the terrestrial term Co-Lat is also used with reference to the celestial sphere, just as latitude of the subpoint is considered to be the Dec amount from the equator. Latitude is used when referring to the observer or zenith, and Dec is used when referring to the star or its subpoint. The distance between the subpoint and the assumed position is generally referred to as zenith distance (Co-Alt) rather than the segment of the vertical circle joining the subpoint and the assumed position. These angular distance terms are interchangeable on the celestial and terrestrial spheres.

The values of the Zn and the interior angle (Z) are listed in the Pub. No. 249 tables depending upon whether or not a Dec solution is desired. Pub. No. 249, Volume 1, lists the Zn rather than the interior angle. Pub. No. 249, Volumes 2 and 3, list the interior angle (Z). It is necessary to follow rules printed on each page to convert the interior angle (Z) to true azimuth (Zn).

## Pub. No. 249, Volume 1

Volume 1 of Pub. No. 249 deals solely with the solution concerning selected stars and is considered separately from Volumes 2 and 3. Volume 1 provides complete worldwide coverage from pole to pole for each degree of latitude. The LHA of Aries is listed in 1° increments from latitudes of 0° to 69° North and South inclusive. From 70° through 89° of latitude, the meridians are so close together that it is only necessary to tabulate the values of the LHA of Aries in

even 2° increments. There are two pages devoted to each whole degree of latitude between latitudes 69° N and 69° S inclusive. From there to the pole, only one page is devoted to each whole degree of latitude. The three stars marked by diamonds on each page provide sets for fixing purposes that are favorably situated in altitude and azimuth.

The entering arguments are the assumed latitude and the LHA of Aries (to whole degrees). At any one time, the navigator has the choice of the seven listed stars for that latitude plus Polaris. The names of the stars are in capital letters if the star is of first magnitude or brighter; the second magnitude stars are printed in small letters. The names of the stars are listed every 15° of LHA of Aries (every 30° in the polar latitudes). For the time the navigator expects to make an observation, commonly called a shot, they look up the GHA of Aries and apply the approximate longitude to get a whole degree LHA of Aries. The navigator then enters Pub. No. 249, Volume 1, with the latitude closest to the dead reckoning (DR) latitude and the LHA of Aries to select the stars that will be shot.

Since single celestial observation results in only one LOP, it is necessary to shoot two or more bodies to obtain a fix. Suppose the navigator wants to shoot at approximately 0230 Greenwich Mean Time (GMT), he or she looks up the GHA of Aries (in the Air Almanac) and finds it to be 196°. The DR position for this time is 31° 48' N, 075° 26' W. A quick calculation shows the LHA of Aries is approximately 121°, and the closest latitude is 32° N. Notice in the portion of the tables reproduced in *Figure 9-5* the available stars at this position are Alkaid, Regulus, Alphard, Sirius, Rigel, Aldebaran, and Capella. Using Sirius, a shot is taken at 0231 and the Ho obtained is 37° 50'.

LAT. 32° N

| LHA ♈ | *Alkard | | Regulus | | *Alphard | | Sirius | | Rigel | | Aldebaran | | *Capella | |
|---|---|---|---|---|---|---|---|---|---|---|---|---|---|---|
| | Hc | Zn | Hc | Zn | Hc | Zn | Hc | Zn | Hc | Zn | Hc | Zn | Hc | Zn |
| 120 | 25 51 | 046 | 54 49 | 117 | 44 36 | 150 | 38 00 | 203 | 33 21 | 232 | 41 00 | 264 | 55 21 | 306 |
| 121 | 26 28 | 046 | 55 24 | 118 | 45 01 | 15[illegible] | 37 40 | 205 | 32 41 | 233 | 40 09 | 265 | 54 40 | 306 |
| 122 | 27 35 | 04[illegible] | 56 19 | 119 | 45 26 | 152 | 37 18 | 206 | 32 00 | 234 | 39 19 | 265 | 53 58 | 306 |
| 123 | 27 42 | 047 | 57 03 | 121 | 45 49 | 153 | 36 55 | 207 | 31 19 | 235 | 38 28 | 266 | 53 17 | 306 |
| 124 | 28 19 | 047 | 57 47 | 122 | 46 12 | 155 | 36 32 | 208 | 30 37 | 236 | 37 37 | 267 | 52 35 | 305 |
| 125 | 28 57 | 047 | 58 30 | 123 | 46 33 | 156 | 36 08 | 209 | 29 55 | 236 | 36 46 | 267 | 51 54 | 305 |
| 126 | 29 34 | 047 | 59 12 | 124 | 46 53 | 157 | 35 43 | 210 | 29 12 | 237 | 35 56 | 268 | 51 12 | 305 |
| 127 | 30 11 | 048 | 59 54 | 126 | 47 12 | 159 | 35 17 | 211 | 28 29 | 238 | 35 05 | 268 | 50 31 | 305 |
| 128 | 30 49 | 048 | 60 35 | 127 | 47 30 | 160 | 34 50 | 212 | 27 46 | 239 | 34 14 | 269 | 49 49 | 305 |
| 129 | 31 27 | 048 | 61 15 | 129 | 47 47 | 162 | 34 23 | 213 | 27 03 | 239 | 33 23 | 269 | 49 07 | 305 |
| 130 | 32 05 | 048 | 61 54 | 130 | 48 02 | 163 | 33 55 | 214 | 26 19 | 240 | 32 32 | 270 | 48 26 | 305 |
| 131 | 32 43 | 048 | 62 33 | 132 | 48 17 | 164 | 33 26 | 215 | 25 34 | 241 | 31 41 | 270 | 47 44 | 305 |
| 132 | 33 21 | 049 | 63 10 | 133 | 48 30 | 166 | 32 56 | 216 | 24 50 | 242 | 30 50 | 271 | 47 02 | 305 |
| 133 | 33 59 | 049 | 63 47 | 135 | 48 42 | 167 | 32 26 | 217 | 24 05 | 242 | 29 59 | 272 | 46 21 | 305 |
| 134 | 34 37 | 049 | 64 22 | 137 | 48 52 | 169 | 31 55 | 218 | 23 20 | 243 | 29 09 | 272 | 45 39 | 305 |

**Figure 9-5.** *Enter tables with LHA Aries and latitude.*

| | |
|---|---|
| GHA Aries for 0230 GMT | 196° 06' |
| Correction for 1 minute | 15' |
| GHA Aries for 0231 GMT | 196° 21' |
| Closest longitude to DR for whole LHA (assumed longitude) | W075° 21' |
| LHA Aries for 0231 GMT | 121° |

The closest whole degree of latitude is 32° N; therefore, it is used as the assumed latitude. The assumed longitude is selected as the closest point, resulting in an LHA of Aries that is a whole degree (no minutes). The Hc of Sirius is listed as 37° 40'. The Zn is 205°.

The second shot was taken at 0234 using Regulus, the Ho being 55° 30'. A new DR position could be obtained for 0234 GMT, but the 0230Z DR position will suffice for this determination of Hc and Zn.

| | |
|---|---|
| GHA Aries for 0230 GMT | 196° 06' |
| Correction for 4 minutes | 1° 00' |
| GHA Aries for 0234Z | 197° 06' |
| Closest longitude for whole LHA (assumed longitude) | W075° 06' |
| LHA Aries for 0234Z | 122° |

The assumed latitude is still 32° N and, in this case, 075° 06' W is the assumed longitude since this is the closest longitude to the DR longitude that results in the LHA of Aries being a whole degree. The Hc of Regulus is listed as 56° 19', and the Zn is 119°. The various corrections that must be applied, as well as the plotting of the fix, are discussed later.

### Postcomputation Method

The steps in the precomputation method are as follows:

1. Determine the GHA of Aries for the time of observation from the Air Almanac.
2. Assume a position as close as possible to the DR position at the time of the shot so the latitude and LHA of Aries in whole degrees may be determined.
3. Turn to the page in Pub. No. 249 for the assumed latitude and, opposite the LHA of Aries, select the stars to be shot. In making the selection, assume the LHA of Aries will change 1° every 4 minutes of time.
4. Shoot the body and record the time, Ho, and name of the body.
5. Obtain the GHA of Aries for the time of the observation, and apply the assumed longitude to determine the LHA of Aries.

6.   Turn to the pages for the assumed latitude and, opposite the LHA of Aries in the column headed by the name of the star, find and record the Hc and Zn.

## Pub. No. 249, Volumes 2 and 3

Volume 1 consists of tables of Hc and Zn for selected stars. Because the Dec and SHA of each star change slowly, these tables may be used for many years with only small corrections. The Dec and SHA of a nonstellar body change rapidly, making a permanent format similar to Volume 1 impossible for the sun, moon, and planets.

Volumes 2 and 3 have Dec tables adequate for determining the Hc and Zn of any celestial body within the Dec range of 30° N to 30° S. They are intended primarily for use when observing nonstellar (solar system) bodies. Volume 2 provides latitudes between 39° N and 39° S, and Volume 3 provides for latitudes from 40° N or S to the poles.

Provision is made for observed altitudes from 90° above to 3° below the horizon (7° from latitudes 70° to the pole). In view of refraction and of possible long intercepts, the tables are actually extended 2° below these limits.

## Entering Argument

Volumes 2 and 3 are entered with the LHA of the body, in contrast to Volume 1, which is entered with the LHA of Aries. The range extends from 0° through all LHAs applicable within the altitude limits of the body. Between latitude 70° and the pole, the LHA interval is 2°; for latitudes below 70°, the interval is 1°. Arguments of LHA of the body less than 180° appear on the left margin, and arguments greater than 180° appear on the right.

Several pages are devoted to each degree of latitude. Each page has 15 declination (Dec) columns and is labeled with its value at the top and bottom. Each page is also marked Declination Contrary Name to Latitude or Declination Same Name as Latitude.

The entering arguments of LHA of the body, for declination of contrary name to latitude, always increase from the bottom of the page on the left side and decrease on the right. The opposite arrangement exists on pages where Dec and latitude has the same name. Occasionally, one page is blank in the middle and the top half covers Declination Same Name as Latitude, while the bottom half is Declination Contrary Name to Latitude.

Azimuth angle (Z) is listed instead of true azimuth (Zn). Since Zn is used for plotting, it is necessary to convert Z to Zn. The rules for conversion are listed on the left-hand side at the top and bottom of every page. Notice that LHA and Zn will never occur on the same side of 180°.

In addition to Hc and Z, a value of d is also listed. This d-value is the change in altitude (Hc) with a 1° increase in Dec. If the LHA and Dec of the body and the latitude of the assumed position are each a whole number of degrees, the Hc and Z are found in the correct Dec column opposite the LHA of the body on the page marked by the proper latitude value.

For example, refer to the portion of the table shown in *Figure 9-6*. At the latitude 40° N, if the LHA of a body is 86° and its Dec is 5° N, the Hc is 06° 16' and the azimuth angle (Z) is 089°. The rule in the upper left-hand corner of the page applies for the conversion of Z to Zn. Zn = 360° – Z or 360° – 089° = 271°. Here again the position is assumed so that latitude and LHA are whole numbers.

## Interpolation for Declination (Dec)

When the Dec of a body is a number of minutes in addition to a whole number of degrees, the altitude (Hc) is extracted for the whole number of degrees and corrected by interpolation for the additional minutes. There is rarely a need for interpolation of Z, which is given only to the nearest degree.

Interpolation for Hc should always be made in the direction of increasing Dec in accordance with the sign of the d-value. Not all of the signs are printed; the sign is given at least once in each block of five entries and can always be found by looking either up or down the column from the value of d in question. The correction to altitude for additional minutes of Dec is proportional to d and proportional to the number of additional minutes.

In the previous example, the latitude was 40°, the LHA of the body was 086°, and the Dec was 5° N. Suppose the Dec had been 5° 17' N. The basic figures obtained would be 06° 16' Hc and 089° Z as before, and the true azimuth (Zn) would still be 271°. The Hc of 06° 16' is not correct for a Dec of 5° 17' N, but is correct for 5° N. The Hc change for an additional 1° of Dec (d-value) is +39 minutes of altitude. However, the correction needed in this case is for 17 minutes of Dec, not a whole degree. Consequently, the additional correction is 17/60 of 39'. To the closest whole number, this would be +11 minutes of altitude.

This multiplication can be done on the slide rule face of the DR computer or by means of a table found in back of Pub. No. 249, Volumes 2 and 3. A portion of this table is shown in *Figure 9-7*. Notice that there are no signs listed. The proper sign for the answer from this table is the same sign as the basic d-value. A rule of thumb is the correction is a plus (+) for Declination of Same Name as Latitude and a negative (–) for Declination of Contrary Name as Latitude. Values of d are given across the top of the table and additional minutes of Dec are given down the side of the table. In the table, the correction

N. Lat {  LHA greater than 180° . . . Zn = Z
          LHA less than 180° . . . Zn = 360 – Z

**Declination (0°–14°) Same Name as Latitude**

| LHA | 0° (Hc d Z) | 1° | 2° | 3° | 4° | 5° | 6° | 7° | 8° | 9° | 10° | 11° |
|---|---|---|---|---|---|---|---|---|---|---|---|---|
| 70 | 15 11 40 103 | 15 51 40 102 | 16 31 40 102 | 17 11 39 102 | 17 50 39 100 | 18 29 39 100 | 19 08 39 99 | 19 47 38 98 | 20 25 39 97 | 21 05 38 96 | 21 42 37 95 | 22 19 38 94 |
| 71 | 14 27 39 103 | 15 06 40 102 | 15 46 39 101 | 16 25 40 100 | 17 05 39 99 | 17 44 39 99 | 18 23 38 98 | 19 01 39 97 | 19 40 38 96 | 20 18 38 95 | 20 56 38 95 | 21 34 37 94 |
| 72 | 13 42 39 102 | 14 21 40 101 | 15 01 39 100 | 15 40 39 100 | 16 19 39 99 | 16 58 39 98 | 17 37 39 97 | 18 16 38 96 | 18 54 38 96 | 19 32 38 94 | 20 10 38 94 | 20 38 37 93 |
| 73 | 12 57 39 101 | 13 36 39 100 | 14 15 40 100 | 14 55 39 99 | 15 34 39 98 | 16 13 38 97 | 16 51 39 96 | 17 30 38 96 | 18 08 38 95 | 18 46 38 94 | 19 24 38 93 | 20 02 37 92 |
| 74 | 12 11 40 100 | 12 51 39 100 | 13 30 39 99 | 14 09 39 98 | 14 48 39 97 | 15 27 39 97 | 16 06 38 96 | 16 44 38 95 | 17 22 38 94 | 18 00 38 93 | 18 28 38 93 | 19 16 37 92 |
| 75 | 11 26 39 100 | 12 05 39 99 | 12 45 39 98 | 13 24 39 97 | 14 03 38 97 | 14 41 39 95 | 15 20 38 95 | 15 58 38 94 | 16 36 38 93 | 17 14 38 93 | 17 52 38 92 | 18 30 37 91 |
| 76 | 10 41 39 99 | 11 20 39 98 | 11 59 39 98 | 12 38 39 97 | 13 17 39 96 | 13 58 38 95 | 14 34 39 94 | 15 12 39 94 | 15 51 38 93 | 16 29 37 92 | 17 06 38 91 | 17 44 37 90 |
| 77 | 09 55 40 98 | 10 35 39 98 | 11 14 38 97 | 11 52 39 96 | 12 31 39 95 | 13 10 38 95 | 13 48 38 94 | 14 26 39 93 | 15 05 38 92 | 15 43 37 91 | 16 20 38 90 | 16 58 37 90 |
| 78 | 09 10 39 98 | 09 49 39 97 | 10 28 39 96 | 11 07 38 96 | 11 45 39 95 | 12 24 38 94 | 13 02 38 93 | 13 41 38 92 | 14 19 38 92 | 14 57 37 91 | 15 34 38 90 | 16 12 37 89 |
| 79 | 08 24 39 97 | 09 03 39 96 | 09 42 39 96 | 10 21 38 95 | 10 59 39 94 | 11 38 38 93 | 12 16 38 93 | 12 55 38 92 | 13 33 38 91 | 14 11 37 90 | 14 48 38 89 | 15 26 38 89 |
| 80 | 07 39 39 97 | 08 18 38 96 | 08 56 39 95 | 09 35 39 94 | 10 14 38 93 | 10 52 38 93 | 11 30 39 92 | 12 09 38 91 | 12 47 38 90 | 13 25 38 89 | 14 03 37 89 | 14 40 38 88 |
| 81 | 06 53 39 96 | 07 32 39 95 | 08 11 38 94 | 08 49 39 94 | 09 28 38 93 | 10 06 39 92 | 10 45 38 91 | 11 23 38 91 | 12 01 38 90 | 12 39 38 89 | 13 17 37 88 | 13 54 38 87 |
| 82 | 06 07 39 95 | 06 46 39 94 | 07 25 38 94 | 08 03 38 93 | 08 42 38 92 | 09 20 39 91 | 09 59 38 91 | 10 37 38 90 | 11 15 38 89 | 11 53 38 88 | 12 31 37 87 | 13 08 38 87 |
| 83 | 05 21 39 95 | 06 00 39 94 | 06 39 38 93 | 07 17 39 92 | 07 56 38 91 | 08 34 39 91 | 09 13 38 90 | 09 51 38 89 | 10 29 38 88 | 11 07 38 88 | 11 45 37 87 | 12 22 38 86 |
| 84 | 04 36 38 94 | 05 14 39 93 | 05 53 38 92 | 06 31 39 92 | 07 10 38 91 | 07 48 38 90 | 08 27 38 89 | 09 05 38 89 | 09 43 38 88 | 10 21 38 87 | 10 59 38 86 | 11 37 37 85 |
| 85 | 03 50 39 93 | 04 28 39 92 | 05 07 39 92 | 05 45 39 91 | 06 24 38 90 | 07 02 39 89 | 07 41 38 89 | 08 19 38 88 | 08 57 38 88 | 09 35 38 86 | 10 13 38 86 | 10 51 38 85 |
| 86 | 03 04 38 93 | 03 42 39 92 | 04 21 38 91 | 04 59 39 90 | 05 38 38 90 | 06 16 39 89 | 06 55 38 88 | 07 33 38 87 | 08 11 38 86 | 08 49 38 86 | 09 27 38 85 | 10 05 38 84 |
| 87 | 02 18 38 92 | 02 56 39 91 | 03 35 39 90 | 04 14 38 90 | 04 52 38 89 | 05 30 39 88 | 06 09 38 88 | 06 47 38 87 | 07 25 38 86 | 08 03 38 85 | 08 41 38 84 | 09 19 38 83 |
| 88 | 01 32 39 91 | 02 10 39 91 | 02 49 39 90 | 03 28 38 89 | 04 06 38 88 | 04 44 39 87 | 05 23 38 87 | 06 01 38 86 | 06 39 38 85 | 07 17 38 84 | 07 56 38 84 | 08 34 38 83 |
| 89 | 00 46 39 91 | 01 25 39 90 | 02 03 39 89 | 02 42 38 88 | 03 20 39 88 | 03 59 38 87 | 04 37 38 86 | 05 15 39 85 | 05 54 38 85 | 06 32 38 84 | 07 10 38 83 | 07 48 38 82 |
| 90 | 00 00 39 90 | -0 00 38 89 | 01 17 38 88 | 01 56 38 88 | 02 34 39 87 | 03 13 38 86 | 03 51 39 85 | 04 30 38 85 | 05 08 38 84 | 05 46 39 83 | 06 25 38 82 | 07 03 38 82 |
| 91 | -0 46 39 89 | -0 07 38 89 | 00 31 39 88 | 01 10 38 87 | 01 48 39 86 | 02 27 38 86 | 03 05 39 85 | 03 44 38 84 | 04 22 39 83 | 05 01 38 82 | 05 39 38 82 | 06 17 39 81 |
| 92 | -1 32 39 89 | -0 53 38 88 | -0 15 38 87 | 00 24 38 86 | 01 02 39 86 | 01 41 39 85 | 02 20 38 84 | 02 58 39 83 | 03 37 38 83 | 04 15 39 82 | 04 54 38 81 | 05 32 38 80 |
| 93 | -2 18 38 88 | -1 39 38 88 | -1 01 39 87 | -0 22 39 86 | 00 17 38 85 | 00 55 39 84 | 01 34 39 83 | 02 12 38 83 | 02 51 39 82 | 03 30 38 81 | 04 08 39 80 | 04 47 38 80 |
| 94 | | -2 35 39 87 | -1 46 39 86 | -1 08 39 | 00 29 38 84 | | 00 48 39 83 | 01 27 39 | | 02 44 39 81 | 03 23 39 80 | 04 02 39 79 |

---

| LHA | 0° (Hc d Z) | 1° | 2° | 3° | 4° | 5° | 6° | | 11° | 12° | 13° | 14° | Zn |
|---|---|---|---|---|---|---|---|---|---|---|---|---|---|
| 89 | 00 46 39 91 | 00 07 39 91 | -0 31 39 92 | -1 10 38 93 | -1 48 39 94 | -2 27 38 94 | -3 05 39 95 | 5 17 39 99 | | | | | 271 |
| 88 | 01 32 39 91 | 00 53 38 92 | 00 15 39 93 | -0 24 38 94 | -1 02 39 94 | -1 41 39 95 | -2 20 38 96 | 5 32 38 100 | -6 10 39 100 | | | | 272 |
| 87 | 02 18 38 92 | 01 39 38 93 | 01 01 39 93 | 00 22 39 94 | -0 17 38 95 | -0 55 39 95 | -1 34 39 97 | 4 47 38 100 | -5 25 39 101 | -6 04 38 102 | | | 273 |
| 86 | 03 04 38 93 | 02 25 39 93 | 01 46 38 94 | 01 08 39 95 | 00 29 39 96 | -0 10 38 96 | -0 48 39 97 | 4 02 38 101 | -4 40 39 102 | -5 19 38 102 | -5 57 38 103 | | 274 |
| 85 | 03 50 39 | 03 11 39 94 | 02 32 38 95 | 01 54 39 95 | 01 15 39 96 | 00 36 39 97 | -0 03 38 | 3 16 39 102 | -3 55 39 102 | -4 34 38 103 | -5 12 39 104 | | 275 |
| 84 | 04 36 39 94 | 03 57 39 95 | 03 18 39 95 | 02 39 39 96 | 02 00 40 97 | 01 22 39 98 | 00 43 39 98 | 31 39 102 | -3 10 39 103 | -3 49 39 104 | -4 28 38 105 | | 276 |
| 83 | 05 21 38 95 | 04 43 39 95 | 04 04 39 96 | 03 25 39 97 | 02 46 39 98 | 02 07 39 98 | 01 28 39 99 | 47 39 103 | -2 26 38 104 | -3 04 39 104 | -3 43 39 105 | | 277 |
| 82 | 06 07 39 95 | 05 28 38 96 | 04 50 39 97 | 04 11 39 97 | 03 32 39 99 | 02 54 39 99 | 02 14 40 100 | 17 39 104 | -1 41 39 104 | -2 20 39 105 | -2 59 39 106 | | 278 |
| 81 | 06 53 39 96 | 06 14 39 97 | 05 35 39 97 | 04 56 39 98 | 04 17 39 99 | 03 38 39 100 | 02 59 39 101 | 17 39 104 | -0 56 40 105 | -1 36 39 106 | -2 15 39 106 | | 279 |
| 80 | 07 39 39 97 | 07 00 39 97 | 06 21 39 98 | 05 42 40 99 | 05 02 39 100 | 04 24 39 100 | 03 44 39 101 | 27 39 105 | -0 12 40 106 | -0 52 39 106 | -1 31 39 107 | | 280 |
| 79 | 08 24 39 97 | 07 45 39 98 | 07 06 39 99 | 06 27 39 99 | 05 48 40 101 | 05 09 39 101 | 04 30 39 102 | 12 40 106 | 00 32 39 106 | -0 07 40 107 | -0 47 40 108 | | 281 |
| 78 | 09 10 39 98 | 08 31 39 99 | 07 52 40 99 | 07 12 39 100 | 06 33 40 101 | 05 54 40 102 | 05 14 40 103 | 56 40 106 | 01 16 40 107 | 00 36 39 108 | -0 03 40 108 | | 282 |
| 77 | 09 55 39 99 | 09 16 39 99 | 08 37 39 100 | 07 58 40 101 | 07 18 40 102 | 06 38 38 102 | 06 00 39 103 | 40 40 107 | 02 00 40 108 | 01 20 40 108 | 00 40 40 109 | | 283 |
| 76 | 10 41 39 99 | 10 02 40 100 | 09 22 39 101 | 08 43 40 101 | 08 03 40 102 | 07 23 39 103 | 06 44 40 104 | 24 40 107 | 02 44 40 108 | 02 04 40 109 | 01 24 41 110 | | 284 |
| 75 | 11 26 39 100 | 10 47 40 101 | 10 07 39 101 | 09 28 40 102 | 08 48 40 103 | 08 08 40 104 | 07 28 40 104 | 08 41 108 | 03 27 40 109 | 02 47 40 110 | 02 07 41 110 | | 285 |
| 74 | 12 11 39 100 | 11 32 40 101 | 10 52 39 102 | 10 13 40 103 | 09 33 40 104 | 08 53 40 104 | 08 13 40 105 | 4 51 40 109 | 04 11 40 110 | 02 50 41 111 | | | 286 |
| 73 | 12 57 40 101 | 12 17 40 102 | 11 37 40 103 | 10 57 40 103 | 10 17 40 104 | 09 37 40 105 | 08 57 40 106 | 05 35 41 109 | 04 54 41 110 | 04 13 40 111 | | | 287 |
| 72 | 13 42 40 102 | 13 02 40 103 | 12 22 40 103 | 11 42 40 104 | 11 01 40 105 | 10 21 41 106 | 09 41 41 106 | 16 18 41 110 | 05 37 41 111 | 04 56 41 112 | 04 15 41 112 | | 288 |
| 71 | 14 27 39 103 | 13 47 40 103 | 13 07 41 104 | 12 26 40 105 | 11 46 40 106 | 11 06 41 106 | 10 25 41 107 | 7 01 41 111 | 06 20 41 112 | 05 39 41 112 | 04 58 42 113 | | 289 |
| 70 | 15 11 40 103 | 14 31 40 104 | 13 51 40 105 | 13 11 41 106 | 12 30 40 106 | 11 50 41 107 | 11 09 41 108 | 7 44 41 111 | 07 03 42 112 | 06 21 41 113 | 05 40 42 114 | | 290 |

**LAT. 40°**

S. Lat {  LHA greater than 180° . . . Zn = 180 – Z
          LHA less than 180° . . . Zn = 180 + Z

**Declination (0°–14°) CONTRARY Name to Latitude**    **LAT. 40°**

**Figure 9-6.** *Enter tables with latitude, Dec, and LHA.*

| d / r | 1 2 3 | 4 5 6 | 12 | 13 14 15 | 16 17 18 | 19 20 21 | 22 23 24 | 25 26 27 | 28 29 30 | 31 32 33 | 34 35 36 | 37 38 39 | 40 |
|---|---|---|---|---|---|---|---|---|---|---|---|---|---|
| 0 | 0 0 0 | 0 0 0 | 0 | 0 0 0 | 0 0 0 | 0 0 0 | 0 0 0 | 0 0 0 | 0 0 0 | 0 0 0 | 0 0 0 | 0 0 0 | 0 |
| 1 | 0 0 0 | 0 0 0 | 0 | 0 0 0 | 0 0 0 | 0 0 0 | 0 0 0 | 0 0 0 | 0 0 0 | 1 1 1 | 1 1 1 | 1 1 1 | 1 |
| 2 | 0 0 0 | 0 0 0 | 0 | 1 1 1 | 1 1 1 | 1 1 1 | 1 1 1 | 1 1 1 | 1 1 1 | 1 1 1 | 1 1 1 | 1 1 1 | 1 |
| 3 | 0 0 0 | 0 0 0 | 1 | 1 1 1 | 1 1 1 | 1 1 1 | 1 1 1 | 1 1 2 | 2 2 2 | 2 2 2 | 2 2 2 | 2 2 2 | 2 |
| 4 | 0 0 0 | 0 0 0 | | 1 1 1 | 1 1 1 | 1 1 2 | 2 2 2 | 2 2 2 | 2 2 2 | 2 2 2 | 2 3 3 | 2 3 3 | 3 |
| 5 | 0 0 0 | 0 0 0 | 1 | 1 1 1 | 1 2 2 | 2 2 2 | 2 2 2 | 2 2 3 | 3 3 3 | 3 3 3 | 3 3 3 | 3 3 | 3 |
| 6 | 0 0 0 | 0 0 0 | 1 | 1 2 2 | 2 2 2 | 2 2 2 | 3 3 3 | 3 3 3 | 3 4 4 | 4 4 4 | 4 4 4 | 4 | 4 |
| 7 | 0 0 0 | 0 1 1 | 1 | 2 2 2 | 2 2 3 | 3 3 3 | 3 3 4 | 4 4 4 | 4 4 4 | 4 5 5 | 5 | 5 |
| 8 | 0 0 0 | 1 1 1 | 2 | 2 2 3 | 3 3 3 | 3 4 4 | 4 4 4 | 4 5 5 | 5 5 5 | 5 | 5 |
| 9 | 0 0 0 | 1 1 1 | 2 | 2 3 3 | 3 3 3 | 4 4 4 | 4 5 5 | 5 5 5 | 5 | 5 |
| 10 | 0 0 0 | 1 1 1 | | 2 3 3 | 3 3 4 | 4 4 5 | 5 5 6 | 6 6 6 | 6 6 | 6 |
| 11 | 0 0 1 | 1 1 1 | | 3 3 3 | 4 4 4 | 5 5 6 | 6 6 7 | 7 7 7 | 7 | 7 |
| 12 | 0 0 1 | 1 1 1 | 2 | 3 3 3 | 4 4 5 | 5 5 6 | 6 7 7 | 7 8 8 | 8 | 8 |
| 13 | 0 0 1 | 2 2 | 3 3 4 | 4 5 5 | 6 6 7 | 7 7 8 | 8 9 | 8 |
| 14 | 0 0 1 | 2 2 2 | 4 4 4 | 5 5 5 | 6 6 7 | 7 7 8 | 9 9 | 9 |
| 15 | 0 0 1 | 1 2 | 2 2 | 4 4 4 | 5 5 6 | 6 6 7 | 7 8 8 | 9 10 10 | 10 |
| 16 | 0 0 1 | 1 2 | 2 2 2 | 4 4 5 | 5 6 6 | 7 7 7 | 8 9 9 | 9 10 | 10 11 |
| 17 | 0 1 1 | 2 2 2 | 5 5 5 | 6 6 6 | 7 8 | 9 10 10 | 10 11 11 | 11 |
| 18 | 0 1 1 | 2 2 2 | 5 6 6 | 6 7 | 8 8 | 10 10 11 | 11 11 11 | 11 |

**Figure 9-7.** *Table performs the multiplication.*

11' is found by looking across 17' for Dec and down 39' for d to their intersection at 11'. Since the sign of the d-value is plus, this correction is added to the tabulated Hc. The correct Hc value then becomes 06° 16' + 11' or 06° 27'.

Following is a sample problem illustrating the solution. Refer to the portion of the tables in *Figure 9-8* for the solution.

Suppose the sun is observed at 1005 GMT. The DR position is 38° 12' N, 101° 47' E, and the Ho of the sun is 10° 52'.

| | |
|---|---|
| Dec of the Sun for 1000Z | S7° 37' |
| GHA Sun for 1000Z | 326° 53' |
| Correction to GHA for 5 minutes | +1° 15' |
| GHA Sun for l005Z | 328° 08' |
| Closest longitude for whole degree LHA (assumed longitude) | +E101° 52' |
| | 430° 00' |
| | −360° 00' |
| LHA Sun for 1005Z | 070° 00' |

The closest whole degree of latitude is 38° N and is used as the assumed latitude. Since the assumed latitude is north and Dec is south, the navigator must use Pub. 249, Volume 2, page for 38° latitude, which is headed Declination (0°–14°) Contrary Name to Latitude. Following LHA 070° across the page to 7° Dec, the navigator extracts:

| | |
|---|---|
| Tab Hc | 11° −06' |
| d-value | −40' |
| Z | 108° |
| d-correction from Pub. No. 249, Volume 2 | −25' |

| | |
|---|---|
| Corrected Hc | 10° −41' |
| Zn using rule in the upper left-hand corner of the page | 252° |

## Postcomputation Summary

Before proceeding, review the procedures for finding the Hc and Zn of a body whose Dec lies between 30° N and 30° S, using Pub. No. 249, Volume 2 or 3.

1. Shoot the body and record the time of observation, the body's name, and the Ho.

2. From the Air Almanac, extract GHA and Dec of the body for the time of the observation.

3. Assume a position close to the DR position so that the latitude is a whole number of degrees and the longitude combined with the GHA of the body gives a whole number of degrees of LHA of the body. Find the LHA of the body for this position.

4. Select the correct volume (2 or 3) and page that contains the correct arguments of Dec and LHA of the body, temporarily disregarding the odd additional minutes of Dec. Thus, if the Dec were N 19° 55', use the column for 19°. Select the table labeled Declination Same Name as Latitude, if Dec and latitude are both north or both south, or select the table labeled Declination Contrary Name to Latitude, if one is north and the other south. Opposite the LHA of the body, read the tabulated altitude, d-value, and Zn in the column headed by the whole degrees of Dec.

5. If the Dec is not a whole number of degrees, determine the correction for the additional minutes of Dec. Enter the table in the Pub. No. 249 volume with the value d and the number of additional minutes of Dec. Apply the correction to the tabulated altitude (Hc) according to the sign of d. This is the corrected Hc.

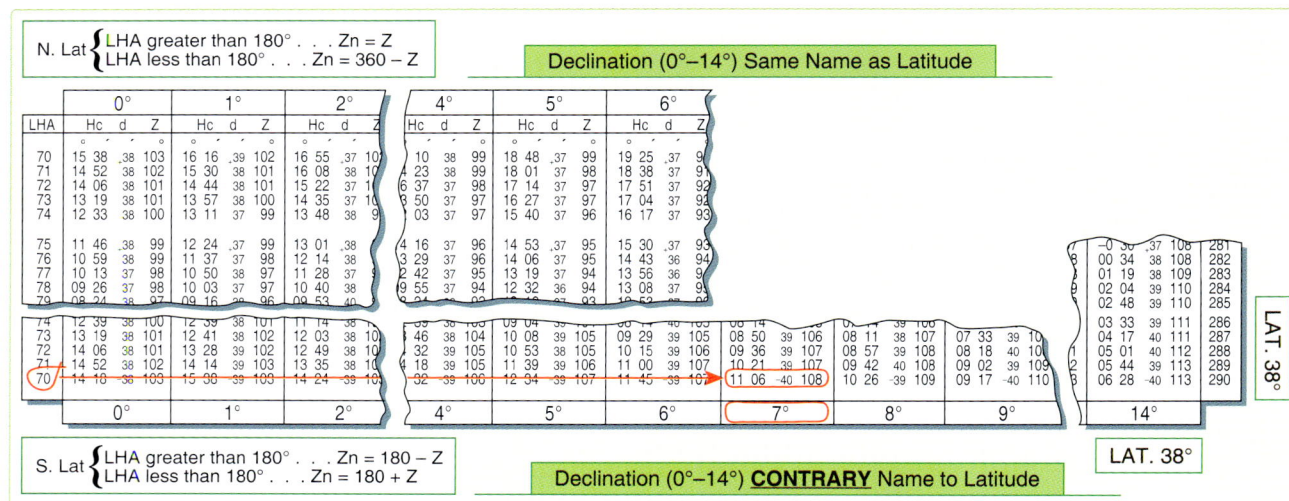

**Figure 9-8.** *Declination 0–14° Contrary Name to Latitude.*

6.  Convert azimuth angle (Z) to true azimuth (Zn) by means of the rule at the top or bottom of the page.

7.  This completes the solution for the Dec tables. Keep in mind this solution is computed after the observation. Because of the speeds involved in air navigation, we will explain a way to compute the solution before the shot in the next chapter.

## Precession and Nutation

The earth's axis does not maintain a fixed direction in space. Actually, the earth is like a slow running gyro that is wobbling. There are several separate patterns that the wobble makes. Some of those patterns have short cycles, while others take hundreds of years to complete. Two of the many patterns are shown in *Figure 9-9*. One involves small nodding motions while at the same time completing a larger circular path. You must use a correction called precession and nutation to account for these variations in the apparent position of the stars. This correction is applied only to celestial LOPs determined with Pub. No. 249, Volume 1.

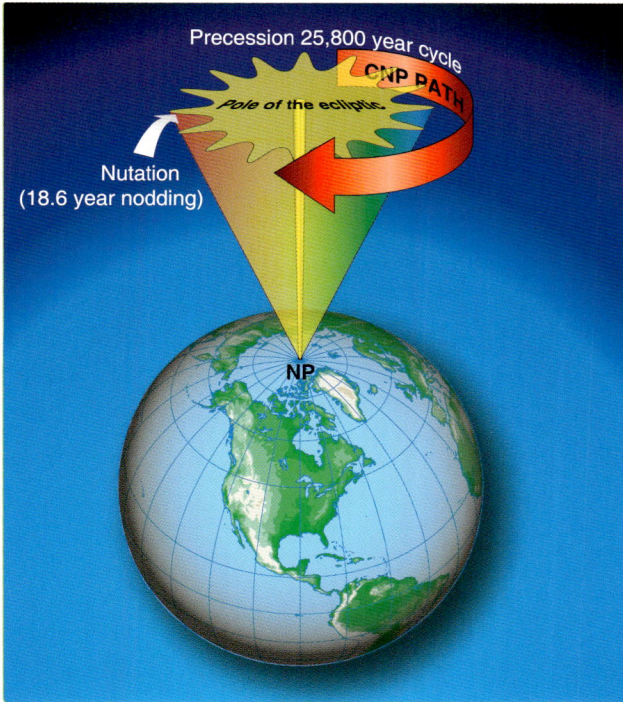

**Figure 9-9.** *Earth's axis wobble.*

## Precession

Because of the equatorial bulge, the attractive forces of other solar system bodies, principally the moon, are unbalanced about the center of the earth. The imbalance is directed toward aligning the equator with the plane of the ecliptic. However, the rotation of the earth transforms this force into an effect acting 90° away in the direction of rotation—a precessional effect. The result is that the poles travel in a conical path westward around the ecliptic poles, as shown in *Figure 9-10* (the point 90° from the ecliptic). Consequently, the points of intersection of the equator with the ecliptic, or the equinoxes, travel in a westerly direction along the ecliptic. This travel is called precession of the equinoxes, and it amounts to approximately 5/6 of a minute (50.26") annually. The equinoxes complete one revolution along the ecliptic in approximately 25,800 years. The equator is used as a reference for Dec and its movement, due to precession of the equinoxes, causes slight changes in the celestial coordinates of the stars that otherwise appear fixed in space.

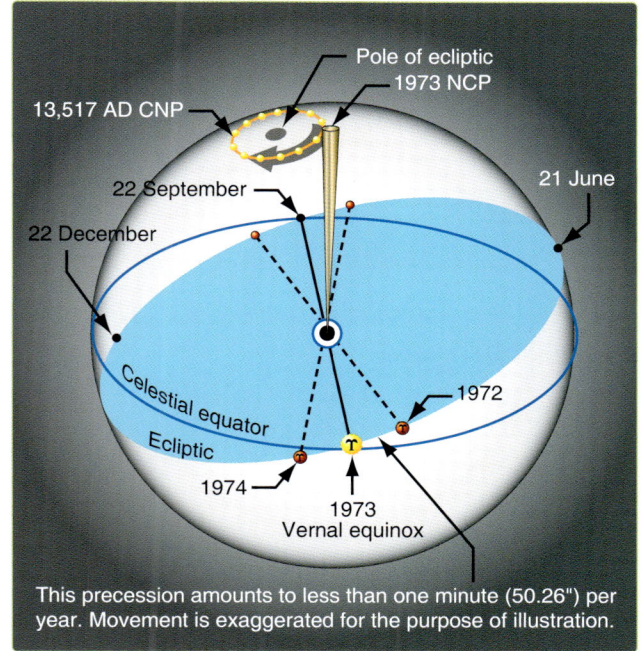

**Figure 9-10.** *Precession of the equinoxes.*

## Nutation

As the relative positions and distances from the earth to the sun, moon, and planets vary, so does the rate of precession. The only variation of importance in navigation is nutation. Nutation is a nodding of the poles, one oscillation occurring in about 18.6 years.

In *Figure 9-11*, you can see that if the stars remain fixed and the equinoctial moves up and down, the Dec of these bodies is changing.

Nutation, being approximately perpendicular to the ecliptic, has an appreciable influence on Dec. It is caused by complex gravitational forces among the sun, moon, and earth because the moon's orbit does not always lie in the plane of the ecliptic. The change in Dec of the celestial bodies caused by the resulting wobble of the earth's axis is called nutation.

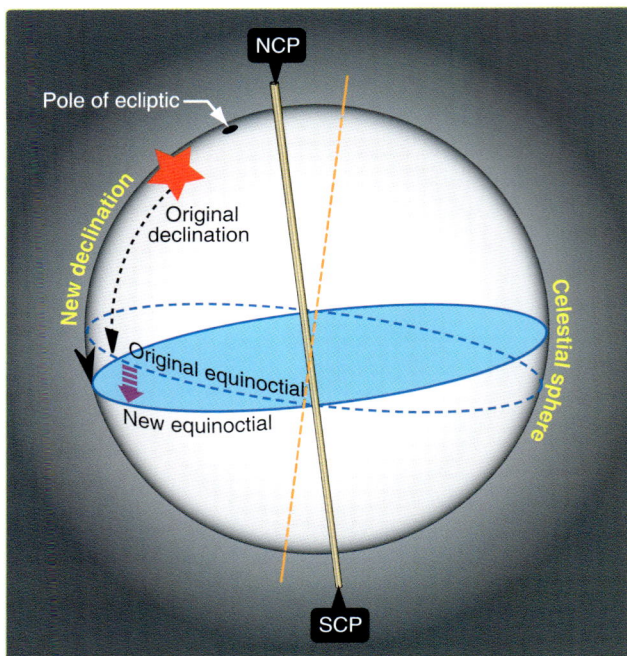

**Figure 9-11.** *Nutation changes the declination.*

## Position Corrections

Because of precession and nutation, Hc and Zn for a star are accurate only at the instant, or epoch, at which the LHA and Dec for the computations are correct. A position obtained at any other time with that Hc and Zn requires a correction. Pub. No. 249, Volume 1, contains Hc and Zn calculated for an epoch year (midnight, 1 January, of that year) so, if the volume is used in any other year, the resultant position must be corrected. The precession and nutation corrections are combined and given in Pub. No. 249, Volume 1, Table 5.

Entering arguments for the table is year, latitude, and LHA of Aries, and the correction is presented in the form of a distance and direction to move the fix. The tabulated values show the distance, parallel to the ecliptic, between the observer's position in the year of the fix and the position in the epoch year at the latitude and LHA of Aries.

Directions for using Table 5 are printed in the introduction of Pub. No. 249, Volume 1. One point needs emphasis here: the table is to be used only for observations plotted with the aid of Volume 1, never in conjunction with Volumes 2 or 3.

## Chapter Summary

This chapter introduces the astronomical triangle and how the Pub. No. 249 volumes you use in resolving the astronomical triangle. We have discussed obtaining solutions involving celestial bodies using Pub. No. 249, Volumes 1, 2, and 3. Succeeding chapters discuss plotting of the celestial LOPs and techniques of pre-computation.

# Chapter 10
# Celestial Precomputation

## Introduction

Celestial precomputation is neither new nor revolutionary. The tables necessary to do precomputation have been available since 1940; however, there was no operational requirement for precomputation at that time. With present day high-speed aircraft, however, the picture has changed radically. By postcomping, a great deal of work must be done after the last celestial observation. The fix could easily be 15 minutes old by the time it is plotted on the chart. At 450 knots groundspeed (GS), a fix that is 15 minutes old is over 100 miles behind the aircraft and is of questionable value. Another factor necessitating precomputation in high-speed aircraft lies in the method of shooting celestial. With the limited field of view of the sextant, the correct star is difficult to find unless you know where to look.

## Presetting the Sextant

Precomputation greatly reduces both of the problems just mentioned. By completing most of the computations before shooting, the navigator reduces the time necessary to plot the fix after the last observation. Also, the problem of finding the star in the optics of the sextant is simplified. The procedure for finding the star is similar to the heading check performed with the periscopic sextant, using the true bearing (TB) method as explained in Chapter 12, Special Celestial Techniques. In this case, the true azimuth (Zn) is set into the sextant mount and the computed altitude (Hc), which approximates the sextant altitude (Hs), is set into the sextant. Now, instead of sighting the body to determine the true heading (TH), set the TH under the vertical crosshair to find the selected body, hopefully very close to the crosshairs in the sextant field of view. Use the inverse relative bearing (IRB) method to avoid erroneous settings in the azimuth window and to increase speed in setting up the sextant. In this method, the azimuth window remains permanently at 000.0° and the IRB is computed by the formula: IRB = TH – Zn. The body should be found at its computed altitude when its IRB appears under the crosshairs.

## Precomputation Techniques

There are many acceptable methods of precomputation in general usage. However, these methods are basically either graphical, mathematical, or a combination of both methods. Selection is largely based on individual navigator preference. Celestial corrections that are used in precomputation include atmospheric refraction, parallax of the moon, instrument and acceleration errors, Coriolis and rhumb line, precession and nutation, motion of the observer, and wander. With precomputation, new corrections and terminology are introduced that include fix time, solution time, observation time, scheduled time, and motion of the body adjustment.

Fix time is the time for which the lines of position (LOP) are resolved and plotted on the chart. Solution time is the time for which the astronomical triangle is solved. Observation time is the midtime of the actual observation for each celestial body. Scheduled time is the time for which the astronomical triangle is solved for each LOP in the graphic method. Motion of the body correction is used to correct for the changing altitude of the selected bodies from shot to fix time and may be applied either graphically or mathematically.

### Motion of the Body Correction

Motion of the body correction can be applied graphically by moving the assumed position eastward or westward for time. This is possible because the Greenwich hour angle (GHA) and the subpoint of the body move westward at the rate of 1° of longitude per 4 minutes of time. In the graphic method, a scheduled time of observation is given to each body. If shooting is off schedule, the following rules apply:

1. For every minute of time that the shot is taken early, move the assumed position 15' of longitude to the east; for every minute of time that the shot is taken late, move the assumed position 15' of longitude to the west.

2. When the latitude of the assumed position and the Zn of the body are known, the motion of the body can be computed mathematically. For 1 minute, the formula is: 15(cos lat)/(sin Zn). This correction is shown in tabular form in *Figure 10-1*.

The National Imagery and Mapping Agency has published the Sight Reduction Tables for Air Navigation in a publication referred to as Pub. No. 249. These tables are published in three volumes. Volume 1, used by both the marine and air navigator, contains the altitude and azimuth values of seven selected stars for the complete ranges of latitude and hour angle of Aries. These seven stars represent the best selection for observation at any given position and time, and provide the data for presetting instruments before observation and for sight reduction afterwards. Volumes 2 and 3 cover latitudes 0-40 and 39-89 respectively and are primarily used by the air navigator in conjunction with observations of celestial bodies to calculate the geographic position of the observer.

In Publication No. 249, the local hour angle (LHA) increases 1° in 4 minutes of time. Thus, the Hc for an LHA that is 1° less than the LHA used for precomputation is the Hc for 4 minutes of time earlier than the solution time. The difference between the two Hcs is the value to apply to the Hc or Hs to advance or retard the LOP for 4 minutes of time. If the Hc decreases (Zn greater than 180°), the body is setting and the sign is minus (–) to advance the LOP if the value is applied to the Hs. If the Hc increases (Zn less than 180°), the body is rising and the sign is plus (+) to advance the LOP if the value is applied to the Hs.

In addition, motion corrections may be determined by using a modified MB-4 computer. This modification allows for greater accuracy and speed in computation of combined motions (motion of the observer and motion of the body) than the Pub. No. 249 tables. For a discussion of this modification, see chapter 12.

### Special Celestial Techniques

The main difference between the basic methods of precomputation is the manner in which the motion of the observer and the motion of the body corrections are applied. In the graphic method, both corrections are applied graphically by movement of the assumed position or the LOP. In the mathematical method, both corrections are applied mathematically to the Hc, the Hs, or the intercept after being obtained from tables, a modified MB-4 computer, or the Pub. No. 249.

| True Zn | Correction for 4 minutes of time | | | | | | | | | | | | | | | | | | | | | | | | | | | True Zn |
|---|---|---|---|---|---|---|---|---|---|---|---|---|---|---|---|---|---|---|---|---|---|---|---|---|---|---|---|---|
| | Latitude | | | | | | | | | | | | | | | | | | | | | | | | | | | |
| | 0 | 8 | 16 | 20 | 24 | 28 | 30 | 32 | 34 | 36 | 38 | 40 | 42 | 44 | 46 | 48 | 50 | 52 | 54 | 56 | 60 | 64 | 68 | 72 | 76 | 80 | 84 | |
| 090 | +60 | +59 | +58 | +56 | +55 | +53 | +52 | +51 | +50 | +49 | +47 | +46 | +45 | +43 | +42 | +40 | +39 | +37 | +35 | +34 | +30 | +26 | +22 | +19 | +15 | +10 | +6 | 090 |
| 095 | 60 | 59 | 57 | 56 | 55 | 53 | 52 | 51 | 50 | 48 | 47 | 46 | 44 | 43 | 42 | 40 | 38 | 37 | 35 | 33 | 30 | 26 | 22 | 18 | 14 | 10 | 6 | 085 |
| 100 | 59 | 59 | 57 | 56 | 54 | 52 | 51 | 50 | 49 | 48 | 47 | 45 | 44 | 43 | 41 | 40 | 38 | 36 | 35 | 33 | 30 | 26 | 22 | 18 | 14 | 10 | 6 | 080 |
| 105 | 58 | 57 | 56 | 54 | 53 | 51 | 50 | 49 | 48 | 47 | 46 | 44 | 43 | 42 | 40 | 39 | 37 | 36 | 34 | 32 | 29 | 25 | 22 | 18 | 14 | 10 | 6 | 075 |
| 110 | 56 | 56 | 54 | 53 | 52 | 50 | 49 | 48 | 47 | 46 | 44 | 43 | 42 | 41 | 39 | 38 | 36 | 35 | 33 | 32 | 28 | 25 | 21 | 17 | 14 | 10 | 6 | 070 |
| 115 | 54 | 54 | 52 | 51 | 50 | 48 | 47 | 46 | 45 | 44 | 43 | 42 | 40 | 39 | 38 | 36 | 35 | 33 | 32 | 30 | 27 | 24 | 20 | 17 | 13 | 9 | 6 | 065 |
| 120 | +52 | +51 | +50 | +49 | +47 | +46 | +45 | +44 | +43 | +42 | +41 | +40 | +39 | +37 | +36 | +35 | +33 | +32 | +31 | +29 | +26 | +23 | +19 | +16 | +13 | +9 | +5 | 060 |
| 125 | 49 | 49 | 47 | 46 | 45 | 43 | 43 | 42 | 41 | 40 | 39 | 38 | 37 | 35 | 34 | 33 | 32 | 30 | 29 | 27 | 25 | 22 | 18 | 15 | 12 | 9 | 5 | 055 |
| 130 | 46 | 45 | 44 | 43 | 42 | 41 | 40 | 39 | 38 | 37 | 36 | 35 | 34 | 33 | 32 | 31 | 30 | 28 | 27 | 26 | 23 | 20 | 17 | 14 | 11 | 8 | 5 | 050 |
| 135 | 42 | 42 | 41 | 40 | 39 | 37 | 37 | 36 | 35 | 34 | 33 | 33 | 32 | 31 | 29 | 28 | 27 | 26 | 25 | 24 | 21 | 19 | 16 | 13 | 10 | 7 | 4 | 045 |
| 140 | 39 | 38 | 37 | 36 | 35 | 34 | 33 | 33 | 32 | 31 | 30 | 30 | 29 | 28 | 27 | 25 | 25 | 24 | 23 | 22 | 19 | 17 | 14 | 12 | 9 | 7 | 4 | 040 |
| 145 | 34 | 34 | 33 | 32 | 31 | 30 | 30 | 29 | 29 | 28 | 27 | 26 | 26 | 25 | 24 | 23 | 22 | 21 | 20 | 19 | 17 | 15 | 13 | 11 | 8 | 6 | 4 | 035 |
| 150 | +30 | +30 | +29 | +28 | +27 | +26 | +26 | +25 | +25 | +24 | +24 | +23 | +22 | +22 | +21 | +20 | +19 | +18 | +18 | +17 | +15 | +13 | +11 | +9 | +7 | +5 | +3 | 030 |
| 155 | 25 | 25 | 24 | 24 | 23 | 22 | 22 | 22 | 21 | 21 | 20 | 19 | 19 | 18 | 18 | 17 | 16 | 16 | 15 | 14 | 13 | 11 | 9 | 8 | 6 | 4 | 3 | 025 |
| 160 | 21 | 20 | 20 | 19 | 19 | 18 | 18 | 17 | 17 | 17 | 16 | 16 | 15 | 15 | 14 | 14 | 13 | 13 | 12 | 11 | 10 | 9 | 8 | 6 | 5 | 4 | 2 | 020 |
| 165 | 16 | 15 | 15 | 15 | 14 | 14 | 13 | 13 | 13 | 13 | 12 | 12 | 12 | 11 | 11 | 10 | 10 | 10 | 9 | 9 | 8 | 7 | 6 | 5 | 4 | 3 | 2 | 015 |
| 170 | 10 | 10 | 10 | 10 | 10 | 9 | 9 | 9 | 9 | 8 | 8 | 8 | 8 | 7 | 7 | 7 | 7 | 6 | 6 | 6 | 5 | 5 | 4 | 3 | 3 | 2 | 1 | 010 |
| 175 | +5 | +5 | +5 | +5 | +5 | +5 | +5 | +4 | +4 | +4 | +4 | +4 | +4 | +4 | +4 | +3 | +3 | +3 | +3 | +3 | +3 | +2 | +2 | +2 | +1 | +1 | +1 | 005 |
| 180 | 0 | 0 | 0 | 0 | 0 | 0 | 0 | 0 | 0 | 0 | 0 | 0 | 0 | 0 | 0 | 0 | 0 | 0 | 0 | 0 | 0 | 0 | 0 | 0 | 0 | 0 | 0 | 000 |
| 185 | -5 | -5 | -5 | -5 | -5 | -5 | -5 | -4 | -4 | -4 | -4 | -4 | -4 | -4 | -4 | -3 | -3 | -3 | -3 | -3 | -3 | -2 | -2 | -2 | -1 | -1 | -1 | 355 |
| 190 | 10 | 10 | 10 | 10 | 10 | 9 | 9 | 9 | 9 | 8 | 8 | 8 | 8 | 7 | 7 | 7 | 7 | 6 | 6 | 6 | 5 | 5 | 4 | 3 | 3 | 2 | 1 | 350 |
| 195 | 16 | 15 | 15 | 15 | 14 | 14 | 13 | 13 | 13 | 13 | 12 | 12 | 12 | 11 | 11 | 10 | 10 | 10 | 9 | 9 | 8 | 7 | 6 | 5 | 4 | 3 | 2 | 345 |
| 200 | 21 | 20 | 20 | 19 | 19 | 18 | 18 | 17 | 17 | 17 | 16 | 16 | 15 | 15 | 14 | 14 | 13 | 13 | 12 | 11 | 10 | 9 | 8 | 6 | 5 | 4 | 2 | 340 |
| 205 | 25 | 25 | 24 | 24 | 23 | 22 | 22 | 22 | 21 | 21 | 20 | 19 | 19 | 18 | 18 | 17 | 16 | 16 | 15 | 14 | 13 | 11 | 9 | 8 | 6 | 4 | 3 | 335 |
| 210 | 30 | 30 | 29 | 28 | 27 | 26 | 26 | 25 | 25 | 24 | 24 | 23 | 22 | 22 | 21 | 20 | 19 | 18 | 18 | 17 | 15 | 13 | 11 | 9 | 7 | 5 | 3 | 330 |
| 215 | -34 | -34 | -33 | -32 | -31 | -30 | -30 | -29 | -29 | -28 | -27 | -26 | -26 | -25 | -24 | -23 | -22 | -21 | -20 | -19 | -17 | -15 | -13 | -11 | -8 | -6 | -4 | 325 |
| 220 | 39 | 38 | 37 | 36 | 35 | 34 | 33 | 33 | 32 | 31 | 30 | 30 | 29 | 28 | 27 | 26 | 25 | 24 | 23 | 22 | 19 | 17 | 14 | 12 | 9 | 7 | 4 | 320 |
| 225 | 42 | 42 | 41 | 40 | 39 | 37 | 37 | 36 | 35 | 34 | 33 | 33 | 32 | 31 | 29 | 28 | 27 | 26 | 25 | 24 | 21 | 19 | 16 | 13 | 10 | 7 | 4 | 315 |
| 230 | 46 | 46 | 44 | 43 | 42 | 41 | 40 | 39 | 38 | 37 | 36 | 35 | 34 | 33 | 32 | 31 | 30 | 28 | 27 | 26 | 23 | 20 | 17 | 14 | 11 | 8 | 5 | 310 |
| 235 | 49 | 49 | 47 | 46 | 45 | 43 | 43 | 42 | 41 | 40 | 39 | 38 | 37 | 35 | 34 | 33 | 32 | 30 | 29 | 27 | 25 | 22 | 18 | 15 | 12 | 9 | 5 | 305 |
| 240 | 52 | 51 | 50 | 49 | 47 | 46 | 45 | 44 | 43 | 42 | 41 | 40 | 39 | 37 | 36 | 35 | 33 | 32 | 31 | 29 | 26 | 23 | 19 | 16 | 13 | 9 | 5 | 300 |
| 245 | -54 | -54 | -52 | -51 | -50 | -48 | -47 | -46 | -45 | -44 | -43 | -42 | -40 | -39 | -38 | -36 | -35 | -33 | -32 | -30 | -27 | -24 | -20 | -17 | -13 | -9 | -6 | 295 |
| 250 | 56 | 56 | 54 | 53 | 52 | 50 | 49 | 48 | 47 | 46 | 44 | 43 | 42 | 41 | 39 | 38 | 36 | 35 | 33 | 32 | 28 | 25 | 21 | 17 | 14 | 10 | 6 | 290 |
| 255 | 58 | 57 | 56 | 54 | 53 | 51 | 50 | 49 | 48 | 47 | 46 | 44 | 43 | 42 | 40 | 39 | 37 | 36 | 34 | 32 | 29 | 25 | 22 | 18 | 14 | 10 | 6 | 285 |
| 260 | 59 | 59 | 57 | 56 | 54 | 52 | 51 | 50 | 49 | 48 | 47 | 45 | 44 | 43 | 41 | 40 | 38 | 36 | 35 | 33 | 30 | 26 | 22 | 18 | 14 | 10 | 6 | 280 |
| 265 | 60 | 59 | 57 | 56 | 55 | 53 | 52 | 51 | 50 | 48 | 47 | 46 | 44 | 43 | 42 | 40 | 38 | 37 | 35 | 33 | 30 | 26 | 22 | 18 | 14 | 10 | 6 | 275 |
| 270 | -60 | -59 | -58 | -56 | -55 | -53 | -52 | -51 | -50 | -49 | -47 | -46 | -45 | -43 | -42 | -40 | -39 | -37 | -35 | -34 | -30 | -26 | -22 | -19 | -15 | -10 | -6 | 270 |

Observations earlier than solution time   signs as given
Observations later than solution time   signs reversed

**Figure 10-1.** *Correction for motion of the body.*

## Celestial Computation Sheets

The format in *Figure 10-2* is a typical celestial precomputation and illustrates one acceptable method of completing a precomputation. The explanation is numbered to help locate the various blocks on the celestial sheets. *[Figure 10-2]*

NOTE: Not all blocks apply on every precomputation.

1. DATE—place the Zulu date of the Air Almanac page used in this block.

2. FIX TIME—GMT (coordinated universal time) of the computation.

3. BODY—the celestial body being observed.

4. DR LAT LONG—the dead reckoning (DR) position for the time of the observation.

5. GHA—the value of GHA extracted from Air Almanac (10-minute intervals).

6. CORR—the GHA correction for additional minutes of time added to the GHA in block 5 and, if necessary, the 360° addition required establishing the LHA. SHA–When a star is precomped with Pub. No. 249, Volume 2 or 3, SHA is placed in this block.

7. GHA—corrected GHA (sum of blocks 5 and 6).

8. ASSUM LONG (–W/+E)—the assumed longitude required to obtain a whole degree of LHA.

9. LHA—LHA of the body (or Aries).

10. ASSUME LAT—the whole degree of latitude nearest the DR position.

11. DEC—the declination of the celestial body (not used with Pub. No. 249, Volume 1).

12. TAB Hc—the Hc from the appropriate page of Pub. No. 249, Volume 2 or 3.

13. D—the d correction factor found with previous Hc. Include + or –, as appropriate. The value is used to interpolate between whole degrees of Dec.

14. DEC—minutes of declination from block 11.

15. CORR—the correction from the Correction to Tabulated Altitude for Minutes of Declination table in Volume 2 or 3, using blocks 13 and 14 for entering arguments.

# CELESTIAL PRECOMPUTATION

SHEET NUMBER

## PRECOMPUTATION—PERISCOPIC SEXTANT

| NAVIGATOR | | | | ALT MSL. | 20 | DATE(Z) | 1 | FIX TIME | 2 | Z |

| STAR SELECTOR BY AZIMUTH | | | |
|---|---|---|---|

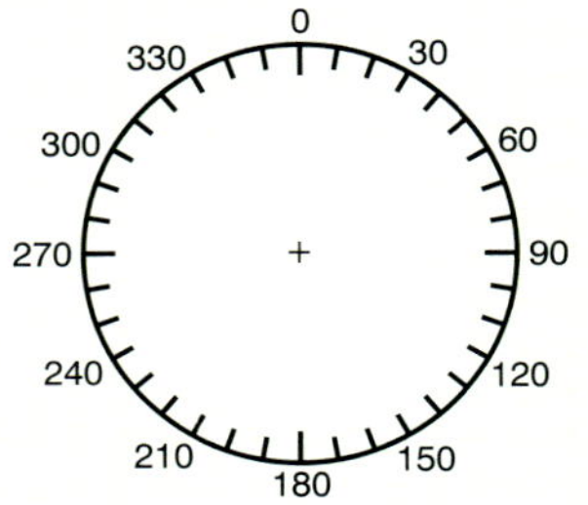

(compass diagram: 0, 30, 60, 90, 120, 150, 180, 210, 240, 270, 300, 330)

| | | | | | |
|---|---|---|---|---|---|
| TRACK | 18 | BODY | 3 | Polaris | |
| GS | 19 | BASE GHA | 5 | | |
| CORIOLIS | 21 (R/L) | CORR | 6 | | |
| PREC/ NUT | 22 | +360 | 6 | | |
| DR LAT | 4 (N/S) | GHA | 7 | | |
| DR LONG | 4 (E/W) | ASSUM LONG (W) +E | 8 | | |

| | | | | | | | | | |
|---|---|---|---|---|---|---|---|---|---|
| MOTION OF OBSERVER | 24 | | | LHA | 9 | | | | |
| MOTION OF BODY | 25 | | | ASSUM LAT | 10 | | | | |
| 4 MIN ADJUST | 26 | | | DEC | 11 N/S | N/S | N/S | N/S | N/S |
| OFF-FIX TIME | 27 E/L | E/L | E/L | E/L | E/L | PLANNED MID-TIME | | | |
| TOTAL MOT. ADJUST | 28 | | | ACTUAL MID-TIME | | | | | |
| POLARIS Q | 32 | 33 | | TAB Hc | 12 | | | | |
| MOON PA SD | 31 | | | D 13 / DEC 14  CORR | 15 | | | | |
| REF (–) | 29 | | | | | | | | |
| PERS/SEXT | 30 | | | CORR Hc | 16 | | | | |
| TOTAL → ADJ | 34 | | | TOTAL → ADJ | 34 | | | | |
| TH/GH | | | | ADJ Hc | 16±34 | | | | |
| Zn/GZa(–) | | | | OFF TIME MOTION | 35 | | | | |
| IRB | | | | Hc | 16±34 ±35 | | | | |
| IRB | | | | Ho | 36 | | | | |

REFRACTION TABLE (condensed)

| Ro | Altitude MSL (thousands of feet) | | | | | |
|---|---|---|---|---|---|---|
| | 0 | 20 | 25 | 30 | 35 | 40 |
| 1 | 53 | 46 | 41 | 36 | 33 | 26 |
| 2 | 33 | 19 | 16 | 14 | 11 | 9 |
| 3 | 23 | 12 | 10 | 8 | 7 | 5 |
| 4 | 16 | 8 | 7 | 6 | 5 | 3-10 |
| | 12 | 7 | 5 | 4 | 3-10 | 2-10 |

| | | | | | | | | | |
|---|---|---|---|---|---|---|---|---|---|
| Za/GZa(+) | | | | | | INT | 37 T/A | T/A | T/A | T/A | T/A |
| TH/GH | | | | | | Zn | 17 | | | | |
| TRACK T/G | 18 | | | CONV +W ANGLE –E | 39 | | | | |
| Za | 17 | | | GRID Zn | 40 | | | | |
| REL Za | 23 | | | | | | | | |

| | TK | DC | TH | VAR | MH | DEV | CH |
|---|---|---|---|---|---|---|---|
| | | | | | | | |

**Figure 10-2.** *Typical celestial precomputation format.*

16. CORR Hc—this is the corrected Hc—sum of blocks 12 and 15 or extracted from Pub. No. 249, Volume 1.

17. Zn—true azimuth of the celestial body from the formula in Pub. No. 249, Volume 2 or 3, or directly from Volume 1.

18. TRACK—the true course (track) of the aircraft.

19. GS—the groundspeed of the aircraft.

20. ALT MSL—aircraft altitude.

21. CORIOLIS—the Coriolis correction extracted from Pub. No. 249, the Air Almanac, or a Coriolis/rhumb line table.

22. PREC/NUT—precession and nutation correction computed from the table in Pub. No. 249, Volume 1.

23. REL Zn or Zn—the difference between Zn and track, used to determine motion of the observer correction.

24. MOTION OF OBSERVER (MOO)—motion of the observer correction for either 1 minute (using 1-minute motion correction table) or 4 minutes (using 4-minute correction table in Pub. No. 249) of time.

25. MOTION OF BODY (MOB)—motion of the body correction for either 1 minute (using 1-minute motion correction table) or 4 minutes (using tabulated Hc change for 1° of LHA or 4-minutes correction table in Pub. No. 249) of time.

26. 4-MINUTE ADJUST—algebraic sum of 24 and 25; for use of 4-minute motion corrections extracted from Pub. No. 249.

27. X-Time—time in minutes between planned shot time and fix time.

28. TOTAL MOT ADJUST/ADV/RET—correction based on combined motion of observer and body, for the difference between the time of the shot and fix time. The sign of this correction is the same as the sign in block 26 if the observation was taken prior to the computation time. If it was taken later, the sign is reversed.

29. REFR—correction for atmospheric refraction.

30. PERS/SEXT—sextant correction or personal error.

31. SD—semidiameter correction for Sun or Moon.

32. PA—parallax correction for Moon observation.

33. POLARIS/Q CORR—the Q correction for the time of the Polaris observation (extracted from Pub. No. 249 or the Air Almanac).

34. Total ADJ—algebraic sum of blocks 28–33 as applicable.

35. OFF-TIME MOTION—motion adjustment for observation other than at planned time.

36. Ho—height observed (sextant reading).

37. INT—intercept distance (NM) is the difference between the final Hc and Ho. Apply the HOMOTO rule to determine direction (T or A) along the Zn.

38. LAT—polaris latitude.

39. CONV ANGLE (W/–E)—convergence angle used in grid navigation.

40. GRID Zn—the sum of blocks 17 and 39.

## Corrections Applied to Hc

In some methods of precomputation, corrections are applied in advance to the Hc to derive an adjusted Hc. When using corrections that are normally applied to Hs, the signs of the corrections are reversed if applied to Hc. For example:

### Corrections Applied to Hs

| | |
|---|---|
| Hs | 31° 05 |
| REFR | –01 |
| PERS/SEXT | –05 |
| Ho | 30° 59 |
| Hc | 30° 40 |
| INT | 19T |

### Corrections Applied to Hc

| | |
|---|---|
| Hc | 30° 40 |
| REFR | +01 |
| PERS/SEXT | +05 |
| ADJ Hc | 30° 46 |
| Hs | 31° 05 |
| INT | 19T |

This demonstrates that corrections may be applied to either Hs or Hc. As long as they are applied with the proper sign, the intercept remains the same. The following sample precomp uses a common fix time (though computation times are different) and common observation times to facilitate comparison.

NOTE: Atmospheric refraction correction must be extracted for the actual Hs. It may then be applied to either Hc or Hs using the proper sign. Extracting the value for Hc may cause large errors, especially when the body is near the horizon. *Figure 10-3* is a sample three-star precomputation using the mathematical format. Corrections to altitude of the body are applied to the Hc and the sign of the correction has been reversed in this process, so the fix can be plotted prior to the

# CELESTIAL PRECOMPUTATION

| | SHEET NUMBER |
|---|---|

## PRECOMPUTATION—PERISCOPIC SEXTANT

| NAVIGATOR | P. HENRY | | | ALT MSL | 330 | DATE(Z) | 20 APR 79 | FIX TIME | 0740 Z |
|---|---|---|---|---|---|---|---|---|---|

| STAR SELECTOR BY AZIMUTH | | | | | | | | | |
|---|---|---|---|---|---|---|---|---|---|
| | TRACK | 120 | | BODY | VEGA | Spica | Pollux | | |
| | GS | 450 | | BASE GHA | 312–46 | | | | |
| | CORIOLIS | 10 | Ⓡ L | CORR | 1–00 | | | | |
| | PREC/NUT | 0/000 | | +360 | ---- | | | | |
| | DR LAT | 38–14 | Ⓝ S | GHA | 313–46 | | | | |
| | DR LONG | 120–50 | E Ⓦ | ASSUM LONG +E | Ⓦ 120–46 | | | | |

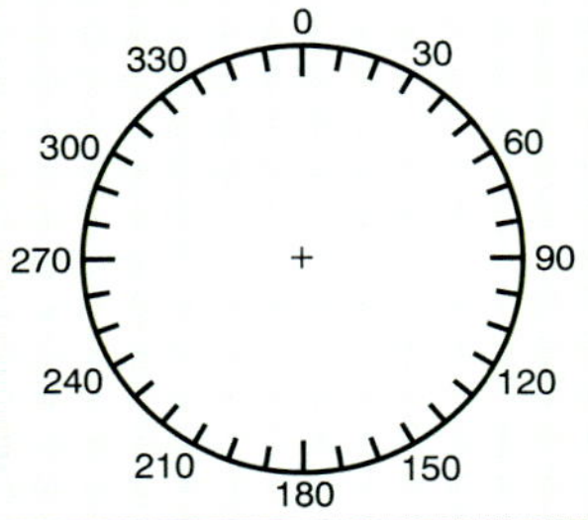

| MOTION OF OBSERVER | +15 | +19 | –29 | | LHA | 193 | | | | |
|---|---|---|---|---|---|---|---|---|---|---|
| MOTION OF BODY | +41 | +09 | –45 | | ASSUM LAT | 38 N S | | | | |
| 4 MIN ADJUST | +56 | +28 | –1–14 | | DEC | N S | N S | N S | N S | N S |
| OFF-FIX TIME | 12 Ⓔ L | 8 Ⓔ L | 4 Ⓔ L | E L | E L | PLANNED MID-TIME | 0652 | 0656 | 0700 | |
| TOTAL MOT. ADJUST | +2–48 | +56 | –1–14 | | ACTUAL MID-TIME | | | | | |
| POLARIS Q MOON PA SD | | | | | TAB Hc | | | | | |
| REF (–) | –01 | 0 | –01 | | D DEC CORR | | | | | |
| PERS/SEXT | 0 | 0 | 0 | | CORR Hc | 25–19 | 40–22 | 26–31 | | |
| TOTAL →ADJ | +2–47 | +56 | –1–15 | | TOTAL →ADJ | –2–47 | –56 | +1–15 | | |
| TH/GH | | | | | ADJ Hc | 22–32 | 39–26 | 27–46 | | |
| Zn/GZa(-) | | | | | OFF TIME MOTION | | | | | |
| IRB | | | | | Hc | 22–32 | 39–26 | 27–46 | | |
| IRB | | | | | Ho | 23–04 | 39–20 | 27–24 | | |
| Za/GZa(+) | | | | | INT | 32 Ⓣ A | 6 Ⓣ A | 22 Ⓣ A | T A | T A |
| TH/GH | | | | | Zn | 059 | 170 | 286 | | |
| TRACK T/G | 120 | 170 | 286 | | CONV +W ANGLE –E | | | | | |
| Za | 059 | 120 | 120 | | GRID Zn | | | | | |
| REL Za | 061 | 050 | 166 | | TK | DC | TH | VAR | MH | DEV | CH |

REFRACTION TABLE (condensed)

| Ro | Altitude MSL (thousands of feet) | | | | | |
|---|---|---|---|---|---|---|
| | 0 | 20 | 25 | 30 | 35 | 40 |
| 1 | 53 | 46 | 41 | 36 | 33 | 26 |
| 2 | 33 | 19 | 16 | 14 | 11 | 9 |
| 3 | 23 | 12 | 10 | 8 | 7 | 5 |
| 4 | 16 | 8 | 7 | 6 | 5 | 3–10 |
| | 12 | 7 | 5 | 4 | 3–10 | 2–10 |

**Figure 10-3.** *Mathematical solution.*

computation time. All shots are early shots, allowing the navigator to resolve the fix and alter at fix time. However, any minor errors in interpolation for motions are multiplied for the two earliest shots and may cause inaccuracies in the fix. Refer to Appendix D for a blank celestial precomputation sheet.

*Figure 10-4* shows a three-star precomputation using a three-LHA or graphical solution. The assumed position is then moved for track and GS to accommodate LOPs shot off time. Each observation is taken on time and then plotted out of its own plotting position. This precomp is easier and faster to accomplish with relatively few opportunities for math errors to occur. The three assumed positions required for this solution, on the other hand, often cause large intercepts and may make star identification difficult if care is not taken in choosing the precomp assumed position.

## Limitations

Precomputational methods lose accuracy when the assumed position and the actual position differ by large distances. Another limiting factor is the difference in time between the scheduled and actual observation time. The motion of the body correction is intended to correct for this difference. The rate of change of the correction for motion of the body changes very slowly within 40° of 090° and 270° Zn, and the observation may be advanced or retarded for a limited period of time with little or no error. When the body is near the observer's meridian, however, the correction for motion of the body changes rapidly due in part to the fast azimuth change and it is inadvisable to adjust such observations for long (over 6 minutes) periods of time.

NOTE: Errors in altitude and azimuth creep into the solution if adjustments are made for too long an interval of time. Because of these errors, the navigator should attempt to keep observation time as close as possible to computation time.

## Preplotting True Azimuth (Zn)

To speed up fix resolution, some navigators preplot the Zn of the bodies. This technique works best when used on a constant scale chart and using a technique of precomputation that gives one assumed position. Before making any observations, plot the assumed position, correct it for Coriolis and precession and/or nutation (if required), and draw the Zn of the bodies through this point. Label each Zn as the 1st, 2d, or 3d as shown in *Figure 10-5,* or use the name of the bodies. Use arrowheads to identify the direction of the body. Suppose the corrected assumed position is 30°40' N, 117° 10' W and the following Zn were computed for the bodies:

1st shot ZN 020°

2d shot ZN 135°

3d shot ZN 270°

| BODY | | | |
|---|---|---|---|
| TIME | | | |
| LAT | | | |
| DEC | | | |
| GHA | | | |
| ± 360 | | | |
| GHA | | | |
| LONG $^{-W}_{+E}$ | | | |
| LHA | | | |
| HC | | | |
| CORR | | | |
| HC | | | |
| HO | | | |
| INT ($^T_A$) | | | |
| ZN | | | |
| LONG/CONV | | | |
| GRIO ZN | | | |
| BODY | VERA | SPICA | ALLOW |
| TIME | 0652 | 0656 | 0700 |
| LAT | | | 39N |
| DEC | | | |
| GHA | | | 312-46 |
| ± 360 | | | —— |
| GHA | | | 312-46 |
| LONG $^{-W}_{+E}$ | | | 120-46W |
| LHA | 190 | 191 | 192 |
| HC | 23-49 | 39-05 | 27-32 |
| CORR | +01 | 0 | +01 |
| HC | 23-50 | 39-05 | 27-33 |
| HO | 23-04 | 39-20 | 27-24 |
| INT ($^T_A$) | 46 A | 15 T | 9 A |
| ZN | 058 | 167 | 285 |
| LONG/CONV | | | |
| GRIO ZN | | | |
| **DRIFT DATA** | | | |
| TIME | | | |
| GROSS WT | | | |
| PAGE NO. | | | |
| FUEL MEM | | | |
| O/M FUEL | | | |
| DIFF | | | |
| FUEL ETE | | | |
| ETE DEST | | | |
| EXTM TIME | | | |
| EHOUR | | | |
| | | | |

**Figure 10-4.** *Graphical solution.*

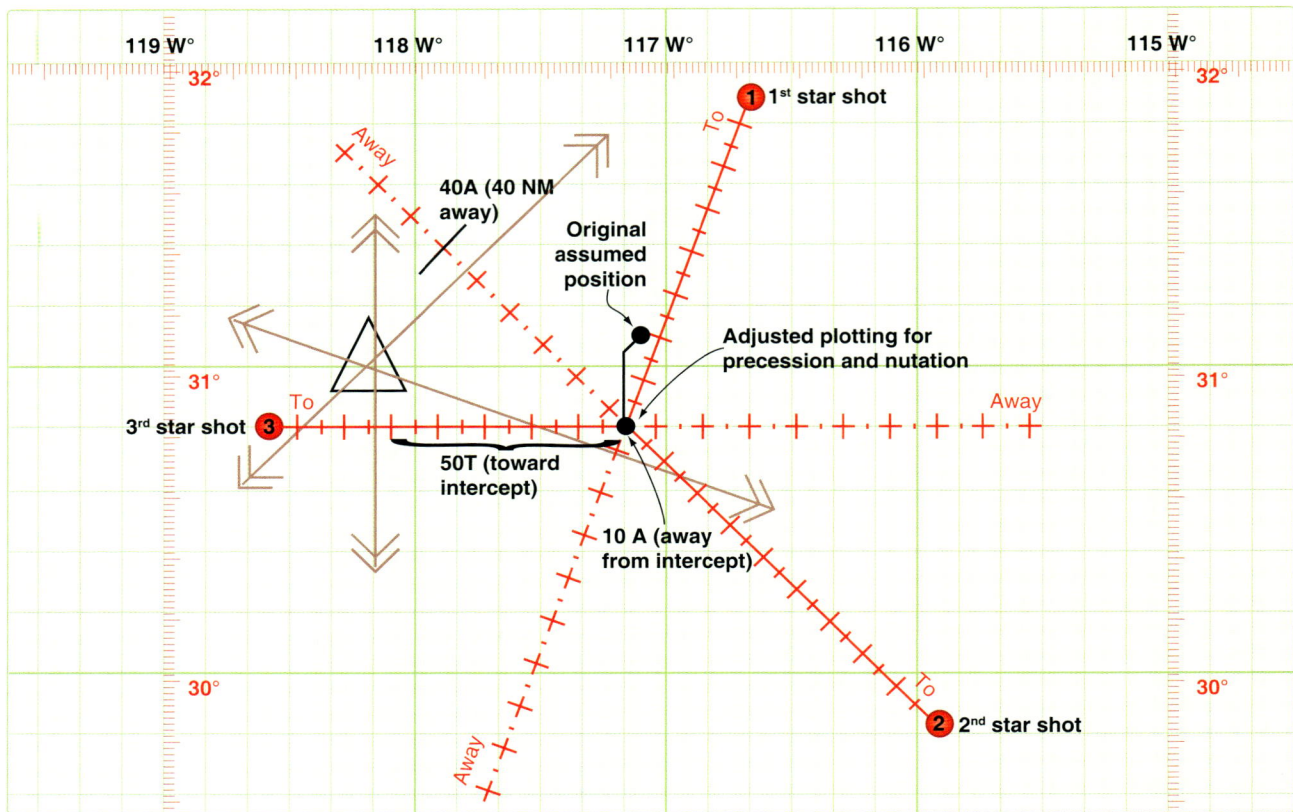

**Figure 10-5.** *The fix can be plotted quickly.*

The original assumed position of 31° N; 117° 08' W has been corrected for precession and/or nutation and for Coriolis or rhumb line error to obtain the plotting position. When the first intercept is found to be 10A, second intercept 40A, and the third intercept 50T, the fix may be plotted quickly by constructing perpendicular lines at the correct point on the respective Zn line. This greatly reduces the time necessary to plot the fix.

## Chapter Summary

Celestial precomputation methods have been brought to the forefront with the proliferation of high-speed aircraft. Aircraft speeds make it necessary to minimize the time between shooting and fixing. Since the sextant may be the only means of viewing the body, it is necessary to precompute the altitude and azimuth of a body in order to locate it. Remember corrections may be applied to the Hc, Ho, or intercept, and pay close attention to the sign of the correction. In addition to precomputation, the fix may be resolved faster by preplotting the true azimuths of the bodies.

# Chapter 11

# Plotting and Interpreting the Celestial Line of Position

## Introduction

This chapter explains the methods that transform the tabulated and inflight observation values into an aircraft position. The navigator is faced with two tasks: plotting the resultant information onto a chart and resolving this information into an aircraft position. There are two basic methods of obtaining a line of position (LOP): the subpoint method and the intercept method.

Step 4: Flop plotter plotter is now perpendicular to Zn

Step 3: Flip di

Step 5: Draw lop

123°W

121°W

120°W

Observer's assumed position

TN

Toward the stars

Hc (altitude)

Zn

LOP

LOP

Intercept if Ho less than Hc

90°-Ho = Complement angle (90°-60°=30°)
1°= 60 Nautical miles on the earths surface
30°=1,800 NM (zenith distance)

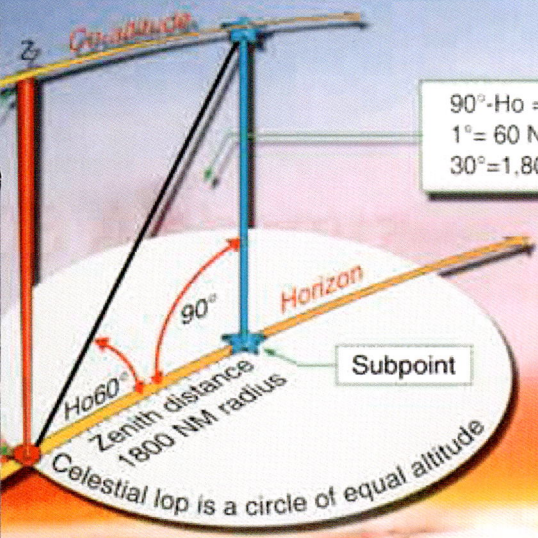

Zd

Co-altitude

90°

Horizon

Ho60°

Zenith distance
1800 NM radius

Subpoint

Celestial lop is a circle of equal altitude

Great circle thru observer's position and subpoint of star

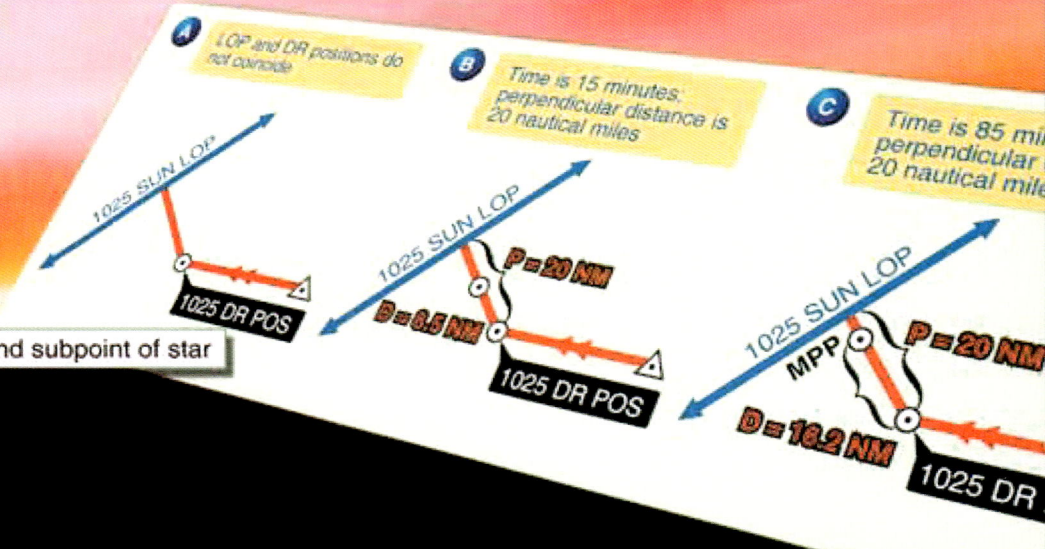

**A** LOP and DR positions do not coincide

**B** Time is 15 minutes. perpendicular distance is 20 nautical miles

**C** Time is 85 mi perpendicular 20 nautical mile

1025 SUN LOP

1025 DR POS

1025 SUN LOP

P = 20 NM

D = 6.5 NM

1025 DR POS

1025 SUN LOP

MPP

P = 20 NM

D = 18.2 NM

1025 DR

## Subpoint Method

A detailed explanation of the theory concerning the subpoint method is in Chapter 9, Computing Altitude and True Azimuth, and Chapter 10, Celestial Precomputation. *[Figure 11-1]* Following is a summary of the steps involved:

1. Positively identify the body and measure the altitude using a sextant.

2. Because no tabulated information for azimuth or elevation is required for this method, corrections for refraction, parallax, semidiameter, wander error, and sextant correction are applied directly to the Ho.

3. The resultant measurement is subtracted from 90° to obtain the co-altitude (co-alt). To convert to NM (1°= 60 NM), multiply the number of degrees times 60. Any fractional portion of degrees is added to the previous value.

Example: Vega is observed at an altitude (Ho) of 88° 23'. Sextant correction is –03'.

$$88° \ 23' - 03' = 88° \ 20'$$

$$90° - 88° \ 20' = 1° \ 40'$$

$$1° \ 40' = 60' + 40' = 100 \ NM$$

In this example, 100 NM represents the distance from the observer's position to the subpoint of the body. The coordinates of the body are its corresponding declination (Dec) and Greenwich hour angle (GHA). For this example, Vega's Dec is N38° 46'. The GHA is obtained by applying the sidereal hour angle (SHA) of Vega to the GHA of Aries.

Example:

SHA = 080° 59'

GHA Aries = 039° 18'

GHA Vega = 120° 17'

Subpoint of Vega is located at 38° 46' N 120° 17' W. The observer is now ready to apply the information:

1. Plot the subpoint on an appropriate chart.

2. With dividers or compass, span the co-alt distance; in this case 100 NM.

3. Use the body's subpoint (38° 46' N 120° 17' W) as the center and 100 NM (co-alt) as the radius. The circle is called the circle of equal altitude and the observer is located on that portion of the circle nearest the dead reckoning (DR) position. There are definite advantages to this method. It requires no precomputation values and plotting is very simple if the observer and body are reasonably close together. When the observer and body are separated by great distances, some disadvantages appear.

4. If a body is observed at 20° above the horizon, the observer is 4,200 NM from its subpoint. To swing a LOP from this subpoint, the subpoint and the arc must be plotted on the same chart. To permit plotting of any LOP, the chart must cover an area extending more than 4,000 miles in every direction from the DR position. This means that the chart must be either of such large size that it cannot be spread out on a table in the aircraft, or of such small scale that plotting on it is inaccurate. To cover an area 8,000 miles across, a chart 4 feet square must be drawn to a scale of about 1:10,000,000. Furthermore, measuring would be difficult because of distortion.

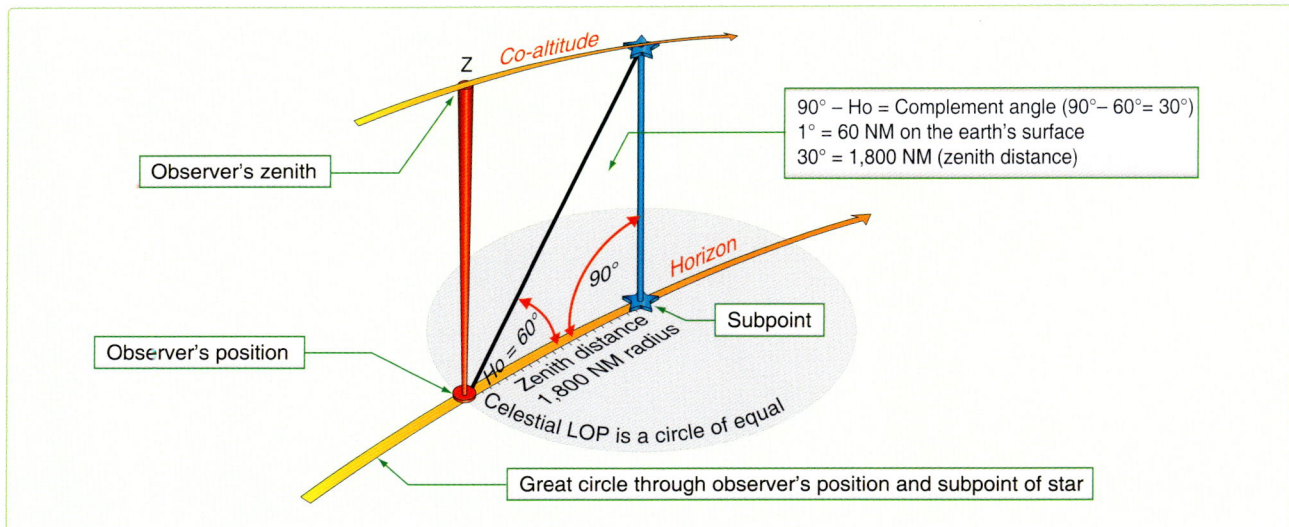

**Figure 11-1.** *The subpoint method.*

5. Since a celestial LOP cannot always be drawn by the subpoint method, the intercept method, based on the same principles, is often used.

## Intercept Method

You can eliminate the need for plotting the body's subpoint and still draw the arc representing the circle of equal altitude. *[Figure 11-2]* By using the following formula, you can calculate the altitude and azimuth of the body for the DR position:

$$Hc = SIN^{-1} [SIN (DEC') SIN (LDr) +COS (DEC')COS (LDr) COS (LHA)]$$

$$Z = COS (Z) = [SIN (DEC') - SIN (LDR) SIN (HC)]/[COS (Hc COS (LDr)]$$

$$Zn = Z \text{ if } SIN (LHA) < 0$$

$$Zn = 360 - Z \text{ if } SIN (LHA) \geq 0$$

The calculations may be performed quickly using a programmable calculator, or they may be extracted from the appropriate volume of the National Imagery and Mapping Agency's Sight Reduction Tables for Air Navigation in a publication referred to as Pub. No. 249. This method enables the observer to use any of the navigational bodies available at the appropriate fix time. Here is a brief review:

- Compute a DR for the time of the position, using preflight or inflight data.

- Determine the necessary entering values for the Pub. 249 volume being used (Lat, LHA, Dec contrary, or same) and extract all the necessary values of computed altitude (Hc) and azimuth angle (Z).

- After making all the necessary conversions and corrections (Chapter 10), compare the Ho and corrected Hc. This difference is the intercept. If the Ho equals the corrected Hc, then the circle of equal altitude passed through the plotting position. If the Ho is greater than the Hc, the difference is plotted in the direction of the true azimuth (Zn). The Zn represents the azimuth from the observer's position to the subpoint. If the Ho is less than the Hc, plot the difference 180° from the Zn.

- NOTE: If HO is MOre, plot TOward the subpoint (HO MO TO)

Example: The assumed position is 38° N, 121° 30' W for a shot taken at 1015Z on Aldebaran. The Ho is 32° 14'. The Hc is determined to be 32° 29' and the Zn is 120°. A comparison of Ho and Hc determines the intercept to be 15 NM away (15A).

## Plotting LOP Using Zn Method

1. Plot the assumed position and set the intercept distance on the dividers. *[Figure 11-3]*

2. Draw a dashed line through the assumed position toward the subpoint.

3. Span intercept distance along dashed Zn line.

4. Place plotter perpendicular to Zn.

5. Draw LOP along plotter as shown in *Figure 11-3*.

## Plotting LOP Using Flip-Flop Method

1. Plot the assumed position and set the intercept distance on the dividers. *[Figure 11-4]*

2. Measure 120° of the Zn with point A of the dividers on the assumed position and place point B of the dividers down. In this case, away from 120° or in the direction

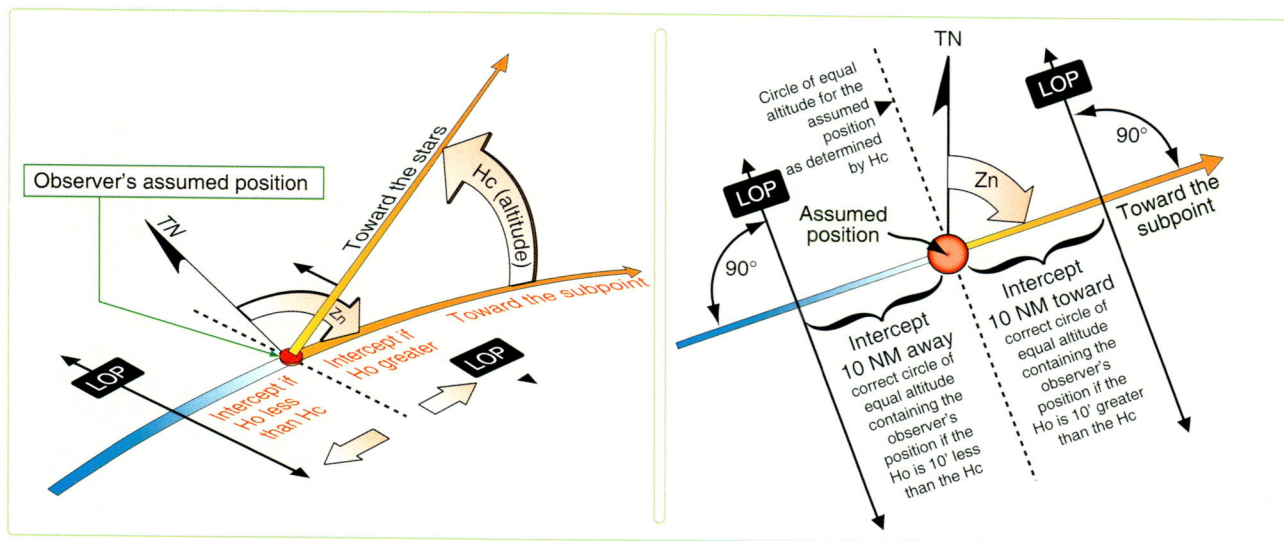

**Figure 11-2.** *Line of position computed by intercept method.*

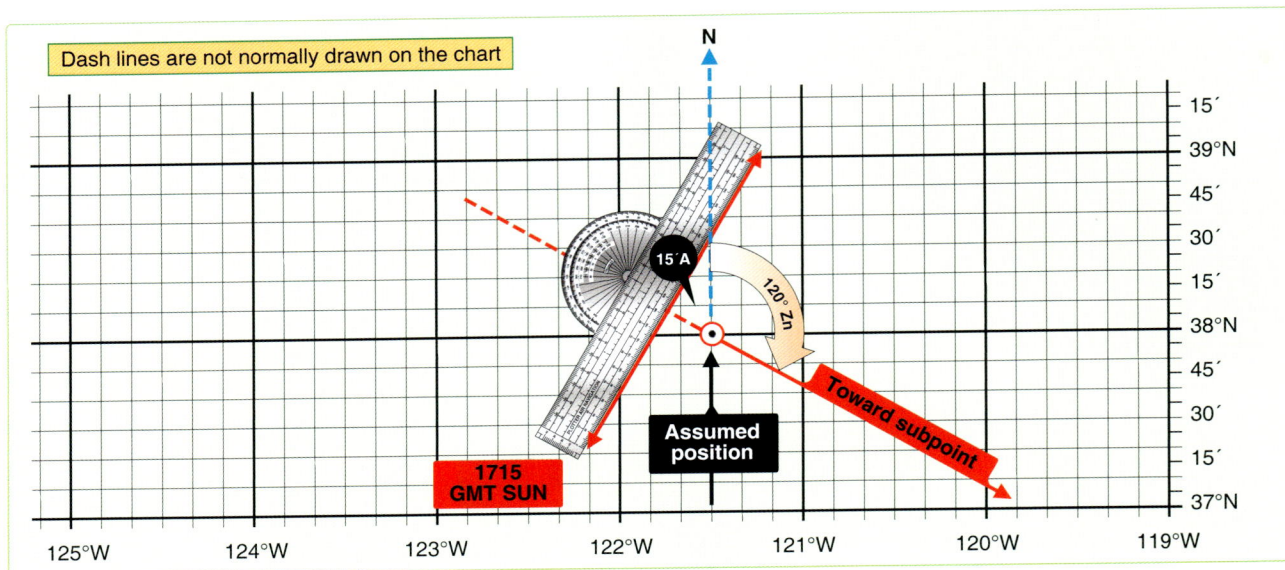

**Figure 11-3.** *Celestial line of position using true azimuth method.*

of 300° from the assumed position. Slide the plotter along the dividers until the center grommet and the 100/200-mile mark are lined up directly over point B of the dividers marking the intercept point.

3.  Remove point A of the dividers from the assumed position, keeping point B in place. Flip point A (that was on the assumed position) across the plotter, at the same time expanding the dividers so that point A can be placed on the chart at the 90°/270° mark of the plotter.

4.  Flop the plotter around and place the straight edge against the perpendicular, which is established by the dividers.

5.  Draw LOP along the plotter as shown in *Figure 11-4.*

When using the intercept method, remember:

*   For some assumed position near the DR position, find the Hc and Zn of this body for the time of the observation. This is done with the aid of celestial tables, such as Pub. No. 249 or a programmable calculator.

*   Obtain needed corrections, sextant correction, refraction, etc., and apply these to the Hc by reversing the sign. Remember, we are striving to derive a precomputed value to ensure the correct body is shot. Measure the altitude (Ho) of the celestial body with the sextant and record the midtime of the observation.

*   Find the intercept, which is the difference between Ho and Hc. Intercept is toward the subpoint if Ho is greater than Hc, and away from the subpoint if Ho is smaller than Hc.

*   From the assumed position, measure the intercept toward or away from the subpoint (in the direction of Zn

or its reciprocal) and locate a point on the LOP. Through this point, draw the LOP perpendicular to the Zn.

## Additional Plotting Techniques

The preceding techniques involve the basic plotting procedures used on most stars and the bodies of the solar system. However, there are certain techniques of plotting that are peculiar to their own celestial methods; for example, the plotting of LOPs obtained by using Polaris, which is discussed later. Also, certain precomputation techniques lend themselves more readily to other plotting techniques, such as preplotting the true azimuths or plotting the fix on the DR computer.

These last plotting techniques are discussed in Pub. No. 249 in the section on precomputation. Other special techniques are discussed in the section on curves, in which the celestial observation is plotted on a graph rather than on the chart.

## Interpretation of an LOP

Navigation has two aspects—the mechanical and the interpretive. The mechanical aspect includes operation and reading of instruments, simple arithmetical calculations, plotting, and log keeping. The interpretive aspect is the analysis of the data that have been gathered mechanically. These data are variable and subject to error. You must convert them into probabilities as to the position, track, and GS of the aircraft and the direction and speed of the wind. The more these data are subject to error, the more careful the interpretations must be and the less mechanical the work can be. LOPs and fixes especially require careful interpretation. It is convenient to think of a fix as the true position of the aircraft and of the LOP as a line passing through this position, but these definitions are optimistic. It is almost impossible

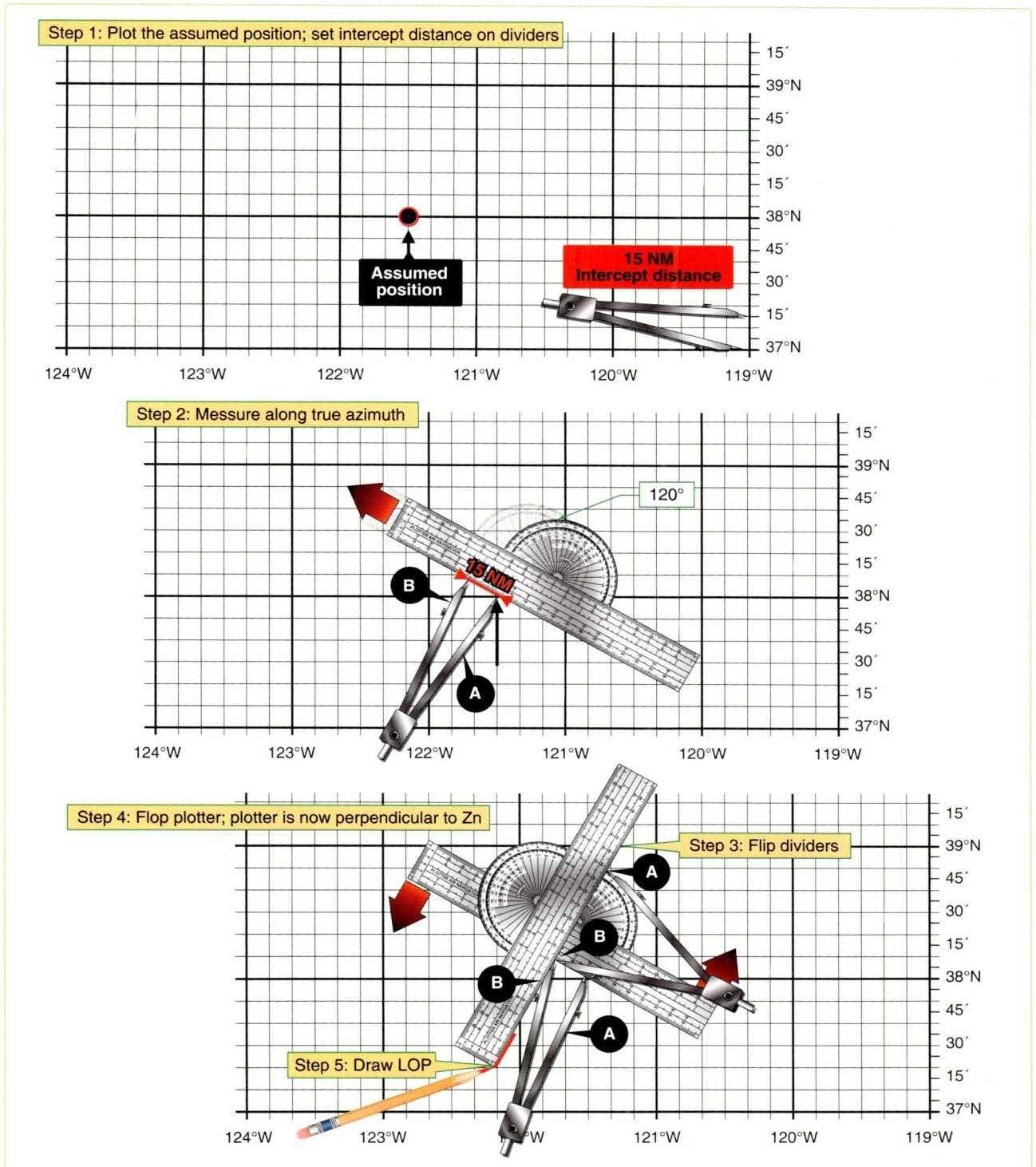

**Figure 11-4.** *Plotting celestial line of position using flip-flop method.*

to make a perfect observation and plot a perfect LOP. Therefore, a LOP passes some place near this position, but not necessarily through it, and a fix determined by the intersection of LOPs is simply the best estimate of this position on the basis of one set of observations. In reality, a fix is a most probable position (MPP) and a LOP is a line of MPP.

The best interpretation of LOPs and fixes means they are used, to the best advantage, with DR. But good interpretation cannot compensate for poor LOPs, nor can good LOPs compensate for careless DR. To get good results, every precaution must be taken to ensure the accuracy of LOPs and exact DR calculations.

Intelligent interpretation requires fine judgment, which can be acquired from experience. You can be guided, however, by certain well-established, though flexible, rules. The following discussion pertains especially to celestial LOPs and fixes. It also applies to LOPs and fixes established by radio and, to some extent, to those obtained by map reading.

## Single LOP

Previous discussions dealt with the basic plotting of a LOP and errors in LOPs, but they did not show the actual mechanics of the plotted corrections that must be applied. The LOP must be corrected for Coriolis or rhumb line correction and also for precession and/or nutation correction if it is based on a Volume 1 star shot. Coriolis or rhumb line correction becomes a very significant correction at higher speeds and latitudes. For example, suppose the correction determined from the Coriolis or rhumb line correction table is 9 NM right (of the track). The LOP must be moved a distance of 9 NM to the right of track. This can be done either by moving the assumed position prior to plotting or by moving the LOP

itself after it is plotted. (Remember, the assumed position is not used in the plotting of the LOP obtained from a Polaris observation.) Consider *Figure 11-5*, which shows a track of 90°. Notice that, in both methods, the corrected LOP is in the same place with respect to the original assumed position and that the intercept value is the same. The resultant LOP is the same regardless of the method used.

If, in addition to the Coriolis or rhumb line correction, a precession and/or nutation correction of 3 NM in the direction of 60° is required, it would have been further applied as shown in *Figure 11-6*. Again, the corrected LOP is the same, using either method, because the intercept and resultant position of the corrected LOP to the original assumed position are the same. The corrected LOP alone gives very little information; hence, a position must be arrived at only after considering the LOP and the DR position for the same time.

## Most Probable Position (MPP) by C-Plot

The MPP is just what the name implies. It is not a fix; however, since it is the best information available, it is treated as such. Notice in *Figure 11-7A* that the DR position and celestial LOP (for the same time) do not coincide.

Obviously, the DR information, or celestial information, or both is in error. Notice that the prior fix has no time on it. Suppose this prior fix had been for the time of 1010. It would then be very likely that most of the error is in the celestial information and the probable position is closer to the DR position than to the celestial LOP. On the other hand, suppose the prior fix had been for the time of 0900. Since the accuracy of the celestial information is unaffected by the time from the last fix, it would, in this case, be most likely that the actual position is closer to the LOP than to the DR position.

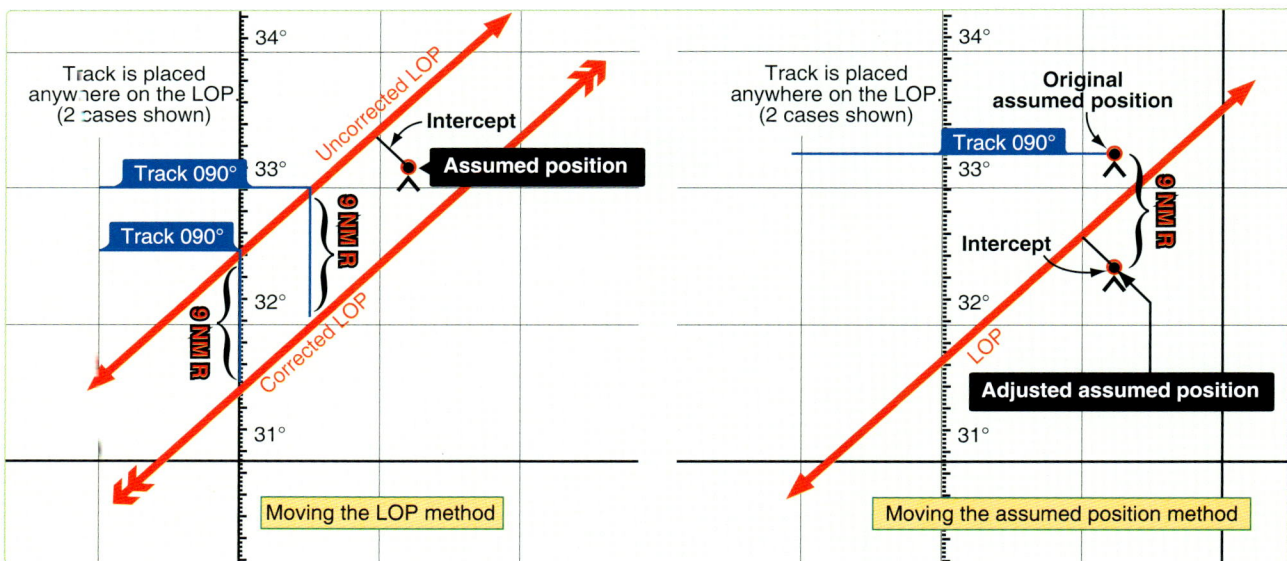

**Figure 11-5.** *Two methods of coriolis/rhumb line correction.*

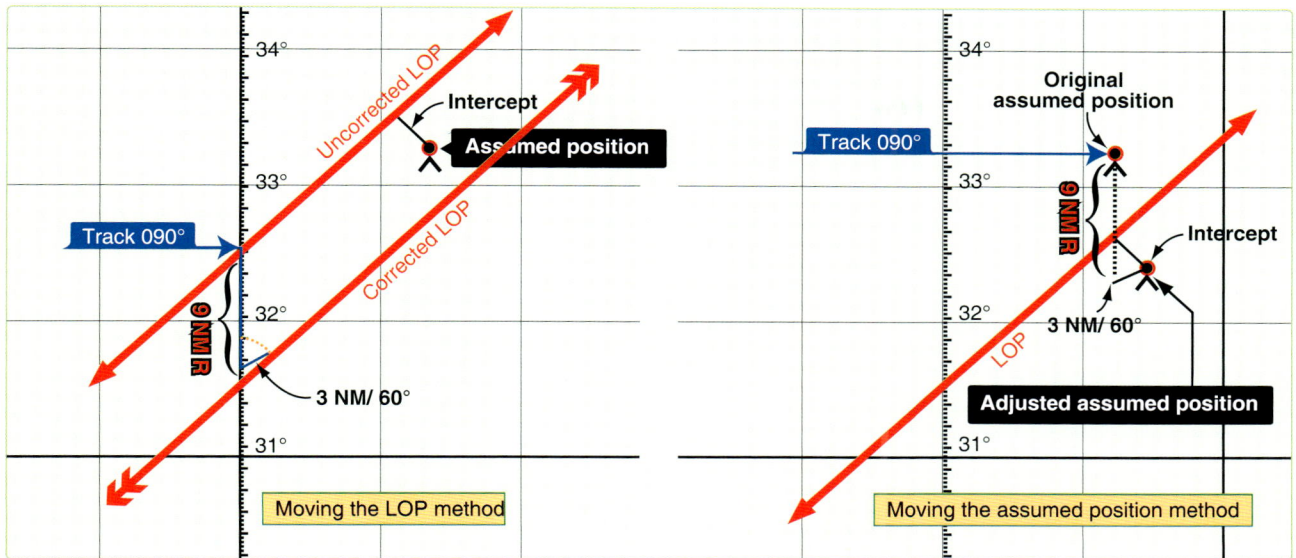

**Figure 11-6.** *Two methods of coriolis/rhumb line and precession/nutation correction.*

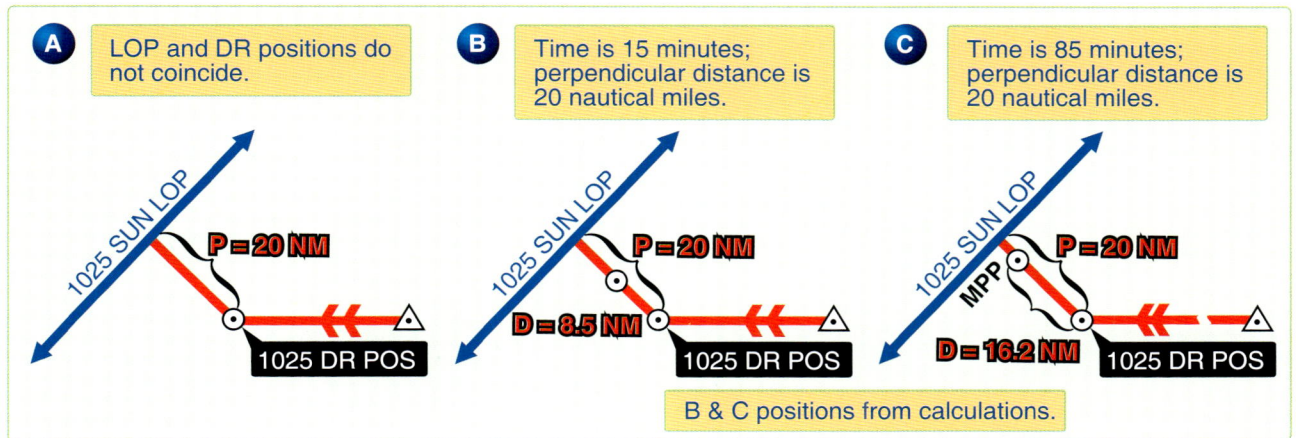

**Figure 11-7.** *Most probable position by C-plot.*

A formula has been devised to position the observer along the perpendicular to the LOP according to the time factor. The formula is:

$$\frac{d}{t} = \frac{p}{t + p}$$

where t is time in minutes, p is the perpendicular distance between the DR position and the LOP, and d is the distance from the DR position for the time of the MPP measured along the perpendicular to the LOP. Look at *Figure 11-7B* and *C* and see how the formula works for the two problems cited above if the perpendicular is 20 NM in length. In *Figure 11-7B*, t is 15 minutes and p is 20 NM, so the MPP would be located along the perpendicular about 8½ NM from the DR position.

$$\frac{d}{15} = \frac{20}{15 + 20} \qquad d = \frac{300}{35} = 8.57 \text{ NM}$$

Now, consider *Figure 11-7C* where t is 1 hour 25 minutes or 85 minutes, p is 20 NM and, in this case, the MPP would be over 16 NM away from the DR position along the perpendicular to the LOP.

$$\frac{d}{85} = \frac{20}{85 + 20} \qquad d = \frac{1700}{105} = 16.2 \text{ NM}$$

If you prefer not to use the formula, a simple table can be easily constructed to solve for d with entering arguments of t and p. *[Figure 11-8]* The table could easily be enlarged to handle larger values of t and p. In most fixes, the DR position is so close to the LOP that the midpoint between these two can be considered the MPP. A good rule to use is to take the midpoint of the perpendicular if the total distance between the DR position and the LOP is 10 NM or less. If the value of p is greater than 10 NM, use a table or the formula to determine

| P ▼ | "t" time (minutes) | | | |
|---|---|---|---|---|
| | 10m | 15m | 30m | 60m |
| 12 | 5 | 7 | 9 | 10 |
| 14 | 6 | 7 | 10 | 11 |
| 16 | 6 | 8 | 10 | 13 |
| 18 | 6 | 8 | 11 | 14 |
| 20 | 7 | 9 | 12 | 15 |
| 25 | 7 | 9 | 14 | 18 |
| 30 | 8 | 10 | 15 | 20 |
| 35 | 8 | 10 | 16 | 22 |
| 40 | 8 | 11 | 17 | 24 |
| 45 | 8 | 11 | 18 | 26 |
| 50 | 8 | 12 | 19 | 27 |
| 55 | 9 | 12 | 19 | 29 |
| 60 | 9 | 12 | 20 | 30 |
| To the closest NM values of "d" in the formula | | | | |

**Figure 11-8.** *To solve for distance.*

the MPP. Up to this point, determination of the MPP has been rather mechanical. Experienced navigators frequently further adjust the position of the MPP for other factors not yet considered. For example, if the LOP is carefully obtained under good conditions or if it is the average of several LOPs, you may further weight the MPP in the direction of the LOP by an amount that judgment dictates. However, the reverse may be true if the LOP is obtained under adverse conditions of rough air. In the latter case, you might move the MPP closer to the DR position by some amount determined by sound judgment.

Further, consider the validity of the DR position in relation to factors other than time. A DR position at the end of 40 minutes would be more reliable with Doppler drift and GS versus one based on metro information. These factors may also adjust the original MPP closer to or farther away from the DR position, along the perpendicular. However, these last mentioned factors are judgment values that come only with experience. In fact, with experience you may mentally calculate all the factors involved and arrive at the final position of the MPP without recourse to a formula or table.

### Finding a Celestial Fix Point

Up to this point, only the single celestial LOP and what to do with it have been considered. Now, the celestial fix should be considered. To establish a fix, two or more LOPs must be obtained. Since, in most cases, two or more LOPs cannot be obtained simultaneously, they must be converted to a common time. For example, a LOP obtained at 1010 must be converted to the LOP obtained at the fix time of 1014. There are several methods for making this conversion that are discussed in this chapter. Consideration is also given to the planning of the fix and the final interpretation of the fix itself.

## Conversion of LOPs to a Common Time

### Moving the LOP

One method of converting LOPs to a common time is to move the LOP along the best-known track for the number of minutes of GS necessary for the time conversions. This method is similar to that used in correcting for Coriolis or rhumb line and precession or nutation. For example, suppose the track is 110° and the GS is 300 knots. LOPs are for 1500, 1504, and 1508, and a fix is desired at 1508. This means the 1500 LOP must be moved to the time of the fix using the track and 8 minutes of the best known GS. The 1504 LOP must be moved to the time of the fix using the track and 4 minutes of GS. The 1508 LOP is already at the fix time, so it requires no movement. *Figure 11-9* shows the method of conversion as it is completed on the chart.

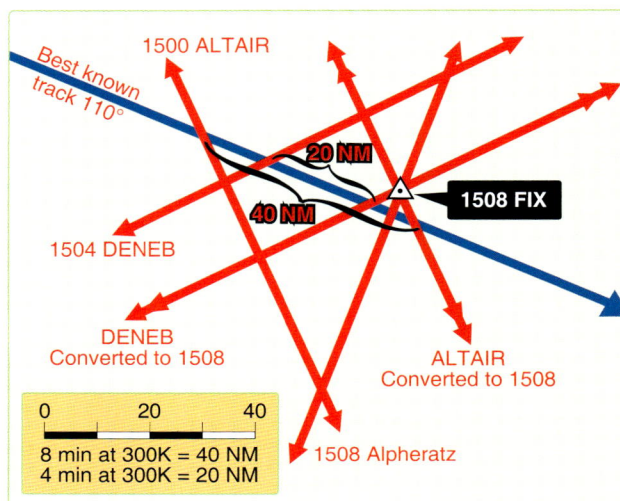

**Figure 11-9.** *Conversion of lines of position to a common time.*

If, at any time, the LOP has to be retarded (moved back) to the time of the fix, use the following procedures. Using the reciprocal track and GS, obtain the correction in the regular manner for the number of minutes of difference. For example, suppose the fix is at 1800 and the last shot is at 1802.

Retarding the LOP 2 minutes of GS on a track of 70° would be the same as advancing it 2 minutes of GS on a track of 250°.

### Motion of Observer Tables

A second method of conversion of LOPs to a common time is with a Motion of the Observer table such as the one in Pub. No. 249. This table gives a correction to be applied to the Ho or Hc so that the LOP plots in its converted position. The correction obtained from Table 1 in all volumes of Pub. No. 249 is for 4 minutes of time. An additional table allows you to get the correction for the number of minutes needed. For example, suppose the LOP needs to be advanced for 11 minutes and the Ho of the body is 33° 29' and Zn is 080°.

The track of the aircraft is 020º and the GS is 240 knots. In Table 1, Correction for Motion of the Observer for 4 minutes of Time *[Figure 11-10]*, the entering arguments is Rel Zn and GS. Rel Zn is azimuth relative to course (Zn minus track or track minus Zn). Subtract the smaller angle from the larger and enter the table with the answer. In this case, Zn – track = 080° – 020° = 060° (Rel Zn) and GS is 240 knots. Entering this table with these arguments, the correction listed is +08' for 4 minutes of time.

Use the whiz wheel to calculate the total motion for 11 minutes. In this case, the 11-minute correction totals 22'. By applying any other correction (refraction, sextant correction, etc.), a total adjustment is derived. By changing the sign, this total may be applied to the Hc. To apply the correction to the Ho, the sign of the adjustment would remain the same. Apply the adjustment to the intercept as the rules state in Table 1. In each case, the resultant intercept would be the same.

Suppose the Hc was 33° 57'. Applying the correction –22 yields 33° 35'. Comparing this with our Ho 33°29' results in an intercept of 6 NM away. If you decide to apply the correction to the Ho, 33° 29' + 22' yields 33° 5l'. Comparing this to the Hc 33° 57' yields the same result, 6 NM away. When using the Motion of the Observer table and when the fix time is earlier than the observation (LOP to be retarded), the rule for the sign of the correction is also printed below Table 1.

## Moving the Assumed Position

Another method of converting LOPs to a common time is to move the assumed position. This method is recommended for shots 4 minutes apart computed to give all three bodies a single assumed position. However, it is not limited to that type of computation. The assumed position is moved along the best-known track at the best-known GS. For example, again suppose the track is 330° and the GS 300 knots. LOPs are for 1500, 1504, and 1508 and a fix is desired at 1508. *[Figure 11-11]* Since the first LOP would have to be advanced

| Rel. Zn | Correction for 4 minutes of time | | | | | | | | | | | | | | | | | | | | | | | | | | | | | | Rel. Zn |
|---|---|---|---|---|---|---|---|---|---|---|---|---|---|---|---|---|---|---|---|---|---|---|---|---|---|---|---|---|---|---|---|
| | Groundspeed (knots) | | | | | | | | | | | | | | | | | | | | | | | | | | | | | | |
| | 90 | 120 | 150 | 180 | 210 | 240 | 270 | 300 | 330 | 360 | 390 | 420 | 450 | 480 | 510 | 540 | 570 | 600 | 630 | 660 | 690 | 720 | 750 | 780 | 810 | 840 | 870 | 900 | |
| 000 | +6 | +8 | +10 | +12 | +14 | +16 | +18 | +20 | +22 | +24 | +26 | +28 | +30 | +32 | +34 | +36 | +38 | +40 | +42 | +44 | +45 | +48 | +50 | +52 | +54 | +56 | +58 | +60 | 000 |
| 005 | 6 | 8 | 10 | 12 | 14 | 16 | 18 | 20 | 22 | 24 | 26 | 28 | 30 | 32 | 34 | 36 | 38 | 40 | 42 | 44 | 45 | 48 | 50 | 52 | 54 | 56 | 58 | 60 | 355 |
| 010 | 6 | 8 | 10 | 12 | 14 | 16 | 18 | 20 | 22 | 24 | 26 | 28 | 30 | 32 | 33 | 35 | 37 | 39 | 41 | 43 | 45 | 47 | 49 | 51 | 53 | 55 | 57 | 59 | 350 |
| 015 | 6 | 8 | 10 | 12 | 14 | 15 | 17 | 19 | 21 | 23 | 25 | 27 | 29 | 31 | 33 | 35 | 37 | 39 | 41 | 43 | 44 | 46 | 48 | 50 | 52 | 54 | 56 | 58 | 345 |
| 020 | 6 | 8 | 9 | 11 | 13 | 15 | 17 | 19 | 21 | 23 | 24 | 26 | 28 | 30 | 32 | 34 | 36 | 38 | 39 | 41 | 43 | 45 | 47 | 49 | 51 | 53 | 55 | 56 | 340 |
| 025 | 5 | 7 | 9 | 11 | 13 | 15 | 16 | 18 | 20 | 22 | 24 | 25 | 27 | 29 | 31 | 33 | 34 | 36 | 38 | 40 | 42 | 44 | 45 | 47 | 49 | 51 | 53 | 54 | 335 |
| 030 | +5 | +7 | +9 | +10 | +12 | +14 | +16 | +17 | +19 | +21 | +23 | +24 | +26 | +28 | +29 | +31 | +33 | +35 | +35 | +38 | +40 | +42 | +43 | +45 | +47 | +48 | +50 | +52 | 330 |
| 035 | 5 | 7 | 8 | 10 | 11 | 13 | 15 | 16 | 18 | 20 | 21 | 23 | 25 | 26 | 28 | 29 | 31 | 33 | 34 | 36 | 38 | 39 | 41 | 43 | 44 | 46 | 48 | 49 | 325 |
| 040 | 5 | 6 | 8 | 9 | 11 | 12 | 14 | 15 | 17 | 18 | 20 | 21 | 23 | 25 | 26 | 28 | 29 | 31 | 32 | 34 | 35 | 37 | 38 | 40 | 41 | 43 | 44 | 46 | 320 |
| 045 | 4 | 6 | 7 | 8 | 10 | 11 | 13 | 14 | 16 | 17 | 18 | 20 | 21 | 23 | 24 | 25 | 27 | 28 | 30 | 31 | 33 | 34 | 35 | 37 | 38 | 40 | 41 | 42 | 315 |
| 050 | 4 | 5 | 6 | 8 | 9 | 10 | 12 | 13 | 14 | 15 | 17 | 18 | 19 | 21 | 22 | 23 | 24 | 26 | 27 | 28 | 30 | 31 | 32 | 33 | 35 | 36 | 37 | 39 | 310 |
| 055 | 3 | 5 | 6 | 7 | 8 | 9 | 10 | 11 | 13 | 14 | 15 | 16 | 17 | 18 | 20 | 21 | 22 | 23 | 24 | 25 | 25 | 28 | 29 | 30 | 31 | 32 | 33 | 34 | 305 |
| 060 | +3 | +4 | +6 | +6 | +7 | +8 | +9 | +10 | +11 | +12 | +13 | +14 | +15 | +15 | +17 | +18 | +19 | +20 | +21 | +22 | +23 | +24 | +25 | +26 | +27 | +28 | +29 | +30 | 300 |
| 065 | 3 | 3 | 4 | 5 | 6 | 7 | 8 | 8 | 9 | 10 | 11 | 12 | 13 | 14 | 14 | 15 | 16 | 17 | 18 | 19 | 19 | 20 | 21 | 22 | 23 | 24 | 25 | 25 | 295 |
| 070 | 2 | 3 | 3 | 4 | 5 | 5 | 6 | 7 | 8 | 8 | 9 | 10 | 10 | 11 | 12 | 12 | 13 | 14 | 14 | 15 | 16 | 16 | 17 | 18 | 18 | 19 | 20 | 21 | 290 |
| 075 | 2 | 2 | 3 | 3 | 4 | 4 | 5 | 5 | 6 | 6 | 7 | 7 | 8 | 8 | 9 | 9 | 10 | 10 | 11 | 11 | 12 | 12 | 13 | 13 | 14 | 14 | 15 | 16 | 285 |
| 080 | 1 | 1 | 2 | 2 | 2 | 3 | 3 | 3 | 4 | 4 | 5 | 5 | 5 | 6 | 6 | 6 | 7 | 7 | 7 | 8 | 8 | 8 | 9 | 9 | 9 | 10 | 10 | 10 | 280 |
| 085 | +1 | +1 | +1 | +1 | +1 | +1 | +2 | +2 | +2 | +2 | +2 | +2 | +3 | +3 | +3 | +3 | +3 | +3 | +4 | +4 | +4 | +4 | +4 | +5 | +5 | +5 | +5 | +5 | 275 |
| 090 | 0 | 0 | 0 | 0 | 0 | 0 | 0 | 0 | 0 | 0 | 0 | 0 | 0 | 0 | 0 | 0 | 0 | 0 | 0 | 0 | 0 | 0 | 0 | 0 | 0 | 0 | 0 | 0 | 270 |
| 095 | −1 | −1 | −1 | −1 | −1 | −1 | −2 | −2 | −2 | −2 | −2 | −2 | −3 | −3 | −3 | −3 | −3 | −3 | −4 | −4 | −4 | −4 | −4 | −5 | −5 | −5 | −5 | −5 | 265 |
| 100 | 1 | 1 | 2 | 2 | 2 | 3 | 3 | 3 | 4 | 4 | 5 | 5 | 5 | 6 | 6 | 6 | 7 | 7 | 7 | 8 | 8 | 8 | 9 | 9 | 9 | 10 | 10 | 10 | 260 |
| 105 | 2 | 2 | 3 | 3 | 4 | 4 | 5 | 5 | 6 | 6 | 7 | 7 | 8 | 8 | 9 | 9 | 10 | 10 | 11 | 11 | 12 | 12 | 13 | 13 | 14 | 14 | 15 | 16 | 255 |
| 110 | 2 | 3 | 3 | 4 | 5 | 5 | 6 | 7 | 8 | 8 | 9 | 10 | 10 | 11 | 12 | 12 | 13 | 14 | 14 | 15 | 16 | 16 | 17 | 18 | 18 | 19 | 20 | 21 | 250 |
| 115 | 3 | 3 | 4 | 5 | 6 | 7 | 8 | 8 | 9 | 10 | 11 | 12 | 13 | 14 | 14 | 15 | 16 | 17 | 18 | 19 | 19 | 20 | 21 | 22 | 23 | 24 | 25 | 25 | 245 |
| 120 | 3 | 4 | 5 | 6 | 7 | 8 | 9 | 10 | 11 | 12 | 13 | 14 | 15 | 16 | 17 | 18 | 19 | 20 | 21 | 22 | 23 | 24 | 25 | 26 | 27 | 28 | 29 | 30 | 240 |
| 125 | −3 | −5 | −6 | −7 | −8 | −9 | −10 | −11 | −13 | −14 | −15 | −16 | −17 | −18 | −20 | −21 | −22 | −23 | −24 | −25 | −26 | −28 | −29 | −30 | −31 | −32 | −33 | −34 | 235 |
| 130 | 4 | 5 | 6 | 8 | 9 | 10 | 12 | 13 | 14 | 15 | 17 | 18 | 19 | 21 | 22 | 23 | 24 | 25 | 27 | 28 | 30 | 31 | 32 | 33 | 35 | 36 | 37 | 39 | 230 |
| 135 | 4 | 6 | 7 | 8 | 10 | 11 | 12 | 14 | 16 | 17 | 18 | 20 | 21 | 23 | 24 | 25 | 27 | 28 | 30 | 31 | 33 | 34 | 35 | 37 | 38 | 40 | 41 | 42 | 225 |
| 140 | 5 | 6 | 8 | 9 | 11 | 12 | 13 | 15 | 17 | 18 | 20 | 21 | 23 | 25 | 26 | 28 | 29 | 31 | 32 | 34 | 35 | 37 | 38 | 40 | 41 | 43 | 44 | 46 | 220 |
| 145 | 5 | 7 | 8 | 10 | 11 | 13 | 15 | 16 | 18 | 20 | 21 | 23 | 25 | 26 | 28 | 29 | 31 | 33 | 34 | 35 | 38 | 39 | 41 | 43 | 44 | 46 | 48 | 49 | 215 |
| 150 | 5 | 7 | 9 | 10 | 12 | 14 | 16 | 17 | 19 | 21 | 23 | 24 | 26 | 28 | 29 | 31 | 33 | 35 | 36 | 38 | 40 | 42 | 43 | 45 | 47 | 48 | 50 | 52 | 210 |
| 155 | −5 | −7 | −9 | −11 | −13 | −15 | −16 | −18 | −20 | −22 | −24 | −25 | −27 | −29 | −31 | −33 | −34 | −36 | −38 | −40 | −42 | −44 | −45 | −47 | −49 | −51 | −53 | −54 | 205 |
| 160 | 6 | 8 | 9 | 11 | 13 | 15 | 17 | 19 | 21 | 23 | 24 | 26 | 28 | 30 | 32 | 34 | 35 | 38 | 39 | 41 | 43 | 45 | 47 | 49 | 51 | 53 | 55 | 56 | 200 |
| 165 | 6 | 8 | 10 | 12 | 14 | 15 | 17 | 19 | 21 | 23 | 25 | 27 | 29 | 31 | 33 | 35 | 37 | 39 | 41 | 43 | 44 | 46 | 48 | 50 | 52 | 54 | 56 | 58 | 195 |
| 170 | 6 | 8 | 10 | 12 | 14 | 16 | 18 | 20 | 22 | 24 | 26 | 28 | 30 | 32 | 33 | 35 | 37 | 39 | 41 | 43 | 45 | 47 | 49 | 51 | 53 | 55 | 57 | 59 | 190 |
| 175 | 6 | 8 | 10 | 12 | 14 | 16 | 18 | 20 | 22 | 24 | 26 | 28 | 30 | 32 | 34 | 36 | 38 | 40 | 42 | 44 | 45 | 48 | 50 | 52 | 54 | 56 | 58 | 60 | 185 |
| 180 | −6 | −8 | −10 | −12 | −14 | −16 | −18 | −20 | −22 | −24 | −26 | −28 | −30 | −32 | −34 | −36 | −38 | −40 | −42 | −44 | −46 | −48 | −50 | −52 | −54 | −56 | −58 | −60 | 180 |

**Figure 11-10.** *Entering arguments are relative true azimuth and groundspeed.*

# Correction for less than 4 minutes of time

Value from 4 minute motion tables (for values greater than 60')

| Interval of time (m s) | 2 | 4 | 6 | 8 | 10 | 12 | 14 | 16 | 18 | 20 | 22 | 24 | 26 | 28 | 30 | 32 | 34 | 36 | 38 | 40 | 42 | 44 | 46 | 48 | 50 | 52 | 54 | 56 | 58 | 60 | Interval of time (m s) |
|---|---|---|---|---|---|---|---|---|---|---|---|---|---|---|---|---|---|---|---|---|---|---|---|---|---|---|---|---|---|---|---|
| 0 10 | 0 | 0 | 0 | 0 | 0 | 0 | 1 | 1 | 1 | 1 | 1 | 1 | 1 | 1 | 1 | 1 | 1 | 2 | 2 | 2 | 2 | 2 | 2 | 2 | 2 | 2 | 2 | 2 | 2 | 2 | 0 10 |
| 0 20 | 0 | 0 | 0 | 1 | 1 | 1 | 1 | 1 | 2 | 2 | 2 | 2 | 2 | 2 | 2 | 3 | 3 | 3 | 3 | 3 | 4 | 4 | 4 | 4 | 4 | 4 | 5 | 5 | 5 | 5 | 0 20 |
| 0 30 | 0 | 0 | 1 | 1 | 1 | 2 | 2 | 2 | 2 | 2 | 3 | 3 | 3 | 4 | 4 | 4 | 4 | 4 | 5 | 5 | 5 | 6 | 6 | 6 | 6 | 6 | 7 | 7 | 7 | 8 | 0 30 |
| 0 40 | 0 | 1 | 1 | 1 | 2 | 2 | 2 | 3 | 3 | 3 | 4 | 4 | 4 | 5 | 5 | 5 | 6 | 6 | 6 | 7 | 7 | 7 | 8 | 8 | 8 | 9 | 9 | 9 | 10 | 10 | 0 40 |
| 0 50 | 0 | 1 | 1 | 2 | 2 | 2 | 3 | 3 | 4 | 4 | 5 | 5 | 5 | 6 | 6 | 7 | 7 | 8 | 8 | 8 | 9 | 9 | 10 | 10 | 10 | 11 | 11 | 12 | 12 | 12 | 0 50 |
| 1 00 | 0 | 1 | 2 | 2 | 2 | 3 | 4 | 4 | 4 | 5 | 6 | 6 | 6 | 7 | 8 | 8 | 8 | 9 | 10 | 10 | 10 | 11 | 12 | 12 | 12 | 13 | 14 | 14 | 14 | 15 | 1 00 |
| 1 10 | 1 | 1 | 2 | 2 | 3 | 4 | 4 | 5 | 5 | 6 | 6 | 7 | 8 | 8 | 9 | 9 | 10 | 10 | 11 | 12 | 12 | 13 | 13 | 14 | 15 | 15 | 16 | 16 | 17 | 18 | 1 10 |
| 1 20 | 1 | 1 | 2 | 3 | 3 | 4 | 5 | 5 | 6 | 7 | 7 | 8 | 9 | 9 | 10 | 11 | 11 | 12 | 13 | 13 | 14 | 15 | 15 | 16 | 17 | 17 | 18 | 19 | 19 | 20 | 1 20 |
| 1 30 | 1 | 2 | 2 | 3 | 4 | 4 | 5 | 6 | 7 | 8 | 8 | 9 | 10 | 10 | 11 | 12 | 13 | 14 | 14 | 15 | 16 | 16 | 17 | 18 | 19 | 20 | 20 | 21 | 22 | 22 | 1 30 |
| 1 40 | 1 | 2 | 2 | 3 | 4 | 5 | 6 | 7 | 8 | 8 | 9 | 10 | 11 | 12 | 12 | 13 | 14 | 15 | 16 | 17 | 18 | 18 | 19 | 20 | 21 | 22 | 22 | 23 | 24 | 25 | 1 40 |
| 1 50 | 1 | 2 | 3 | 4 | 5 | 6 | 6 | 7 | 8 | 9 | 10 | 11 | 12 | 13 | 14 | 15 | 16 | 16 | 17 | 18 | 19 | 20 | 21 | 22 | 23 | 24 | 25 | 26 | 27 | 28 | 1 50 |
| 2 00 | 1 | 2 | 3 | 4 | 5 | 6 | 7 | 8 | 9 | 10 | 11 | 12 | 13 | 14 | 15 | 16 | 17 | 18 | 19 | 20 | 21 | 22 | 23 | 24 | 25 | 26 | 27 | 28 | 29 | 30 | 2 00 |
| 2 10 | 1 | 2 | 3 | 4 | 5 | 6 | 8 | 9 | 10 | 11 | 12 | 13 | 14 | 15 | 16 | 17 | 18 | 20 | 21 | 22 | 23 | 24 | 25 | 26 | 27 | 28 | 29 | 30 | 31 | 32 | 2 10 |
| 2 20 | 1 | 2 | 4 | 5 | 6 | 7 | 8 | 9 | 10 | 12 | 13 | 14 | 15 | 16 | 18 | 19 | 20 | 21 | 22 | 23 | 24 | 26 | 27 | 28 | 29 | 30 | 32 | 33 | 34 | 35 | 2 20 |
| 2 30 | 1 | 2 | 4 | 5 | 6 | 8 | 9 | 10 | 11 | 12 | 14 | 15 | 16 | 18 | 19 | 20 | 21 | 22 | 24 | 25 | 26 | 28 | 29 | 30 | 31 | 32 | 34 | 35 | 36 | 38 | 2 30 |
| 2 40 | 1 | 3 | 4 | 5 | 7 | 8 | 9 | 11 | 12 | 13 | 15 | 16 | 17 | 19 | 20 | 21 | 23 | 24 | 25 | 27 | 28 | 29 | 31 | 32 | 33 | 35 | 36 | 37 | 39 | 40 | 2 40 |
| 2 50 | 1 | 3 | 4 | 6 | 7 | 8 | 10 | 11 | 13 | 14 | 16 | 17 | 18 | 20 | 21 | 23 | 24 | 26 | 27 | 28 | 30 | 31 | 33 | 34 | 35 | 37 | 38 | 40 | 41 | 42 | 2 50 |
| 3 00 | 2 | 3 | 4 | 6 | 8 | 9 | 10 | 12 | 14 | 15 | 16 | 18 | 20 | 21 | 22 | 24 | 26 | 27 | 28 | 30 | 32 | 33 | 34 | 36 | 38 | 39 | 40 | 42 | 44 | 45 | 3 00 |
| 3 10 | 2 | 3 | 5 | 6 | 8 | 10 | 11 | 13 | 14 | 16 | 17 | 19 | 21 | 22 | 24 | 25 | 27 | 28 | 30 | 32 | 33 | 35 | 36 | 38 | 40 | 41 | 43 | 44 | 46 | 48 | 3 10 |
| 3 20 | 2 | 3 | 5 | 7 | 8 | 10 | 12 | 13 | 15 | 17 | 18 | 20 | 22 | 23 | 25 | 27 | 28 | 30 | 32 | 33 | 35 | 37 | 38 | 40 | 42 | 43 | 45 | 47 | 48 | 50 | 3 20 |
| 3 30 | 2 | 4 | 5 | 7 | 9 | 10 | 12 | 14 | 16 | 18 | 19 | 21 | 23 | 24 | 26 | 28 | 30 | 32 | 33 | 35 | 37 | 38 | 40 | 42 | 44 | 46 | 47 | 49 | 51 | 52 | 3 30 |
| 3 40 | 2 | 4 | 6 | 7 | 9 | 11 | 13 | 15 | 16 | 18 | 20 | 22 | 24 | 26 | 28 | 29 | 31 | 33 | 35 | 37 | 38 | 40 | 42 | 44 | 46 | 48 | 50 | 51 | 53 | 55 | 3 40 |
| 3 50 | 2 | 4 | 6 | 8 | 10 | 12 | 13 | 15 | 17 | 19 | 21 | 23 | 25 | 27 | 29 | 31 | 33 | 34 | 36 | 38 | 40 | 42 | 44 | 46 | 48 | 50 | 52 | 54 | 56 | 58 | 3 50 |
| 4 00 | 2 | 4 | 6 | 8 | 10 | 12 | 14 | 16 | 18 | 20 | 22 | 24 | 26 | 28 | 30 | 32 | 34 | 36 | 38 | 40 | 42 | 44 | 46 | 48 | 50 | 52 | 54 | 56 | 58 | 60 | 4 00 |

| Time of fix | Sign from 4 min. table | To observed altitude | To tabulated altitude | To intercept |
|---|---|---|---|---|
| Later than observation | + | Add | Subtract | Toward |
| | − | Subtract | Add | Away |
| Earlier than observation | + | Subtract | Add | Away |
| | − | Add | Subtract | Toward |

*Figure 11-10. Entering arguments are relative true azimuth and groundspeed (continued).*

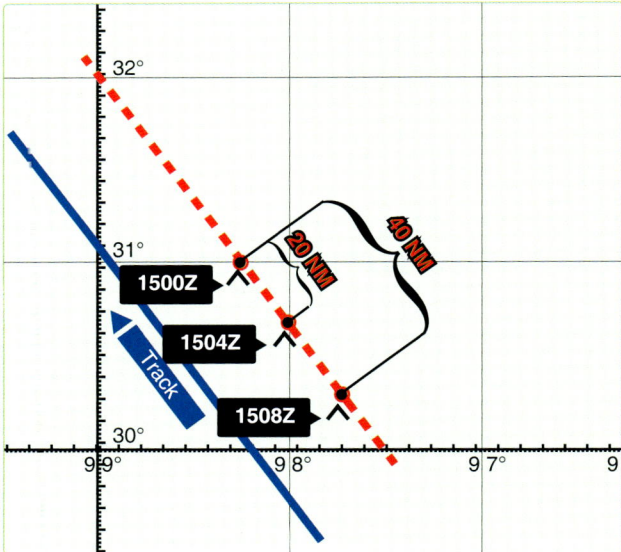

*Figure 11-11. Moving assumed positions.*

40 NM (8 minutes at 300 knots), the same result is realized by advancing the assumed position 40 NM parallel to the best-known track. The 1504 LOP must be advanced 20 NM; therefore, the assumed position is advanced 20 NM miles parallel to the best-known track. The third shot requires no movement, and it is plotted from the original assumed position.

It should be noted that the first shot is always plotted from the assumed position, which is closest to destination. In this method, if observations are precomputed and the assumed position is moved prior to shooting, the following procedure is used when shooting is off schedule. For every minute of time that the shot is taken early, move the assumed position 15 minutes of longitude to the east. For every minute of time that the shot is taken late, move the assumed position 15 minutes of longitude to the west. In addition, the affected LOP must be moved along the best-known track for the number of minutes of GS the observation was early or late. If the shot was early, advance the LOP; if the shot was late, retard the LOP.

## Planning the Fix

In selecting bodies for observation, one should generally consider azimuth primarily and such factors as brightness, altitude, etc., secondarily. If all observations were precisely correct in every detail, the resulting LOPs would meet at a point. However, this is rarely the case. Three observations generally result in LOPs forming a triangle. If this triangle is not more than 2 or 3 miles on a side under good conditions and 5 to 10 miles under unfavorable conditions, there is normally no reason to suppose that a mistake has been made. Even a point fix, however, is not necessarily accurate. An uncorrected error in time, for instance, would require the entire fix to be moved eastward if observations were early and westward if observations were late, at the rate of 1 minute of longitude for each 4 seconds of time.

In a two-LOP fix, the ideal cut of the LOPs is 90°. In *Figure 11-12,* a 90° cut with a 5 NM error in one LOP causes a 5 NM error in the fix. If the acute angle between the LOPs is 30°, a 5 NM error in one LOP causes a 10 NM error in the fix. Thus, with a two-LOP fix, an error in one LOP causes at least an equal error in the fix; the smaller the acute angle between the LOPs, the greater the fix error caused by a given error in one LOP. Of course, if both LOPs are in error, the fix may be thrown off even more. In a three-LOP fix, the ideal cut of the LOPs is 60° (star azimuths 120° apart). With this cut, a 3 NM error in any one LOP causes a 2 NM error in the fix. With any other cut, a 3 NM error in any one LOP causes more than a 2 NM error in the fix. In a three-star fix, the cut will be 60° if the azimuths of the stars differ by 60° or if they differ by 120°. If there is any unknown constant error in the observations, all the Hos will be either too great or too small.

Notice in *Figure 11-13* that, if stars are selected whose azimuths differ by 120°, this constant error of the Hos causes a displacement of the three LOPs, either all toward the center

or all away from the center of the triangle. In either case, the position of the center of the triangle is not affected. If you use any three stars with azimuths outside a 180° range, any constant error in observations tends to cancel out.

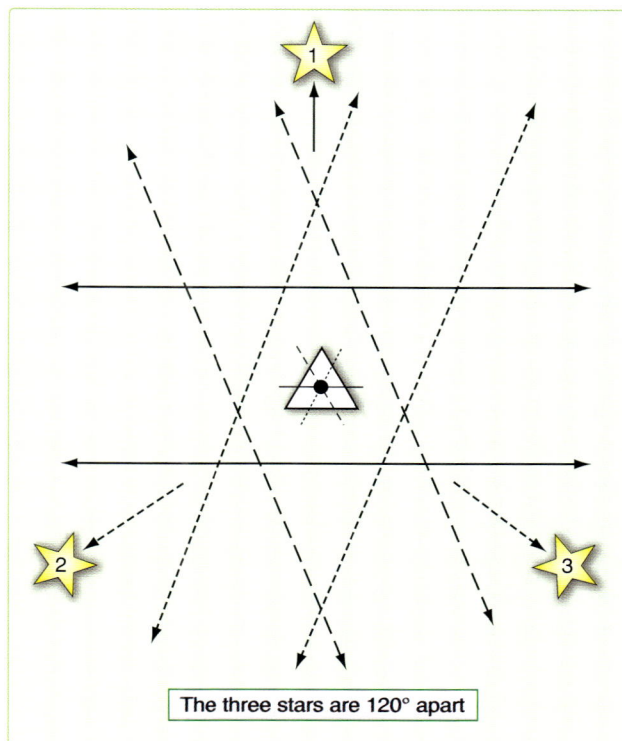

The three stars are 120° apart

**Figure 11-13.** *Effect of azimuth on accuracy of fix.*

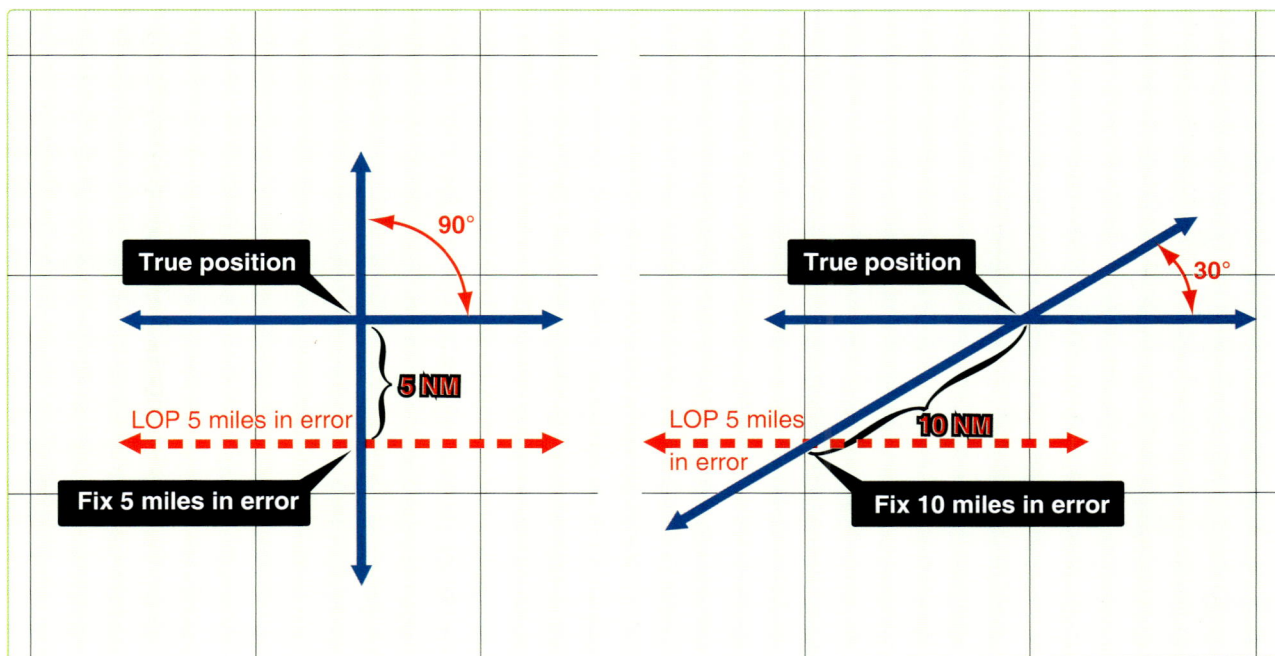

**Figure 11-12.** *Effect of cut on accuracy of a fix.*

The three-star fix has two distinct advantages over the two-star fix. First, it is the average of three observations. Second, selecting the stars carefully can counteract the effect of constant errors of observation. There is also a third advantage. Each pair of two LOPs furnishes a rough check on the third. In resolving an observation into a LOP, you might possibly make a gross error; for example, obtaining an LHA that is in error by a whole degree. Such an error might not be immediately apparent. Neither would such a discrepancy come to immediate attention in a two-LOP fix. However, this third advantage does not apply when a single LHA is used in solving all LOPs, such as is done when precomputing and using motion corrections to resolve all LOPs to a common time. Because of these three advantages, it is evident that a three-star fix should be used, rather than a two-star fix, when possible.

Whatever the number of observations, common practice, backed by logic, is to take the center of the figure formed unless there is reason for deviating from this procedure. Center is meant as the point representing the least total error of all lines considered reliable. With three LOPs, the center is considered that point within the triangle equidistant from the three sides. It may be found by bisecting the angles, but is usually located by eye.

## Chapter Summary

Because of all the factors involved, a certain amount of judgment is necessary, along with the proper use of the mechanics comprising celestial navigation. When using a single LOP or a fix, you have to take into consideration the existing conditions and weigh the DR information against the information obtained from the LOP. An accurate DR position should always be computed.

The C-Plot formula helps place the MPP with a single LOP, but you might want to make further adjustments to the final position. The formula is:

$$\frac{d}{t} = \frac{p}{t + p}$$

Remember, d is the distance measured along a perpendicular from the DR position to the LOP. In the case of the two- or three-star fix, planning plays a very important part. Selecting stars whose azimuths differ by 120° for a three-star fix minimizes errors in the fix position. In two-star fixes, the ideal azimuth separation is 90°. Also, when dealing with more than one LOP, it is necessary to resolve the LOPs to a common time. This adjustment can be accomplished by moving the assumed position, by moving the LOPs, or by applying a correction factor to the Hc or Ho.

## Chapter 12

# Special Celestial Techniques

## Introduction

This chapter describes some techniques that may not be used every day and under all circumstances but are valuable alternatives from normal precomping procedures. Most of these techniques save time by eliminating either some extractions or computations. Some navigational techniques and planning procedures are also discussed.

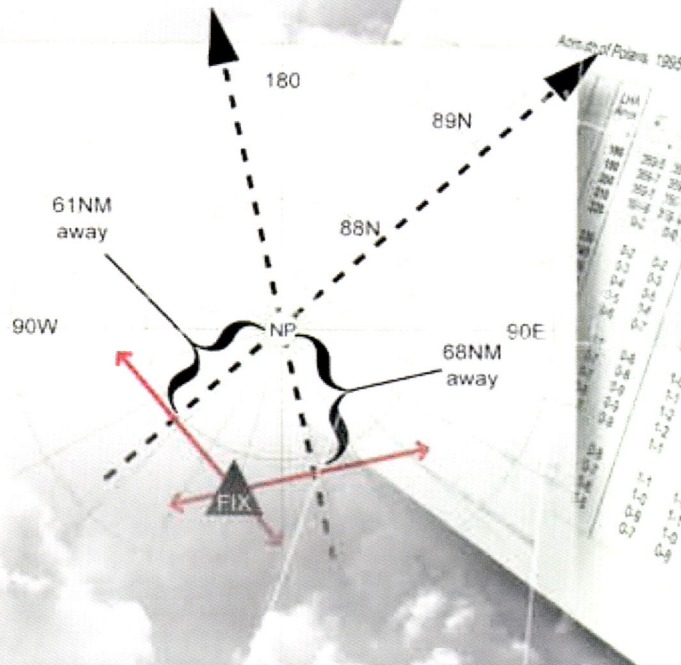

## Determining Availability of Celestial Bodies

By doing a quick comparison of Greenwich hour angle (GHA) to the observer's position, it is easy to determine the availability of celestial bodies. For example, the observer anticipates being at 18°N 135° W at 0015Z on 28 September 1995. There are several bodies listed in the Air Almanac, but not all of them are available for observation. To determine availability, take the observer's longitude and look 80° either side of it. Within this range, compare the GHA of a body. Looking at *Figure 12-1,* we see that the sun, moon, Venus, and Jupiter are within the 80° range and are therefore usable. Saturn is outside of the 80° range, so it is not usable. The declination (Dec) of a body is not normally a factor; however, at high latitudes a body may not be available when its subpoint is near the pole opposite the observer.

## Latitude by Polaris

Polaris is the polestar, or North Star. Because Polaris is approximately 1° from the North Pole, it makes a small diurnal circle and seemingly stays in about the same place all night. This fact makes Polaris very useful in navigation. With certain corrections, it serves as a reference point for direction and for latitude in the Northern Hemisphere. Latitude by Polaris is a quick method of obtaining a latitude line of position (LOP); only the tables given in the Air Almanac are needed.

## Obtaining Latitude by Polaris

A latitude by Polaris LOP is obtained by applying the Q correction to the corrected observed altitude. *[Figure 12-2]* This adjusts the altitude of the pole, which is equal to the navigator's latitude. The Q correction table is in the back of the Air Almanac. The entering argument for the table is exact local hour angle (LHA) of Aries. The effect of refraction is not included in Q correction, so the observed altitude must be fully corrected. When refraction is used for a latitude by Polaris LOP, it is applied to the observed altitude and the sign of the correction is negative. A Polaris LOP can also be plotted using the intercept method. In this case, the Hc is computed by reversing the sign of the Q correction and applying it to the assumed latitude (rounded off to the nearest degree). Refraction is positive when applied to get an Hc for the intercept method.

## Obtaining Azimuth of Polaris

For either method, the azimuth of Polaris is obtained from the Azimuth of Polaris table found in the Air Almanac or in the Pub. No. 249. *[Figure 12-2]* Whether plotted as an intercept or a latitude, the assumed position should be corrected for Coriolis, or rhumb line, and precession, or nutation. The resulting LOPs should fall in the same place for either method. To plot the LOP using the latitude method, choose the longitude line closest to the DR and plot perpendicular to the longitude line. For the intercept method, use the assumed latitude and plot the intercept normally using the azimuth of Polaris.

## Latitude by Polaris Example

On 18 April 1995 for Greenwich mean time (GMT) 1600 at 23° 10' N 120° W, with an observed altitude 23° –06' at 31,000'. When doing a latitude by Polaris you must use the exact latitude and longitude. See *Figure 12-3* for plotting.

| | |
|---|---|
| GHA | 086° –18' |
| Longitude (West) | –120° –00' |
| LHA | 326° –18' |
| True Course (TC) | 090° |
| Groundspeed (GS) | 400 knots |
| Coriolis/rhumb line | 7R |
| Corrected Observed Altitude | 23° –06' |

### (Day 271) Greenwich A. M. 1995 September 28 (Thursday)  541

| UT (GMT) | ⊙ Sun GHA | Dec. | Aries GHA ♈ | Venus 1.4 GHA | Dec. | Jupiter 2.0 GHA | Dec. | Saturn 0.7 GHA | Dec. | ● Moon GHA | Dec. | Lat. | Moon rise | Diff. |
|---|---|---|---|---|---|---|---|---|---|---|---|---|---|---|
| h m | D ' | ° ' | ° ' | D ' | ° ' | ° ' | ° ' | ° ' | ° ' | ° ' | ° ' | N | h m | m |
| 00 00 | 82 15.7 | S 1 45.0 | 6 18.1 | 172 23 | S 4 48 | 117 52 | S21 31 | 14 16 | S 6 00 | 141 29 | S 15 06 | ° | h m | * |
| 10 | 184 45.8 | 45.1 | 8 48.6 | 174 53 | | 120 22 | | 16 46 | | 143 53 | 08 | 72 | 13 34 | |
| 20 | 187 15.8 | 45.3 | 11 19.0 | 177 23 | | 122 53 | | 19 17 | | 146 18 | 09 | 70 | 12 43 | +54 |
| 30 | 189 45.8 | 45.5 | 13 49.4 | 179 53 | | 125 23 | | 21 47 | | 148 42 | 10 | 68 | 12 11 | 48 |
| 40 | 192 15.9 | 45.6 | 16 19.8 | 182 23 | | 127 53 | | 24 18 | | 151 06 | 11 | 66 | 11 48 | 45 |
| 50 | 194 45.9 | 45.8 | 18 50.2 | 184 53 | | 130 24 | | 26 48 | | 153 31 | 12 | 64 | 11 28 | 42 |
| 01 00 | 197 15.9 | S 1 46.0 | 21 20.6 | 187 23 | S 4 49 | 132 54 | S21 31 | 29 18 | S 6 00 | 155 55 | S 15 13 | 62 | 11 14 | 40 |
| 10 | 199 46.0 | 46.1 | 23 51.0 | 189 53 | | 135 25 | | 31 49 | | 158 20 | 15 | 60 | | 39 |
| 20 | 202 16.0 | 46.3 | 26 21.4 | 192 23 | | 137 55 | | 34 19 | | 160 44 | 16 | 58 | 11 02 | 38 |
| 30 | 204 46.0 | 46.4 | 28 51.8 | 184 52 | | 140 25 | | 36 50 | | 163 09 | 17 | 56 | 10 51 | 37 |
| 40 | 207 16.1 | 46.6 | 31 22.3 | 197 22 | | 142 55 | | 39 20 | | 165 33 | 18 | 54 | 10 42 | 36 |
| 50 | 209 46.1 | 46.8 | 33 52.7 | 199 52 | | 145 26 | | 41 51 | | 167 57 | 19 | 52 | 10 33 | |
| 02 00 | 212 16.1 | S 1 46.9 | 36 23.1 | 202 22 | S 4 50 | 147 56 | S21 32 | 44 21 | S 6 00 | 170 22 | S 15 20 | 50 | 10 26 | 36 |
| 10 | 214 46.2 | 47.1 | 38 53.5 | 204 52 | | 150 27 | | 46 52 | | 172 46 | 21 | 45 | 10 19 | 35 |
| 20 | 217 16.2 | 47.2 | 41 23.9 | 207 22 | | 152 57 | | 49 22 | | 175 11 | 23 | 40 | 10 05 | 34 |
| 30 | 219 46.3 | 47.4 | 43 54.3 | 209 52 | | 155 27 | | 51 52 | | 177 35 | 24 | 35 | 09 53 | 33 |
| 40 | 222 16.3 | 47.6 | 46 24.7 | 212 22 | | 157 58 | | 54 23 | | 179 59 | 25 | 30 | 09 43 | 32 |
| 50 | 224 46.3 | 47.7 | 48 55.1 | 214 52 | | 160 28 | | 56 53 | | 182 24 | 26 | | 09 35 | 31 |

**Figure 12-1.** *A quick check of body availability.*

## Polaris (Pole Star) Table, 1995

For determining the latitude from a sextant altitude

| LHA Aries | Q | LHA Aries | Q | LHA Aries | Q | LHA Aries | Q | LHA Aries | Q | LHA Aries | Q | LHA Aries | Q | LHA Aries | Q |
|---|---|---|---|---|---|---|---|---|---|---|---|---|---|---|---|
| ° ′ | ′ | ° ′ | ′ | ° ′ | ′ | ° ′ | ′ | ° ′ | ′ | ° ′ | ′ | ° ′ | ′ | ° ′ | ′ |
| 358 46 | −36 | 86 00 | −29 | 120 50 | −4 | 153 30 | +21 | 227 46 | +44 | 281 46 | +19 | 314 17 | −6 | 349 32 | −31 |
| 0 50 | −37 | 87 38 | −28 | 122 07 | −3 | 154 56 | +22 | 233 14 | +43 | 283 10 | +18 | 345 33 | −7 | 351 16 | −32 |
| 3 01 | −38 | 89 14 | −27 | 123 23 | −2 | 156 23 | +23 | 237 18 | +42 | 284 32 | +17 | 316 50 | −8 | 353 00 | −33 |
| 5 20 | −39 | 90 49 | −26 | 124 39 | −1 | 157 51 | +24 | 240 42 | +41 | 285 55 | +16 | 318 07 | −9 | 354 53 | −34 |
| 7 48 | −40 | 92 21 | −25 | 125 54 | −0 | 159 21 | +25 | 243 42 | +40 | 287 16 | +15 | 319 24 | −10 | 356 47 | −35 |
| 10 29 | −41 | 93 52 | −24 | 127 11 | −1 | 160 53 | +26 | 246 25 | +39 | 288 36 | +14 | 320 42 | −11 | 358 46 | −36 |
| 13 27 | −42 | 95 21 | −23 | 128 27 | −2 | 162 26 | +27 | 248 55 | +38 | 289 56 | +13 | 322 00 | −12 | 0 50 | −37 |
| 16 49 | −43 | 96 49 | −22 | 129 43 | −3 | 164 01 | +28 | 251 15 | +37 | 291 16 | +12 | 323 19 | −13 | 3 01 | −38 |
| 20 50 | −44 | 98 15 | −21 | 130 59 | −4 | 165 37 | +29 | 253 28 | +36 | 292 35 | +11 | 324 38 | −14 | 5 20 | −39 |
| 26 13 | −45 | 99 40 | −20 | 132 16 | −5 | 167 17 | +30 | 255 33 | +35 | 293 53 | +10 | 325 57 | −15 | 7 48 | −40 |
| 47 38 | −44 | 101 05 | −19 | 133 32 | −6 | 168 58 | +31 | 257 34 | +34 | 295 11 | +9 | 327 18 | −16 | 10 29 | −41 |
| 53 01 | −43 | 102 28 | −18 | 134 49 | −7 | 170 43 | +32 | 259 29 | +33 | 296 28 | +8 | 328 39 | −17 | 13 27 | −42 |
| 57 02 | −42 | 103 51 | −17 | 136 05 | −8 | 172 31 | +33 | 261 20 | +32 | 297 46 | +7 | 330 00 | −18 | 16 49 | −43 |
| 60 24 | −41 | 105 12 | −16 | 137 23 | −9 | 174 22 | +34 | 263 08 | +31 | 299 02 | +6 | 331 23 | −19 | 20 50 | −44 |
| 63 22 | −40 | 106 33 | −15 | 138 40 | −10 | 176 17 | +35 | 264 53 | +30 | 300 19 | +5 | 332 46 | −20 | 26 13 | −45 |
| 66 03 | −39 | 107 54 | −14 | 139 58 | −11 | 178 18 | +36 | 266 34 | +29 | 301 35 | +4 | 334 11 | −21 | 47 38 | −44 |
| 68 31 | −38 | 109 13 | −13 | 141 16 | −12 | 180 23 | +37 | 268 14 | +28 | 302 52 | +3 | 335 36 | −22 | 53 01 | −43 |
| 70 50 | −37 | 110 32 | −12 | 142 35 | −13 | 182 36 | +38 | 269 50 | +27 | 304 08 | +2 | 337 02 | −23 | 57 02 | −42 |
| 73 01 | −36 | 111 51 | −11 | 143 55 | −14 | 184 56 | +39 | 271 25 | +26 | 305 24 | +1 | 338 30 | −24 | 60 24 | −41 |
| 75 05 | −35 | 113 09 | −10 | 145 15 | −15 | 187 26 | +40 | 272 58 | +25 | 306 40 | 0 | 339 59 | −25 | 63 22 | −40 |
| 77 04 | −34 | 114 27 | −9 | 146 35 | −16 | 190 09 | +41 | 274 30 | +24 | 307 57 | −1 | 341 30 | −26 | 66 03 | −39 |
| 78 58 | −33 | 115 44 | −8 | 147 56 | −17 | 193 09 | +42 | 276 00 | +23 | 309 12 | −2 | 343 02 | −27 | 68 31 | −38 |
| 80 49 | −32 | 117 01 | −7 | 149 19 | −18 | 196 33 | +43 | 277 28 | +22 | 310 28 | −3 | 344 37 | −28 | 70 50 | −37 |
| 82 35 | −31 | 118 18 | −6 | 150 41 | −19 | 200 37 | +44 | 278 55 | +21 | 311 44 | −4 | 346 13 | −29 | 73 01 | −36 |
| 84 19 | −30 | 119 34 | −5 | 152 05 | −20 | 206 05 | +45 | 280 21 | +20 | 313 01 | −5 | 347 51 | −30 | 75 05 | 35 |
| 86 00 | | 120 50 | | 153 30 | | 227 46 | | 281 46 | | 314 17 | | 349 32 | | 77 04 | |

*In critical cases, ascend Q, which does not include refraction, is to be applied to the corrected sextant altitude of Polaris.*
*Polaris: Mag. 2.1, SHA 323° 04', Dec N89° 14:7*

## Azimuth of Polaris, 1995

| LHA Aries | Latitude | | | | | | | LHA Aries | Latitude | | | | | | |
|---|---|---|---|---|---|---|---|---|---|---|---|---|---|---|---|
| | 0° | 30° | 50° | 55° | 60° | 65° | 70° | | 0° | 30° | 50° | 55° | 60° | 65° | 70° |
| | ° | ° | ° | ° | ° | ° | ° | | ° | ° | ° | ° | ° | ° | ° |
| 0 | 0-5 | 0-5 | 0-7 | 0-8 | 0-9 | 1-1 | 1-4 | 180 | 359-5 | 359-5 | 359-3 | 359-2 | 359-1 | 359-0 | 358-7 |
| 10 | 0-3 | 0-4 | 0-5 | 0-6 | 0-7 | 0-8 | 1-0 | 190 | 359-7 | 359-6 | 359-5 | 359-4 | 359-3 | 359-2 | 359-0 |
| 20 | 0-2 | 0-3 | 0-3 | 0-4 | 0-4 | 0-5 | 0-7 | 200 | 359-8 | 359-7 | 359-7 | 359-6 | 359-6 | 359-5 | 359-4 |
| 30 | 0-1 | 0-1 | 0-1 | 0-2 | 0-2 | 0-2 | 0-3 | 210 | 359-9 | 359-9 | 359-9 | 359-8 | 359-8 | 359-8 | 359-8 |
| 40 | 0-0 | 0-0 | 359-9 | 359-9 | 359-9 | 359-9 | 359-9 | 220 | 0-0 | 0-0 | 0-1 | 0-1 | 0-1 | 0-1 | 0-1 |
| 50 | 359-8 | 359-8 | 359-7 | 359-7 | 359-7 | 359-6 | 359-5 | 230 | 0-2 | 0-2 | 0-3 | 0-3 | 0-3 | 0-4 | 0-5 |
| 60 | 359-7 | 359-7 | 359-5 | 359-5 | 359-4 | 359-3 | 359-1 | 240 | 0-3 | 0-3 | 0-5 | 0-5 | 0-6 | 0-7 | 0-8 |
| 70 | 359-6 | 359-5 | 359-4 | 359-3 | 359-2 | 359-0 | 358-8 | 250 | 0-4 | 0-5 | 0-6 | 0-7 | 0-8 | 1-0 | 1-2 |
| 80 | 359-5 | 359-4 | 359-2 | 359-1 | 359-0 | 358-8 | 358-5 | 260 | 0-5 | 0-6 | 0-8 | 0-9 | 1-0 | 1-2 | 1-5 |
| 90 | 359-4 | 359-3 | 359-1 | 358-9 | 358-8 | 358-5 | 358-2 | 270 | 0-6 | 0-7 | 0-9 | 1-0 | 1-2 | 1-4 | 1-7 |
| 100 | 359-3 | 359-2 | 358-9 | 358-8 | 358-6 | 358-4 | 358-0 | 280 | 0-7 | 0-8 | 1-0 | 1-2 | 1-3 | 1-6 | 1-9 |
| 110 | 359-3 | 359-2 | 358-9 | 358-7 | 358-5 | 358-3 | 357-9 | 290 | 0-7 | 0-8 | 1-1 | 1-3 | 1-4 | 1-7 | 2-1 |
| 120 | 359-2 | 359-1 | 358-8 | 358-7 | 358-5 | 358-2 | 357-8 | 300 | 0-7 | 0-9 | 1-2 | 1-3 | 1-5 | 1-8 | 2-2 |
| 130 | 359-2 | 359-1 | 358-8 | 358-7 | 358-5 | 358-2 | 357-8 | 310 | 0-8 | 0-9 | 1-2 | 1-3 | 1-5 | 1-8 | 2-2 |
| 140 | 359-3 | 359-2 | 358-9 | 358-7 | 358-5 | 358-3 | 357-9 | 320 | 0-7 | 0-9 | 1-1 | 1-3 | 1-5 | 1-8 | 2-2 |
| 150 | 359-3 | 359-2 | 358-9 | 358-8 | 358-6 | 358-4 | 358-0 | 330 | 0-7 | 0-8 | 1-1 | 1-2 | 1-4 | 1-7 | 2-1 |
| 160 | 359-4 | 359-3 | 359-0 | 358-8 | 358-8 | 358-5 | 358-2 | 340 | 0-6 | 0-7 | 1-0 | 1-1 | 1-3 | 1-5 | 1-9 |
| 170 | 359-4 | 359-4 | 359-2 | 358-1 | 358-9 | 358-7 | 358-4 | 350 | 0-6 | 0-6 | 0-9 | 1-0 | 1-1 | 1-3 | 1-7 |
| 180 | 359-5 | 359-5 | 359-3 | 359-2 | 359-1 | 359-0 | 358-7 | 360 | 0-5 | 0-5 | 0-7 | 0-8 | 0-9 | 1-1 | 1-4 |

*When Cassiopeia is left (right), Polaris is west (east).*

**Figure 12-2.** *Polaris Q correction and azimuth tables from the Air Almanac.*

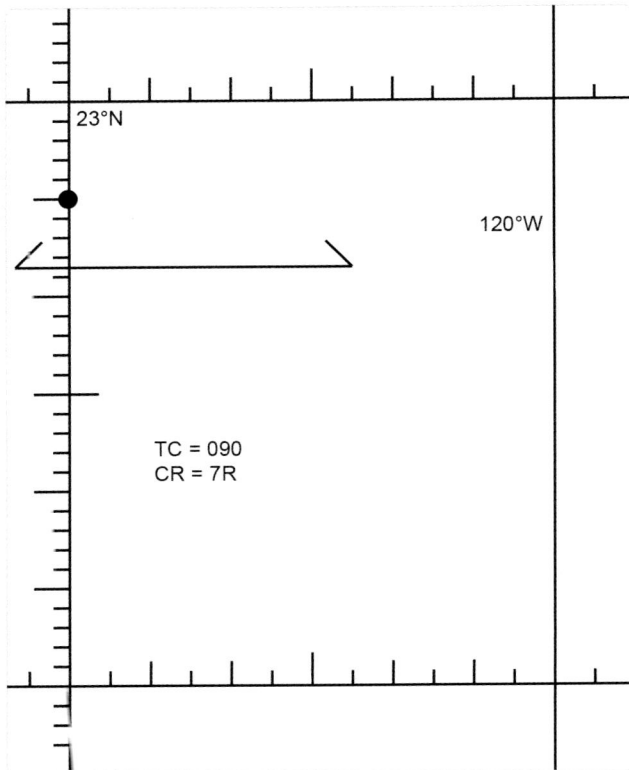

**Figure 12-3.** *Plotting the Polaris LOP.*

| | |
|---|---|
| Q (based on LHA 072-44) | −15' |
| Refraction | −01' |
| Latitude | 22° −50' |
| Azimuth (LHA 326° −18', Latitude 23° N) | 000.8° |

NOTE: If the Q correction table in Volume 1 is used, precession and nutation (P/N) and Coriolis, or rhumb line, must be used in plotting the LOP. This is because the Pub. No. 249 covers a 5-year period, and the further the years get from the Epoch year, the greater the error is when using the Polaris table. P/N compensates for this error.

## Intercept Method Example

Refer to the previous problem and *Figure 12-3* for plotting. NOTE: Applying 10A to assumed latitude gives 22° −50' N, which is the same the answer in the latitude by Polaris example.

| | |
|---|---|
| Azimuth of Polaris | 359.5 |
| Coriolis/rhumb line | 7R |
| Assumed Lat (rounded off) | 23° −00' N |
| Q (reversed sign) | +15' |
| Refraction | +01' |
| Hc Polaris | 23° −16' |
| Ho Polaris | 23° −06' |
| Intercept | 10A |

NOTE: In these examples, all information was taken from the Air Almanac. No P/N is required.

## LHA Method of Obtaining Three-Star Fix

The LHA technique allows you to solve the motion problem for a three-star fix by applying a correction to the assumed position rather than computing a numerical solution on the precomp. This eliminates mathematical motion calculations, therefore reducing the chance of math errors on the precomp. To accomplish a three-LHA fix, you must plan 4 minutes between the midtime of each shot. *[Figures 12-4* and *12-5]* Because LHA changes 1 degree for every 4 minutes, the precomp has three successive LHAs, 1 degree apart. To correct for off-time motion, adjust the assumed position based on true course (TC) and groundspeed (GS). If a shot is planned earlier than fix time, the assumed position is advanced (down-track). For shots planned later than fix time, the assumed position is retarded (up-track).

The example in *Figures 12-4* and *12-5* shows the LHA method for a 12-8-4 early shooting schedule. This shooting schedule allows the fix and/or MPP to be resolved before the fix time. To adjust the assumed positions, plot the fix time assumed position and then advance it for 4 minutes of track and GS for each body. This satisfies motion of the observer. When shooting the selected bodies, take care to shoot them exactly on the prescribed times. This eliminates motion of the body.

A variation of advancing the assumed position is to use half motions. This enables you to plot all three LOPs from one assumed position. Table 1 from Pub. No. 249 lists corrections to position of the observer. Each correction is for 4 minutes of time. To use it, enter with your relative Zn (Zn-track) and GS. Now, look at the bottom of the table and note you can apply this correction to your tabulated altitude or observed altitude. It does not matter which you choose, but note that the sign changes dependent on where you apply it. Now, take the number and multiply it by the 4-minute increment of the shot. For example, *Figure 12-6* shows the precomp for a 0300 fix using 3 LHAs and half motions. The 0248 shot, Alpheratz, relative Zn, and groundspeed were used to extract a +20 correction from Table 1. Because this shot is 12 minutes early, we need to multiply +20 by three before we apply it to the shot. Note the +60 correction was applied to the observed altitude and, therefore, kept its positive sign. The benefit of doing this is a reduction in plotting. See *Figure 12-7* for the plotted LOPs. This technique can be applied to day celestial as well.

# CELESTIAL PRECOMPUTATION

SHEET NUMBER 14

### PRECOMPUTATION   PERISCOPIC SEXTANT

| NAVIGATOR | | | ALT MSL. | FL 200 | DATE(Z) | 1 Jan 1993 | FIX TIME | 0702 Z |

**STAR SELECTOR BY AZIMUTH**

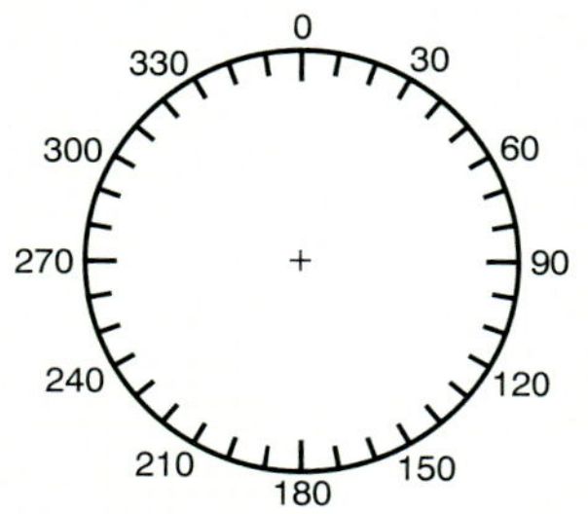

(Azimuth dial: 0, 30, 60, 90, 120, 150, 180, 210, 240, 270, 300, 330 with center +)

| | | | | BODY | Dubhe | Sirius | Capella |
|---|---|---|---|---|---|---|---|
| TRACK | 090 | | | BASE GHA | | | 205-57 |
| GS | 400 | | | | | | |
| CORIOLIS | 7 (R/L) | | | CORR | | | +30 |
| PREC/NUT | 1/270 | | | +360 | | | |
| DR LAT | 33-02 (N/S) | | | GHA | | | 206-27 |
| DR LONG | 100-30 (E/W) | | | ASSUM LONG (W) +E | | | 100-27 |
| MOTION OF OBSERVER | | | | LHA | 103 | 104 | 105 / 106 |
| MOTION OF BODY | | | | ASSUM LAT | 33 (N/S) | | |
| 4 MIN ADJUST | | | | DEC | N S | N S | N S | N S | N S |
| OFF-FIX TIME | 12 (E/L) | 8 (E/L) | 4 (E/L) | E L | E L | PLANNED MID-TIME | 0650 | 0654 | 0658 |
| TOTAL MOT. ADJUST | | | | ACTUAL MID-TIME | | | |
| POLARIS  Q | | | | TAB Hc | | | |
| MOON  PA  SD | | | | | | | |
| REF ( ) | 01 | 01 | | D/DEC CORR | | | |
| PERS/SEXT | | | | CORR Hc | 41-22 | 40-13 | 66-19 |
| TOTAL → ADJ | 01 | 01 | | TOTAL → ADJ | +01 | +01 | |
| TH/GH | | | | ADJ Hc | 41-32 | 40-14 | 66-19 |
| Zn/GZa( ) | | | | OFF TIME MOTION | | | |
| IRB | | | | Hc | 41-23 | 40-14 | 66-19 |
| IRB | | | | Ho | 41-15 | 39-54 | 65-59 |
| Za/GZa(+) | | | | INT | 8 T/A | 20 T/A | 20 T/A | T A | T A |
| TH/GH | | | | Zn | 034 | 183 | 311 |
| TRACK  T/G | 18 | | | CONV +W ANGLE E | | | |
| Za | 17 | | | GRID Zn | | | |
| REL Za | 23 | | | | | | |

**REFRACTION TABLE (condensed)**

| Ro | Altitude MSL (thousands of feet) | | | | | |
|---|---|---|---|---|---|---|
| | 0 | 20 | 25 | 30 | 35 | 40 |
| 1 | 63 | 46 | 41 | 36 | 31 | 26 |
| 2 | 33 | 19 | 16 | 14 | 11 | 9 |
| 3 | 21 | 12 | 10 | 8 | 7 | 5 |
| 4 | 16 | 8 | 7 | 6 | 5 | 3-10 |
| | 12 | 7 | 5 | 4 | 3-10 | 2-10 |

| TK | DC | TH | VAR | MH | DEV | CH |
|---|---|---|---|---|---|---|
| | | | | | | |
| | | | | | | |

**Figure 12-4.** *Typical example of the three LHA method.*

12-5

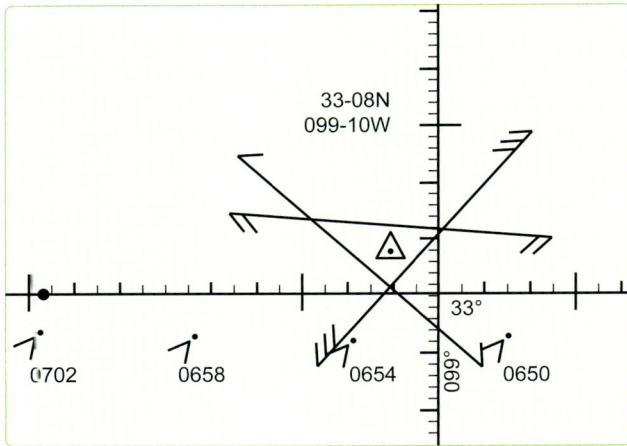

33-08N
099-10W

33°

060°

0702 0658 0654 0650

**Figure 12-5.** *Plotting three LHA.*

| | | 28 Sept 95 | | | | |
|---|---|---|---|---|---|---|
| **D R** | TIME | | | 0300 | | |
| | LAT | | | N27-18 | | |
| | LONG | | | W121-04 | | |
| **P R E C O M P** | Body | Alphazet | Antares | Alkaid | | |
| | GHA | | | 51-25 | | |
| | CORR | | | - | | |
| | SHA | | | - | | |
| | GHA | | | 51-25 | | |
| | +360 | | | 411-25 | | |
| | ASSUM LONG -W +E | | | W121-25 | | |
| | LHA | 287 | 288 | 289/290 | | |
| | ASSUM LAT | | | N27- | | |
| | DEC | | | - | | |
| | TAS HC | 24-56 | 23-45 | 25-03 | | |
| | d | — - | — - | — - | — - | — - |
| | DEC | I | I | I | I | I |
| **C O R R S & M O T S** | CORR MC | 24-56 | 23-45 | 25-03 | | |
| | FS TIME | 0248 | 0252 | 0256 | | |
| | FS | 24-19 | 23-55 | 25-16 | | |
| | SEXT | 0 | 0 | 0 | | |
| | REFR | -1 | -1 | -1 | | |
| | PA/sp/O | - | - | - | | |
| | MOT DSS | +60 | -34 | -8 | | |
| | MOT BODY | - | - | - | | |
| | TCTAL | +59 | -35 | -9 | | |
| | HC | 25-18 | 23-20 | 25-07 | | |
| | INTCPT | 22T | 25A | 4T | | |
| | ZN | 069 | 220 | 315 | | |
| **G R I D H D G S** | CAMPS | | | | | |
| | GZN/TH | | | | | |
| | MAG VAR | | | | | |
| | MAG HDG | | | | | |
| | DEV CORR | | | | | |
| | CH | | | | | |
| **M I S C** | TRACK | 068 | 068 | 068 | | |
| | ZN-TR | 001 | 152 | 247 | | |
| | GS | 310 | | | | |
| | ALT | 220 | | | | |
| | CORIOLUS | 4R | | | | |
| | P-N | N/A | | | | |

**Figure 12-6.** *Half motions three LHA format.*

# Daytime Celestial Techniques

Daytime fixing, using celestial techniques, is rather limited because often only one body, the sun, is visible. Ordinarily, three LOPs cannot be obtained for a fix from one body, because the LOPs plot nearly parallel to each other.

## The Sun Heading Shot at High Noon

The azimuth of the sun changes very rapidly when the subpoint of the sun is directly over the longitude of the observer, which is called the time of transit. The LHA at transit time is 360°. This phenomenon is more pronounced at lower latitudes as the subpoint of the sun passes closer to the observer. This makes it extremely difficult to get an accurate celestial heading shot at the transit time. Therefore, if you need a heading shot near the time of transit, you must take extra precaution to get the heading observation exactly at the precomputed fix time. If the moon or Venus is available, consider using these bodies for an accurate celestial heading. If using the sun, you should weigh the increased possibility of an inaccurate heading shot. If the accuracy is questionable, get another heading shot as the sun's rate of azimuth change slows enough to allow a more accurate shot.

## Intercept Method

The intercept method is normally used in obtaining a noon day fix. If the sun passes close to the observer's position, within about 4°, the subpoint method of plotting the fix may be used. This method differs from normal procedures in that three different precomps for three different times are computed. Because of the rapid change of the sun's azimuth at or near transit, this variation is necessary. The procedure is:

1. Determine the time of transit.

2. Select the LHA before and after transit for which the change in azimuth is 30° or more. Since 1° of LHA is equal to 4 minutes of time, the difference in transit LHA and the new LHA can be converted to time in minutes. Thus, the time preceding and following transit can be determined.

3. Plot the DR positions for times determined in 12.7.2. Select the appropriate assumed positions necessary for the computation and plotting of the LOPs. The assumed position for time of transit is also plotted.

4. Determine the intercepts and azimuth for each LOP. Plot these data from the respective assumed positions.

5. Resolve the LOPs to a common time, preferably that of the transit LOP.

NOTE: At 30° N latitude, the linear speed of the sun is approximately 780 knots. Thus, on westerly headings in high-speed aircraft, the DR distance involved before encountering a 30° change in azimuth is considerable.

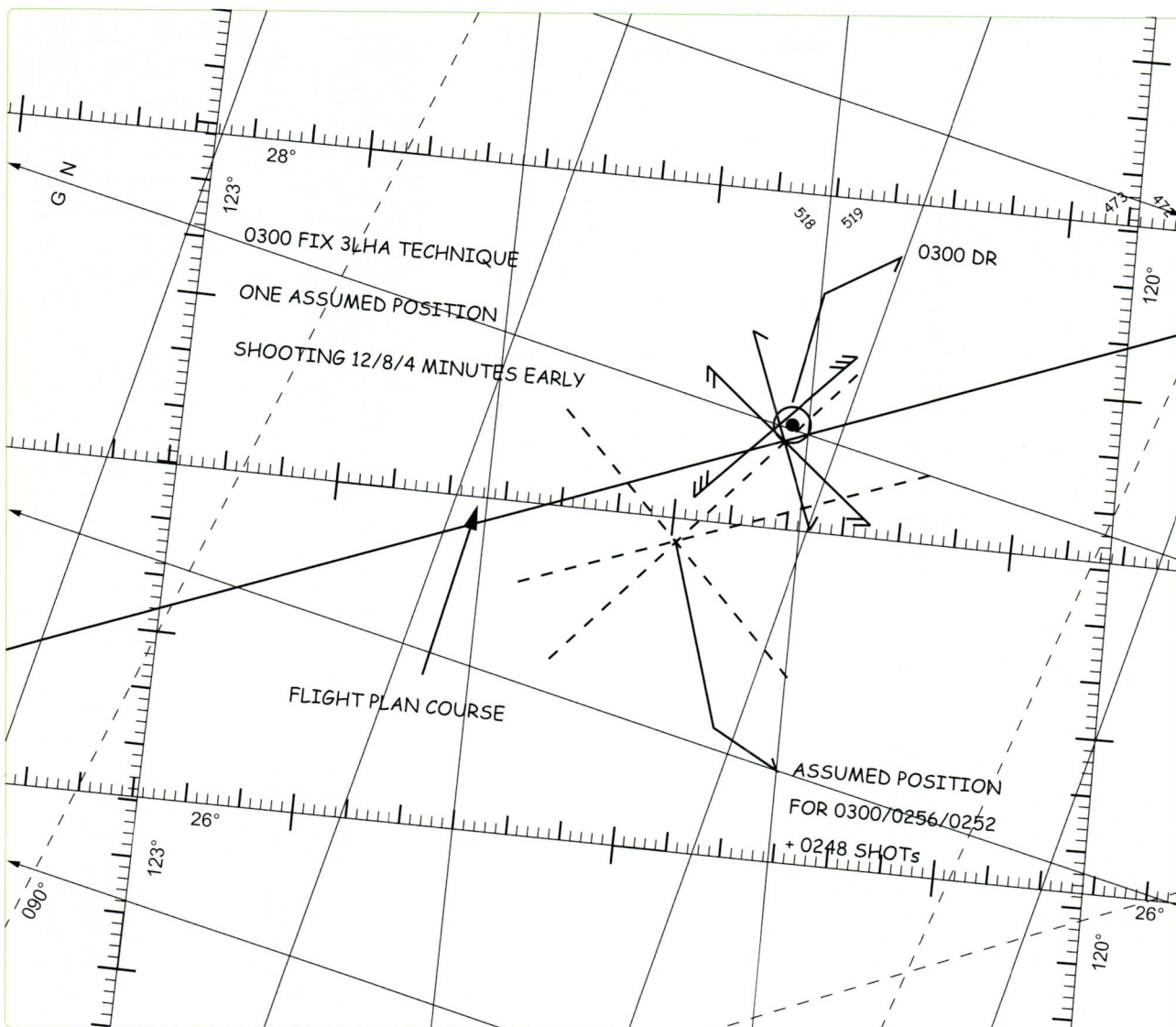

**Figure 12-7.** *Plotting a half motions observation.*

## Subpoint Method

When the observer is within approximately 4° of the subpoint of the body, the subpoint method of solution is normally used. This is because the radius of the circle of equal altitude is so small that a straight line does not approximate the arc and a straight line does not give an accurate LOP. The procedure is:

1. Plot the subpoints of the body for the time of the observations (using GHA and/or Dec).

2. Find the co-altitude of the shots and convert it to NM (90° − Alt × 60 NM).

3. Advance the first subpoint and retard the third along the DR track, using best-known track and GS.

4. Set the distance found from the co-altitude and strike it off from the resolved subpoints (with a compass or pair of dividers). Do this for each observation.

NOTE: The resulting intersection, or triangle, gives one on-time fix. If the LOPs form a triangle, the aircraft position is probably within the triangle.

The subpoint method is convenient because Pub. No. 249 is not used—only the Air Almanac. This method can also be used with a star near your assumed position and may be necessary if, for some reason, your Volume 1 is unavailable. The stars Dec and GHA are needed to determine if the observer is within 4° of the subpoint. The Air Almanac may be used to find the Dec and sidereal hour angle (SHA) of the star. The SHA of the star is added to the GHA of Aries to find the GHA of the star.

## Eliminating Motions with the Bracket Technique

For sun observations, you can eliminate motion calculations by using a shooting schedule of 3 minutes early, on fix time, and 3-minutes late. With this schedule, the 3-minute early and 3-minute late shots have the same magnitude of motion but an opposite sign. Therefore, these motions cancel each other out and do not need to be computed. The on-time shot has no motions. Therefore, the three intercepts can be averaged for a single LOP. At night, shooting the same star 4 minutes early and late, with a different star shot on time, can employ a similar method. In this case, the intercepts for the same star's 4-minute early or late shots can be averaged. This reduces workload, but only two LOPs are obtained.

## DR Computer Modification

Rather than eliminating motions, your DR computer can be modified so both observer and body motions can be computed at one time, without entry into the Pub. No. 249. Make a GS and latitude scale. *[Figure 12-8]* After constructing these, the DR computer can be modified for quick and accurate computations of 1-minute motion adjustments.

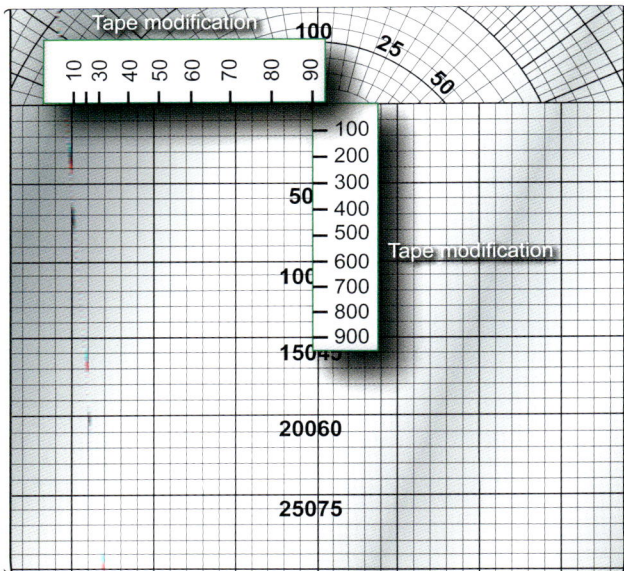

**Figure 12-8.** *MB-4 motions modification.*

Tape the GS scale (0 through 900) along the centerline of the grid scale. Match zero to zero, 300 to 50, and 600 to 100 as shown in *Figure 12-8*. Then, tape the latitude scale along the zero grid line so that 90° falls on the centerline and the scale extends to the left as shown. Check the accuracy of your placement: 30° latitude should fall 13 divisions left of centerline. Juggle the scale as necessary to provide the greatest accuracy between 30° and 45°.

To use the modified MB-4 computer for motion adjustments:

- Set true north under the index. If computing for grid, set polar angle (PA) under the index. In the NW and SE hemisphere quadrants, PA equals convergence angle (CA). In the NE and SW quadrants, PA = 360 − CA. Next, place the grommet over the zero grid line. Mark a cross (+) at the assumed latitude. *[Figure 12-9]*

- Set track (or grid track) under the index and position the slide so the GS is under the grommet. Place a dot on the zero point of the grid scale. *[Figure 12-10]*

- Place the Zn (or grid Zn) of the body under the index. Position the slide so the cross or the dot, whichever is uppermost, is on the zero line of the grid. *[Figure 12-11]*

NOTE: The vertical distance between the zero line and the low mark is the combined 1-minute motion. Each line of the grid equals 1 minute of arc (1 mile). If the cross is on the zero line, the motion is positive. If the dot is on the zero line, the motion is negative. When solving for motions using grid, all directions must be grid directions.

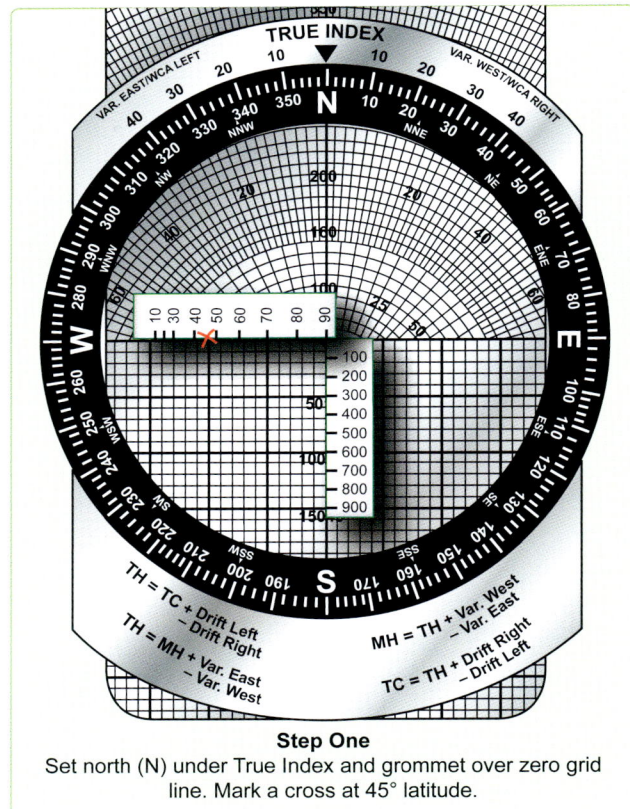

**Step One**
Set north (N) under True Index and grommet over zero grid line. Mark a cross at 45° latitude.

**Figure 12-9.** *Celestial motions–step one.*

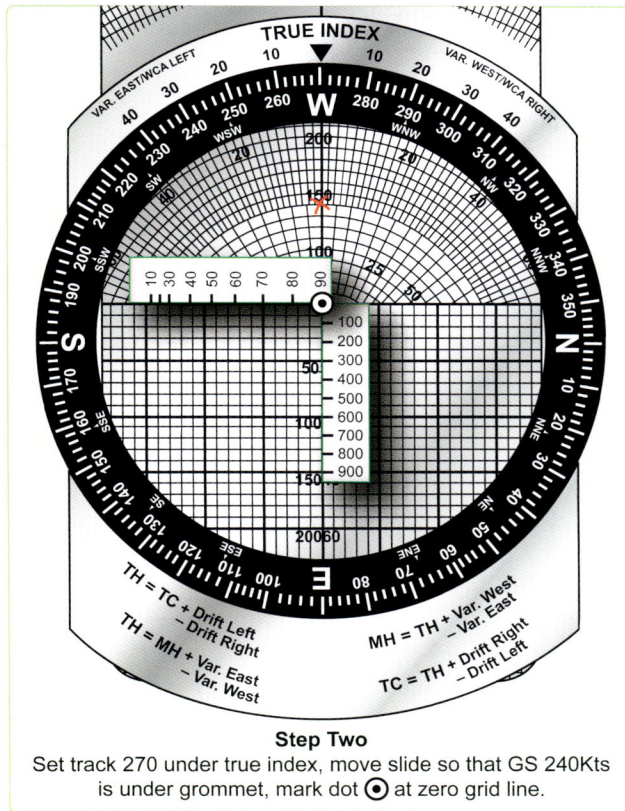

**Step Two**
Set track 270 under true index, move slide so that GS 240Kts is under grommet, mark dot ⊙ at zero grid line.

**Figure 12-10.** *Celestial motions–step two.*

EXAMPLE: Given the following information, find the combined 1-minute motion adjustment.

| | |
|---|---|
| Assumed Latitude | 45° 10′ N |
| True Track | 270° |
| GS | 240 knots |
| True Zn | 171° |
| Answer | +1′ |

## Combinations of Sun, Moon, and Venus

The moon or Venus is often visible during daylight hours and can be used to obtain an LOP. Always consider fixing using these bodies during daylight celestial flights. When planning the flight, use the sky diagrams in the Air Almanac to determine the availability of the moon and Venus. If the bodies are available, they can be readily found by accurately precomputing their altitudes and azimuths.

When looking for Venus, take all the filters out of the sextant and point it at the precise location of the planet. A bright, small pinpoint of light is visible but hard to detect, unless sky conditions and separation from the sun are ideal. With practice, acquisition should become easier and you will be familiar with those conditions conducive to successfully making a Venus shot.

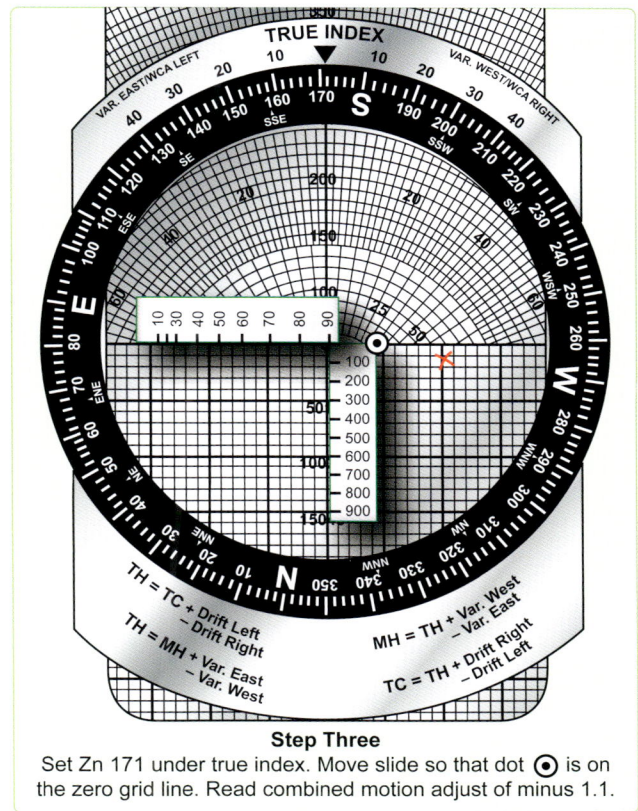

**Step Three**
Set Zn 171 under true index. Move slide so that dot ⊙ is on the zero grid line. Read combined motion adjust of minus 1.1.

**Figure 12-11.** *Celestial motions–step three.*

During the day when the sun is high, the moon or Venus, if they are available, can be used to obtain compass deviation checks. In polar regions during periods of continuous twilight, the moon and Venus are available if their Dec is the same name as the latitude.

### Duration of Light

Sunrise and sunset at sea level and at altitude, moonrise and moonset and semiduration graphs will not be discussed in detail in this chapter. It is imperative; however, to preplan for any flight where twilight occurs during the course of the flight, especially at the higher latitudes where twilight extends over longer periods of time. An excellent discussion, with appropriate examples, is provided in the Air Almanac and should be sufficient for those missions requiring detailed planning.

## True Heading Celestial Observation

The periscopic sextant, in addition to measuring celestial altitudes, can be used to determine true headings (TH) and true bearings (TB). Any celestial body, whose azimuth can be computed, can be used to obtain a TH. Except for Polaris, the appropriate volume of Pub. No. 249 is entered to obtain Zn (true bearing). In the case of Polaris, the Air Almanac has

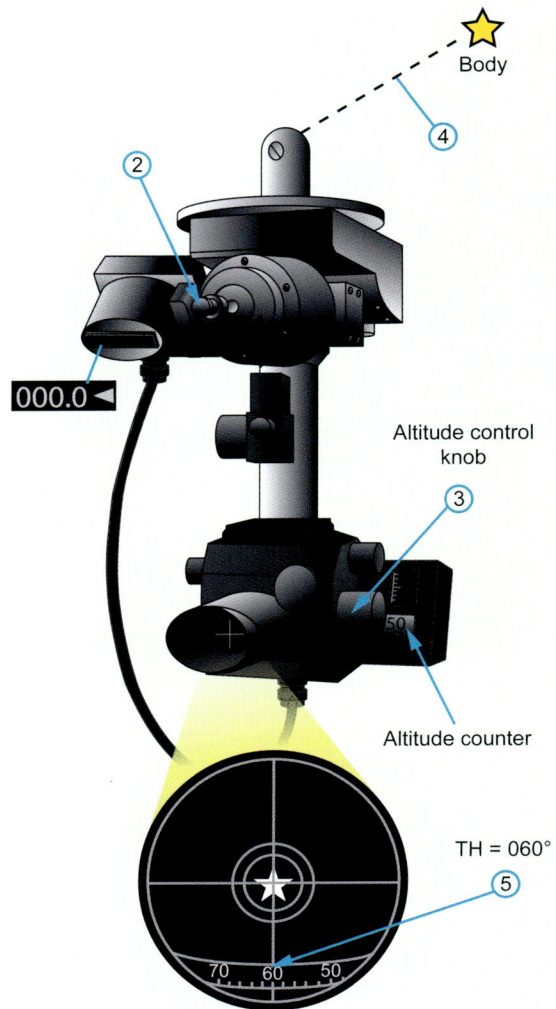

1. Precompute the Zn of the body.

2. Using the azimuth crank, set the Zn of the body in the azimuth counter window.

3. Using the altitude control knob, set Hc in the altitude counter window.

4. Locate the body by turning the sextant until the approximate TH of the aircraft falls under the vertical crosshair. Body should be in the field of vision. Bring body into collimation.

5. Read the exact TH under the vertical crosshair (060°).

**Figure 12-12.** *True bearing method (except Polaris).*

an azimuth of Polaris table. It does not require information from the Pub. No. 249 tables. There are two methods used to obtain TH with the periscopic sextant. The TB method requires precomputation of Zn. Postcomputation of Zn is possible with the inverse relative bearing (IRB) method. The procedures follow.

## True Bearing (TB) Method:

1. Determine GMT and body to be observed.

2. Extract GHA from the Air Almanac.

3. Apply exact longitude, at the time of the shot, to GHA to obtain exact LHA.

4. Enter appropriate Pub. No. 249. table with exact LHA, latitude, and Dec. Interpolate if necessary and extract Zn and Hc. *[Figure 12-12]* If Polaris is used, obtain the azimuth from the Azimuth of Polaris table in the

Air Almanac and use your latitude instead of Hc. *[Figure 12-13]*

5. Set Zn in the azimuth counter window with the azimuth crank, and set Hc in the altitude counter window with the altitude control knob.

6. Collimate the body at the precomputed time and read the TH of the aircraft under the vertical crosshair in the field of vision. If you are using precomputation techniques, a TH is available every time an altitude observation is made.

NOTE: Shot must be taken at precomp time.

## Inverse Relative Bearing (IRB) Method:

1. Set 000° in the azimuth counter window with the azimuth crank. *[Figure 12-14]*

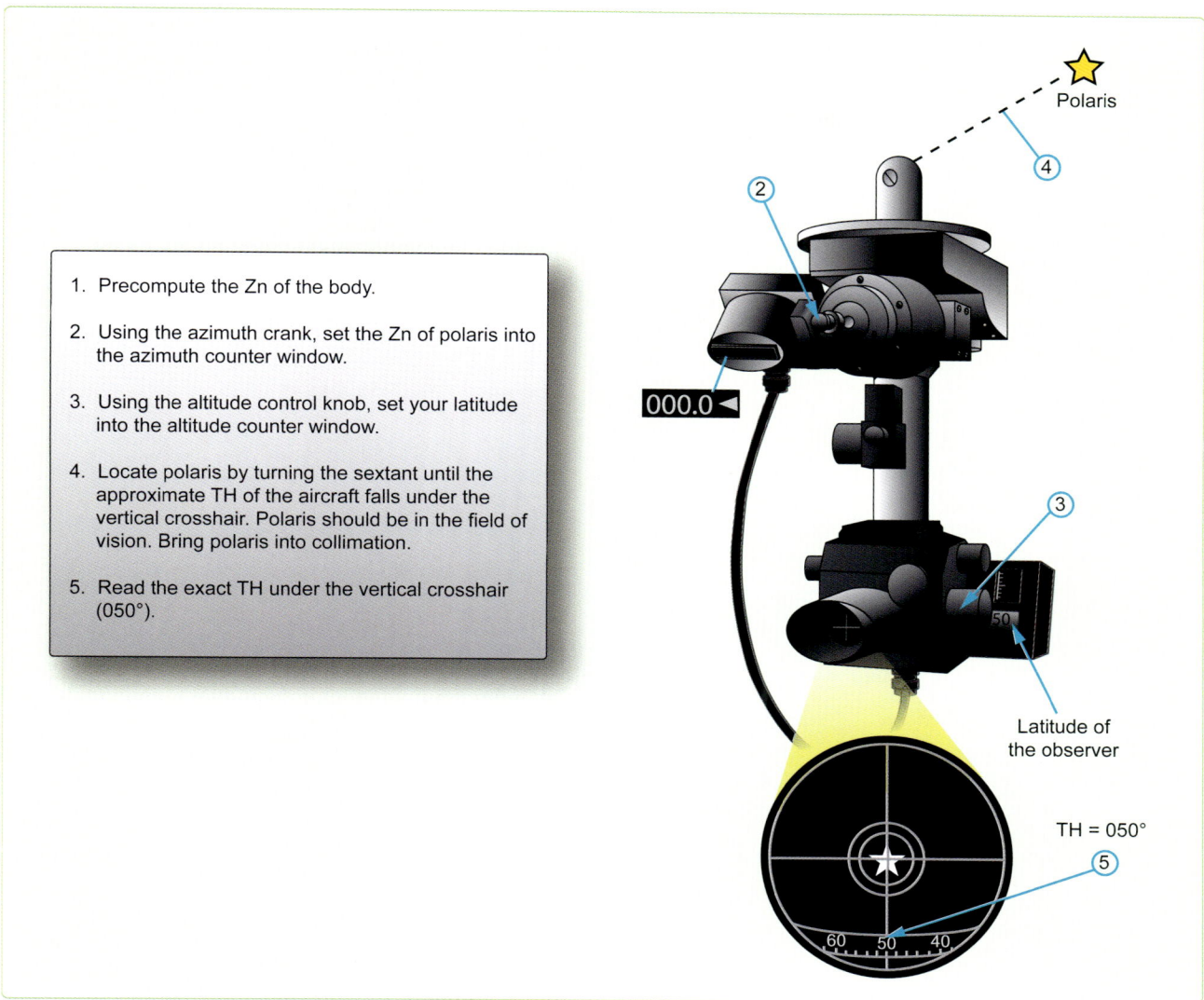

1. Precompute the Zn of the body.

2. Using the azimuth crank, set the Zn of polaris into the azimuth counter window.

3. Using the altitude control knob, set your latitude into the altitude counter window.

4. Locate polaris by turning the sextant until the approximate TH of the aircraft falls under the vertical crosshair. Polaris should be in the field of vision. Bring polaris into collimation.

5. Read the exact TH under the vertical crosshair (050°).

**Figure 12-13.** *True bearing method (including Polaris).*

2. Collimate the body. At the desired time, read the IRB under the vertical crosshair in the field of vision.

3. Compute Zn of the celestial body and use the formula: TH = Zn + IRB

## Celestial Navigation in High Latitudes

Celestial navigation in polar regions is of primary importance because it constitutes a primary method of determining position other than by DR, and it provides a reliable means of establishing direction over much of the polar regions. The magnetic compass and directional gyro (DG) are useful in polar regions, but they require an independent check that can be provided by a celestial body or other automatic system, such as inertial navigation system (INS) or global positioning system (GPS).

At high latitudes, the sun's daily motion is nearly parallel to the horizon. The motion of the aircraft in these regions can easily have greater effect upon altitude and Zn of the sun than the motion of the sun itself.

At latitude 64°, an aircraft flying west at 400 knots keeps pace with the sun, which appears to remain stationary in the sky. At higher latitudes, the altitude of a celestial body might be increasing at any time of day, if the aircraft is flying toward it and a body might rise or set, at any azimuth, depending upon the direction of motion of the aircraft relative to the body.

### Bodies Available for Observation

During the continuous daylight of the polar summer, only the sun is regularly available for observation. The moon is

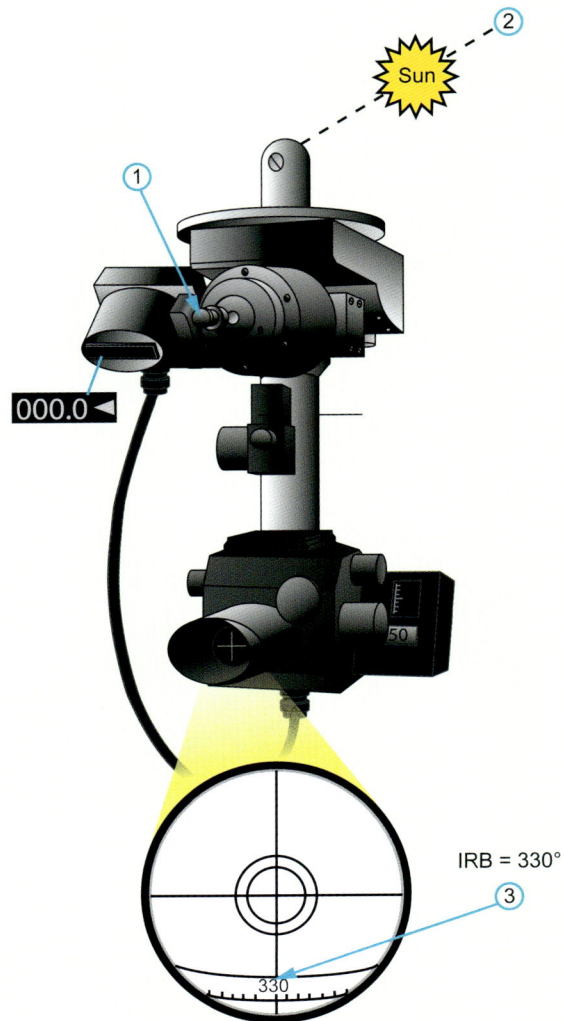

1. Turn crank until 000° is in the azimuth counter window

2. Locate the body and bring into collimation.

3. Read IRB under the vertical crosshair.

4. Solve the formula:

   TH = Zn + IRB

Sun ②

①

000.0 ◄

50

IRB = 330°
③

330

**Figure 12-14.** *Inverse relative bearing method.*

above the horizon about half the time, but generally it is both visible and at a favorable position with respect to the sun for only a few days each month.

During the long polar twilight, no celestial bodies may be available for observation. As in lower latitudes, the first celestial bodies to appear after sunset and the last to remain visible before sunrise are those brighter planets, which are above the horizon.

The sun, moon, and planets are never high in polar skies, thus making low altitude observations routine. Particularly with the sun, observations are made when any part of the celestial body is visible. If it is partly below the horizon, the upper limb is observed and a correction of –16' for semidiameter (SD) is used in the SD block of the precomputation form.

During the polar night, stars are available. Polaris is not generally used, because it is too near the zenith in the arctic and not visible in the Antarctic. A number of good stars are in favorable positions for observation. Because of large refractions near the horizon avoid low altitudes (below about 20°) when higher bodies are visible.

**Sight Reduction**

Sight reduction in polar regions presents some slightly different problems from those at lower latitudes. Remember, for latitudes greater than 69° N or 69° S, Pub. No. 249 tables have tabulated Hc and azimuths for only even degrees of LHA. This concerns you in two ways. First, it is necessary to adjust assumed longitude to achieve a whole, even LHA for extractions. This precludes interpolating. Second, the difference between successive, tabulated Hc is for 2° of LHA, or 8 minutes of time, so this difference must be divided in half when computing motion of the body for 4 minutes of time.

For ease of plotting, all azimuths can be converted to grid. To convert, use the longitude of the assumed position to determine convergence, because the Zn is for the assumed position, not the DR position. On polar charts, convergence is equal to longitude.

In computing motion of the observer, it is imperative that you use the difference between grid azimuth and grid track, or Zn and true track, since this computation is based on relative bearing (RB). Zn minus grid course does not give RB.

Since low altitudes and low temperatures are normal in polar regions, refer to the refraction correction table and use the temperature correction factor for all observations.

In polar regions, Coriolis corrections reach maximum values and should be carefully computed.

## Poles as Assumed Positions

Within approximately 2° of the pole, it is possible to use the pole as the assumed position. With this method, no tabulated celestial computation is necessary and the position may be determined by use of the Air Almanac alone.

At either of the poles of the earth, the zenith and the elevated poles are coincident or the plane of the horizon is coincident with the plane of the equator. Vertical circles coincide with the meridians and parallels of latitude coincide with Dec circles. Therefore, the altitude of the body is equal to its Dec and the azimuth is equal to its hour angle.

To plot any LOP, an intercept and the azimuth of the body are needed. In this solution, the elevated pole is the assumed position. The azimuth is plotted as the GHA of the body or the longitude of the subpoint. The intercept is found by comparing the Dec of the body, as taken from the Air Almanac, with the observed altitude of the body. To summarize, the pole is the assumed position, the Dec is the Hc, and the GHA equals the azimuth.

For ease of plotting, convert the GHA of the body to grid azimuth by adding or subtracting 180° when using the North Pole as the assumed position. When at the South Pole, 360° − GHA of the body equals grid azimuth. The result allows the use of the grid lines for plotting the LOPs. When using grid azimuth for plotting, apply Coriolis to the assumed position (in this case, the pole). Precession or nutation corrections are not necessary since current SHA and Dec are used. Motion of the observer tables may also be used in precomputation, since grid azimuth relative to grid course may be determined. Motion of the body is zero at the poles.

Note the exact GMT of the celestial observation. From the Air Almanac, extract the proper Dec and GHA. Plot the azimuth. Compare Ho and Hc to obtain the intercept. When the observed altitude (Ho) is greater than the Dec (Hc), it is necessary to go from the pole toward the celestial body along the azimuth. If the observed altitude is less than the Dec, as is the case with the sun in *Figure 12-15*, it is necessary to go from the pole away from the body along the azimuth. Draw the LOPs perpendicular to the azimuth line in the usual manner. Do not be concerned about large intercepts; they have no bearing on the accuracy of this type of fix. Observations on well-separated bearings give a fix that is as good close to the pole as it is anywhere else.

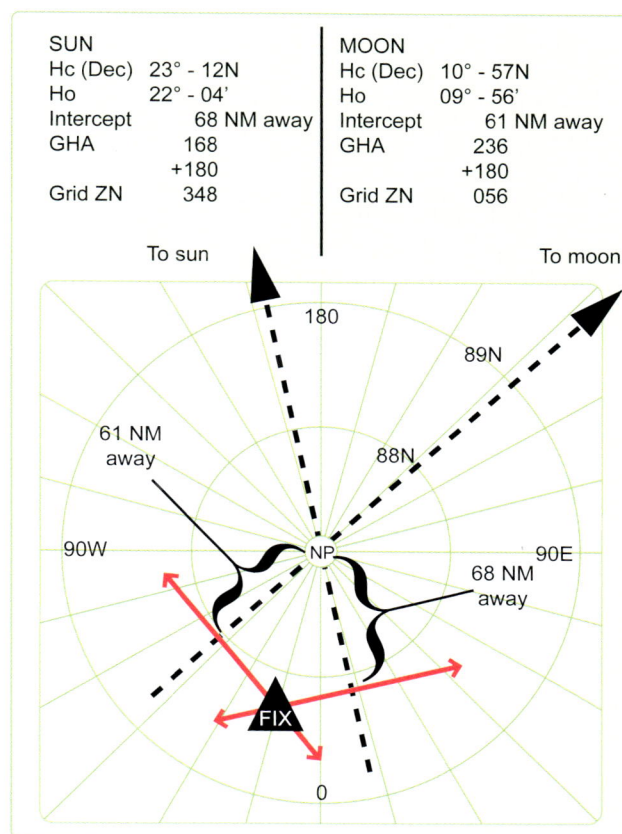

**Figure 12-15.** *Using pole as assumed position.*

## Adjusting Assumed Position

### Adjusting Assumed Position for Off-Time Shot

There are times when the observer does not start the shot at the prescribed time for various reasons. For example, the observer may struggle to find the body due to cloud cover. If a shot is taken off time, you can use the FEAST (Fast EAST) rule: a shot taken too fast or too early has the assumed position moved 15' of longitude east for each minute early to compensate for body motion. *[Figure 12-16]* Apply the reverse of the FEAST rule for late shots (move the assumed position west). This adjusted position is then advanced or retarded for track and GS to account for motion of the observer, applying the same concept used in the three-LHA method. *[Figure 12-5]* This technique for solving motions is also discussed in Chapter 10.

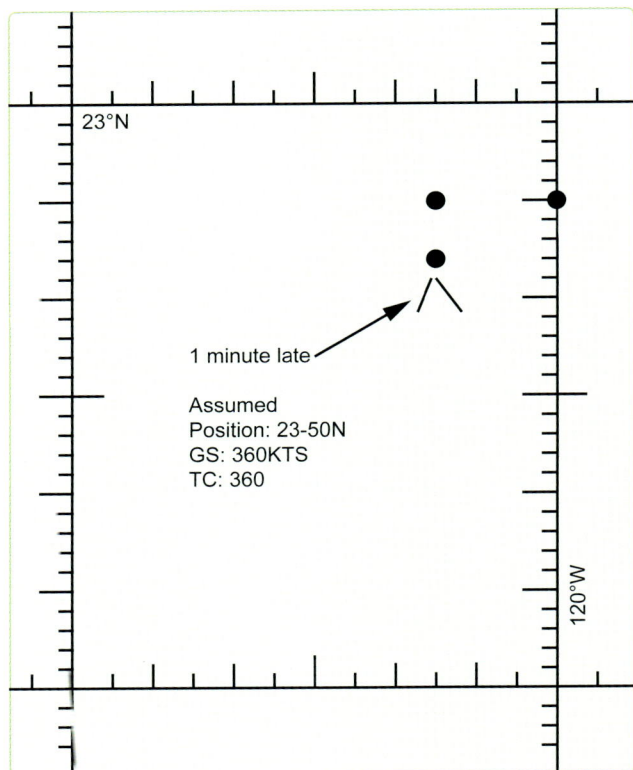

**Figure 12-16.** *Corrections for off-time shooting.*

EXAMPLE:

Original assumed position:          23°–50' N 120° –00' W
Move 15' of longitude west for 1 minute late
Retard 6 NM from track of 360° TH
New assumed position:          23°–44' N 120 ° –15' W

## Longitude Adjustment Principle

You will occasionally make errors in your precomputations. Possibly the most common would be an extraction error of the GHA or math error while computing the LHA. If one of these numbers is incorrect, then all the extractions from the Pub. No. 249 would be based on erroneous information, and the result would be an LOP error. Fortunately there is a way of compensating for this type of error without having to reenter the table and retrieve the correct data. This method is called the Longitude Adjustment Principle (LAP). You need only adjust the assumed longitude (up to 2½°) to correct for a GHA extraction error or a math error. Moving the assumed position beyond the 2½° induces some error in the plotting LOP. Suppose you wanted the GHA for

1410Z, you extracted the value for 1400Z and applied it to the longitude. *[Figure 12-17]* The resultant LHA was used and the precomp completed before you realized your error. To do the LAP first, extract the correct GHA (031–20), keep the old LHA, and adjust the longitude so that the math is correct. *[Figure 12-18]* A math error can occur in solving for the LHA. *[Figure 12-19]* Once you have corrected the precomp, use the adjusted longitude for your assumed longitude to plot the LOP.

| Body | Sun | |
|---|---|---|
| Base GHA | 028-50 | Incorrect value for the GHA. |
| CORR | | |
| (+360) | 360 | 1. Is this error within 2½°? |
| GHA | 388-50 | 031-20<br>028-50<br>―――――<br>2-30 |
| ASSUM -W<br>LONG +E | 130-50 | |
| LHA | 258 | 2. 2° 30' of longitude is less than 150 NM. |
| ASSUM<br>LAT | 42 N S | |

**Figure 12-17.** *LAP using incorrect GHA.*

| Body | Sun | | |
|---|---|---|---|
| Base GHA | 028-50 | 031-20 | **1410 GHA** |
| CORR | | | |
| (+360) | 360 | 360 | |
| GHA | 388-50 | 391-20 | |
| ASSUM -W<br>LONG +E | 130-50 | 133-20 | **Longitude is adjusted** |
| LHA | 258 | 258 | |
| ASSUM<br>LAT | 42 N S | | **LHA remains the same** |

**Figure 12-18.** *LAP using correct GHA.*

## Chapter Summary

Any of the techniques discussed here, if used on a regular basis, can be just as accurate as normal precomping procedures and save some time as well. These techniques are not all inclusive. There are many commercial publications available as a source for celestial navigators: for example, American Practical Navigator by Bowditch (available through the NGA) and the Journal of the Institute of Navigation (available through the Institute of Navigation).

| | Body | Sun | | |
|---|---|---|---|---|
| | Base GHA | 320-30 | | |
| LHA should be 199 | CORR | | | |
| | (+360) | | | |
| | GHA | 320-30 | 320-30 | Adjust longitude |
| | ASSUM (-W) LONG +E | 121-30 | 122-30 | |
| | LHA | 198 | 198 | LHA remains the same |
| | ASSUM | 40 (N) S | | |

**Figure 12-19.** *LAP correcting a math error.*

# Chapter 13

# Sextants and Errors of Observation

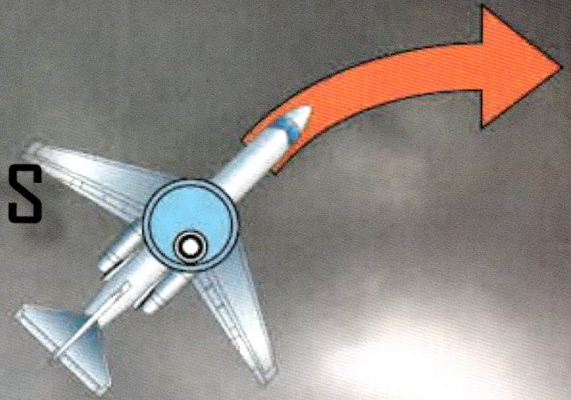

## Sextants

For hundreds of years, mariners have navigated the seas keeping track of their positions by use of the sextant. This instrument measured the altitude of celestial bodies (their angular distance above the horizon), and the information derived from this measurement was used to determine the position of the vessel. All celestial navigation follows this rule. Today's navigator measures the altitude of the celestial bodies in much the same manner as Magellan or Columbus. However, there is a difference between air and marine celestial navigation. Because marine navigators are on the surface of the ocean, they can establish their horizon by referring to the natural horizon. In an aircraft, this is impossible because altitude and aircraft attitude induce error. In the sextant designed for air navigation, a bubble, like the one in a carpenter's level, determines an artificial horizon, which is parallel to the celestial horizon. The bubble chamber is placed in the sextant so the bubble is superimposed upon the field of view. Both the celestial body and the bubble are viewed simultaneously, making it possible to keep the sextant level while sighting the body.

Sextants are subject to certain errors that must be compensated for when determining a line of position (LOP). Some of these errors are instrument errors while others are induced by the various inflight conditions. The first half of this chapter discusses the sextant and the second half explains sextant errors.

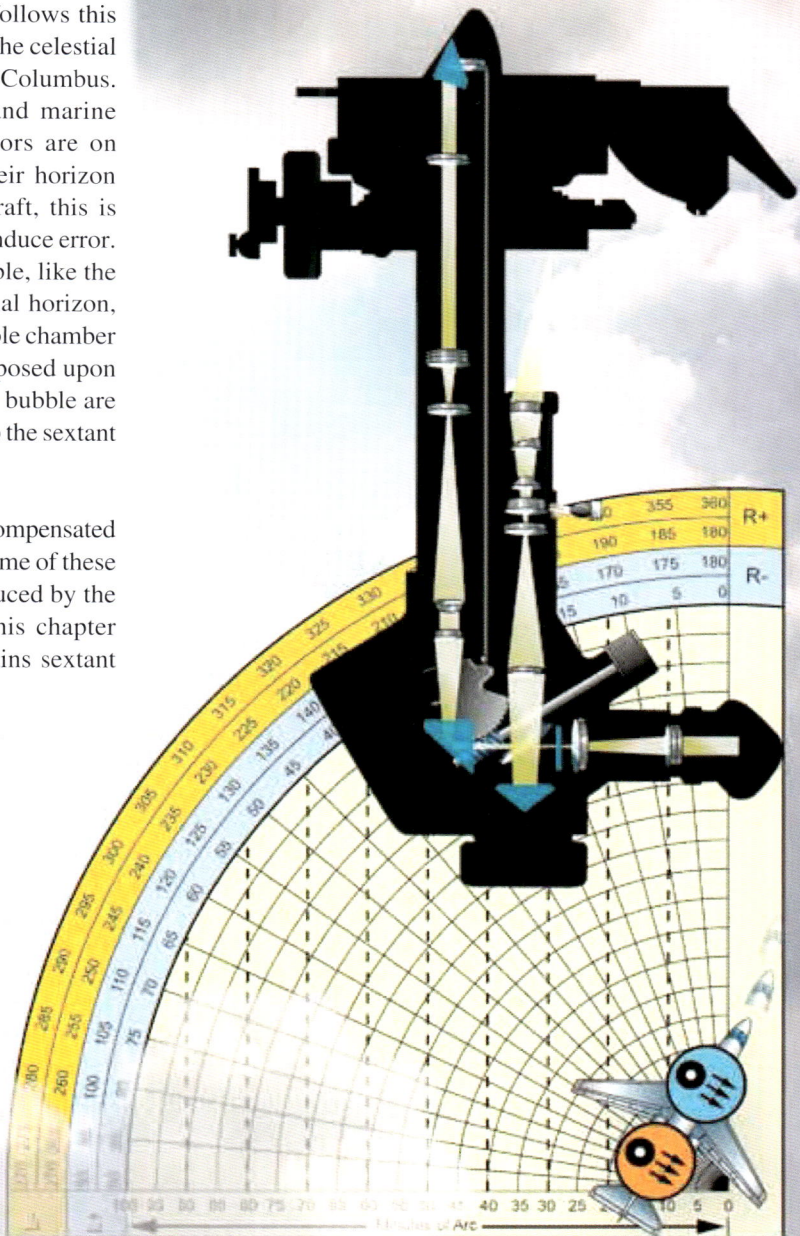

## The Bubble Sextant

The aircraft bubble sextant measures altitude above a horizontal plane established by a bubble. Aviators use several types of bubble sextants, all of which are indirect sighting. This means the navigator does not look directly toward the celestial body, but always looks in a horizontal direction as shown in *Figure 13-1*. The image of the body is reflected into the field of view when the field prism is set at the correct angle. In the bubble sextant, the bubble and body are visible in the same field of view. The sextant system consists of four parts: the mount, the sextant, the electrical cables, and the carrying case.

## The Periscopic Sextant

The periscopic sextant is an optical instrument that enables the navigator to determine true azimuth (Zn), relative bearing (RB), altitude angle of a celestial body, and aircraft true heading (TH). The sextant provides an angle of observation from below the horizon to directly overhead, as compared to an artificial horizon. *[Figure 13-2]*

Proper collimation techniques and the correct size bubble are essential ingredients of accurate celestial observations. Collimation is effected when the body is placed in the center of the bubble. For greatest accuracy, the bubble should be in the center of the field, with the body in the center of the bubble. The error is small if the bubble is anywhere on the vertical line of the field, as long as it does not touch the top or bottom of the bubble chamber. *Figure 13-3* shows examples of collimation from better to worse.

Bubble size affects the accuracy of a sextant observation. The ideal situation for collimation is to have a small bubble for ease in determining the center. A bubble that is too small sticks to the lens, decreasing accuracy. A bubble that is too large moves like a creature from a science fiction movie, making it difficult to find the center. Experience shows that best results are obtained with a bubble approximately one and a half times the apparent diameter of the sun or moon,

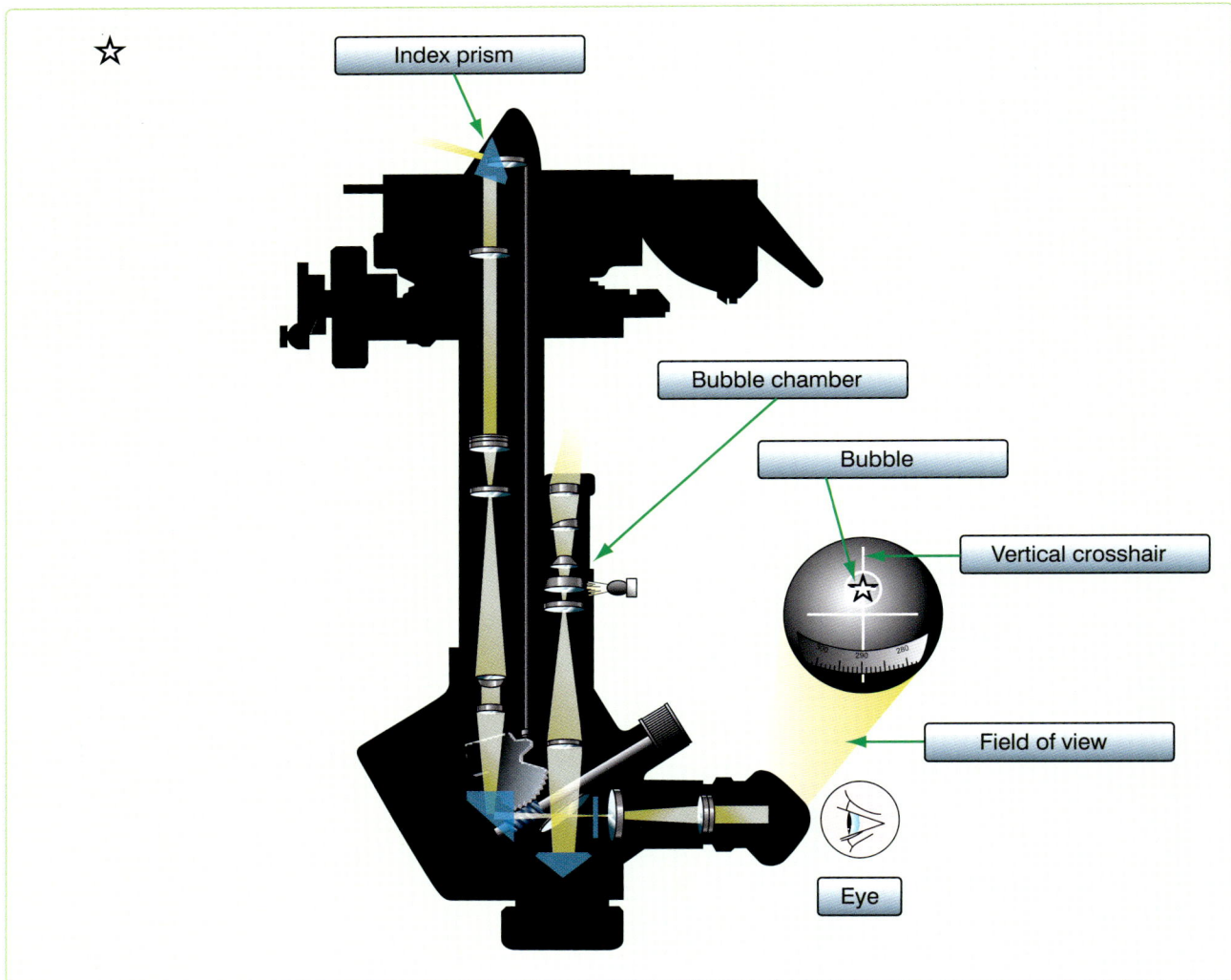

**Figure 13-1.** *Body is not sighted directly.*

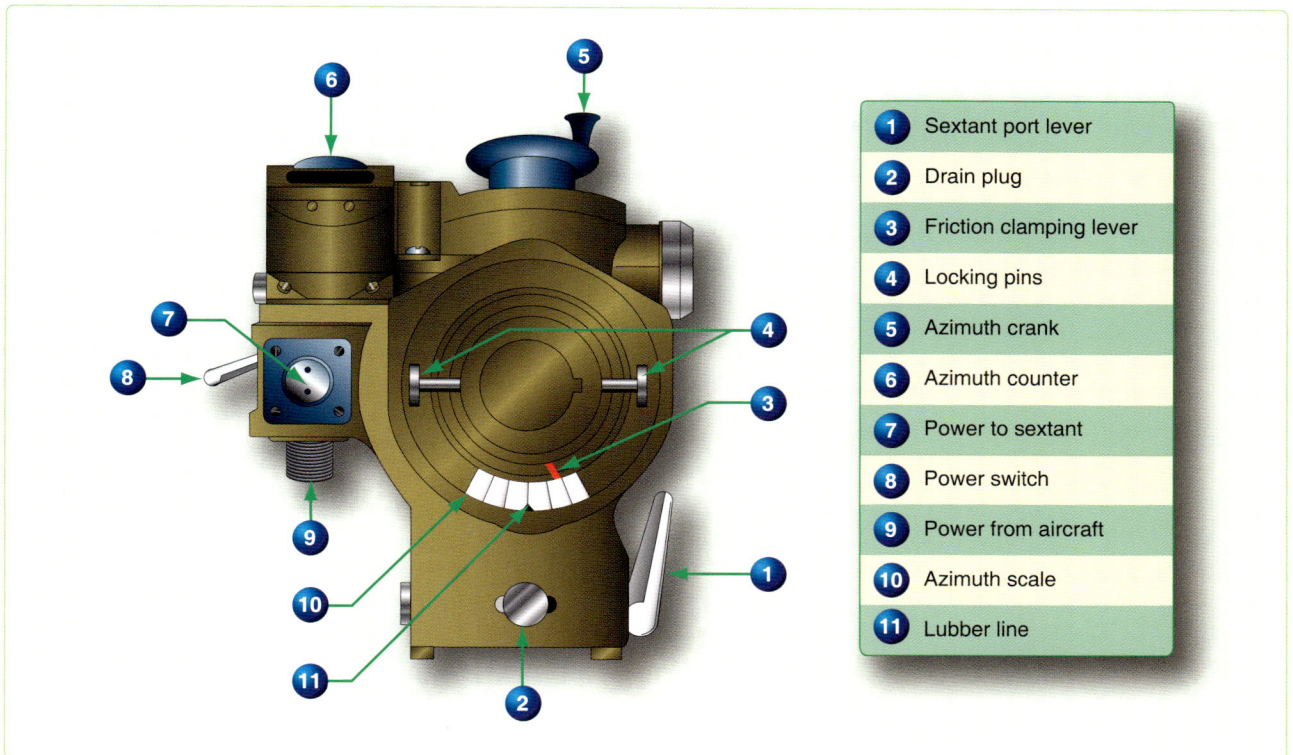

**Figure 13-2.** *Periscopic sextant mount.*

| | |
|---|---|
| 1 | Sextant port lever |
| 2 | Drain plug |
| 3 | Friction clamping lever |
| 4 | Locking pins |
| 5 | Azimuth crank |
| 6 | Azimuth counter |
| 7 | Power to sextant |
| 8 | Power switch |
| 9 | Power from aircraft |
| 10 | Azimuth scale |
| 11 | Lubber line |

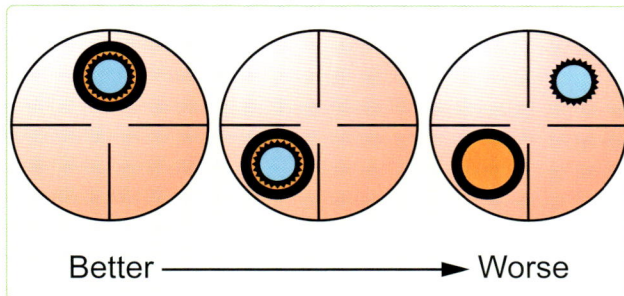

Better ⟶ Worse

**Figure 13-3.** *Correct and incorrect collimation.*

or about the size of a small washer ring. The field prism is geared to an altitude scale so that when the body is collimated, the altitude can be read from the scale.

An averaging mechanism is also incorporated that allows the navigator to take an observation over a period of time. The continuous motion of the aircraft affects the bubble and resultant artificial horizon. This movement resolves itself into a cycle in which the aircraft rolls, yaws, and pitches. To obtain an accurate reading, it is necessary to sight the body for a period of time during this cyclic movement and to average the results of a series of sightings. An averaging device has been incorporated in the sextant allowing an average reading to be obtained.

The sextant is actually a low-power periscope with a 15° field of view. *[Figure 13-4]* All lens surfaces in the sextant are coated to minimize light loss. To prevent condensation when the tip of the sextant is extended into cold air, the tube is filled with a dry gas and sealed. A desiccant, composed of silica gel, is used to remove moisture and check on the dryness of the gas inside the tube and is visible in the periscopic end of the sextant, or in some models, on the sextant body. When the silica gel is pink, there is moisture in the tube and the sextant should be replaced before flight.

Note: The numbers in parentheses refer to the parts indicated in *Figure 13-4.*

An eyepiece (1) rotates to correct the eyesight of the individual observer. Filters (2) are provided for selective use in the optical system so that the intensity of the sun's light might be adequately reduced. The filter control (2) is located on the left of the sextant.

Most sextants currently in use have been modified with an electronic device for accomplishing all the functions of the averaging mechanism. General differences in these and the unmodified sextants are addressed in this discussion.

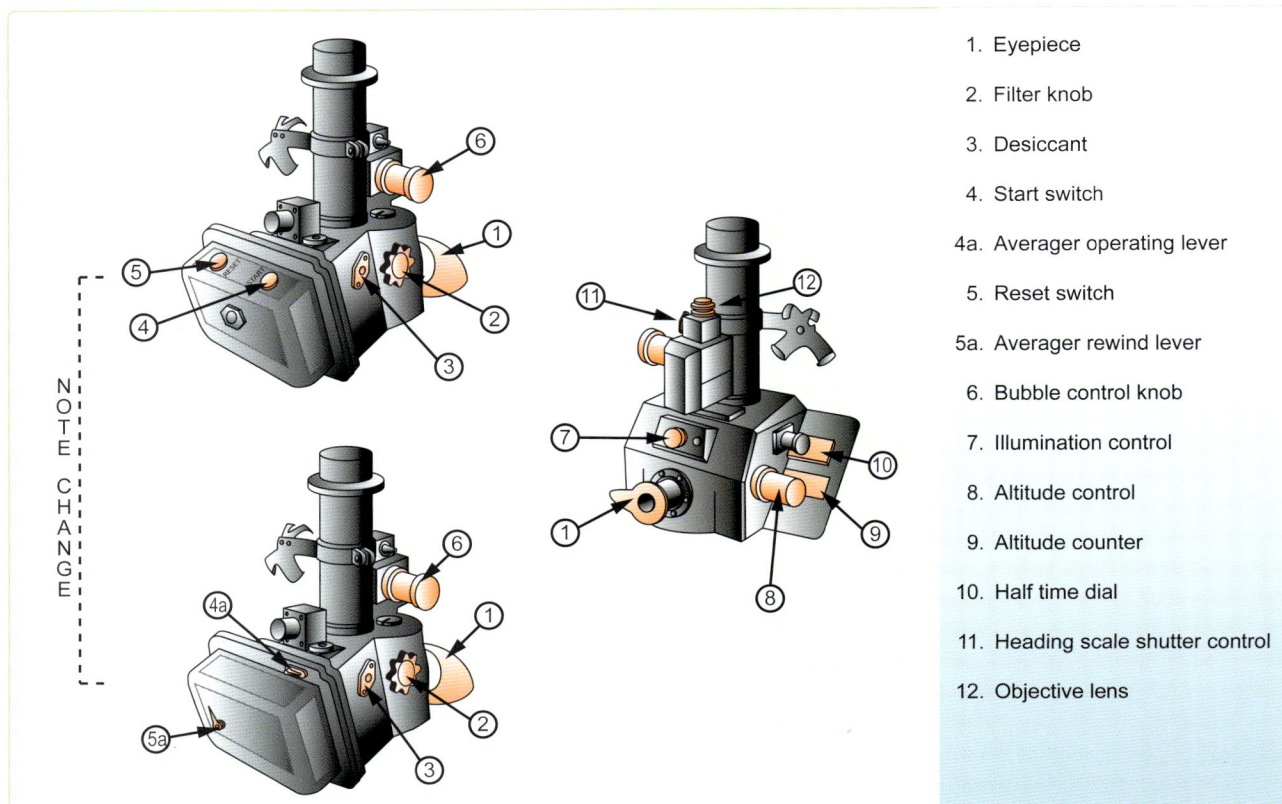

1. Eyepiece
2. Filter knob
3. Desiccant
4. Start switch
4a. Averager operating lever
5. Reset switch
5a. Averager rewind lever
6. Bubble control knob
7. Illumination control
8. Altitude control
9. Altitude counter
10. Half time dial
11. Heading scale shutter control
12. Objective lens

**Figure 13-4.** *Periscopic sextant.*

A start switch (4) (a start and stop or averager operating lever (4A) on unmodified sextants) starts and stops the operation of the sextant. Adjacent to this switch is the reset switch (5) (the averager rewind lever, if unmodified (5A), located below the averager operating lever). The reset switch, or averager rewind lever, has four functions. When depressed and released, it does the following:

1. Removes the shutter from the field of vision,

2. Zeroes and resets (rewinds if unmodified) the timer,

3. Zeroes the averager and places initial values in registers and data memory (realigns indices on unmodified sextants), and

4. Disconnects the altitude control knob from the averager.

The bubble control knob (6) should be left in the maximum increase position after adjustments have been made. With the control in the maximum increase position, an aneroid is locked to the bubble chamber to compensate for changes in ambient pressure and temperature.

On the front of the sextant, there is a rheostat control (7) that varies the intensity of the light in the bubble chamber. The altitude knob (8) is located on the right side of the sextant. It keeps the observed body in vertical collimation during the period of the observation. At the end of the scheduled observation, it adjusts the altitude counter until the exact average indication appears, or to align the indices on unmodified sextants. The body's altitude is read in the altitude counter (9). Directly behind the altitude knob is the averager display (10) (half-time dial and indices if unmodified). The averager display, or half-time dial, is graduated from 0–60 and indicates the half time of the observation. The indices, when aligned, permit the direct reading of the observed altitude on the altitude dial.

In the periscope sextant, the averaging is accomplished by microprocessor (Deimel-Black ball integrator if unmodified), which effects a continuous moving averager over any observation period up to 2 minutes. This system is very simple to operate and has many advantages over other known averaging devices. A single switch, or lever, sets or winds the mechanism and no other presetting of the sextant, timing mechanism, or averaging is necessary. It is continuously integrating altitude against elapsed time. After at least 30 seconds, it may be stopped at any time up to 2 minutes. The average altitude is read directly from the counter. A half-time clock indicates the half time of the observation. The time indication may be added directly to the time of starting the observation to compute the mean time of the observation. At the end of the observation, the averager energizes a solenoid (actuates a lever if unmodified) that drops a shutter across the field of view, indicating the end of the observation. Although it is possible to utilize an instantaneous shot, the normal timed observation lasts for 2 minutes. It is impossible to time any observation for less than 30 seconds using the sextant timer.

A heading scale shutter (diffuser lever) control (11) provides a convenient means of blocking out the bright illumination on the azimuth scale for night celestial observations. The objective lens (12) is located just above the heading scale shutter control. The lens aligns the azimuth scale of the sextant with the longitudinal axis of the aircraft. The lens can be rotated with the fingers in order to calibrate the azimuth scale on a known bearing while looking through the eyepiece. The objectives lens can remove up to 2° azimuth error in the azimuth ring. A locking ring beneath the lens prevents accidental movement. A dial lamp located on the right side of the sextant provides three beams of light to illuminate the averager indicators, the altitude counter, and the watch clip. The watch clip is made to hold an old-fashioned pocket watch.

### Electrical Cables

Cables provide power for sextant operation and illumination. One Y cable provides power from the mount to the sextant for illumination and averager operation.

### Sextant Case

The case provides shock-absorbent storage for the sextant when it is not in use. The sextant fits into form-fitting foam blocks and is secured by straps. The case also contains spare bulbs for sextant illumination and provides storage for the electrical cable.

## Errors of Sextant Observation

If collimation of the body with the bubble and reading the sextant were all that had to be done, celestial navigation would be simple. This would mean LOPs that are accurate to within 1 or 2 miles could be obtained without any further effort. Unfortunately, considerable errors are encountered in every sextant observation made from an aircraft. A thorough understanding of the cause and magnitude of these errors, as well as the proper application of corrections to either computed altitude (Hc) or Hs, helps minimize their effects. Remember that any correction applied to the Hs may be applied to the Hc with a reverse sign. Accuracy of celestial navigation depends upon thorough application of these corrections, together with proper shooting techniques. The errors of sextant observation may be classified into four groups: parallax, refraction, acceleration, and instrument.

### Parallax Error

Parallax in altitude is the difference between the altitude of a body above a bubble horizon at the surface of the earth and its calculated altitude above the celestial horizon at the center of the earth. All Hc are given for the center of the earth. If the light rays reaching the earth from a celestial body are parallel, the body has the same altitude at both the center and the surface of the earth. For most celestial bodies, parallax is negligible for purposes of navigation.

### Parallax Correction for the Moon

The moon is so close to the earth that its light rays are not parallel. The parallax of the moon may be as great as 1° thus, when observing the moon, a parallax correction must be applied to the Hs. This correction is always positive (+) and varies with the altitude and with the distance of the moon from the earth. The correction varies from day to day because the distance of the moon from the earth varies. Corrections for the moon's parallax in altitude are given on the daily pages of the Air Almanac and are always added, algebraically, to sextant altitudes. The values of parallax for negative altitudes are obtained from the Air Almanac for the equivalent positive altitudes.

### Semidiameter Correction

Semidiameter correction is found on the daily pages of the Air Almanac. Apply it when shooting the upper or lower limb of the moon or the sun. It is more likely to occur on observations of the moon because, when the moon is not full (completely round), the center is difficult to estimate. Shoot either the upper or lower limb and apply the semidiameter correction listed on the Air Almanac page for the time and date of the observation. Subtract the correction from the Hs when shooting the upper limb; add the correction to the Hs when shooting the lower limb. Reverse the sign if applying the correction to the Hc. Listed on the same page is the semidiameter correction for the sun, which is applied the same way as for the moon.

Example: Using *Figure 13-5*, extract the corrections for the upper limb of the moon as observed on 11 August 1995 at 1100Z is 33° 41'.

Apply these corrections as:

| | |
|---|---|
| Hs | 33° 41' |
| Parallax | +49' |
| Semidiameter | −16' |
| Ho | 34° 14' |

### Atmospheric Refraction Error

Still another factor to be taken into consideration is atmospheric refraction. If a fishing pole is partly submerged under water, it appears to bend at the surface. The bending of light rays as they pass from the water into the air causes this appearance. This bending of the light rays, as they pass from one medium into another, is called refraction. The

| UT (GMT) | | ☉ Sun GHA | Dec. | Aries GHA ♈ | Mars 1.4 GHA | Dec. | Jupiter 2.3 GHA | Dec. | Saturn 0.9 GHA | Dec. | ○ Moon GHA | Dec. | Lat. | Moon rise | Diff. |
|---|---|---|---|---|---|---|---|---|---|---|---|---|---|---|---|
| h | m | D ′ | ° ′ | ° ′ | D ′ | S ° ′ | ° ′ | S ° ′ | ° ′ | S ° ′ | ° ′ | S ° ′ | N ° | h m | m |
| **00** | 00 | 178 40.2 | N15 28.3 | 318 59.5 | 127 27 | S 4 49 | 75 09 | S20 40 | 323 50 | S 4 35 | 356 59 | S 10 01 | | | |
| | 10 | 181 10.3 | 28.2 | 321 29.9 | 129 57 | | 77 40 | | 326 21 | | 359 24 | 9 59 | 72 | 20 18 | -04 |
| | 20 | 183 40.3 | 28.1 | 324 00.3 | 132 27 | | 80 10 | | 328 51 | | 1 48 | 57 | 70 | 20 09 | -01 |
| | 30 | 186 10.3 | 27.9 | 326 30.8 | 134 57 | | 82 40 | | 331 22 | | 4 13 | 56 | 68 | 20 02 | +02 |
| | 40 | 188 40.3 | 27.8 | 329 01.2 | 137 28 | | 85 11 | | 333 52 | | 6 37 | 54 | 66 | 19 55 | 04 |
| | 50 | 191 10.3 | 27.7 | 331 31.6 | 139 58 | | 87 41 | | 336 22 | | 9 02 | 52 | 64 | 19 50 | 06 |
| **01** | 00 | 193 40.3 | N15 27.6 | 334 02.0 | 142 28 | S 4 50 | 90 12 | S20 40 | 338 53 | S 4 35 | 11 26 | S 9 50 | 62 | 19 45 | 08 |
| | 10 | 196 10.3 | 27.4 | 336 32.4 | 144 58 | | 92 42 | | 341 23 | | 13 51 | 49 | | | |
| | 20 | 198 40.4 | 27.3 | 339 02.8 | 147 28 | | 95 12 | | 343 54 | | 16 15 | 47 | 60 | 19 41 | 10 |
| | 30 | 201 10.4 | 07.2 | 341 33.2 | 149 59 | | 97 43 | | 346 24 | | 18 40 | 45 | 58 | 19 37 | 11 |
| | 40 | 203 40.4 | 27.1 | 344 03.6 | 152 29 | | 100 13 | | 348 55 | | 21 04 | 43 | 56 | 19 34 | 12 |
| | 50 | 206 10.4 | 27.0 | 346 34.0 | 154 59 | | 102 44 | | 351 25 | | 23 29 | 42 | 54 | 19 31 | 13 |
| **02** | 00 | 208 40.4 | N15 26.8 | 349 04.5 | 157 29 | S 4 50 | 105 14 | S20 40 | 353 55 | S 4 35 | 25 54 | S 9 40 | 52 | 19 29 | 14 |
| | 10 | 211 10.4 | 26.7 | 351 34.9 | 159 59 | | 107 44 | | 356 26 | | 28 18 | 38 | | | |
| | 20 | 213 40.5 | 26.6 | 354 05.3 | 162 29 | | 110 15 | | 358 56 | | 30 43 | 37 | 50 | 19 26 | 15 |
| | 30 | 216 10.5 | 26.5 | 356 35.7 | 165 00 | | 112 45 | | 1 27 | | 33 07 | 35 | 45 | 19 24 | 17 |
| | 40 | 218 40.5 | 26.3 | 359 06.1 | 167 30 | | 115 16 | | 3 57 | | 35 32 | 33 | 40 | 19 16 | 18 |
| | 50 | 221 10.5 | 26.2 | 1 36.5 | 170 00 | | 117 46 | | 6 28 | | 37 56 | 31 | 35 | 19 13 | 20 |
| **03** | 00 | 223 40.5 | N15 26.1 | 4 06.9 | 172 30 | S 4 51 | 120 16 | S20 40 | 8 58 | S 4 35 | 40 21 | S 9 30 | 30 | 19 09 | 21 |
| | 10 | 226 10.5 | 26.0 | 6 37.3 | 175 00 | | 122 47 | | 11 29 | | 42 46 | 28 | | | |
| | 20 | 228 40.6 | 25.9 | 9 07.7 | 177 30 | | 125 17 | | 13 59 | | 45 10 | 26 | 20 | 19 03 | 23 |
| | 30 | 231 10.6 | 25.7 | 11 38.1 | 180 01 | | 127 48 | | 16 29 | | 47 35 | 24 | 10 | 18 58 | 24 |
| | 40 | 233 40.6 | 25.6 | 14 08.6 | 182 31 | | 130 18 | | 19 00 | | 49 59 | 23 | 0 | 18 53 | 26 |
| | 50 | 236 10.6 | 25.5 | 16 39.0 | 185 01 | | 132 48 | | 21 30 | | 52 24 | 21 | 10 | 18 48 | 28 |
| **04** | 00 | 238 40.6 | N15 25.4 | 19 09.4 | 187 31 | S 4 51 | 135 19 | S20 40 | 24 01 | S 4 35 | 54 48 | S 9 19 | 20 | 18 43 | 30 |
| | 10 | 241 10.6 | 25.2 | 21 39.8 | 190 01 | | 137 49 | | 26 31 | | 57 13 | 17 | | | |
| | 20 | 243 40.7 | 25.1 | 24 10.2 | 192 32 | | 140 20 | | 29 02 | | 59 38 | 16 | 30 | 18 37 | 32 |
| | 30 | 246 10.7 | 25.0 | 26 40.6 | 195 02 | | 142 50 | | 31 32 | | 62 02 | 14 | 35 | 18 34 | 33 |
| | 40 | 248 40.7 | 24.9 | 29 11.0 | 197 32 | | 145 20 | | 34 02 | | 64 27 | 12 | 40 | 18 30 | 34 |
| | 50 | 251 10.7 | 24.8 | 31 41.4 | 200 02 | | 147 51 | | 36 33 | | 66 51 | 10 | 45 | 18 25 | 36 |
| **05** | 00 | 253 40.7 | N15 24.6 | 34 11.8 | 202 32 | S 4 52 | 150 21 | S20 40 | 39 03 | S 4 35 | 69 16 | S 9 09 | 50 | 18 20 | 38 |
| | 10 | 256 10.7 | 24.5 | 36 42.3 | 205 02 | | 152 51 | | 41 34 | | 71 41 | 07 | | | |
| | 20 | 258 10.8 | 24.4 | 39 12.7 | 207 33 | | 155 22 | | 44 04 | | 74 05 | 05 | 52 | 18 17 | 39 |
| | 30 | 261 10.8 | 24.3 | 41 43.1 | 210 03 | | 157 52 | | 46 34 | | 76 30 | 03 | 54 | 18 15 | 40 |
| | 40 | 263 40.8 | 24.1 | 44 13.5 | 212 33 | | 160 26 | | 49 05 | | 78 54 | 02 | 56 | 18 11 | 41 |
| | 50 | 266 10.8 | 24.0 | 46 43.9 | 215 03 | | 162 53 | | 51 35 | | 81 19 | 9 00 | 58 | 18 08 | 42 |
| **06** | 00 | 268 40.8 | N15 23.9 | 49 14.3 | 217 33 | S 4 53 | 165 23 | S20 40 | 54 06 | S 4 35 | 83 44 | S 8 58 | 60 S | 18 04 | +44 |
| | 10 | 271 10.8 | 23.8 | 51 44.7 | 220 03 | | 167 54 | | 56 36 | | 86 08 | 56 | | | |
| | 20 | 273 40.9 | 23.6 | 54 15.1 | 222 34 | | 170 24 | | 59 07 | | 88 33 | 55 | | Moon's P. in A. | |
| | 30 | 276 10.9 | 23.5 | 56 45.5 | 225 04 | | 172 55 | | 67 37 | | 90 57 | 53 | | | |
| | 40 | 278 40.9 | 23.4 | 59 16.0 | 227 34 | | 175 25 | | 64 07 | | 93 22 | 51 | | | |
| | 50 | 281 10.9 | 23.3 | 61 46.4 | 230 04 | | 177 55 | | 66 38 | | 95 47 | 49 | | | |

| A l t ° | C o r + ′ | A l t ° | C o r + ′ |
|---|---|---|---|
| 0 | 60 | 53 | 35 |
| 3 | 59 | 54 | 34 |
| 11 | 58 | 55 | 33 |
| 15 | 57 | 56 | 32 |
| 18 | 56 | 58 | 31 |
| 21 | 55 | 59 | 30 |
| 23 | 54 | 60 | 29 |
| 26 | 53 | 61 | 28 |
| 28 | 52 | 62 | 27 |
| 30 | 51 | 63 | 26 |
| 32 | 50 | 64 | 25 |
| 33 | 49 | 65 | 24 |
| 35 | 48 | 66 | 23 |
| 37 | 47 | 67 | 22 |
| 38 | 46 | 68 | 21 |
| 40 | 45 | 69 | 20 |
| 41 | 44 | 70 | 19 |
| 43 | 43 | 71 | 18 |
| 44 | 42 | 72 | 17 |
| 45 | 41 | 73 | 16 |
| 47 | 40 | 74 | 15 |
| 48 | 39 | 75 | 14 |
| 49 | 38 | 76 | 13 |
| 51 | 37 | 77 | 12 |
| 52 | 36 | 78 | 11 |
| 53 | 35 | 79 | 10 |
| 54 | | 80 | |

| UT (GMT) | | ☉ Sun GHA | Dec. | Aries GHA ♈ | Mars 1.4 GHA | Dec. | Jupiter 2.3 GHA | Dec. | Saturn 0.9 GHA | Dec. | ○ Moon GHA | Dec. |
|---|---|---|---|---|---|---|---|---|---|---|---|---|
| **07** | 00 | 283 40.9 | N15 23.2 | 64 16.8 | 232 34 | S 4 53 | 180 26 | S20 40 | 69 08 | S 4 35 | 98 11 | S 8 48 |
| | 10 | 286 10.9 | 23.0 | 66 47.2 | 235 04 | | 182 56 | | 71 39 | | 100 36 | 46 |
| | 20 | 288 40.9 | 22.9 | 69 17.6 | 237 35 | | 185 27 | | 74 09 | | 103 00 | 44 |
| | 30 | 291 11.0 | 22.8 | 71 48.0 | 240 05 | | 187 57 | | 76 40 | | 105 25 | 42 |
| | 40 | 293 41.0 | 22.7 | 74 18.4 | 242 35 | | 190 27 | | 79 10 | | 107 50 | 41 |
| | 50 | 296 11.0 | 22.5 | 76 48.8 | 245 05 | | 192 58 | | 81 41 | | 110 14 | 39 |
| **08** | 00 | 298 41.0 | N15 22.4 | 79 19.2 | 247 35 | S 4 54 | 195 28 | S20 40 | 84 11 | S 4 35 | 112 39 | S 8 37 |
| | 10 | 301 11.0 | 22.3 | 81 49.6 | 250 05 | | 197 59 | | 86 41 | | 115 04 | 35 |
| | 20 | 303 41.0 | 22.2 | 84 20.1 | 252 36 | | 200 29 | | 89 12 | | 117 28 | 33 |
| | 30 | 306 11.1 | 22.1 | 56 50.5 | 255 06 | | 202 59 | | 91 442 | | 119 53 | 32 |
| | 40 | 308 41.1 | 21.9 | 59 20.9 | 257 36 | | 205 30 | | 94 13 | | 122 17 | 30 |
| | 50 | 311 11.1 | 21.8 | 91 51.3 | 260 06 | | 208 00 | | 96 43 | | 124 42 | 28 |
| **09** | 00 | 313 41.1 | N15 21.7 | 94 21.7 | 262 36 | S 4 55 | 210 31 | S20 40 | 99 14 | S 4 35 | 127 07 | S 8 26 |
| | 10 | 316 11.1 | 21.6 | 96 52.1 | 265 07 | | 213 01 | | 101 44 | | 129 31 | 25 |
| | 20 | 318 41.1 | 21.4 | 99 22.5 | 267 37 | | 215 31 | | 104 14 | | 131 56 | 23 |
| | 30 | 321 11.2 | 21.3 | 101 52.9 | 270 07 | | 218 02 | | 106 45 | | 134 21 | 21 |
| | 40 | 323 41.2 | 21.2 | 104 23.3 | 272 37 | | 220 32 | | 109 15 | | 136 45 | 19 |
| | 50 | 326 11.2 | 21.1 | 106 53.8 | 275 07 | | 223 03 | | 111 46 | | 139 10 | 17 |
| **10** | 00 | 328 41.2 | N15 20.9 | 109 24.2 | 277 37 | S 4 55 | 225 33 | S20 41 | 114 16 | S 4 35 | 141 35 | S 8 16 |
| | 10 | 331 11.2 | 20.8 | 111 54.6 | 280 08 | | 228 03 | | 116 47 | | 143 59 | 14 |
| | 20 | 333 41.2 | 20.7 | 114 25.0 | 282 38 | | 230 34 | | 119 17 | | 146 24 | 12 |
| | 30 | 336 11.3 | 20.6 | 116 55.4 | 285 08 | | 233 04 | | 121 47 | | 148 49 | 10 |
| | 40 | 338 41.3 | 20.5 | 119 25.8 | 287 38 | | 235 35 | | 124 18 | | 151 13 | 09 |
| | 50 | 341 11.3 | 20.3 | 121 56.2 | 290 08 | | 238 05 | | 126 48 | | 153 38 | 07 |
| **11** | 00 | 343 41.3 | N15 20.2 | 124 26.6 | 292 38 | S 4 56 | 240 35 | S20 40 | 129 19 | S 4 35 | 156 03 | S 8 05 |
| | 10 | 346 11.3 | 20.1 | 126 57.0 | 295 09 | | 243 06 | | 131 49 | | 158 27 | 03 |
| | 20 | 348 41.3 | 20.0 | 129 27.4 | 297 39 | | 245 36 | | 134 20 | | 160 52 | 01 |
| | 30 | 351 11.4 | 19.8 | 131 57.9 | 300 09 | | 248 07 | | 136 50 | | 163 16 | 8 00 |
| | 40 | 353 41.4 | 19.7 | 134 2.3 | 302 39 | | 250 37 | | 139 20 | | 165 41 | 7 58 |
| | 50 | 356 11.4 | 19.6 | 136 58.7 | 305 09 | | 253 07 | | 141 51 | | 168 06 | 56 |
| | Rate | 15 00.1 | 50 00.7 | | 15 01.0 | 50 00 .6 | 15 02.6 | | 15 02.6 | 50 00 .1 | 14 27.6 | NO 10.5 |

Sun SD 15′.8
Moon SD 16′
Age 15d

**Figure 13-5.** *Correction for moon's parallax.*

refraction of light from a celestial body as it passes through the atmosphere causes an error in sextant observation.

As the light of a celestial body passes from the almost perfect vacuum of outer space into the atmosphere, it is refracted as shown in *Figure 13-6*, so that the body appears a little higher above the horizon than it really is. Therefore, the correction to the Hs for refraction is always negative. The higher the body above the horizon, the smaller the amount of refraction and, consequently, the smaller the refraction correction. Moreover, the greater the altitude of the aircraft, the less dense the layer of atmosphere between the body and the observer; hence, the less the refraction.

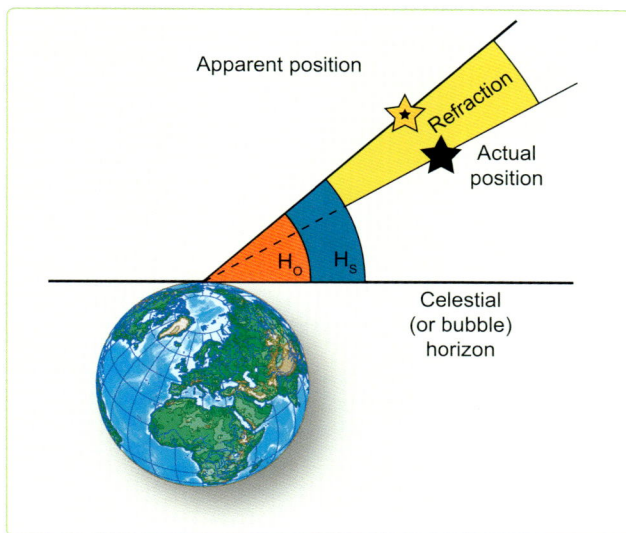

**Figure 13-6.** *Error caused by atmospheric refraction.*

The appropriate correction table for atmospheric refraction is listed inside the back cover of all four books used for celestial computations; namely, the Air Almanac and each of the three volumes of Publication No. 249. It contains the Sight Reduction Tables for Air Navigation published in three volumes. Volume I, used by both the marine and air navigator, contains the altitude and azimuth values of seven selected stars for the complete ranges of latitude and hour angle of Aries. These seven stars represent the best selection for observation at any given position and time, and provide the data for presetting instruments before observation and for sight reduction afterwards. Volumes II and III cover latitudes 0-40 and 39-89, respectively, and are primarily used by the air navigator in conjunction with observations of celestial bodies to calculate the geographic position of the observer. This table, shown in *Figure 13-7*, lists the refraction for different observed altitudes of the body and for different heights of the observer above sea level. The values shown are subtracted from Hs or added to Hc.

## Acceleration Error

Presently, the only practical and continuously available reference datum for the definition of the true vertical is the direction of the gravitational field of the earth. Definition of this vertical establishes the artificial horizon. It is also fundamental that the forces caused by gravity cannot be separated by those caused by accelerations within the sextant. A level or centered bubble in the sextant indicates the true vertical only when the instrument is at rest or moving at a constant velocity in a straight line. Any outside force (changes in GS or changes in track) affect the liquid in the bubble chamber and, consequently, displace the bubble.

When the sextant is moved in a curved path (Coriolis, changes in heading, rhumb line, etc.), or with varying speed, the zenith indicated by the bubble is displaced from the true vertical. This presents a false artificial horizon above which the altitude of the celestial body is measured. Since the horizon used is false, the altitude measured from it is erroneous. Therefore, the accuracy of celestial observations is directly related to changes in track and speed of the aircraft. Acceleration errors have two principal causes: changes in GS and curvature of the aircraft's path in space.

The displacement of the liquid and the bubble in the chamber may be divided into two vectors, and each vector may be considered separately. These vectors may be thought of as a lateral vector (along the wings) and a longitudinal vector (along the nose-tail axis of the aircraft). Any change in GS can cause a longitudinal displacement. This change can be brought about by a change in the airspeed or the wind encountered, or the change in GS brought about by a change in heading due to other factors (gyro precession and rhumb line error). A lateral displacement results from a number of causes, most of which occur in spite of any efforts to hold them in check. These causes are Coriolis, rhumb line, and wander errors.

## Coriolis Force

Any free-moving body traveling at a constant speed above the earth is subject to an apparent force that deflects its path to the right in the Northern Hemisphere and to the left in the Southern Hemisphere. This apparent force, and the resulting acceleration, were first discovered shortly before the middle of the 19th century by Gaspard Gustave de Coriolis (1792–1843) and given quantitative formulation by William Ferrel (1817-1891). The acceleration is known as Coriolis acceleration, or force, or simply, Coriolis, and is expressed in Ferrel's law.

You must realize that the bubble sextant indicates the true vertical only when the instrument is at rest or moving at a constant speed in a straight line as perceived in space. If the earth were motionless, this straight path in space would also

**Refraction**

**To be subtracted from sextant altitude (referred to as observed altitude in AP 3270)**

| | Height above sea level in units of 1,000 feet | | | | | | | | | | | | | $R = R_\circ \times f$ | | | |
|---|---|---|---|---|---|---|---|---|---|---|---|---|---|---|---|---|---|
| | | | | | | | | | | | | | | | $f$ | | |
| | | | | | | | | | | | | | | 0-9 | 1-0 | 1-1 | 1-2 |
| $R_\circ$ | 0 | 5 | 10 | 15 | 20 | 25 | 30 | 35 | 40 | 45 | 50 | 55 | $R_\circ$ | R | | | |
| | | | | | | | Sextant Altitude | | | | | | | | | | |
| 0 | 90 | 90 | 90 | 90 | 90 | 90 | 90 | 90 | 90 | 90 | 90 | 90 | 0 | 0 | 0 | 0 | 0 |
| 1 | 63 | 59 | 55 | 51 | 46 | 41 | 36 | 31 | 26 | 20 | 17 | 13 | 1 | 1 | 1 | 1 | 1 |
| 2 | 33 | 29 | 26 | 22 | 19 | 16 | 14 | 11 | 9 | 7 | 6 | 4 | 2 | 2 | 2 | 2 | 2 |
| 3 | 21 | 19 | 16 | 14 | 12 | 10 | 8 | 7 | 5 | 4 | 2 40 | 1 40 | 3 | 3 | 3 | 3 | 4 |
| 4 | 16 | 14 | 12 | 10 | 8 | 7 | 6 | 5 | 3 10 | 2 20 | 1 30 | 0 40 | 4 | 4 | 4 | 4 | 5 |
| 5 | 12 | 11 | 9 | 8 | 7 | 5 | 4 00 | 3 10 | 2 10 | 1 30 | 0 39 | +0 05 | 5 | 5 | 5 | 5 | 6 |
| 6 | 10 | 9 | 7 | 5 50 | 4 50 | 3 50 | 3 10 | 2 20 | 1 30 | 0 49 | +0 11 | −0 19 | 6 | 5 | 6 | 7 | 7 |
| 7 | 8 10 | 6 50 | 5 50 | 4 50 | 4 00 | 3 00 | 2 20 | 1 50 | 1 10 | 0 24 | −0 11 | −0 38 | 7 | 6 | 7 | 8 | 8 |
| 8 | 6 50 | 5 50 | 5 00 | 4 00 | 3 10 | 2 30 | 1 50 | 1 20 | 0 38 | +0 04 | −0 28 | −0 54 | 8 | 7 | 8 | 9 | 10 |
| 9 | 6 00 | 5 10 | 4 10 | 3 20 | 2 40 | 2 00 | 1 30 | 1 00 | 0 19 | −0 13 | −0 42 | −1 08 | 9 | 8 | 9 | 10 | 11 |
| 10 | 5 20 | 4 30 | 3 40 | 2 50 | 2 10 | 1 40 | 1 10 | 0 35 | +0 03 | −0 27 | −0 53 | −1 18 | 0 | 9 | 10 | 11 | 12 |
| 12 | 4 30 | 3 40 | 2 50 | 2 20 | 1 40 | 1 10 | 0 37 | +0 11 | −0 16 | −0 43 | −1 08 | −1 31 | 2 | 11 | 12 | 13 | 14 |
| 14 | 3 30 | 2 50 | 2 10 | 1 40 | 1 10 | 0 34 | +0 09 | −0 14 | −0 37 | −1 00 | −1 23 | −1 44 | 14 | 13 | 14 | 15 | 17 |
| 16 | 2 50 | 2 10 | 1 40 | 1 10 | 0 37 | +0 10 | −0 13 | −0 34 | −0 53 | −1 14 | −1 35 | −1 56 | 16 | 14 | 16 | 18 | 19 |
| 18 | 2 20 | 1 40 | 1 20 | 0 43 | +0 15 | −0 08 | −0 31 | −0 52 | −1 08 | −1 27 | −1 46 | −2 05 | 18 | 16 | 18 | 20 | 22 |
| 20 | 1 50 | 1 20 | 0 49 | +0 23 | −0 02 | −0 26 | −0 46 | −1 06 | −1 22 | −1 39 | −1 57 | −2 14 | 0 | 18 | 20 | 22 | 24 |
| 25 | 1 12 | 0 44 | +0 19 | −0 06 | −0 28 | −0 48 | −1 09 | −1 27 | −1 42 | −1 58 | −2 14 | −2 30 | 5 | 22 | 25 | 28 | 0 |
| $f$ | 0 | 5 | 10 | 15 | 20 | 25 | 30 | 35 | 40 | 45 | 50 | 55 | $f$ | 0.9 | 1.0 | 1.1 | 1.2 |
| | | | | | Temperature in °C. | | | | | | | | | | $f$ | | |

| $f$ | 0 | 5 | 10 | 15 | 20 | 25 | 30 | 35 | 40 | | | | $f$ | |
|---|---|---|---|---|---|---|---|---|---|---|---|---|---|---|
| 0.9 | +47 | +36 | +27 | +18 | +10 | +3 | −5 | −13 | | For these heights no | | | 0.9 | Where $R_\circ$ is less |
| 1.0 | +26 | +16 | +6 | −4 | −13 | −22 | −31 | −40 | | temperature correction | | | 1.0 | than 10' or the |
| 1.1 | +5 | −5 | +15 | −25 | −36 | −46 | −57 | −68 | | is necessary, so use | | | 1.1 | height greater |
| 1.2 | −16 | −25 | −36 | −46 | −58 | −71 | −83 | −95 | | $R = R_\circ$ | | | 1.2 | than 35000 |
| | −37 | −45 | −56 | −67 | −81 | −95 | | | | | | | | feet, use $R = R_\circ$ |

Choose the column appropriate to height, in units of 1,000 feet, and find the range of altitude in which the sextant altitude lies; the corresponding valve of $R_\circ$ is the refraction, to be subtracted from sextant altitude, unless conditions are extreme. In that case find $f$ from the lower table, with critical argument temperature. Use the table on the right to form the refraction, $R = R_\circ \times f$.

**Figure 13-7.** *Corrections for atmospheric refraction.*

be a straight path over the surface of the earth; conversely, a straight path over the motionless earth would also be a straight path in space.

When the aircraft is flying a path curved in space to the left, the fluid in the bubble chamber is deflected to the right, and the bubble is deflected to the left of the aircraft's path over the earth. When the aircraft is flying a curved path in space to the right, the reverse is true.

In *Figure 13-8,* the aircraft is represented as flying on a curved path to the left. Note that in the inset representing the bubble chamber, the heavy black bubble is indicated in its approximate position representing the true vertical.

The observer always seeks to center the bubble and, on this beam shot facing to the right side of the aircraft to observe the body, tip the sextant up. This would tilt the bubble horizon from its true position, producing a smaller sextant reading

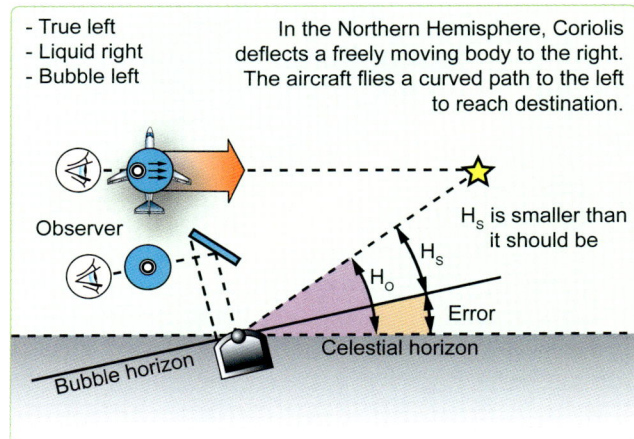

- True left
- Liquid right
- Bubble left

In the Northern Hemisphere, Coriolis deflects a freely moving body to the right. The aircraft flies a curved path to the left to reach destination.

Observer

$H_s$ is smaller than it should be

**Figure 13-8.** *Error caused by Coriolis force.*

than the true value. The smaller the height observed (Ho), the greater the radius of the circle of equal altitude—the LOP falls farther from the subpoint than the true LOP. Obviously, if the erroneous LOP falls farther from the subpoint, it falls to the left of the true LOP and the correction to the right is valid. Corrections for Coriolis error are shown on the inside back cover of the Air Almanac, as well as in all volumes of Pub. No. 249 published by the National Imagery and Mapping Agency.

Coriolis acceleration is directly proportional to the straight-line velocity, directly proportional to the angular velocity of the earth, directly proportional to the sine of the latitude, and at right angles to the direction of flight.

## Rhumb Line Error

The straight Coriolis table in *Figure 13-9,* found in the Air Almanac or Pub. No. 249, has a limited application. As long as a constant TH is flown, the path of the aircraft is a rhumb line. Because a rhumb line on the earth's surface is a curve, it is also a curved line in space. If the aircraft is headed in an easterly direction in the Northern Hemisphere, the apparent curve is to the left and becomes an addition to the Coriolis error. By the same token, if headed in a westerly direction in the Northern Hemisphere, the apparent curve is to the right, or opposite that of Coriolis force. *[Figure 13-10]* There are notable exceptions to this. When flying north or south, the aircraft is flying a great circle and there is no rhumb line error. Also, when steering by a free-running, compensated gyro, the track approximates a great circle and eliminates rhumb line error.

At speeds under 300 knots, the error is negligible. However, at high speeds or high latitudes, rhumb line error is appreciable. For example, at 60° N latitude with a track of 100° and a GS of 650 knots, the Coriolis correction is 15 nautical miles (NM) right, and the rhumb line correction is 10 NM right. Use the following steps and *Figure 13-11* to determine the correction for rhumb line error and Coriolis correction.

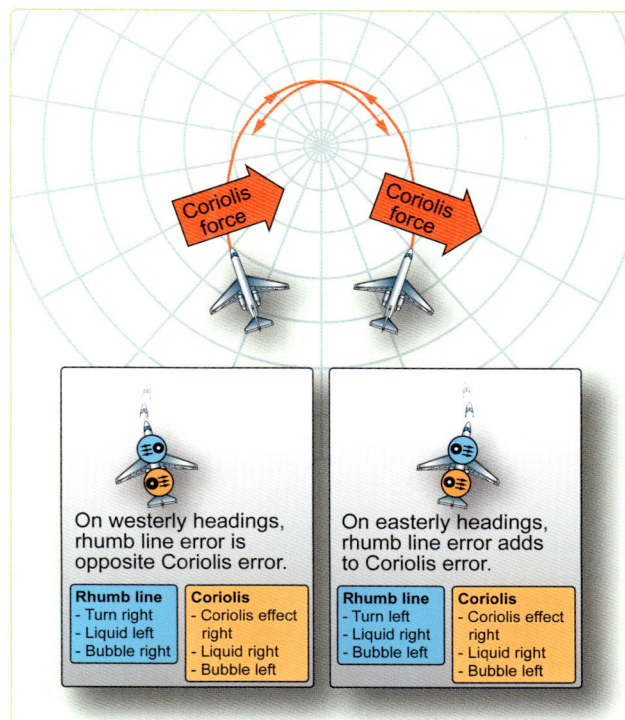

**Figure 13-10.** *Coriolis and/or rhumb line errors in the northern hemisphere.*

1. Enter the nearest latitude on the left side. Interpolate if necessary.

2. Enter the nearest track across the top of the chart. Interpolate if necessary.

3. Choose the closest GS and extract the correction; 50N, track 080°, GS 500 knots = 14.3 Right.

## Groundspeed Acceleration Error

Changes in airspeed or wind velocity cause this error. Prevent changes of airspeed through good crew coordination.

## Coriolis (Z) Correction

To be applied by moving the position line a distance Z to starboard (right) of the track in northern latitudes and to port (left) in southern latitudes.

| G/S Knots | Latitude | | | | | | | G/S Knots | Latitude | | | | | | |
|---|---|---|---|---|---|---|---|---|---|---|---|---|---|---|---|
| | 0° 10° | | 20° 30° | | 40° 50° | | 60° 70° | 80° 90° | | 0° 10° | | 20° 30° | | 40° 50° | | 60° 70° | 80° 90° |

| G/S Knots | 0° | 10° | 20° | 30° | 40° | 50° | 60° | 70° | 80° | 90° | G/S Knots | 0° | 10° | 20° | 30° | 40° | 50° | 60° | 70° | 80° | 90° |
|---|---|---|---|---|---|---|---|---|---|---|---|---|---|---|---|---|---|---|---|---|---|
| 150 | 0 | 1 | 1 | 2 | 3 | 3 | 3 | 4 | 4 | 4 | 550 | 0 | 3 | 5 | 7 | 9 | 11 | 12 | 14 | 14 | 14 |
| 200 | 0 | 1 | 2 | 3 | 3 | 4 | 5 | 5 | 5 | 5 | 600 | 0 | 3 | 5 | 8 | 10 | 12 | 14 | 15 | 16 | 16 |
| 250 | 0 | 1 | 2 | 3 | 4 | 5 | 6 | 6 | 6 | 7 | 650 | 0 | 3 | 6 | 9 | 11 | 13 | 15 | 16 | 17 | 17 |
| 300 | 0 | 1 | 3 | 4 | 5 | 6 | 7 | 7 | 8 | 8 | 700 | 0 | 3 | 6 | 9 | 12 | 14 | 16 | 17 | 18 | 18 |
| 350 | 0 | 2 | 3 | 5 | 6 | 7 | 8 | 9 | 9 | 9 | 750 | 0 | 3 | 7 | 10 | 13 | 15 | 17 | 18 | 19 | 20 |
| 400 | 0 | 2 | 4 | 5 | 7 | 8 | 9 | 10 | 10 | 10 | 800 | 0 | 4 | 7 | 10 | 13 | 16 | 18 | 20 | 21 | 21 |
| 450 | 0 | 2 | 4 | 6 | 8 | 9 | 10 | 11 | 12 | 12 | 850 | 0 | 4 | 8 | 11 | 14 | 17 | 19 | 21 | 22 | 22 |
| 500 | 0 | 2 | 4 | 7 | 8 | 10 | 11 | 12 | 13 | 13 | 900 | 0 | 4 | 8 | 12 | 15 | 18 | 20 | 22 | 23 | 24 |

**Figure 13-9.** *Coriolis correction.*

## Groundspeed 300 knots

| TR → / LAT ↓ | 270 270 | 260 280 | 250 290 | 240 300 | 230 310 | 220 320 | 210 330 | 200 340 | 190 350 | 180 0 | 170 10 | 160 20 | 150 30 | 140 40 | 130 50 | 120 60 | 110 70 | 100 80 | 90 90 |
|---|---|---|---|---|---|---|---|---|---|---|---|---|---|---|---|---|---|---|---|
| 0 | 0.0 | 0.0 | 0 | 0 | 0 | 0 | 0 | 0 | 0 | 0 | 0 | 0 | 0 | 0 | 0 | 0 | 0 | 0 | 0 |
| 10 | 1.1 | 1.1 | 1.1 | 1.2 | 1.2 | 1.2 | 1.2 | 1.3 | 1.3 | 1.4 | 1.4 | 1.4 | 1.5 | 1.5 | 1.6 | 1.6 | 1.6 | 1.6 | 1.6 |
| 20 | 2.2 | 2.2 | 2.2 | 2.3 | 2.3 | 2.4 | 2.4 | 2.5 | 2.6 | 2.7 | 2.8 | 2.8 | 2.9 | 3.0 | 3.1 | 3.1 | 3.1 | 3.2 | 3.2 |
| 30 | 3.2 | 3.2 | 3.2 | 3.3 | 3.4 | 3.5 | 3.6 | 3.7 | 3.8 | 3.9 | 4.1 | 4.2 | 4.3 | 4.4 | 4.5 | 4.6 | 4.6 | 4.7 | 4.7 |
| 40 | 4.0 | 4.0 | 4.0 | 4.1 | 4.2 | 4.4 | 4.5 | 4.7 | 4.9 | 5.1 | 5.2 | 5.4 | 5.6 | 5.8 | 5.9 | 6.0 | 6.1 | 6.1 | 6.2 |
| 50 | 4.5 | 4.5 | 4.5 | 4.7 | 4.8 | 5.0 | 5.2 | 5.5 | 5.7 | 6.0 | 6.3 | 6.6 | 6.8 | 7.0 | 7.2 | 7.4 | 7.5 | 7.6 | 7.6 |
| 60 | 4.6 | 4.6 | 4.7 | 4.9 | 5.1 | 5.4 | 5.7 | 6.0 | 6.4 | 6.8 | 7.2 | 7.6 | 7.9 | 8.2 | 8.5 | 8.8 | 8.9 | 9.0 | 9.0 |
| 70 | 3.8 | 3.8 | 4.0 | 4.3 | 4.6 | 5.1 | 5.6 | 6.1 | 6.8 | 7.4 | 8.0 | 8.6 | 9.2 | 9.7 | 10.2 | 10.5 | 10.8 | 10.9 | 11.0 |
| 80 | 0.3 | 0.4 | 0.8 | 1.3 | 2.0 | 3.0 | 4.0 | 5.2 | 6.4 | 7.7 | 9.0 | 10.3 | 11.5 | 12.5 | 13.5 | 14.2 | 14.7 | 15.0 | 15.1 |
| 89 | 67.2 | 66.1 | 62.6 | 57.4 | 50.1 | 40.2 | 30.0 | 17.9 | 5.3 | 7.9 | 21.0 | 33.6 | 45.7 | 55.9 | 65.8 | 73.1 | 78.3 | 81.8 | 82.9 |

*Interpolate*

## Groundspeed 650 knots

| TR → / LAT ↓ | 270 270 | 260 280 | 250 290 | 240 300 | 230 310 | 220 320 | 210 330 | 200 340 | 190 350 | 180 0 | 170 10 | 160 20 | 150 30 | 140 40 | 130 50 | 120 60 | 110 70 | 100 80 | 90 90 |
|---|---|---|---|---|---|---|---|---|---|---|---|---|---|---|---|---|---|---|---|
| 0 | 0.0 | 0 | 0 | 0 | 0 | 0 | 0 | 0 | 0 | 0 | 0 | 0 | 0 | 0 | 0 | 0 | 0 | 0 | 0 |
| 10 | 1.9 | 1.9 | 1.9 | 2.0 | 2.1 | 2.3 | 2.4 | 2.6 | 2.8 | 3.0 | 3.1 | 3.3 | 3.5 | 3.7 | 3.8 | 3.9 | 4.0 | 4.0 | 4.0 |
| 20 | 3.6 | 3.6 | 3.7 | 3.9 | 4.1 | 4.4 | 4.7 | 5.1 | 5.4 | 5.8 | 6.2 | 6.6 | 7.0 | 7.3 | 4.6 | 7.7 | 7.9 | 8.0 | 8.1 |
| 30 | 5.0 | 5.0 | 5.2 | 5.4 | 5.8 | 6.3 | 6.7 | 7.3 | 7.9 | 8.5 | 9.1 | 9.7 | 10.3 | 10.8 | 11.2 | 11.6 | 11.8 | 12.0 | 12.1 |
| 40 | 5.8 | 5.9 | 6.1 | 6.5 | 7.0 | 7.7 | 8.4 | 9.2 | 10.1 | 11.0 | 11.8 | 12.7 | 13.5 | 14.2 | 14.9 | 15.5 | 15.8 | 16.0 | 16.1 |
| 50 | 5.7 | 5.8 | 6.1 | 6.6 | 7.4 | 8.3 | 9.3 | 10.5 | 11.8 | 13.0 | 14.3 | 15.6 | 16.7 | 17.7 | 18.6 | 19.4 | 19.9 | 20.2 | 20.3 |
| 60 | 4.2 | 4.3 | 4.8 | 5.6 | 6.6 | 8.0 | 9.4 | 11.1 | 12.9 | 14.7 | 16.5 | 18.3 | 20.0 | 21.5 | 22.8 | 23.9 | 24.6 | 25.1 | 25.2 |
| 70 | 0.9 | 0.6 | 0.2 | 1.3 | 3.0 | 5.2 | 7.5 | 10.2 | 13.1 | 16.0 | 19.0 | 21.8 | 24.5 | 26.8 | 29.0 | 30.7 | 31.8 | 32.6 | 32.9 |
| 80 | 18.0 | 17.5 | 15.9 | 13.6 | 10.1 | 5.6 | 0.8 | 4.8 | 10.7 | 16.8 | 22.8 | 28.7 | 34.3 | 39.1 | 43.6 | 47.1 | 49.4 | 51.0 | 51.5 |
| 89 | 334.8 | 329.2 | 313.0 | 288.9 | 254.5 | 208.2 | 160.1 | 103.6 | 44.6 | 17.0 | 78.7 | 137.6 | 194.1 | 242.2 | 288.5 | 322.9 | 347.0 | 363.2 | 368.8 |

Highlighted numbers are plotted in a direction opposite to that of coriolis force.
*Coriolis corrections alone are the figures on the 0 or 180 column.

**Figure 13-11.** *Combined Coriolis and rhumb line correction.*

Changes in wind velocity with resultant changes in GS are more difficult to control. The change in GS causes the liquid to be displaced, with the subsequent shifting of the bubble creating a false horizon. Notice in *Figure 13-12* how the horizon is automatically displaced by keeping the bubble in the center while these changes are taking place. A very simple rule applies to acceleration and deceleration forces. If the aircraft accelerates while a celestial observation is in progress, the resultant LOP falls ahead of the actual position. Accelerate—Ahead. The more the LOP approaches a speed line, the greater the acceleration error becomes. Refer to *Figure 13-13*.

1. Enter with Zn–Track.

2. Extract acceleration error and apply sign.

Example: Track = 080°, Zn = 060°, Beginning GS –500 knots, ending GS –515 knots.

$060° - 080° = 340° = -1.40$

$515 - 500 = 15$ knots

$-1.40 \times 15 = -21$ correction to the Ho

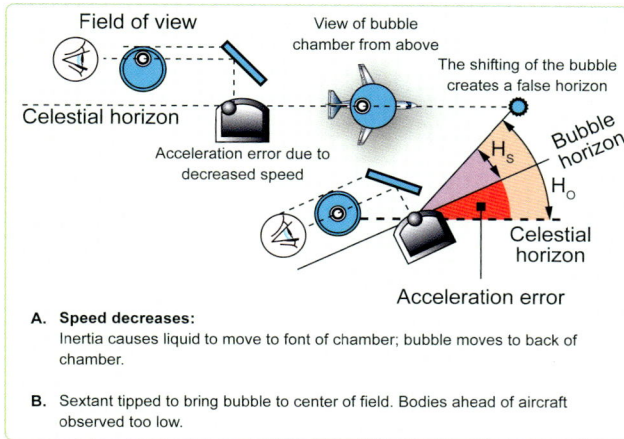

A. **Speed decreases:**
   Inertia causes liquid to move to font of chamber; bubble moves to back of chamber.

B. Sextant tipped to bring bubble to center of field. Bodies ahead of aircraft observed too low.

**Figure 13-12.** *Acceleration/deceleration errors.*

## Wander Error

A change in track can be produced by changes in the wind, heading changes caused by the autopilot, changing magnetic variation, or by heading changes caused by pilot manual steering errors. As with the Coriolis force and rhumb line errors, correction tables have been developed for wander error. Values extracted from the wander correction table, shown in *Figure 13-14*, are to be applied to the Ho.

| ZN–TR | | Groundspeed Acceleration Error/Knot |
|---|---|---|
| +/− | +/− | |
| 000/180 005/175 | 180/360 185/355 | 1.50 |
| 010/170 015/165 | 190/350 195/345 | 1.48 |
| 020/160 025/155 | 200/340 205/335 | 1.40 |
| 030/150 035/145 | 210/330 215/325 | 1.30 |
| 040/140 045/135 | 220/320 225/315 | 1.15 |
| 050/130 055/125 | 230/310 235/305 | 0.97 |
| 060/120 065/115 | 240/300 245/295 | 0.75 |
| 070/110 075/105 | 250/290 255/285 | 0.51 |
| 080/100 085/095 | 260/280 265/275 | 0.26 |
| 090/090 | 270/270 | 0.00 |

**Figure 13-13.** *Groundspeed acceleration error.*

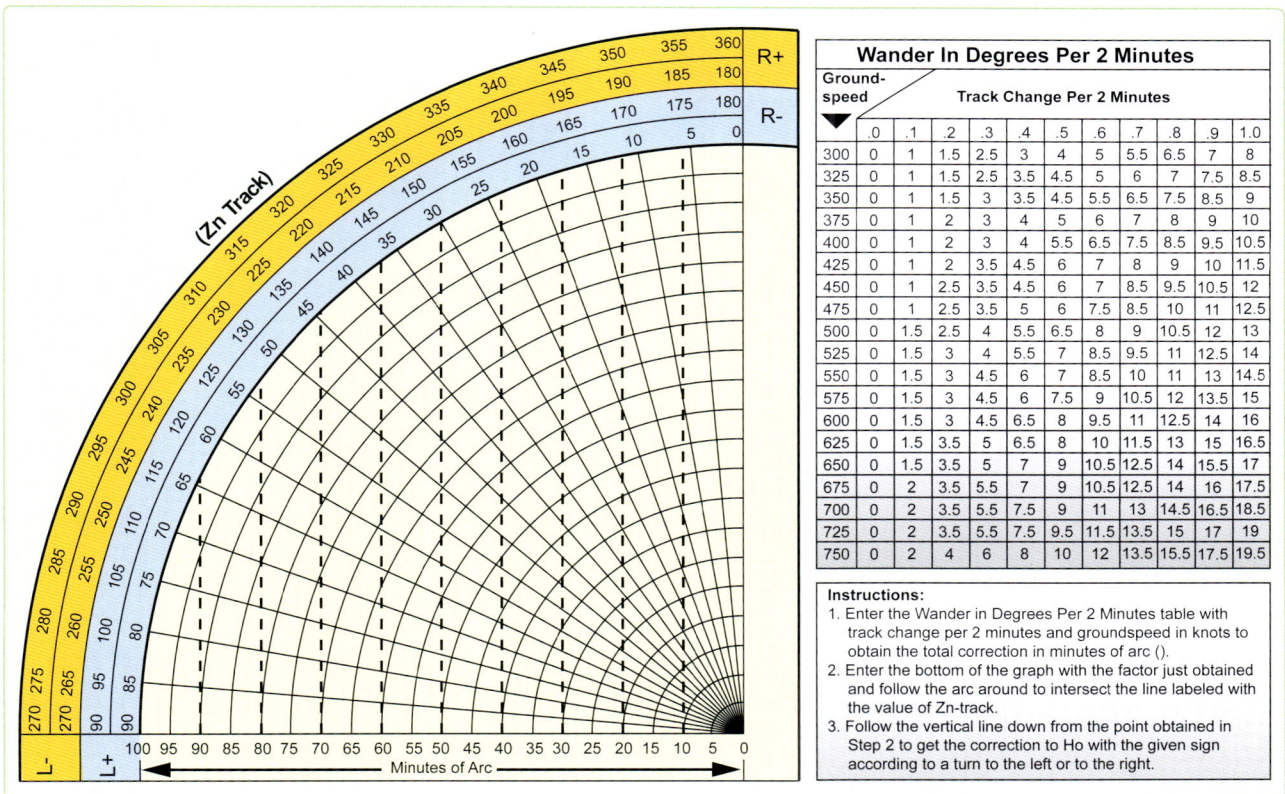

**Wander In Degrees Per 2 Minutes**

| Ground-speed | Track Change Per 2 Minutes | | | | | | | | | | |
|---|---|---|---|---|---|---|---|---|---|---|---|
| | .0 | .1 | .2 | .3 | .4 | .5 | .6 | .7 | .8 | .9 | 1.0 |
| 300 | 0 | 1 | 1.5 | 2.5 | 3 | 4 | 5 | 5.5 | 6.5 | 7 | 8 |
| 325 | 0 | 1 | 1.5 | 2.5 | 3.5 | 4.5 | 5 | 6 | 7 | 7.5 | 8.5 |
| 350 | 0 | 1 | 1.5 | 3 | 3.5 | 4.5 | 5.5 | 6.5 | 7.5 | 8.5 | 9 |
| 375 | 0 | 1 | 2 | 3 | 4 | 5 | 6 | 7 | 8 | 9 | 10 |
| 400 | 0 | 1 | 2 | 3 | 4 | 5.5 | 6.5 | 7.5 | 8.5 | 9.5 | 10.5 |
| 425 | 0 | 1 | 2 | 3.5 | 4.5 | 6 | 7 | 8 | 9 | 10 | 11.5 |
| 450 | 0 | 1 | 2.5 | 3.5 | 4.5 | 6 | 7 | 8.5 | 9.5 | 10.5 | 12 |
| 475 | 0 | 1 | 2.5 | 3.5 | 5 | 6 | 7.5 | 8.5 | 10 | 11 | 12.5 |
| 500 | 0 | 1.5 | 2.5 | 4 | 5.5 | 6.5 | 8 | 9 | 10.5 | 12 | 13 |
| 525 | 0 | 1.5 | 3 | 4 | 5.5 | 7 | 8.5 | 9.5 | 11 | 12.5 | 14 |
| 550 | 0 | 1.5 | 3 | 4.5 | 6 | 7 | 8.5 | 10 | 11 | 13 | 14.5 |
| 575 | 0 | 1.5 | 3 | 4.5 | 6 | 7.5 | 9 | 10.5 | 12 | 13.5 | 15 |
| 600 | 0 | 1.5 | 3 | 4.5 | 6.5 | 8 | 9.5 | 11 | 12.5 | 14 | 16 |
| 625 | 0 | 1.5 | 3.5 | 5 | 6.5 | 8 | 10 | 11.5 | 13 | 15 | 16.5 |
| 650 | 0 | 1.5 | 3.5 | 5 | 7 | 9 | 10.5 | 12.5 | 14 | 15.5 | 17 |
| 675 | 0 | 2 | 3.5 | 5.5 | 7 | 9 | 10.5 | 12.5 | 14 | 16 | 17.5 |
| 700 | 0 | 2 | 3.5 | 5.5 | 7.5 | 9 | 11 | 13 | 14.5 | 16.5 | 18.5 |
| 725 | 0 | 2 | 3.5 | 5.5 | 7.5 | 9.5 | 11.5 | 13.5 | 15 | 17 | 19 |
| 750 | 0 | 2 | 4 | 6 | 8 | 10 | 12 | 13.5 | 15.5 | 17.5 | 19.5 |

**Instructions:**
1. Enter the Wander in Degrees Per 2 Minutes table with track change per 2 minutes and groundspeed in knots to obtain the total correction in minutes of arc ().
2. Enter the bottom of the graph with the factor just obtained and follow the arc around to intersect the line labeled with the value of Zn-track.
3. Follow the vertical line down from the point obtained in Step 2 to get the correction to Ho with the given sign according to a turn to the left or to the right.

**Figure 13-14.** *Wander correction tables.*

Use the following information as entering arguments for the determination of the correction taken from the table:

1. The heading at the beginning of the observation was 079°.

2. The heading at the end of the observation was 081°.

3. The observation was taken over a 2-minute period.

4. The GS was 450 knots.

5. The Zn of the body was 130°.

Following the instructions shown at the bottom of the table, enter the numerical portion of the table with the values of GS and the change of track per 2 minutes. In this case, the GS is 450 knots and the change in track per 2 minutes is 2°. Since the heading at the end of the observation is greater than the heading at the beginning, the change is 2° to the right. Notice that you must know whether the change is to the right or to the left to determine the sign of the correction. The factor obtained from the table is 12 × 2 = 24.

Next, enter the graph portion of the table with the value of the factor (24) and the value of the azimuth of the body, minus the value of track. The graph is so constructed that it must be entered with Zn – Tr.

Zn – Tr = 130° – 080°; so use 050°

Following the rules in steps two and three in the table; the correction is 19'. Since the change in track is to the right, the correction is subtracted from the Ho. This is determined by referring to the signs shown at the ends of the arc in the table. *Figure 13-15* shows the effect of this correction.

If the track and groundspeed are the same at the beginning and the end of a shooting period, there is no wander error.

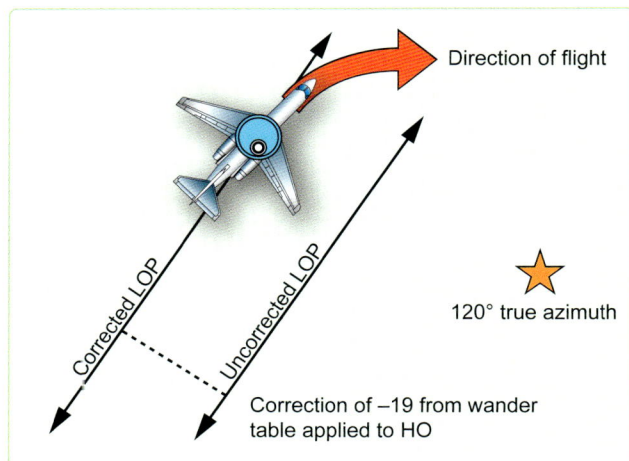

**Figure 13-15.** *Wander correction applied to Ho.*

## Instrument Error

Index error is usually the largest mechanical error in the sextant. This error is caused by improper alignment of the index prism with the altitude counter. No matter how carefully a sextant is handled, it is likely to have some index error. If the error is small, the sextant need not be readjusted; each Hs can be corrected by the amount of the error. This means that the index error of the sextant must be known to obtain an accurate celestial LOP. Another mechanical error found in sextants is backlash. This is caused by excessive play in the gear train connecting the index prism to the altitude counter.

Usually, index and backlash errors are nearly constant through the altitude range of the sextant. Therefore, if the error at one altitude setting is determined, the correction can be applied to any Hs or Hc. The correction is of equal value to the error, but the opposite sign.

The sextant should be checked on the ground before every celestial flight. Preflighting the sextant can determine the sextant error of an individual instrument. The sextant error can also be determined in-flight and a correction can be applied to the precomp to compensate for the error. To determine the error and correction in-flight, one must have a celestial LOP, a Zn, and the actual, or best-known, position of the aircraft at the same time. *[Figure 13-16]*

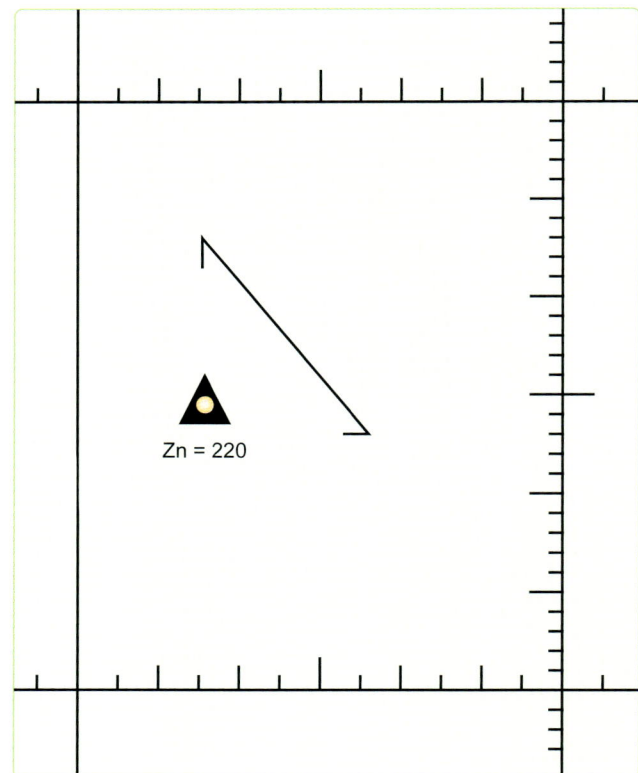

**Figure 13-16.** *Determining sextant error correction.*

The fix symbol represents the best-known position at the time of the celestial LOP. To determine the actual value of the correction, measure the shortest distance between the position and the LOP. This tells you how many minutes of arc (NM) the Ho must be adjusted on subsequent shots to get an accurate LOP. In this case, the value is 10'. To determine whether this value must be added or subtracted, note whether the LOP needs to be adjusted toward the Zn or away from the Zn. Remember the rule HOMOTO? It applies here, too. If the LOP needs to be moved toward the Zn in order to be made more accurate, the Ho needs to be made larger, thus, the correction is added to the Ho to make the Ho value increase. If the LOP needs to be moved away from the Zn, the correction is subtracted from the Ho to make the Ho less. In *Figure 13-16,* the LOP needs to be moved 10 miles toward the Zn in order to be accurate; thus, the sextant error correction is +10 to the Ho and can be used on subsequent shots obtained from the same sextant.

An important thing to remember is that the sextant error correction assumes conditions are consistent. As a technique, it is wise to obtain several LOPs with a sextant, noting the sextant errors on each, before establishing a value to be carried on the precomp. Once using that correction, make sure you use the same sextant.

## Chapter Summary

The first half of this chapter described the parts and operation of the sextant and the second half explained sextant errors. Remember to apply parallax, semidiameter, and refraction errors on every applicable shot. Corrections for acceleration errors can be applied only if you know the track and groundspeed before and after each shot, so be aware of your speed and direction when shooting. Time permitting, always try to evaluate the accuracy of your sextant on the ground.

# Chapter 14
# Grid Navigation

## Introduction

The original purpose of grid navigation was to ease the difficulties facing the navigator during high latitude flights. But grid can be used at all latitudes, particularly on long routes because grid uses a great circle course for a heading reference. Grid is simply a reorientation of the heading reference and does not alter standard fixing techniques.

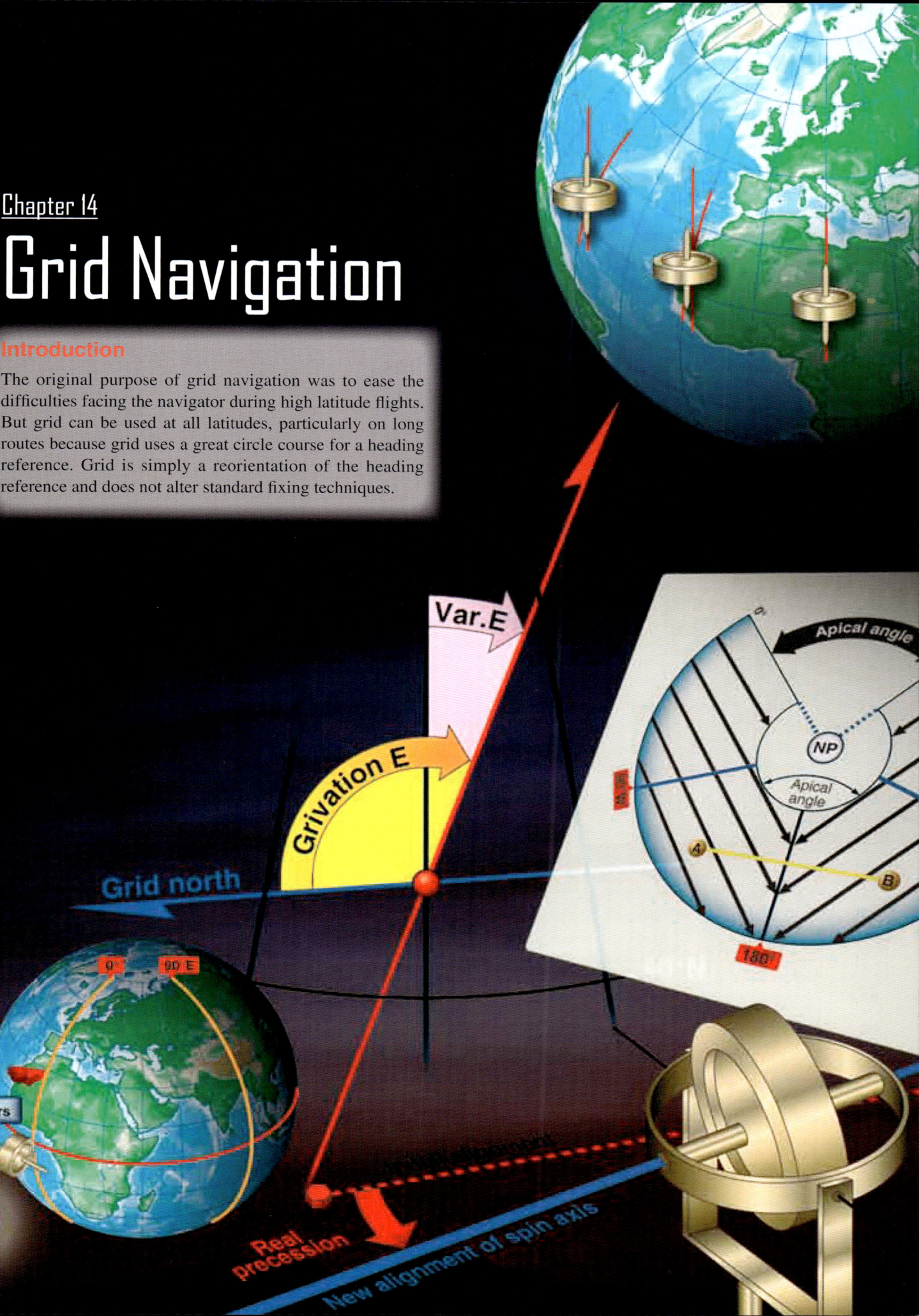

Var.E

Grivation E

Grid north

Apical angle

NP

Apical angle

A

B

0°

90 E

180°

Real precession

New alignment of spin axis

## Problems Encountered in Polar Navigation

Two factors peculiar to polar areas that make steering more difficult than usual are magnetic compass unreliability and geographic meridians converging at acute angles. The combined effect of these two factors makes steering by conventional methods difficult if not impossible. Each factor is examined below.

### Unreliability of Magnetic Compass

Maintaining an accurate heading in high latitudes is difficult when a magnetic compass is used as the heading indicator. Built to align itself with the horizontal component of the earth's magnetic field, the compass instead must react to the strong vertical component that predominates near the magnetic poles. Here, the horizontal component is too weak to provide a reliable indication of direction. As a result, compass performance becomes sluggish and inaccurate. The situation is further aggravated by the frequent magnetic storms in the polar regions that shift the magnetic lines of force.

But even if these conditions did not exist, the mere proximity to the magnetic pole would sharply reduce compass usefulness. While the aircraft may fly a straight course, the compass indicator would swing rapidly, faithfully pointing at a magnetic pole passing off to the left or right. To cope with the unreliable magnetic compass, we use gyro information for our heading inputs.

### Problem of Converging Meridians

The nature of the conventional geographic coordinate system is such that all meridians converge to the pole. Each meridian represents a degree of longitude; each is aligned with true north (TN) and true south. On polar charts, the navigator encounters 1 degree of change in true course for each meridian crossed; thus, the more closely the aircraft approaches a pole, the more rapidly it crosses meridians. Even in straight-and-level flight along a great circle course, true course can change several degrees over a short period of time. You are placed in the peculiar position of constantly altering the aircraft's magnetic heading in order to maintain a straight course. For precision navigation, such a procedure is clearly out of the question. Notice in *Figure 14-1* that the course changes 60° between A and B and much nearer the pole, between C and D, it changes 120°.

The three polar projections most commonly used in polar areas for grid navigation are the transverse Mercator, the polar stereographic, and the polar gnomonic. The transverse Mercator and polar stereographic projections are used in-flight, the polar gnomonic is used only for planning. The Lambert conformal projection is the one most commonly used for grid flight in subpolar areas. The division between

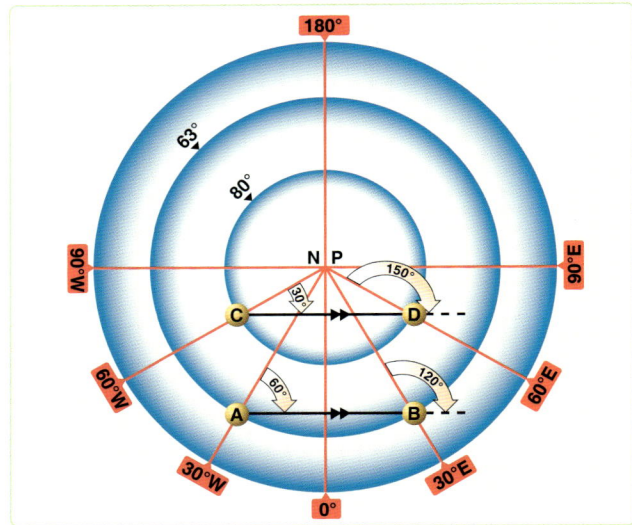

**Figure 14-1.** *Converging meridians.*

polar and subpolar projections varies among the aeronautical chart series. For example, the division is at 70° of latitude for the JN series, and at 80° of latitude for the Operational Navigation Chart (ONC) series charts.

## Grid Overlay

The graticule of the grid overlay eliminates the problem of converging meridians. *[Figure 14-2]* It is a square grid and, though its meridians are aligned with grid north (GN) along the Greenwich meridian, they do not converge at GN. While the grid overlay can be superimposed on any projection, it is most commonly used with the polar stereographic (for flights in polar areas) and the Lambert conformal (for flights in subpolar areas). This is because a straight line on these projections approximates a great circle. As the great circle course crosses the true meridians, its true direction changes but its grid direction remains constant. *[Figures 14-3 and 14-4]* All grid meridians are parallel to the Greenwich meridian and TN along the Greenwich meridian is the direction of GN over the entire chart.

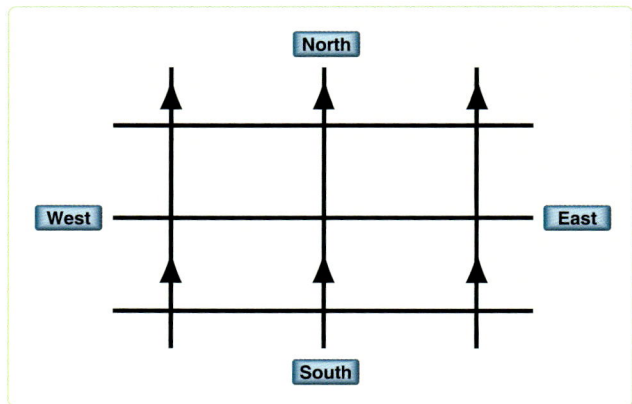

**Figure 14-2.** *Grid overlay.*

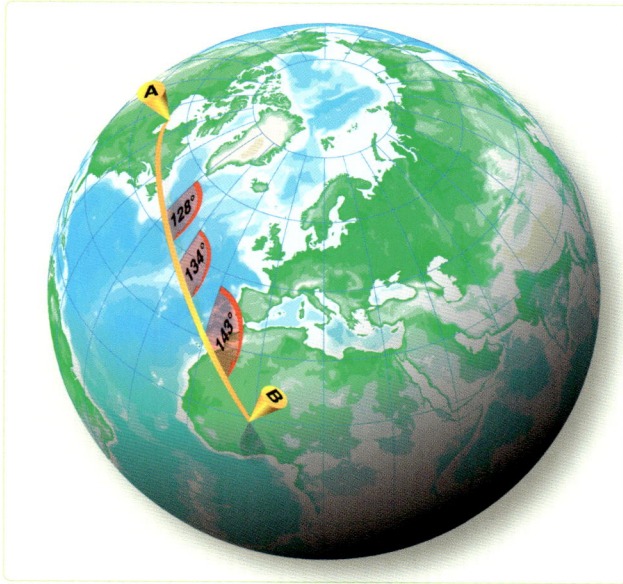

**Figure 14-3.** *Great circle true direction changes.*

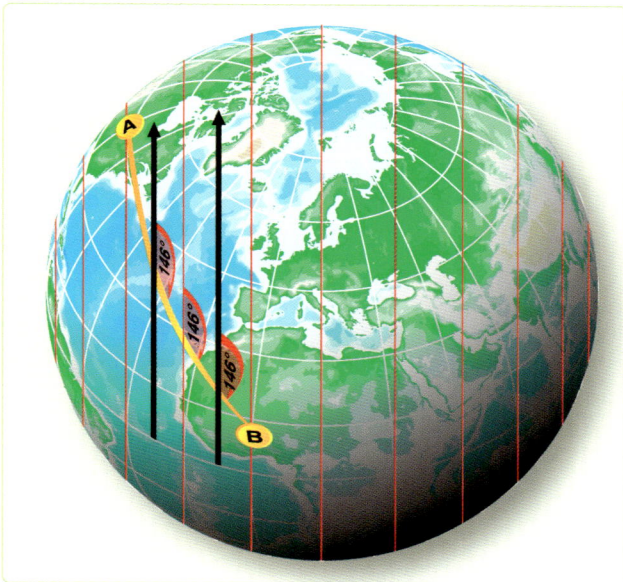

**Figure 14-4.** *Great circle true direction is constant.*

## Relationship of Grid North to True North

Because grid meridians are parallel to the Greenwich meridian, the aircraft longitude and the convergence factor (CF) of the chart govern the angle between GN and TN.

## CF of 1.0

*Figure 14-5* shows that charts having CFs of 1.0 display GN to TN relationship as a direct function of longitude. In the Northern Hemisphere at 30° W, GN is 30° W of TN; at 60° W, GN is 60° W of TN. Similarly, at 130° E longitude, GN is 130° E of TN. In the Southern Hemisphere, the direction of GN with respect to TN is exactly opposite.

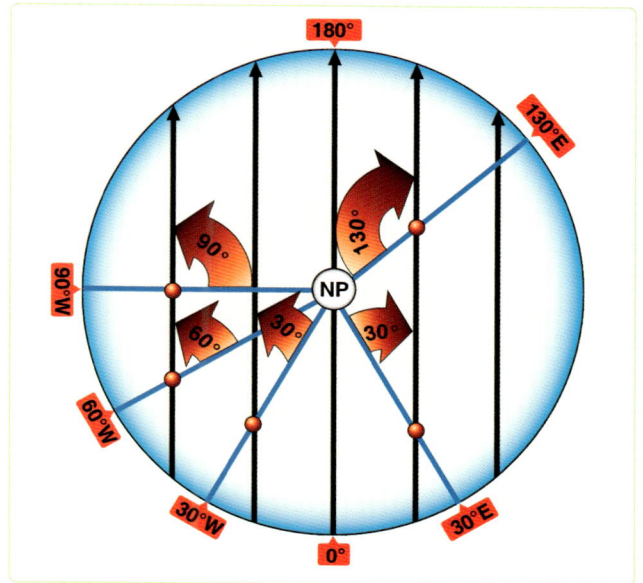

**Figure 14-5.** *Correction for the moon's parallax.*

## CF of Less Than 1.0

*Figure 14-6* shows a chart with a CF of less than 1.0 with a grid overlay superimposed on it. The relationship between GN and TN on this chart is determined in the same manner as on charts with a CF of 1.0. On charts with a CF of less than 1.0, the value of the convergence angle at a given longitude is always smaller than the value of longitude and is equal to the CF times the aircraft longitude.

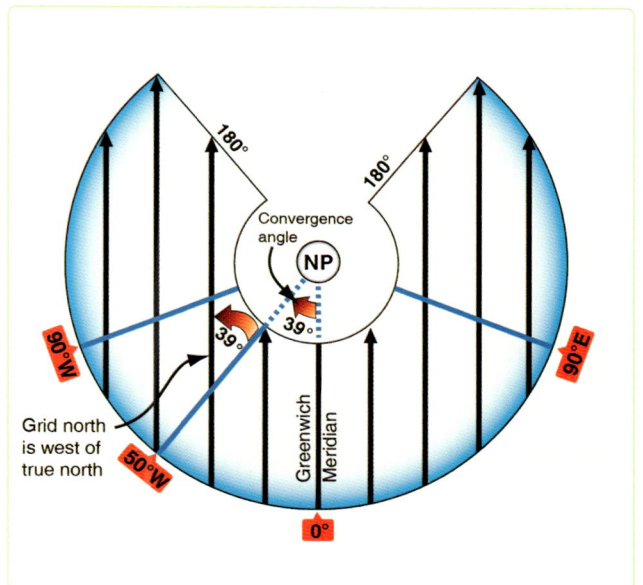

**Figure 14-6.** *Grid overlay superimposed on Lambert conformal (convergence factor 0.785).*

## Relationship of Grid Direction to True Direction

Use the following formulas to determine grid direction.

### In the Northern Hemisphere:

Grid direction = true direction + west longitude × CF

Grid direction = true direction – east longitude × CF

### In the Southern Hemisphere:

Grid direction = true direction – west longitude × CF

Grid direction = true direction + east longitude × CF

Polar angle is used to relate true direction to grid direction. Polar angle is measured clockwise through 360° from GN to TN. It is simple to convert from one directional reference to the other by use of the formula:

Grid direction = true direction + polar angle

To determine polar angle from convergence angle (CA), apply the following formulas:

In the northwest and southeast quadrants, polar angle = CA
In the northeast and southwest quadrants, polar angle = 360° – CA.

## Chart Transition

Since the relationship of the true meridians and the grid overlay on subpolar charts differs from that on polar charts because of different CFs, the overlays do not match when a transition is made from one chart to the other. Therefore, the grid course (GC) of a route on a subpolar chart is different than the GC of the same route on a polar chart. The chart transition problem is best solved during flight planning:

1. Select a transition point common to both charts.

2. Measure the subpolar GC and the polar GC.

3. Compute the difference between the GCs obtained in step two. This is the amount the compass pointer must be changed at the transition point. NOTE: If the GC on the first chart is smaller than the GC on the second chart, add the GC difference to the directional gyro (DG) reading and reposition the DG pointer; if the GC on the first chart is larger, subtract the GC difference.

Example: Chart transition from a subpolar to a polar chart. GC on subpolar chart is 316°. GC on polar chart is 308°. GC difference is 8°. Gyro reading (grid heading (GH)) is 320°. The transition is from a larger GC to a smaller GC; therefore, the GC difference (8°) is subtracted from the GH value read from the DG (320°). The DG pointer is then repositioned to the new GH (312°).

| Computed: | | Applied: | |
|---|---|---|---|
| From (subpolar) GC | 316° | Old (subpolar) GH | 320° |
| GC difference | –8° | GC difference | –8° |
| To (polar) GC | 308° | New (polar) GH | 312° |

Caution: Do not alter the aircraft heading; instead, simply reposition the DG pointer to the new GH.

## Crossing 180th Meridian on Subpolar Chart

When a flight crosses the 180th meridian on a subpolar grid chart, the GH changes because of the convergence of grid meridians along this true meridian. This is very similar to the chart transition procedure described above. When using a subpolar chart that crosses the 180th meridian on an easterly heading [Figure 14-7A to B], the apical angle must be subtracted from the GH. Conversely, the apical angle must be added to the GH when on a westerly heading. [Figure 14-7B to A] The apical angle can be measured on the chart at the 180th meridian between the converging GN references. The angle can also usually be found on the chart border, or computed by use of the following formula:

Apical angle = 360° – (360° × CF)

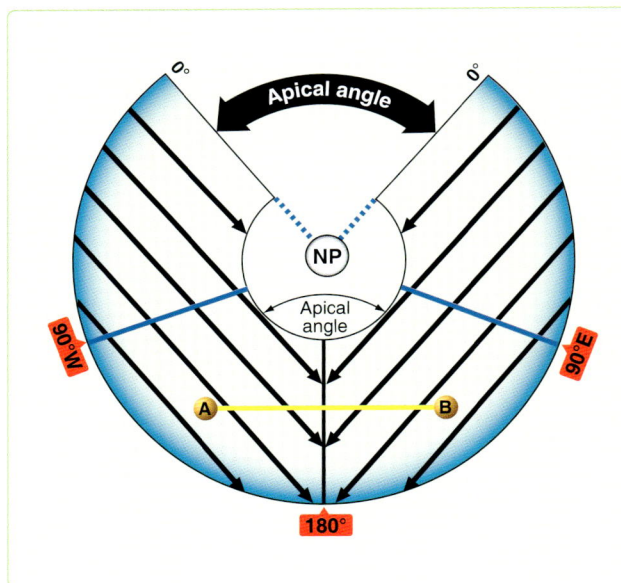

**Figure 14-7.** *Crossing 180th meridian on subpolar chart.*

Example:

Given: Chart CF 0.785

Find: Apical angle

Apical angle = 360° – (360° × 0.785)

Apical angle = 360° – 283°

Apical angle = 77°

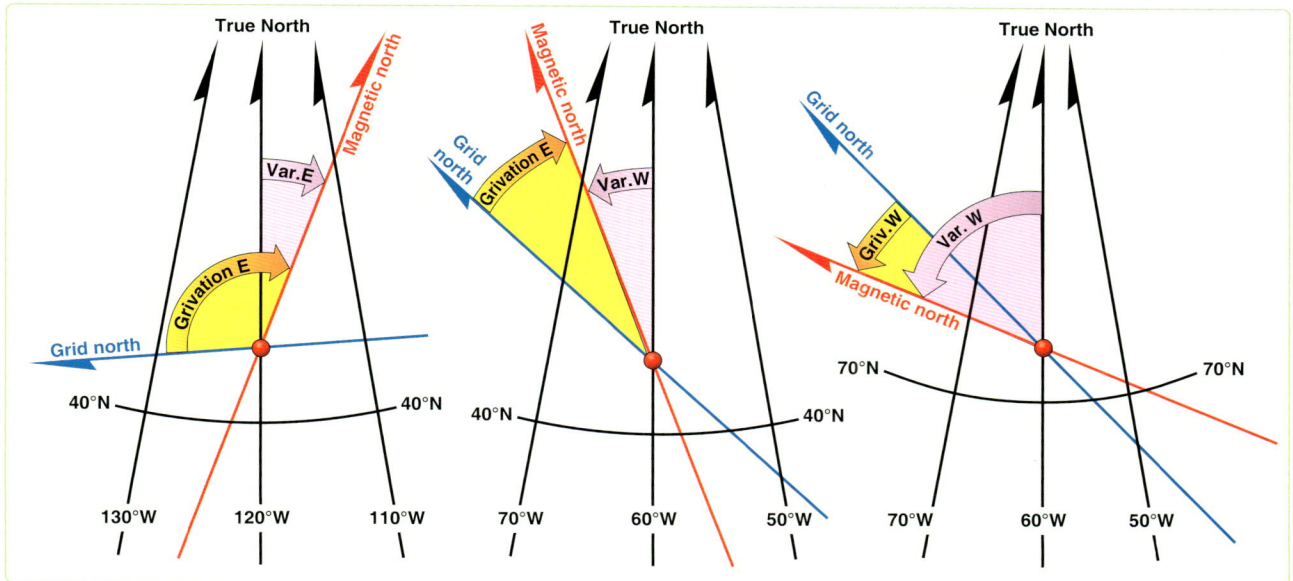

**Figure 14-8.** *Grivation.*

Caution: Do not alter the aircraft heading when crossing the 180<sup>th</sup> meridian; instead, simply reset the DG pointer to the new GH.

## Grivation

The difference between the directions of the magnetic lines of force and GN is called grivation (GV). GV is similar to variation and used to convert MH to GH and vice versa. *Figure 14-8* shows the relationship between GN, TN, and MN. Lines of equal GV (isogrivs) are plotted on grid charts.

The formulas for computing GV in the Northern Hemisphere are:

GV = (–W convergence angle) + W variation

GV = (–W convergence angle) – E variation

GV = (+E convergence angle) + W variation

GV = (+E convergence angle) – E variation

If GV is positive, it is W grivation; if grivation is negative, it is E grivation. For example, if variation is 17° E and convergence angle is 76° W, using the formula:

GV = (–West convergence angle) + (–E variation)

GV = (–76) + (–17) = –93

GV = 93° E

To compute MH from GH, use the formula:

MH = GH + W grivation

MH = GH – E grivation

In the Southern Hemisphere, reverse the signs of west and east convergence angle in the formula above.

## Gyro Precession

To eliminate the difficulties imposed by magnetic compass unreliability in polar areas, you disregard the magnetic compass in favor of a free-running gyro. Gyro steering is used because it is stable and independent of magnetic influence. When used as a steering instrument, the gyro is restricted so its spin axis always remains horizontal to the surface of the earth and is free to turn only in this horizontal plane. Any movement of a gyro spin axis from its initial horizontal alignment is called precession. The two types of precession are real and apparent, with apparent broken into earth rate, transport, and grid transport precession. Total precession is the cumulative effect of real and apparent precession.

### Real Precession

Real precession is the actual movement of a gyro spin axis from its initial alignment in space. *[Figure 14-9]* It is caused by such imperfections as power fluctuation, imbalance of the gyro, friction in gyro gimbal bearings, and acceleration

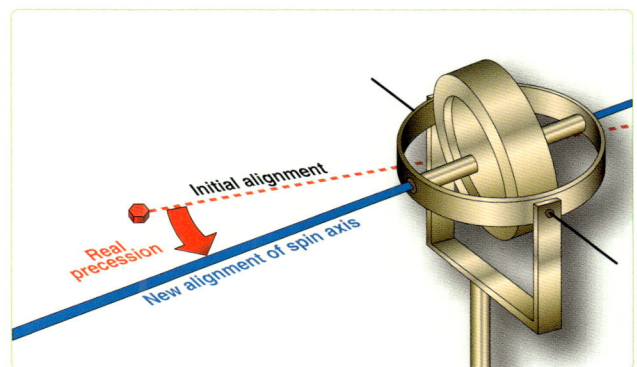

**Figure 14-9.** *Real precession.*

14-5

forces. As a result of the improved quality of equipment now being used, real precession is considered to be negligible. Some compass systems have a real precession rate of less than 1° per hour. Electrical or mechanical forces are intentionally applied by erection or compensation devices to align the gyro spin axis in relation to the earth's surface. In this manner, the effects of apparent precession are eliminated and the gyro can then be used as a reliable reference.

## Apparent Precession

The spin axis of a gyro remains aligned with a fixed point in space, while your plane of reference changes, making it appear that the spin axis has moved. Apparent precession is this apparent movement of the gyro spin axis from its initial alignment.

## Earth Rate Precession

Earth rate precession is caused by the rotation of the earth while the spin axis of the gyro remains aligned with a fixed point in space. Earth rate precession is divided into two components. The tendency of the spin axis to tilt up or down from the horizontal plane of the observer is called the vertical component. The tendency of the spin axis to drift around laterally; that is, to change in azimuth, is called the horizontal component. Generally, when earth rate is mentioned, it is the horizontal component that is referred to, since the vertical component is of little concern.

A gyro located at the North Pole, with its spin axis initially aligned with a meridian, appears to turn 15.04° per hour in the horizontal plane because the earth turns 15.04° per hour. [Figure 14-10] As shown in Figure 14-10A, the apparent relationship between the Greenwich meridian and the gyro spin axis changes by 90° in 6 hours, though the spin axis is still oriented to the same point in space. Thus, apparent precession at the pole equals the rate of earth rotation. At the equator, as shown in Figure 14-10B, no earth rate precession occurs in the horizontal plane if the gyro spin axis is still aligned with a meridian and is parallel to the earth's spin axis.

## Vertical Component

When the gyro spin axis is turned perpendicular to the meridian, maximum earth rate precession occurs in the vertical component. [Figure 14-11] But the directional gyro does not precess vertically because of the internal restriction of the gyro movement in any but the horizontal plane. For practical purposes, earth rate precession is only that precession that occurs in the horizontal plane. Figure 14-11 illustrates earth rate precession at the equator for 6 hours of time.

## Precession Variation

Earth rate precession varies between 15.04°/hour at the poles and 0°/hour at the equator. It is computed for any latitude by

multiplying 15.04° times the sine of the latitude. For example, at 30° N, the sine of latitude is 0.5. The horizontal component of earth rate is, therefore, 15°/hour right × 0.5 or 7.5°/hour right at 30° N. [Figure 14-12]

## Steering by Gyro

Obviously, if the gyro is precessing relative to the steering datum of GN or TN, an aircraft steered by the gyro is led off heading at the same rate. To compensate for this precession, an artificial real precession is induced in the gyro to counteract the earth rate. At 30° N latitude, earth rate precession is equal to 15° × sin lat = 15 × .5 or 7.5° per hour to the right.

## Offsetting Each Rate Effect

Hence, if at 30° N latitude, a real precession of 7.5° left per hour is induced in the gyro, it balances exactly and offsets earth rate effect. In ordinary gyros, a weight is used to produce this effect but, since the rate is fixed for a given latitude, the correction is good for only one latitude. The latitude chosen is normally the mean latitude of the area in which the aircraft operates. The N-1 and AHRS compass systems have a latitude setting knob that you can use to adjust for the earth rate corrections.

## Earth Transport Precession (Horizontal Plane)

Earth transport precession is a form of apparent precession that results from transporting a gyro from one point on the earth's surface to another. The gyro spin axis appears to move because the aircraft, flying over the curved surface of the earth, changes its attitude in relation to the gyro's fixed point in space. [Figure 14-13] Earth transport precession causes the gyro spin axis to move approximately 1° in the horizontal plane for each true meridian crossed. This effect is avoided by using GN as the steering reference.

## Grid Transport Precession

Grid transport precession exists because meridian convergence is not precisely portrayed on charts. The navigator wants to maintain a straight-line track, but the gyro follows a great-circle track, which is a curved line on a chart. The rate at which the great-circle track curves away from a straight-line track is grid transport precession. This is proportional to the difference between convergence of the meridians as they appear on the earth and as they appear on the chart and the rate at which the aircraft crosses these meridians.

## Summary of Precession

Real precession is caused by friction in the gyro gimbal bearings and dynamic unbalance. It is an unpredictable quantity and can be measured only by means of heading checks.

Earth rate precession is caused by the rotation of the earth. It can be computed in degrees per hour with the formula:

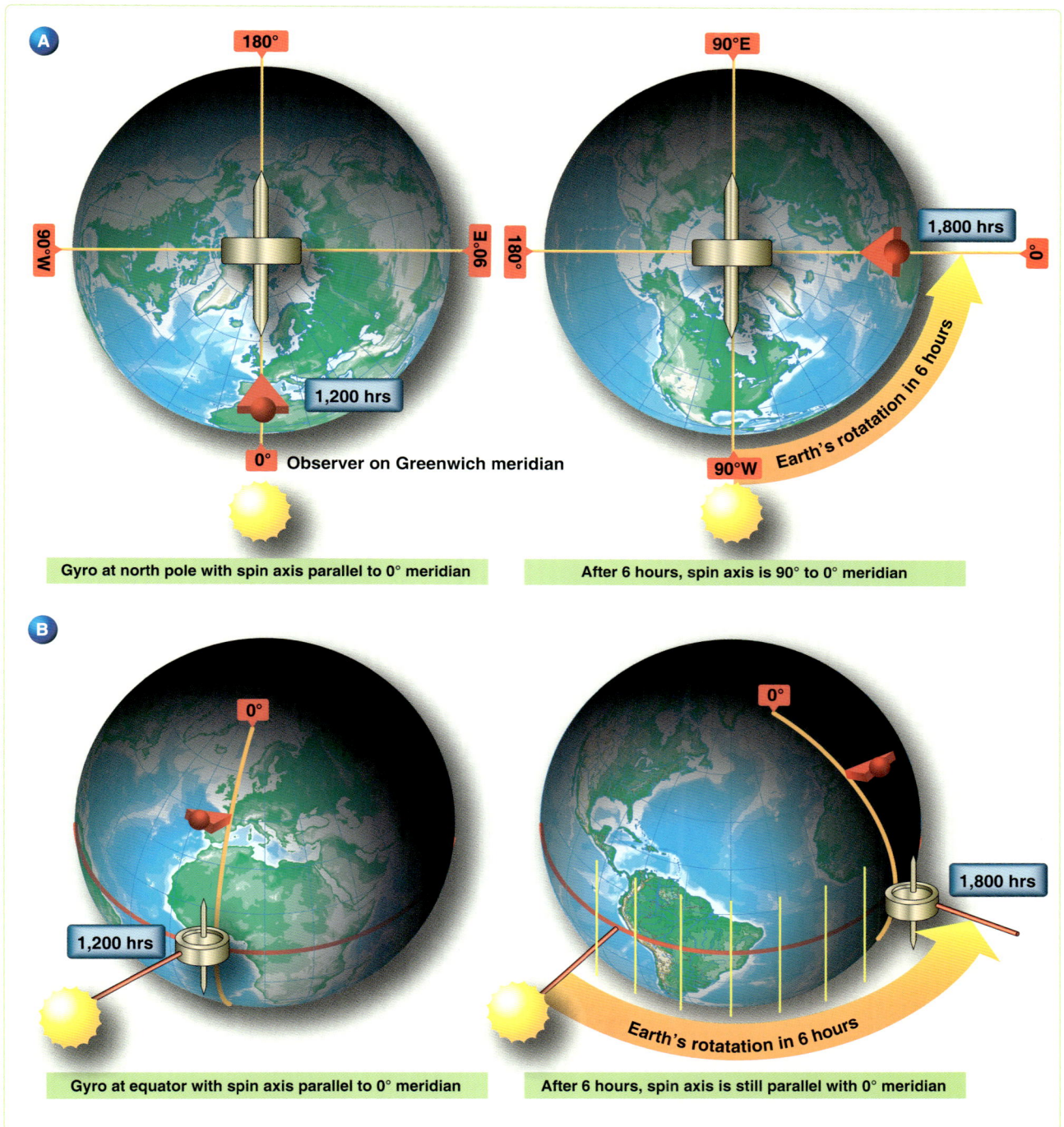

**Figure 14-10.** *Initial location of gyro affects earth rate precession.*

15.04 × sin lat. It is to the right in the Northern Hemisphere and to the left in the Southern Hemisphere. All gyros are corrected to some degree for this precession, many by means of a latitude setting knob.

Earth transport precession (horizontal plane) is an effect caused by using TN as a steering reference. It can be computed by using the formula (change longitude/hour × sine mid latitude). The direction of the precession is a function of the TC of the aircraft. If the course is 0° – 180°, precession is to the right; if the course is 180° – 360°, precession is to the left. This precession effect is avoided by using GN as a steering reference.

Grid transport precession is caused by the fact that the great circles are not portrayed as straight lines on plotting charts.

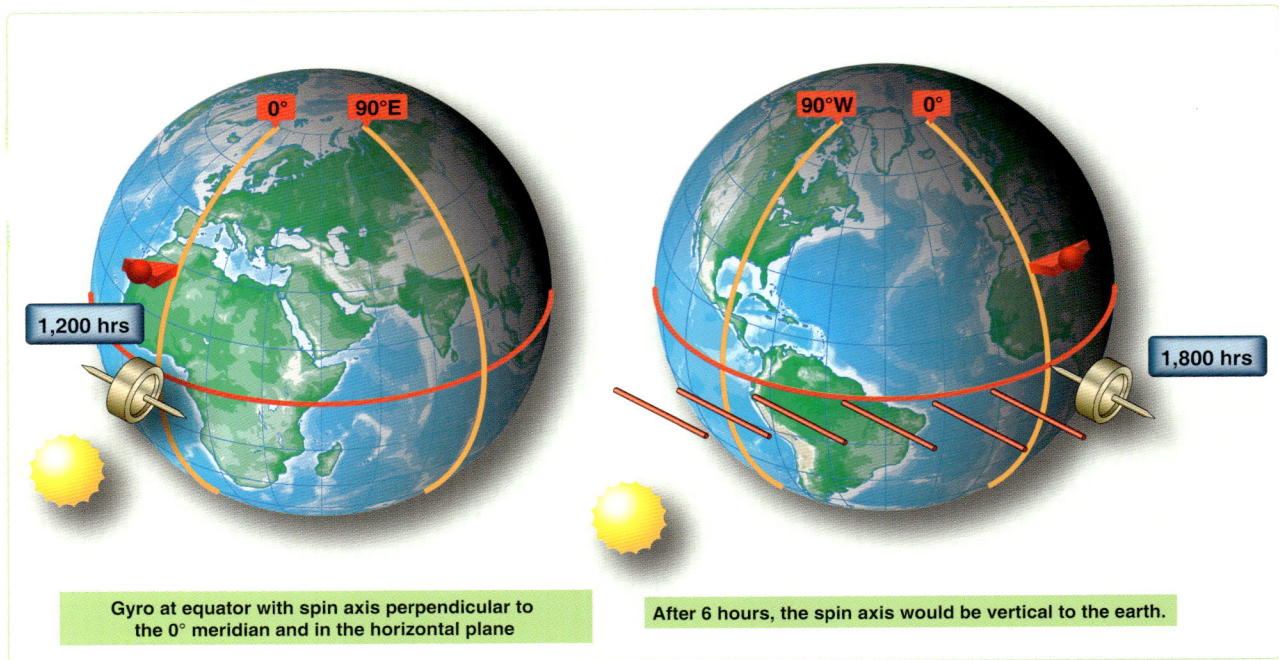

**Gyro at equator with spin axis perpendicular to the 0° meridian and in the horizontal plane**

**After 6 hours, the spin axis would be vertical to the earth.**

1,200 hrs

1,800 hrs

90°W    0°

0°    90°E

**Figure 14-11.** *Direction of spin axis affects earth rate precession.*

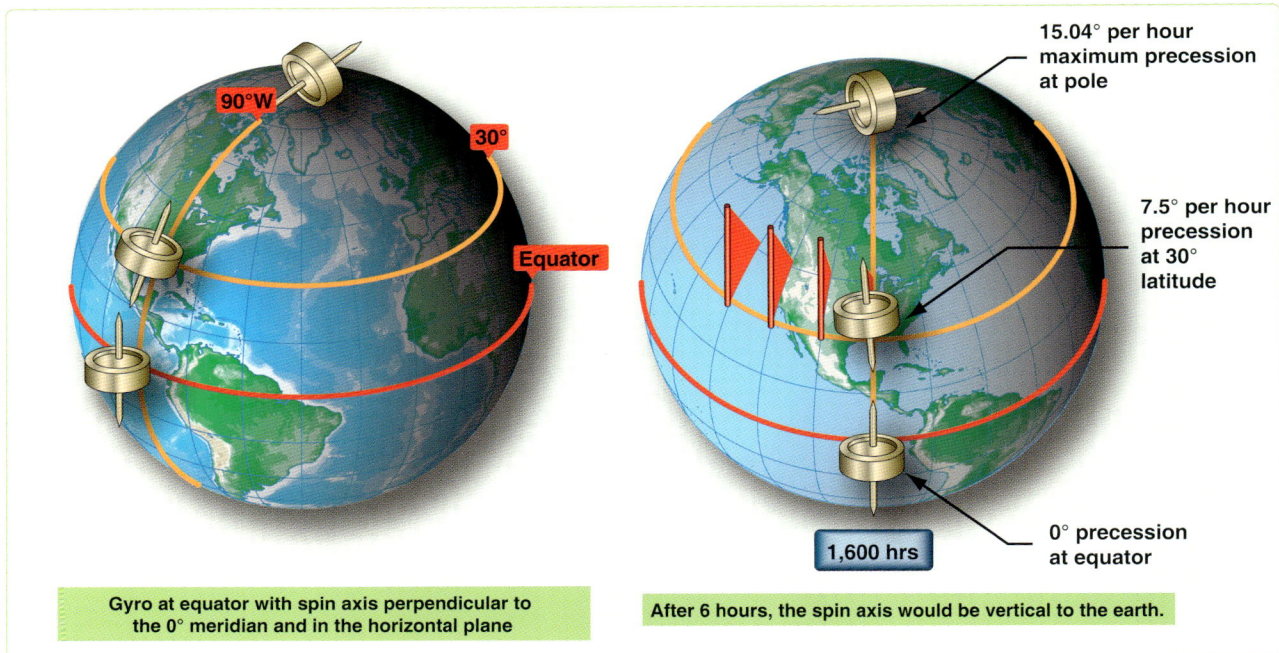

15.04° per hour maximum precession at pole

7.5° per hour precession at 30° latitude

0° precession at equator

90°W

30°

Equator

1,600 hrs

**Gyro at equator with spin axis perpendicular to the 0° meridian and in the horizontal plane**

**After 6 hours, the spin axis would be vertical to the earth.**

**Figure 14-12.** *Earth precession varies according to latitude.*

The navigator tries to fly the straight pencil-line course, the gyro a great circle course. The formula for grid transport precession is change longitude/hour (sin lat − CF), where CF is the chart convergence factor. The direction of this precession is a function of the chart used the latitude and the TC. Direct substitution into the formula produces an answer valid for easterly courses, such as 0° − 180°. For westerly courses, the sign of the answer must be reversed.

## Gyro Steering

Gyro steering is much the same as magnetic steering, except that GH is used in place of true heading (TH). GH has the same relation to GC as TH has to true course (TC). The primary steering gyro in most aircraft provides directional data to the autopilot and maintains the aircraft on a preset heading. When the aircraft alters heading, it turns about the primary gyro while the gyro spin axis remains fixed in

**Figure 14-13.** *Earth transport precession.*

azimuth. If the primary gyro precesses, it causes the aircraft to change its heading by an amount equal to the precession.

## Starting a Grid Navigation Leg

Grid navigation is normally entered while airborne on a constant heading. A constant heading is necessary because grid entry is accomplished by resetting the compass from a magnetic to a grid reference while on the same heading. After obtaining a grid celestial or inertial navigation system (INS) heading check, reset the compass immediately to the correct grid heading to avoid heading errors, because the precomputed grid Zn is only good for shot time. Since the exact grid heading is set at the beginning of the navigation leg, precession is assumed to be zero until subsequent heading checks assess the accuracy of the gyro. The grid heading is normally obtained using a variant of the TH method or the INS TH. Using this method, set a grid Zn in the sextant azimuth counter before collimating on the body. Other heading shot methods can be used, but would delay resetting the gyro to accomplish math computations after the heading shot. Although any celestial body may be used, navigators commonly use the sun or Polaris, depending on the time of day. *[Figure 14-14]*

### Using a Zn Graph

In order to get an accurate grid Zn for daytime grid entry, the navigator must compute Zn of the sun for a time and geographic position where the grid navigation leg begins. If the geographic position for grid entry is known well in advance, you can prepare a Zn graph for a time window. A Zn graph makes grid entry easier, because it is usable for an extended period of time, therefore eliminating the need to precomp for a specific time. The graph can be constructed during flight planning, thus reducing workload in the air.

To construct the graph, precomp and plot Zn on one axis and time on the other. *[Figure 14-15]* Set up the time axis to cover the planned start time and several minutes earlier and later. Plot grid Zn on the other axis using normal precomp procedures and the start point coordinates. Because the time/Zn slope is close to linear, precomping at 20–30 minute intervals and connecting the points gives acceptable accuracy. When the sun is near local noon, precomp Zn at closer intervals because the Zn changes rapidly. To use the graph when it is finished, enter on the time axis. Then extend a line perpendicular to the time axis until reaching the time/Zn line. Finally, read the appropriate Zn on the Zn axis.

*Figure 14-15* demonstrates using a graph to get a grid Zn for the time of 1700Z. Although preparing a Zn graph takes a while, it pays dividends as long as you actually fly over the planned geographic point within the time frame covered by the graph.

## Applying Precession to the DR

The most accurate method for applying precession to the DR is the all behind/half ahead method. This method corrects for the banana effect most commonly associated with precession. Since the full effect of precession does not occur at one time, we have to account for the gradual increase of precession.

Step 1—Determine the Hourly Rate

In *Figure 14-16,* grid entry occurred at 1700. At 1720, the navigator obtained a heading shot or MPP. The heading shot determined precession correction to be −2 and the compass was reset to the GH. On the MB-4 computer, place the −2 correction on the outside scale and the time since grid entry (20 minutes) on the inside scale. The hourly rate now appears above the index (6.0R). To minimize error, the hourly rate has to be computed to the nearest tenth of a degree.

Step 2—Compute All Behind/Half Ahead

Since precession begins at the last time the gyro was reset, for this example we need to start at grid entry 1700. At 1700, all behind would be determined to be 0 minutes and half-ahead to the next dead reckoning (DR) (1706) would be 3 minutes. To determine the amount of precession correction to be used, leave the hourly rate (6.0) over the index and look above 3 minutes. The computed precession correction for the 1706 DR is −0.3° or 0 for use on the log. Next, we need to determine the precession correction for the 1720 DR. At 1706, all behind is 6 minutes and half ahead is 7 minutes. The total time used to compute the precession correction for this DR is 13 minutes. Again, using the hourly rate, precession correction for the 1720 DR is −1.3° or −1° for the log.

### KC-135 Method

Since the all behind/half ahead method tends to keep you behind, the KC-135 method is used by some navigators

# CELESTIAL PRECOMPUTATION

## PRECOMPUTATION   PERISCOPIC SEXTANT

| NAVIGATOR | Major Watters | | ALT MSL. | 35.0 | DATE(Z) | 18 Apr 95 | FIX TIME | 2 | (Z) |
|---|---|---|---|---|---|---|---|---|---|

STAR SELECTOR BY AZIMUTH

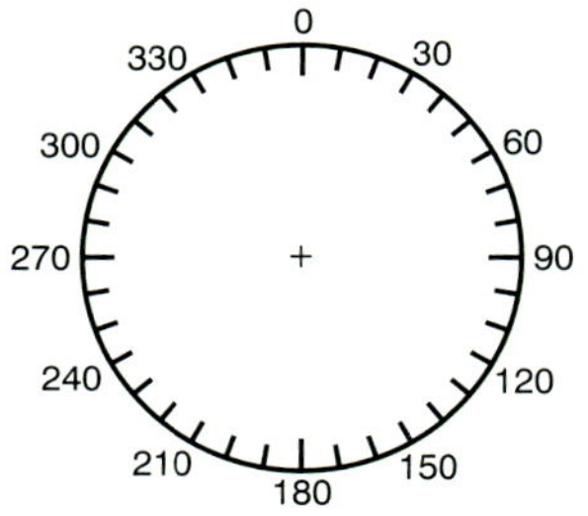

(compass rose diagram: 0 at top, 30, 60, 90, 120, 150, 180, 210, 240, 270, 300, 330, with + in center)

| | | | BODY | SUN | | | | |
|---|---|---|---|---|---|---|---|---|
| TRACK | | | | | | | | |
| GS | | | BASE GHA | 070-09 | 075-09 | 080-09 | 085-09 | |
| CORIOLIS | R / L | | CORR | | | | | |
| PREC/ NUT | | | +360 | 360-00 | | | | |
| DR LAT | 35-00 (N/S) | | GHA | 430-09 | | | | |
| DR LONG | 098-00 (E/W) | | ASSUM LONG (W) +E | 098-00 | | | | |
| MOTION OF OBSERVER | | | LHA | 332-09 | 337-09 | 342-09 | 347-09 | |
| MOTION OF BODY | | | ASSUM LAT | 35 (N)/S | | | | |
| ∠ MIN ADJUST | | | DEC | 10 (N)/S | N/S | N/S | N/S | N/S |
| OFF-FIX TIME | E/L  E/L  E/L  E/L  E/L | | PLANNED MID-TIME | 1640 | 1700 | 1720 | 1740 | |
| TOTAL MOT. ADJUST | | | ACTUAL MID-TIME | | | | | |
| POLARIS  Q | | | TAB Hc | | | | | |
| MOON  PA  SD | | | | | | | | |
| REF ( ) | | | D / DEC  CORR | | | | | |
| PERS/SEXT | | | CORR Hc | | | | | |
| TOTAL → ADJ | | | TOTAL → ADJ | | | | | |
| TH/GH | | | ADJ Hc | | | | | |
| Zn/GZa( ) | | | OFF TIME MOTION | | | | | |
| IRB | | | Hc | | | | | |
| IRB | | | Ho | | | | | |
| Za/GZa(+) | | | INT | T/A | T/A | T/A | T/A | T/A |
| TH/GH | | | Zn | 127 | 133 | 141 | 151 | |
| TRACK  T/G | | | CONV +W ANGLE E | +62 | +62 | +62 | +62 | |
| Za | | | GRID Zn | 189 | 195 | 203 | 213 | |
| REL Za | | | TK  DC  TH  VAR  MH  DEV  CH | | | | | |

REFRACTION TABLE (condensed)

| Ro | Altitude MSL (thousands of feet) | | | | | |
|---|---|---|---|---|---|---|
| | 0 | 20 | 25 | 30 | 35 | 40 |
| 1 | 53 | 46 | 41 | 36 | 33 | 26 |
| 2 | 33 | 19 | 16 | 14 | 11 | 9 |
| 3 | 23 | 12 | 10 | 8 | 7 | 5 |
| 4 | 16 | 8 | 7 | 6 | 5 | 3-10 |
| | 12 | 7 | 5 | 4 | 3-10 | 2-10 |

**Figure 14-14.** *Grid precomp.*

14-10

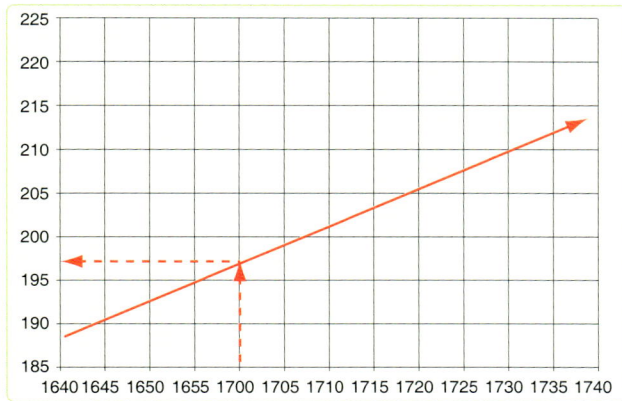

**Figure 14-15.** *Zn graph.*

to predict precession. This method basically uses half of the computed precession correction for future DRs/MPPs when the precession correction is determined between two positions. Though not as accurate as the all behind/half ahead method, the KC-135 method can be effective if used with short DRs. Using the KC-135 method, compensate for precession around the turn by getting a heading shot immediately before and after the turn, resetting the gyro after the heading shot restarts precession.

### False Latitude

A second method of compensating for precession while in-flight involves the use of false latitude inputs into the gyro compass. Most gyro compasses have a latitude control that allows the navigator to compensate for earth rate precession (ERP). Normally, the latitude control is set to the actual latitude of the aircraft. However, other values may be set. For example, if the aircraft is at 30° N and the latitude control knob is set to 70° N, the gyro overcorrects for ERP. Since ERP is right in the Northern Hemisphere, the correction is to the left. Thus, setting a higher than actual latitude corrects for right precession over and above that for ERP. Since ERP 1 5°/hour × sine latitude, a table such as in *Figure 14-17* can be developed to use this procedure.

## Chapter Summary

The grid overlay and the free-running gyro are used to overcome the difficulties of converging meridians and the unreliability of the magnetic compass when navigating in high latitudes. When using gyro steering, maintain a record of the precession of both the primary and secondary gyros. The gyro log provides you with the information necessary to predict values when it is impossible to obtain heading checks because of overcast conditions or twilight. By maintaining a log on the secondary gyro, the navigator can change gyros in case of malfunction of the primary gyro. Use the information recorded in the gyro log in conjunction with the navigator's log to plot position and compute winds, headings, alter headings, and estimated time of arrivals (ETAs). And never forget grid is green.

| | | | | | | | VAR OR | MH | | | | | DIST | | NEXT CHECK PT | | | | TEMP | |
| Time | POS SYM | Present Position | IT OR GC | W/V DC | TH OR GH | & FREE CORR | OR GH | DEV | CH | TAS | AIR DIST | TIME | GS | ACTION PT | DIST | TIME | ETA | ALT | REMARKS |
|---|---|---|---|---|---|---|---|---|---|---|---|---|---|---|---|---|---|---|---|
| 1700 | △ | | | −2 | 090 | | | | | | | | | | | | | ° M | |
| 1706 | ⊙ | | 092 / 092 ✓ | | 090 / 090 | 0 / 0 | 090 | | | 420 | | 40 +06 | 400 | | | | | ° M | |
| 1720 | ⊙ | | 093 / 092 ✓ | | 091 / 090 | −1 / 0 | 090 | | ✓ | | | 133 +20 | ✓ | | | | | ° M | |
| 1800 | △ | | | −3 | 092 | | | | ✓ | | | | ✓ | | | | | ° M | |

Inflight Log

**Figure 14-16.** *Inflight log.*

## FALSE LATITUDE CORRECTION TABLE

| NORM LAT(N) | Observed Precession Rate O/HR | | | | | | | | | | EARTH RATE |
|---|---|---|---|---|---|---|---|---|---|---|---|
| | 1° | | 2° | | 3° | | 4° | | 5° | | |
| | − | + | − | + | − | + | − | + | − | + | |
| 75 | 64 | ↑ | 56 | ↑ | 51 | ↑ | 44 | ↑ | 39 | ↑ | 14.5 |
| 70 | 61 | 90 | 54 | | 48 | | 42 | | 37 | | 14.1 |
| 65 | 57 | 75 | 51 | | 45 | | 40 | | 35 | | 13.6 |
| 60 | 53 | 69 | 47 | 90 | 42 | | 37 | | 32 | | 13.0 |
| 55 | 49 | 63 | 43 | 73 | 38 | 90 | 34 | | 29 | | 12.3 |
| 50 | 44 | 56 | 39 | 64 | 35 | 75 | 30 | 90 | 26 | | 11.5 |
| 45 | 39 | 51 | 35 | 57 | 30 | 65 | 26 | 77 | 22 | 90 | 10.6 |
| 40 | 35 | 45 | 31 | 51 | 26 | 57 | 22 | 65 | 18 | 77 | 9.4 |
| 35 | 30 | 40 | 26 | 45 | 22 | 51 | 18 | 57 | 14 | 65 | 8.6 |
| 30 | 26 | 35 | 22 | 39 | 18 | 44 | 14 | 50 | 10 | 56 | 7.5 |
| 25 | 21 | 29 | 17 | 34 | 13 | 39 | 9 | 44 | 5 | 49 | 6.3 |
| 20 | 16 | 24 | 12 | 28 | 8 | 33 | 4 | 37 | 0 | 43 | 5.1 |
| 15 | 11 | 19 | 7 | 23 | 3 | 27 | 0 | 32 | 4 | 36 | 3.9 |
| 10 | 6 | 14 | 2 | 18 | 2 | 22 | 5 | 26 | 9 | 30 | 2.6 |
| 5 | 1 | 9 | 4 | 13 | 7 | 17 | 10 | 21 | 16 | 25 | 1.3 |
| 0 | 4 | 4 | 8 | 8 | 12 | 12 | 16 | 16 | 20 | 20 | 0 |

Reverse N/S switch in outlined columns

## INSTRUCTIONS

Enter with desired latitude setting and the observed hourly precession rate.

## NOTE

Direction of precession is important:
Select correct column.
In south latitude, reverse +/- column headings.

## EXAMPLE

Normal latitude setting 65° N:
Precession +3° per hour
False latitude setting = 90° N

Normal latitude setting 10° N:
Precession -4° per hour
False latitude setting=5° S

**Figure 14-17.** *False latitude correction table.*

## Chapter 15

# Pressure Pattern Navigation

## Pressure Differential Techniques

Pressure differential flying is based on a mathematically derived formula. The formula predicts windflow based on the fact that air moves from a high pressure system to a low pressure system. This predicted windflow, the geostrophic wind, is the basis for pressure navigation. The formula for the geostrophic wind (modified for a constant pressure surface), combined with inflight information makes available two aids to navigation: Bellamy drift and the pressure line of position (PLOP). Bellamy drift gives information about aircraft track by supplying net drift over a set period of time. Using the same basic information, the PLOP provides a line of position (LOP) as valid as any other type.

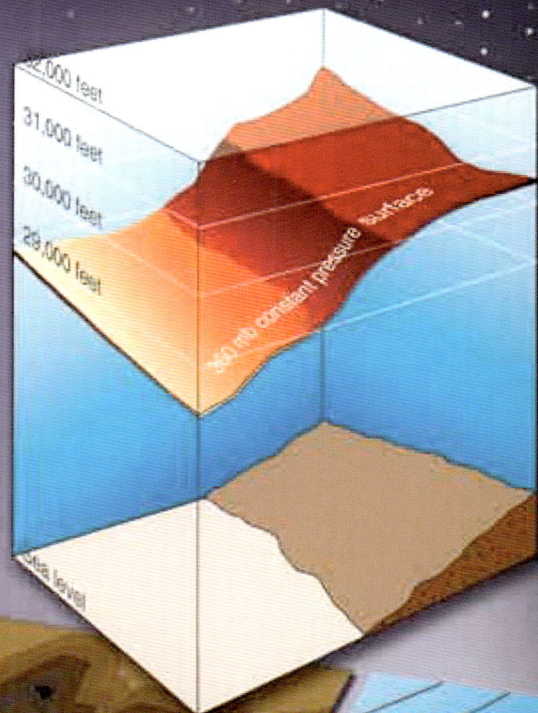

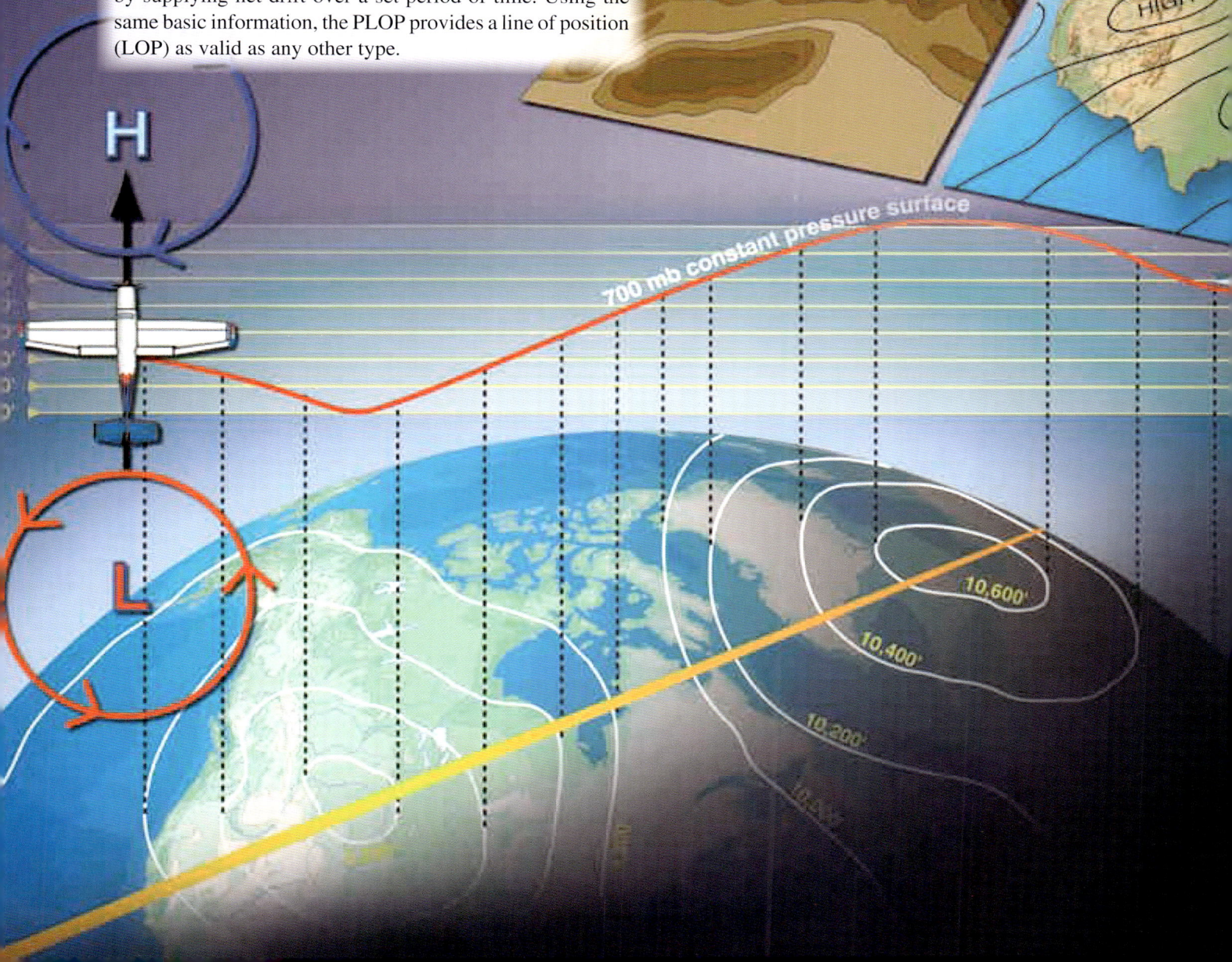

## Constant Pressure Surface

To understand pressure differential navigation, you should know something about the constant pressure surface. The constant pressure surface is one on which the pressure is the same everywhere, even though its height above sea level will vary from point to point as shown in *Figure 15-1*. The pressure altimeter will show a constant reading. A constant pressure surface is shown on a constant pressure chart (CPC) as lines that connect points of equal height above sea level. These lines are referred to as contours and are analogous to contour lines on land maps. *[Figure 15-2]* The intersection of altitude mean sea level (MSL) and constant pressure surfaces form isobars. A comparison of isobars and contours is shown in *Figure 15-2*. The geostrophic wind will blow along and parallel to the contours of a CPC just as it blows along and parallel to the isobars of a constant level chart.

## Geostrophic Wind

The shape and configuration of the constant pressure surface determine the velocity and direction of the geostrophic wind. Flying with 29.92 set in the pressure altimeter will cause the aircraft to follow a constant pressure surface and change its true height as the contours change. *[Figure 15-3]* The slope of the pressure surface, also known as the pressure gradient, is the difference in pressure per unit of distance as shown in *Figure 15-4*. The pressure gradient force (PGF), or slope of the pressure surface, and Coriolis combine to produce the geostrophic wind. The speed of the geostrophic wind is proportional to the spacing of the contours or isobars. Closely spaced contours form a steep slope and produce a stronger wind, while widely spaced contours produce relatively weak winds. According to Buys-Ballots Law, if you stand in the Northern Hemisphere with your back to the wind, the lower pressure is to your left. *[Figure 15-5]* The opposite is true in the Southern Hemisphere where Coriolis deflection is to the left. Further study of *Figure 15-5* shows that as you enter a low or a high system, your drift will be right or left, respectively. The opposite is true as you exit the system. Since the geostrophic wind is based on a constant pressure surface, you must fly a constant pressure altitude. A minimum of 2,000 to 3,000 feet above the surface will usually eliminate distortion introduced through surface friction. Near the equator (20° N to 20° S), Coriolis force approaches zero,

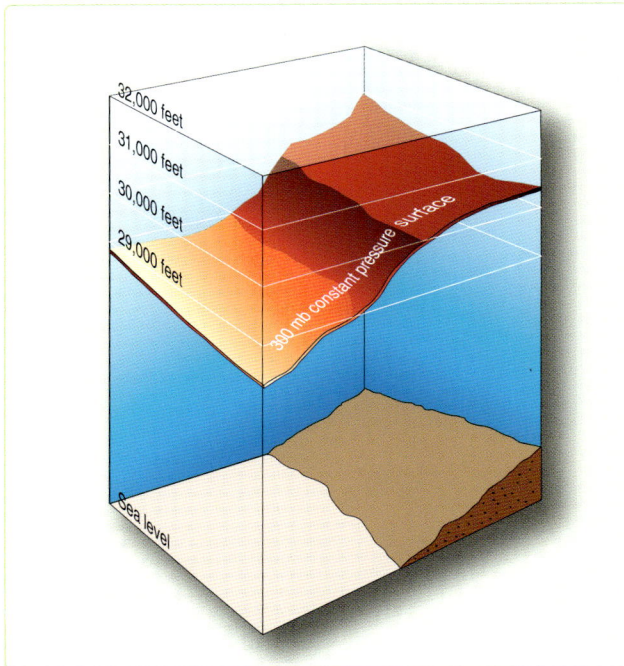

**Figure 15-1.** *Constant pressure surface.*

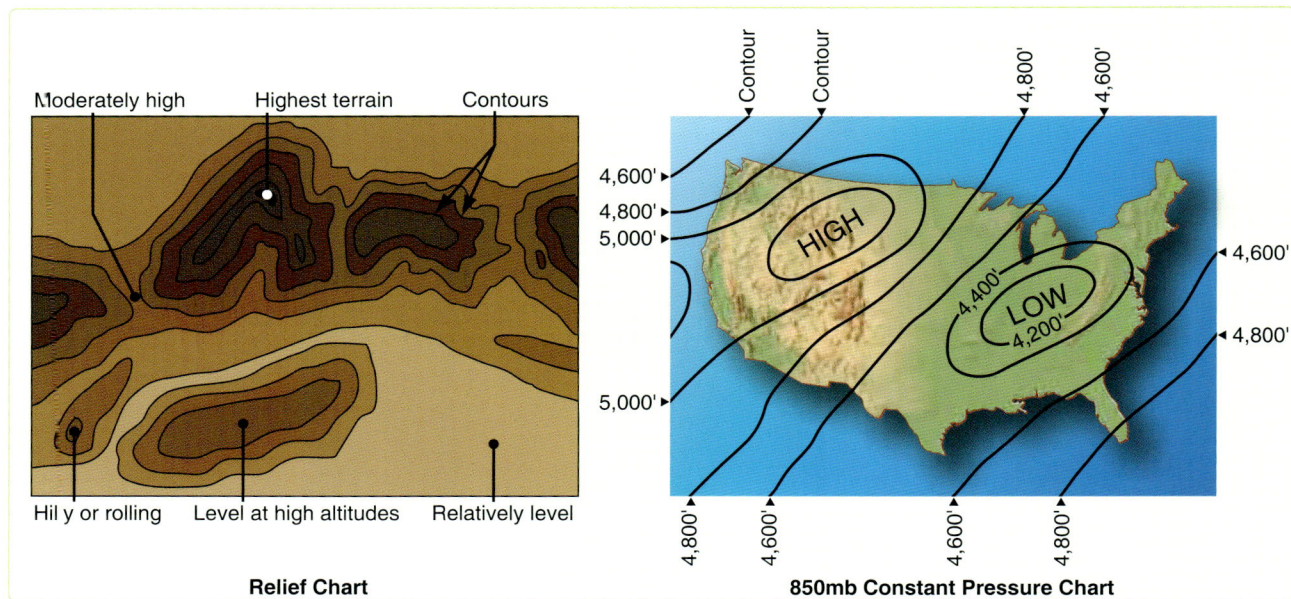

**Figure 15-2.** *Contours.*

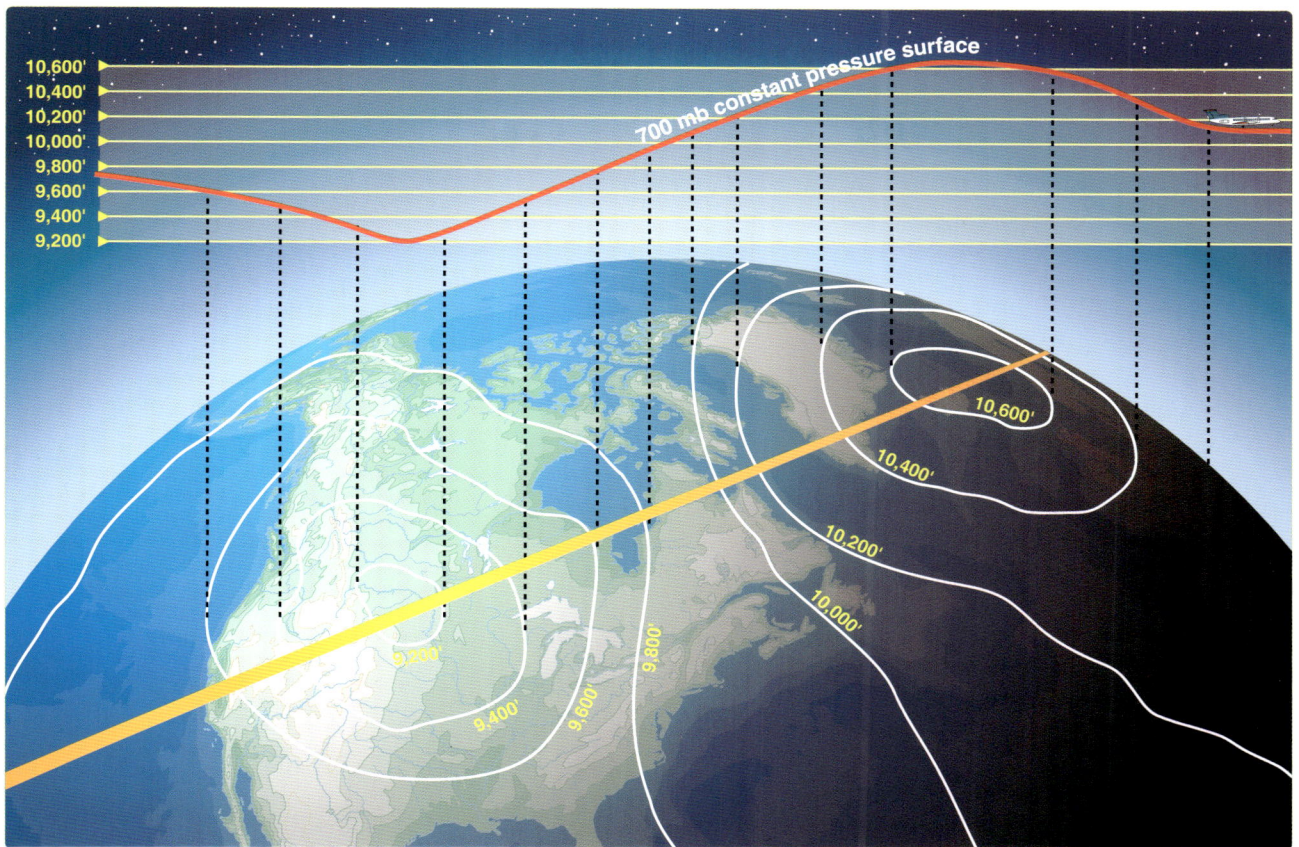

**Figure 15-3.** *Changing contours of constant pressure surface.*

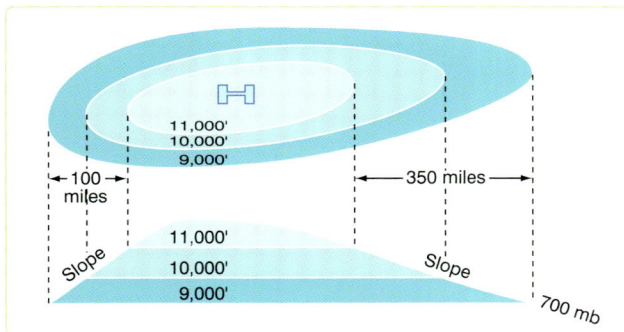

**Figure 15-4.** *Pressure gradient.*

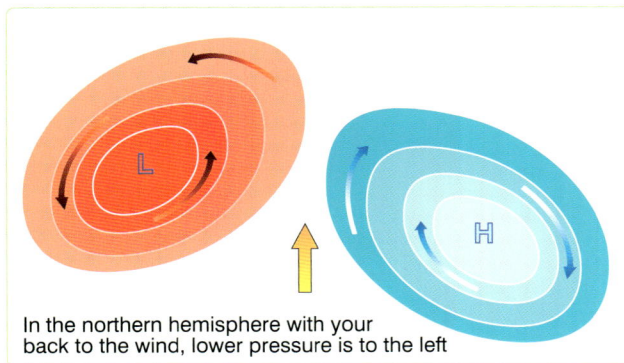

In the northern hemisphere with your back to the wind, lower pressure is to the left

**Figure 15-5.** *Buys-Ballots Law.*

and pressure navigation is unreliable, pressure differential navigation is reliable in midlatitudes.

## Pressure Computations and Plotting

In determining a PLOP or Bellamy drift by pressure differential techniques, use the crosswind component of the geostrophic wind over a given period of time. To determine your pressure pattern displacement (ZN), use the following equation:

$$ZN = \frac{K\,(D_2 - D_1)}{ETAS}$$

This formula gives the direction and crosswind displacement effect of the pressure system you've flown through. To solve for ZN, you must understand how to obtain and apply such special factors as D readings, effective true airspeed (ETAS), effective airpath (EAP), effective air distance (EAD), and K values.

## D Readings

The symbol D stands for the difference between the true altitude (TA) of the aircraft and the pressure altitude (PA) of the aircraft. There are two methods for obtaining D values. The first uses an absolute altimeter to measure TA on overwater flights and the pressure altimeter to measure

15-3

PA. The second method uses outside air temperature (OAT) readings to determine equivalent D values if the absolute altimeter fails. For both methods, the D value is expressed in feet as a plus (+) or minus (–) value. To determine the correct D reading using the altimeter method, assign a plus (+) to TA, a minus (–) to PA, and algebraically add the two. Remember the city in Florida (TAMPA) to keep the signs right. Take the first D reading in conjunction with the initial fix for the pressure navigation leg. This is $D_1$. Take the second reading ($D_2$) at the next fix. Always take the readings at the same time relative to the fix, usually about 4 minutes before fix time. The value, $D_2 - D_1$, is an expression of the slope or pressure gradient experienced by the aircraft. Subtracting $D_1$ from $D_2$ determines the change in aircraft TA between readings. When this altitude change is compared with the distance flown, the resulting value becomes an expression of the slope. The value of $D_2 - D_1$ indicates whether the aircraft has been flying upslope (+) or downslope (–).

Take readings carefully, because an erroneous reading of either altimeter will produce an incorrect D reading and a bad LOP. Gently tap the pressure altimeter before reading it to reduce hysteresis error.

Maintain a constant PA to ensure consistent D readings. If you change altitudes, start with a new D at the new altitude, or correct the previous reading by use of a pastagram. The pastagram will allow you to continue accurately, even though you have changed altitude. The pastagram uses average altitude and average temperature change to determine a correction to the D reading taken before the altitude change. *Figure 15-6* shows a pastagram with instructions for its use and a sample problem.

## Effective True Airspeed (ETAS)

To determine a PLOP, you must compute the ETAS from the last D reading. The ETAS is the TAS that the aircraft flew from the last fix to the next fix air position. *[Figure 15-7]* If the aircraft has maintained a constant true heading (TH) between D readings, the ETAS equals the average TAS. But, if the aircraft has altered heading substantially between the D readings, the effective TAS is derived by drawing a straight line from the fix at the first D reading to the final air position. This line is called the effective airpath (EAP). ETAS is computed by measuring the effective air distance (EAD) and dividing it by the elapsed time. In Figure 15-7, an aircraft flew at 400 knots TAS from the 0820 fix to the 1020 air position via a dogleg route. The EAD is 516 nautical miles (NM); consequently, the ETAS is 258 knots.

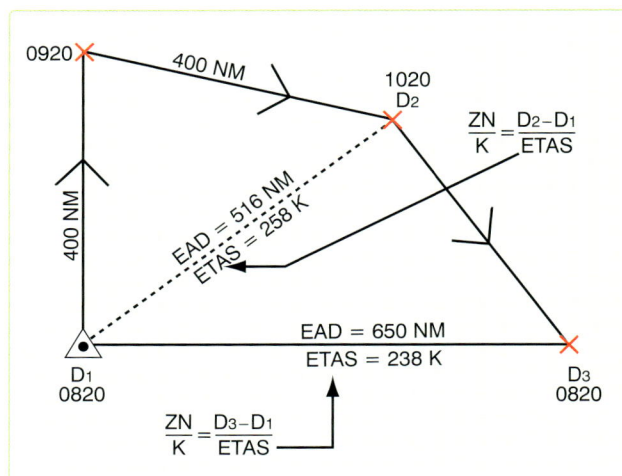

**Figure 15-7.** *Effective true airspeed.*

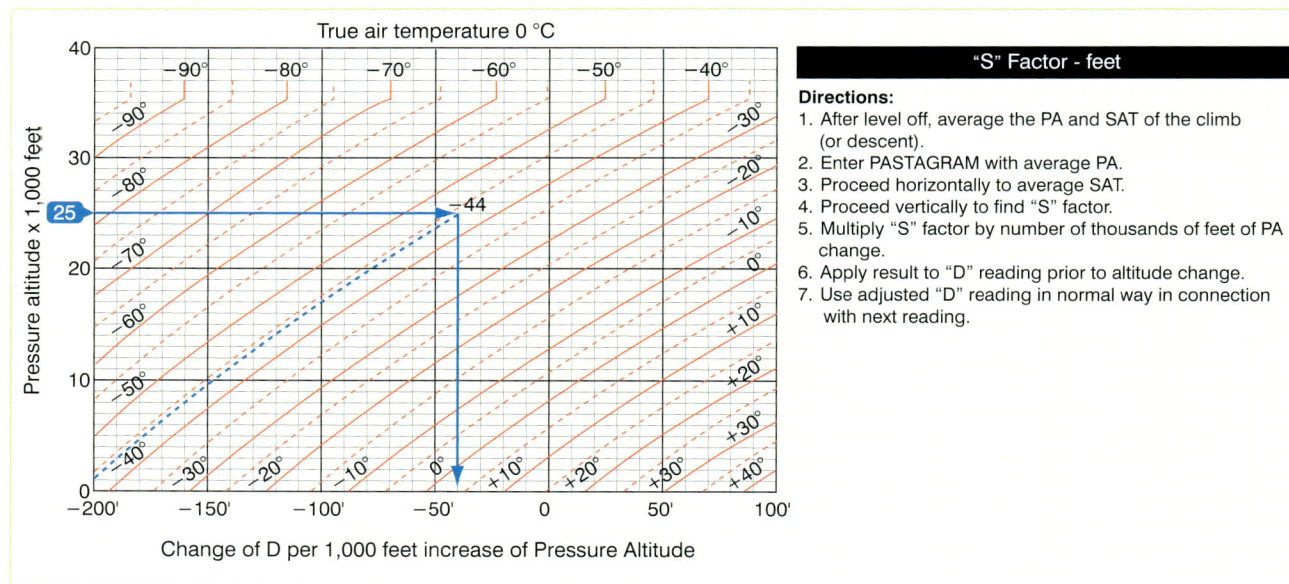

**Figure 15-6.** *Pastagram.*

## K Factor

The constant K takes into account Coriolis and the gravity constant for particular latitudes.

$$K = \frac{21.49}{sin \ \text{midlatitude}}$$

Midlatitude is the average latitude between D1 and D2. It is in tabular form in *Figure 15-8*. In the table, this constant is plotted against latitude since Coriolis force varies with latitude. In using the ZN formula, enter the table with midlatitude and extract the corresponding K factor.

| $\frac{D_2-D_1}{Y} = \frac{BDCA}{I}$ | | $\frac{ZN}{K} = \frac{D_2-D_1}{ETAS}$ | | K FACTORS |
|---|---|---|---|---|
| | | | | LAT — K |
| TA | | | | 20 - 63 |
| PA | | | | 21 - 60 |
| DIFF | | | | 22 - 57.5 |
| TA | | | | 23 - 55 |
| PA | | | | 24 - 53 |
| DIFF | | | | 25 - 51 |
| TA | | | | 26 - 49 |
| PA | | | | 27 - 47.5 |
| DIFF | | | | 28 - 46 |
| TA | | | | 29 - 44 |
| PA | | | | 30 - 43 |
| DIFF | | | | 31 - 42 |
| TA | | | | 32 - 40.5 |
| PA | | | | 33 - 39.5 |
| DIFF | | | | 34 - 38.5 |
| TA | | | | 35 - 37.5 |
| PA | | | | 36 - 36.5 |
| DIFF | | | | 37 - 35.5 |
| TA | | | | 38 - 35 |
| PA | | | | 39 - 34 |
| DIFF | | | | 40 - 33.5 |
| TA | | | | 41 - 33 |
| PA | | | | 42 - 32 |
| DIFF | | | | 43 - 31.5 |
| TA | | | | 44 - 31 |
| PA | | | | 45 - 30.5 |
| DIFF | | | | 47 - 29.5 |
| TA | | | | 49 - 28.5 |
| PA | | | | 51 - 28 |
| DIFF | | | | 53 - 27 |
| TA | | | | 55 - 26 |
| PA | | | | 57 - 25.5 |
| DIFF | | | | 59 - 25 |
| TA | | | | 65 - 24 |
| PA | | | | 70 - 23 |
| DIFF | | | | 75 - 22 |
| TA | | | | 90 - 21.5 |
| PA | | | | |
| DIFF | | | | |

**Figure 15-8.** *Pressure pattern worksheet/K factors table.*

On MB-4 computers, a subscale of latitude appears opposite the values for K factors on the minutes scale. K is computed so that with slope expressed in feet and distance in NM, the geostrophic windspeed is in knots. For training purposes only, the K factors for 20° N or S to 14° N or S are listed in *Figure 15-9*.

| K Factors 20° and below | |
|---|---|
| LAT | K |
| 20° | 63 |
| 19° | 66 |
| 18° | 69.5 |
| 17° | 73.5 |
| 16° | 78 |
| 15° | 83 |
| 14 | 89 |

**Figure 15-9.** *K factors table below 20°.*

## Crosswind Displacement

ZN is the displacement from the straight-line airpath between the readings. Therefore, a PLOP must be drawn parallel to the effective airpath. With all the necessary values available, the ZN formula can be rearranged for convenient solution on the DR computer as follows:

$$\frac{ZN}{K} = \frac{D_2 - D_1}{ETAS}$$

Printed instructions on the face of MB-4 computers specify that to compute crosswind component, set EAD on the minutes scale opposite $D_2 - D_1$ on the miles scale. The crosswind component (V) is not to be confused with ZN. The V is crosswind velocity in knots. V must then be multiplied by the elapsed time between $D_2$ and $D_1$ in order to compute the ZN. Substitute ETAS for EAD on the MB-4 computer, and read the ZN over the K factor (or latitude on the subscale).

## Pressure Line of Position (PLOP)

After you determine ZN, you need to figure out whether to plot it left or right of the EAP. Recall that wind circulation is clockwise around a high and counterclockwise around a low in the Northern Hemisphere; the opposite is true in the Southern Hemisphere. In the Northern Hemisphere, when the value of D increases (a positive $D_2 - D_1$), the aircraft is flying into an area of higher pressure and the drift is left. *[Figure 15-10A]* When the value of D decreases (a negative

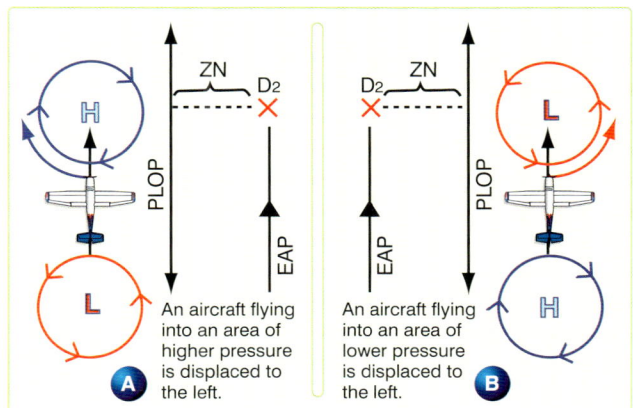

**Figure 15-10.** *Pressure pattern displacement.*

$D_2 - D_1$), the aircraft is flying into an area of lower pressure and the drift is right. *[Figure 15-10B]* Use the memory device PLOP to remember Plot Left On Positive (in the Northern hemisphere) Always plot the PLOP parallel to the EAP, as shown in *Figure 15-11*. Cross the PLOP with another LOP to form a fix, or use it with a DR position to construct an MPP.

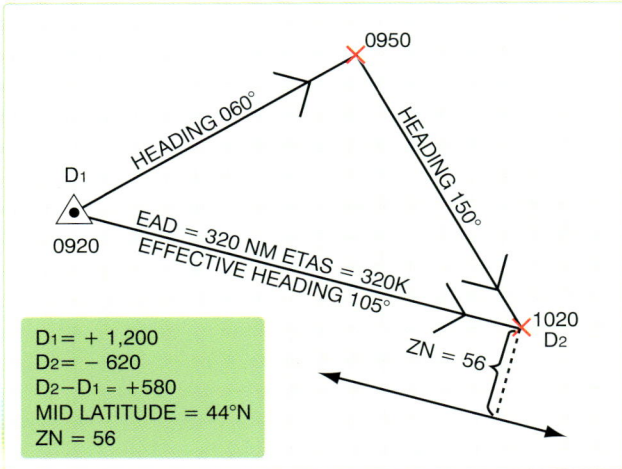

**Figure 15-11.** *Plotting the PLOP.*

## Bellamy Drift

Bellamy drift is a mean drift angle calculated for a past period of time. It is named for Dr. John Bellamy who first demonstrated that drift could be obtained from the use of pressure differential information. Bellamy drift is used in the same way as any other drift reading.

An advantage of Bellamy drift is its independence from external sources. It can serve as a backup if the primary drift source fails, but will not give groundspeed. Bellamy drift is less accurate than Doppler or INS derived sources, but is better than using forecast drift or having none at all.

In *Figure 15-12*, a PLOP has been plotted from the following information:

$D_1$ at a fix at 1000 hrs

$D_2$ at an air position at 1045 hrs

Zn = –20 NM

Constant TH of 90°

Next, construct an MPP on the PLOP. This is done by swinging the arc, with a radius equal to the ground distance traveled, from the fix at the first D reading to intersect the PLOP. The ground distance traveled can be found by multiplying the best known groundspeed (groundspeed by timing, metro groundspeed, etc.) by the time interval between readings. The mean track is shown by the line joining $D_1$ and the MPP. The mean drift is the angle between true heading and the mean track (8°R). Thus, the Bellamy drift is 8° right.

## MB-4 Solution of Bellamy Drift

Compute Bellamy drift on the slide rule side of the DR computer by placing the ZN over the ground distance and reading the Bellamy drift angle opposite 57.3. *[Figures 15-13 and 15-14]* This can be set up in a formula as follows:

$$\frac{\text{Bellamy drift}}{57.3} = \frac{\text{ZN}}{\text{Ground distance (NM)}}$$

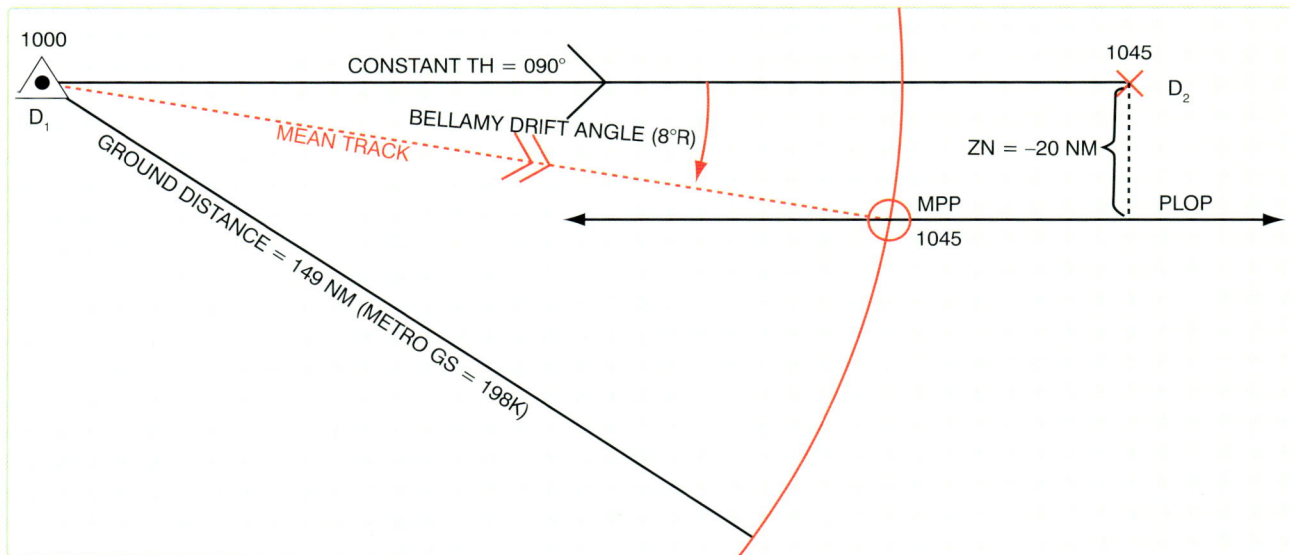

**Figure 15-12.** *Solution of Bellamy drift by using PLOP.*

Given:

ZN = +12.1

Time = 0:30

GS = 190 knots

Find:

Ground distance = 95 NM

Drift = 7° left

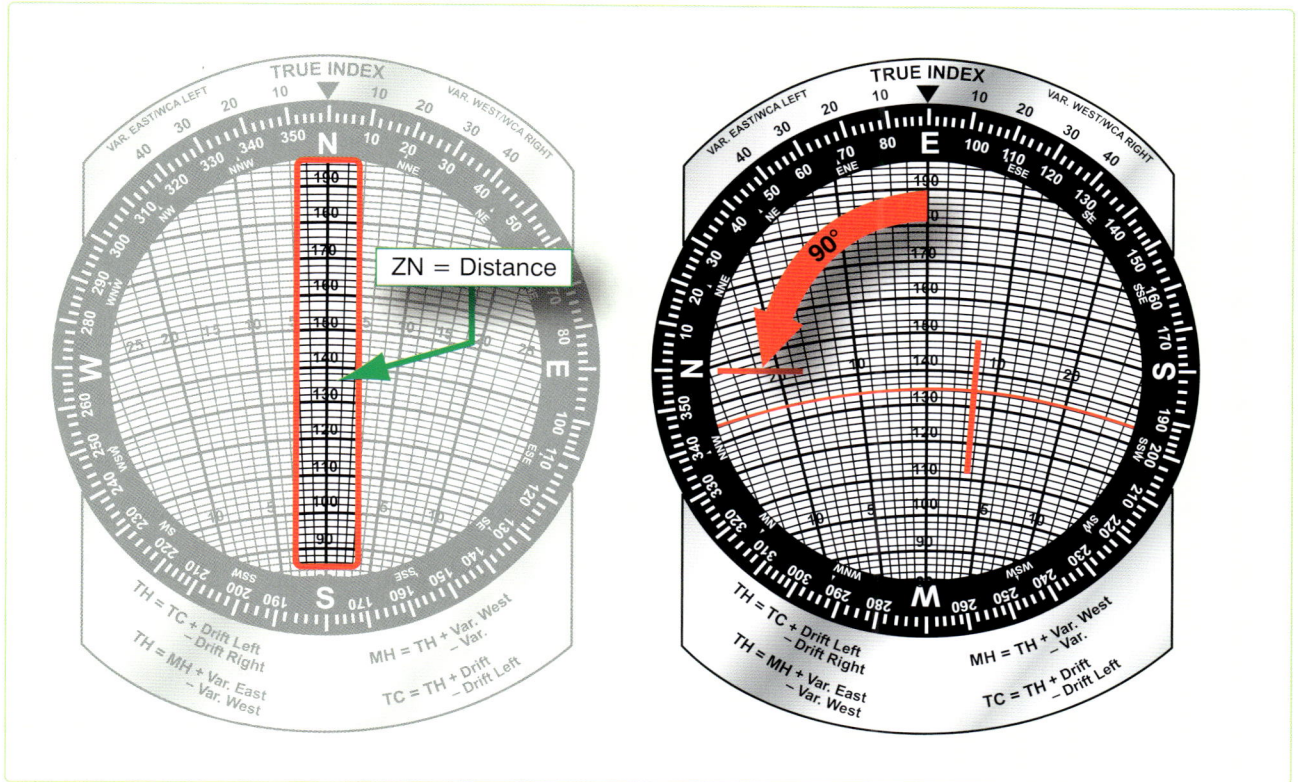

**Figure 15-13.** *Computer solution of Bellamy drift.*

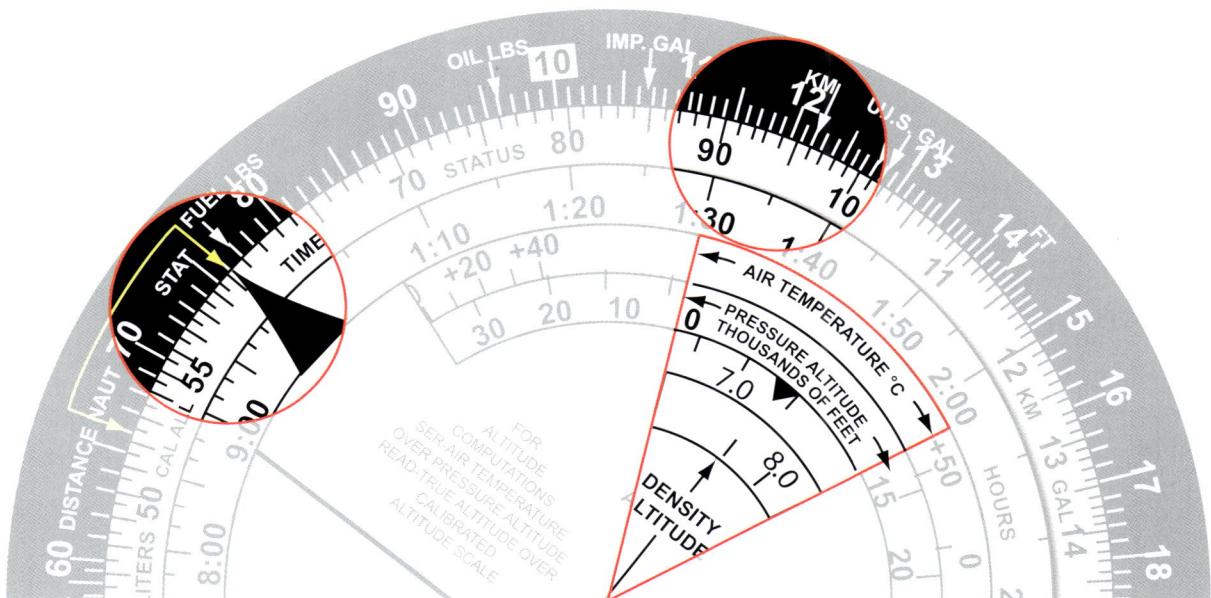

**Figure 15-14.** *Mathematical solution of Bellamy drift.*

## Limitations of Pressure Differential Techniques

Pressure navigation is limited by a few meteorological considerations. The basic accuracy of the LOP in average conditions is about 5 to 10 miles. It will rapidly become worse under the following conditions: tightly circulating pressure systems of highs and lows, flying through a front, or carelessness in reading or computing the information. Bellamy drift has another limitation. To determine drift you must stay on one heading long enough to take two readings about 20 minutes apart.

ZN is a displacement in NM perpendicular to the EAP. Compute ZN on the MB-4 using the equation:

$$ZN = \frac{K\,(D_2 - D_1)}{ETAS}$$

Determine ETAS by using the EAD and time. Measure EAD along a straight line between the two points in question. In the Northern and Southern Hemispheres, the sign of the ZN is the sign of the drift correction. Use airplot in conjunction with a fix position to plot the PLOP, and plot it parallel to the EAP. If the absolute altimeter fails, use pressure by temperature as a backup. With this method, use temperature and pressure altitude to find equivalent D readings. If you change altitudes, restart pressure at the new altitude, or correct the last D reading prior to the altitude change with a pastagram. Another expression of the PLOP is Bellamy drift, used as a backup source of drift angle. *Figure 15-15* shows a fix determined by a PLOP and a celestial LOP.

## Chapter Summary

This chapter discussed pressure differential techniques and how they affect pressure pattern navigation. Topics, such as constant pressure surface, geostophic wind, pressure computations and plotting, D readings, effective true airspeed, K factor, pressure line of position, Bellamy drift, and the MB-4 solution of Bellamy drift are all discussed. Also explained are the limitations of pressure differential techniques and how they are affected my meteorological conditions.

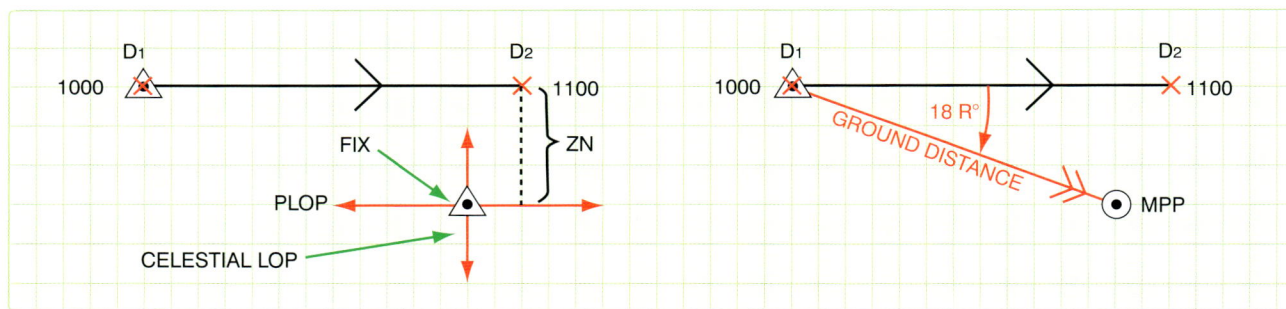

**Figure 15-15.** *Fix using PLOP and celestial line of position.*

## Chapter 16

# Navigation Systems

## Introduction

Navigation systems are computer systems that determine position and calculate navigation information. Simple navigation systems rely on an initial position and basic instrument data, such as doppler groundspeed and drift, magnetic heading, and variation, to compute and constantly update the dead reckoning (DR) position. These systems require the operator to spend time to maintain system accuracy. Complex systems integrate information from a variety of sources using complex statistical algorithms to produce constantly updated, highly accurate position and navigation information. Some components of a navigation system may be stand-alone systems. This chapter examines navigation systems in general and the most common systems in detail—Inertial Navigation Systems (INS) and Global Positioning Systems (GPS).

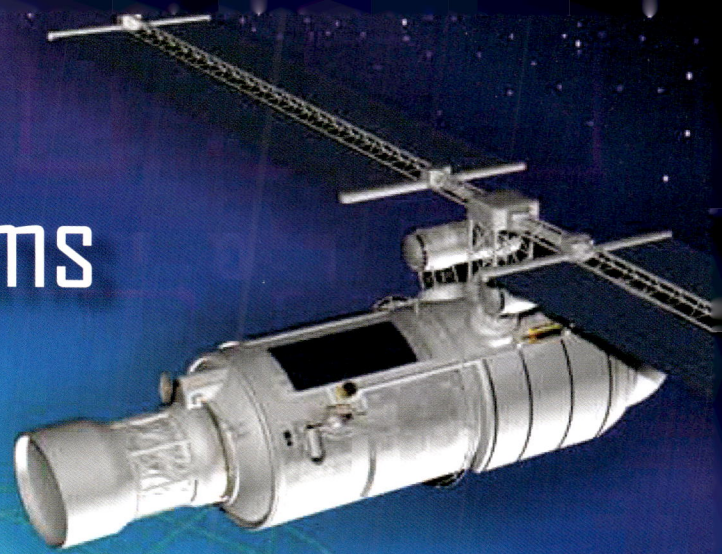

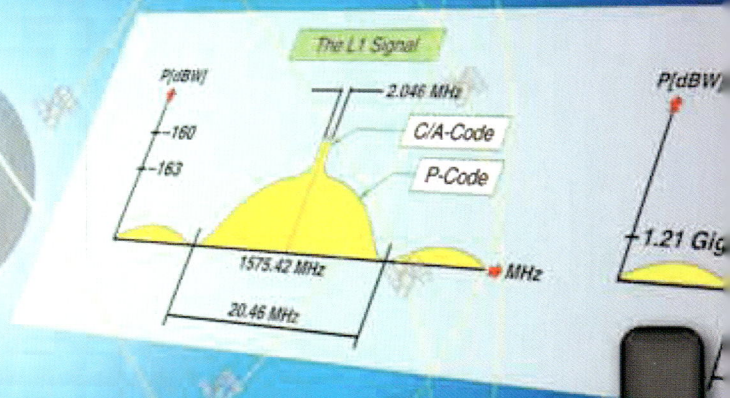

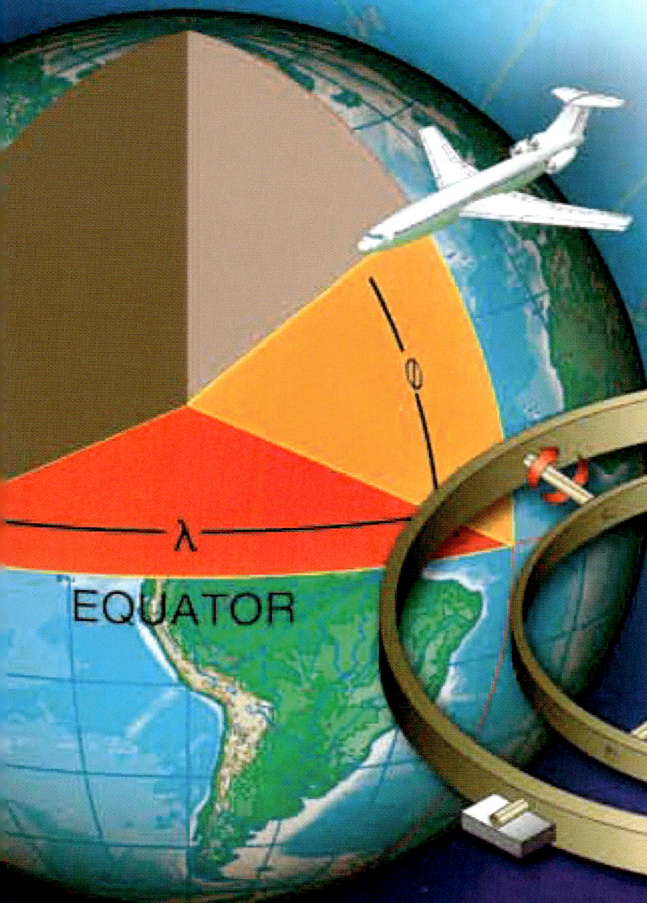

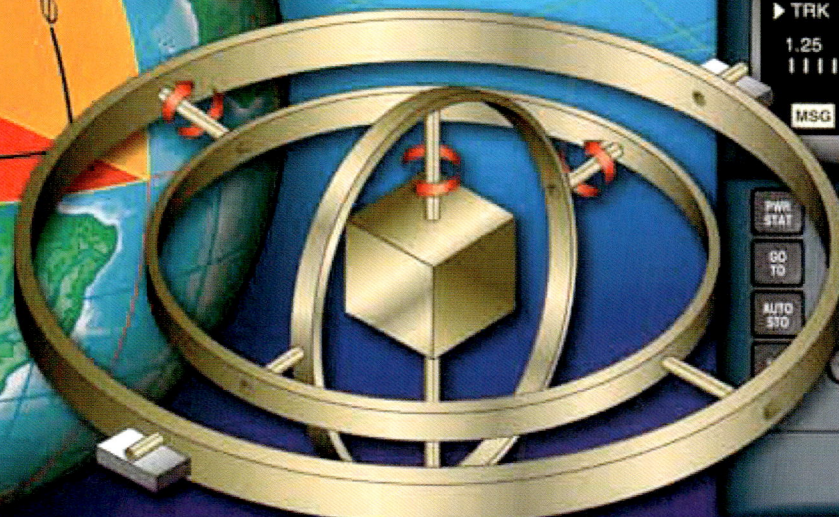

The L1 Signal

EQUATOR

GPS 95

GO TO GARMIN

BRG 045    DIS 2.53

TRK 045    GS 140

1.25    1.25

MSG  MAP  CFG  POSN  NAV

# Navigation Systems

In the same way an autopilot frees a pilot from the manual operations of flying, a navigation system relieves you of many manual operations required to direct the aircraft. When sensors are tied into a navigation system, the system automatically uses their data to compute present position for the navigator. During a flight from the United States to a foreign country, the aircraft may pass over areas of land, water, and icecaps. You may have to deal with conditions of overcast, undercast, day, night, altitude changes, turn points, and air traffic requirements. To handle these conditions at high speeds more effectively, the navigator uses a navigation system.

## Types of Systems

Navigation systems can be classified according to many criteria. Systems can be classified by capability, such as visual flight rules (VFR)-only or all-weather. They can be classified as either self-contained or externally-referenced. Each system has advantages and disadvantages, but this discussion is confined to self-contained and externally referenced systems.

### Self-Contained Navigation Systems

Self-contained systems (radar, celestial, INS, etc.) are complete in that they do not depend upon externally transmitted data.

### Externally-Referenced Navigation Systems

Externally-referenced aids (GPS, NAVAIDs, etc.) include all aids that depend upon transmission of energy or information from an external source to the aircraft. While externally referenced aids have enormous installation and operating costs to the system administrator, they have much lower equipment and maintenance costs to the user.

## The Ideal System

Every navigation system has certain advantages and disadvantages. A particular navigation system is selected for use in an aircraft when its advantages outweigh its disadvantages. In some cases, several components are included in a system to provide adequate, redundant information for all possible flight situations. The ultimate navigation system should have the following characteristics:

- Groundplot DR information—the system must indicate the position and velocity relative to the ground.

- Global coverage—capable of positioning and steering the aircraft accurately and reliably any place in the world.

- Self-contained—must not rely on ground or space transmissions of any kind.

- Flexible—works well despite unplanned deviations. The system must work well at all altitudes and speeds.

## Components

The navigational system consists of three parts:

1. The computer or central processing unit (CPU)

2. Data-gathering sensors, such as astrotrackers, GPS, ground-mapping radar, or NAVAIDS

3. An operator input/output (I/O) interface

The CPU takes in all available data and converts it into usable navigation information. Control panels or computer keyboards allow the operator to control and make inputs to the computer. Data is displayed for the operator on display panels, radar screens, or computer screens. Additional hardware components could include terrain following radar or television cameras.

### Computer Unit

Most navigation systems are hybrids of the two basic computer types: analog and digital.

#### Analog

Analog computers are more specific in design and function than digital computers. While analog computers process vast amounts of similar data, they are not very flexible and cannot be used for multiple purposes. Radar scan converters efficiently process collected radar signals into video images. Video processors collect and process images into video displays. Other examples of analog computers are terrain-avoidance computers and terrain-following computers.

#### Digital

Digital computers are lighter and more compact than analog computers. Hand-held calculators and laptop computers are two examples of the miniaturization possible with digital computers. You can put a great deal of computing power and capability into a small box; the biggest limitation is increased cost. An analog radar scan converter is very efficient at processing radar data, but it cannot be used for other applications. On the other hand, digital computers can be loaded with navigation software, aerial delivery software, and diagnostic programs. These computers can mathematically manipulate data in any way imaginable, because they deal strictly with digital information. The output from digital computers may need to be converted into an analog format for most efficient use by the navigator; however, the digital computer cannot do that. It can display the digital data in an approximation of analog data. While a digital computer can perform any mathematical function, it must first be programmed for that function. Inflight reprogramming is not generally possible.

### Sensors

Many types of sensors are used for inputs to navigation systems.

### Astrotracker

The use of astrotrackers has decreased; however, they are still excellent sources of position information. They automatically track celestial bodies and compute position information using celestial techniques. They are passive but require clear skies.

### Doppler

The Doppler radar measures groundspeed and drift. These two data inputs can be put to several uses in the computer system. Doppler groundspeed is used to determine distance to update the aircraft position. Drift can be used to compute winds and aircraft track. Doppler outputs can be used in platform leveling and verifying inertial groundspeed in an INS. Doppler radar is an essential part of many navigation computer systems.

### Heading System

The gyro-stabilized magnetic heading source is corrected to true heading with the local magnetic variation. This can be applied manually or automatically from a database in the computer. Magnetic or true course can be calculated by applying doppler or inertial drift.

### NAVAIDS

NAVAIDS are easily added to a computer system. Very high frequency (VHF) omnidirectional range (VOR) or tactical air navigation system (TACAN) bearings and distance measuring equipment (DME) provide the same information as a radar fix. The computer needs the location and frequency of the transmitter, which can be programmed into the computer before the flight begins. Some corrections must be applied to bearing data. The computer must correct for magnetic variation and slant range from the station to the aircraft.

### Pressure Altimeter

Pressure altimeter data is an input to the true airspeed computations. Additionally, it can be used with temperature data to compute true altitude.

### Radar

When a ground mapping radar is incorporated into the navigation system, present position can be corrected based on the measurements to surveyed radar returns. The operator identifies radar returns on his radar scope and measures the range and bearing to the return. The operator determines the aircraft position relative to the return and updates the aircraft position. Automatic systems allows the operator to pre-load the coordinates of radar returns in a database, place a movable electronic cursor (or crosshairs) on the return, and push a button to update the system. The computer determines the distance and bearing from the aircraft to the set coordinates. The computer then generates the cursor on the radarscope at the calculated range and bearing. If there is any error in the navigation system position, the cursor will not fall on the radar return. The operator adjusts the cursor or crosshairs onto the radar return. The operator pushes a button to automatically update the system.

### Temperature Sensors

The air data computer uses the information collected by temperature sensors. Temperature gradients can be used with pressure altimeter data to compute true altitude.

### True Airspeed

True airspeed can be calculated from indicated airspeed, temperature, and pressure. True airspeed and winds can be used as a backup for cross-checking groundspeed.

### Independent Systems

INS and GPS can also act as sensors for a navigation system. They are discussed in greater detail later in this chapter.

## Determining Position

The ever-present problem facing the navigator is determining aircraft position. With a navigation system, this problem is solved because the computer converts input data into a constantly updated present position for the aircraft. Advanced systems provide altitude, attitude, heading, and velocity information.

The mathematics of navigation over the surface of a sphere has been known for several centuries. Starting from an initial position, the computer determines the distance and direction traveled since starting navigation. Aircraft direction, or track, may be supplied by INS, GPS, or the heading reference system in combination with doppler drift. Groundspeed may come from INS, GPS, doppler groundspeed, or may be determined from any NAVAID capable of range and bearing fixes. The computer multiplies speed against time interval to determine distance traveled. Distance is projected along the aircraft track to obtain the new position. Track and speed are sampled and present position is updated many times per second. Waypoint navigation is a simple addition to the navigation computer. A database of coordinates can be added to the system to determine distance to go and estimated time of arrival (ETA). If the aircraft changes speed, the ETA is automatically updated using the new groundspeed.

## Decision Algorithm

Simple navigation systems determine position as described above. The operator updates the position for errors that will eventually occur. More complex systems have additional problems. When a system has a variety of sources that provides redundant information, how does the computer decide which source to use? What if the sensors are subject to errors? What if the operator inputs an inaccurate update to the system? How can we get a computer to make simple decisions once left to the navigator? Can we program a computer to analyze and correct for the predictable and unpredictable errors in sensor data? Bias in the accuracy and variability of data are two types of error that navigation systems actually experience and can be solved with the use of statistical software called decision algorithms.

To compensate for these predictable and unpredictable errors in sensor data, we can include statistical measuring software that weigh the accuracy of each data source and the accuracy of the data itself. These programs determine the most likely value for track and velocity in order to compute the most likely present position.

One type of program used to determine the most likely sensor values is called a Kalman filter. Kalman filters are used extensively in computer controlled communications, electronics, and equipment. When used as part of a navigation system, a Kalman filter computes the most likely position of the aircraft and updates the weighing factors with each new position update. The Kalman filter compares the actual sensor data used prior to the update with the data from the update. By comparing the first position with the second position, actual distance and heading can be determined. It then determines the amount of error in the original data and estimates a correction to the data for the next time period. The Kalman filter is an iterative program requiring several updates prior to achieving completely reliable data. If used, the Kalman filter will also be used to evaluate the reliability of operator inputs and weigh how much of each position update to accept. Kalman filtering provides increased reliability in navigation systems so an operator can trust that the information used is valid. Kalman filters protect the operator from inaccurate sensor data and even operator error.

## Inertial Navigation System (INS)

Inertial navigation is accepted as an ideal navigation system because it meets all the criteria of an ideal system. INS provides worldwide ground plot information regardless of flightpath and aircraft performance. An INS can measure groundspeed independently of wind and independently of the operating environment. INS is completely independent of ground transmissions and passive in operation. It is self-contained and portable; most units weigh less than 100 pounds. Some ring laser gyro systems weigh as little as 20 pounds. The need for a system with these properties has spurred development to the point where INS is superior to almost every other navigation system. INS provides accurate velocity information instantaneously for all maneuvers, as well as an accurate attitude and heading reference. INS accuracy decreases as the time between position updates increases. INS maintains its accuracy for short flights without position updates; however, longer flights may require periodic inflight updates.

## Types of Inertial Systems

In the last several years, inertial technology has taken several leaps forward. Early inertials were bulky devices weighing several hundred pounds, whose installation had to be precise and whose operation had to be planned in great detail. Today, there are compact systems that fit in a briefcase and can be bolted to an aircraft in any space available. While some inertial systems still have mechanical gyroscopes, pendulous linear accelerometers, and space stable platforms, most have evolved to keep pace with the advances in technology. Acoustic gyros, ring laser gyros, and electronically suspended gyros have replaced the gimbaled gyroscope. Laser and acoustic accelerometers are replacing the pendulous linear accelerometer. Highly accurate computers and precision sensors have led to modifications of the space stable platform so the INS housing does not need to be accurately aligned with the aircraft. Eventually software will perform all the functions of the space stable platform. Despite all these modern advances, we can better learn about and understand inertial systems by studying the original systems.

The basic principle behind inertial navigation is straightforward. Starting from a known point, you calculate your present position (a continuously running DR) from the direction and speed traveled since starting navigation. The difference between other navigation systems and INS is how it determines direction, distances, and velocities. Accelerations are detected by the three linear accelerometers. These accelerations are integrated over time to determine changes in velocity. Velocity is integrated a second time to determine distance traveled. Changes in vector direction are detected with angular accelerometers. As sensors detect changes in gyroscope orientation, correction signals are generated to reorient the stable platform to the original position and determine new vector direction. INS requires no other inputs. It avoids all environmental inputs, such as indicated or true airspeed, magnetic heading, drift, and winds that are necessary for dead reckoning.

## Components

The five basic components of an INS are:

1. Three linear accelerometers arranged orthogonally to supply X, Y, and Z axis components of acceleration.

2. Gyroscopes to measure and use changes in aircraft vector to maintain and orient the stable platform.

3. A stable platform oriented to keep the X and Y axis linear accelerometers oriented north-south and east-west to provide azimuth orientation and to keep the Z axis aligned with the local gravity vector. The stable platform is necessary to prevent either the X or Y axis accelerometer from picking up the force of gravity and interpreting it as an acceleration on the aircraft.

4. Integrators to convert raw acceleration data into velocity and distance data.

5. A computer to continuously calculate position information.

## Linear Accelerometers

Acceleration-measuring devices are the heart of all inertial systems. It is important that they function reliably for all maneuvers within the capability of the aircraft and that all possible sources of error are minimized. Very slight accelerations and changes in heading in all directions must be detected. Changes in temperature and pressure must not affect INS operation. To do this, INS requires two types of accelerometers: linear and angular.

The simplest type of linear accelerometer consists of a pendulous mass that is free to rotate about a pivot axis in the instrument. There is an electrical pickoff that converts the rotation of the pendulous mass about its pivot axis into an output signal. This output signal is used to torque the pendulum to hold it in the original position and, since the signal is proportional to the measured acceleration, it is sent to the navigation computer as an acceleration output signal. *[Figure 16-1]*

To obtain acceleration in all directions, three accelerometers are mounted mutually perpendicular in a fixed orientation. To convert acceleration into useful information, the acceleration signals must be integrated to produce velocity and then the

velocity information is integrated to get the distance traveled. One of the forces measured by the linear accelerometers is gravity. This acceleration may be incorrectly interpreted as an acceleration of the aircraft if the stabilized platform is tilted relative to the local gravity vector. The accelerometers cannot distinguish between actual acceleration and the force of gravity. This means that the linear accelerometers on the stable platform must be kept level relative to the earth's surface (perpendicular to the local gravity vector). The gyroscopes keep the stabilized platform and the accelerometers level and oriented in a north-south and east-west direction.

## Gyroscopes

Gyroscopes are used in inertial systems to measure angular acceleration and changes in orientation and heading. While the types of gyros are briefly discussed here, the function of the gyro is discussed in great detail in the next section on the stable platform. The original gimbaled gyroscope has been replaced by newer designs.

### Electronically Suspended Gyros

These gimbal-less gyros consist of a ball that is suspended in a magnetic field and spun electronically. Evacuating the air in the gyro cavity further reduces friction. The result is a near frictionless gyro with precession rates measured in years. Optical sensors measure the ball's orientation from symbols etched on the surface of the ball.

### Ring Laser Gyro (RLG)

Accuracy and dependability of first generation systems have greatly improved with the introduction of the ring laser gyro (RLG) INS. The RLG INS replaces the three pendulous mass accelerometers with three RLG accelerometers. Technically, the RLG is not a gyroscope since it has no moving parts, but it gives the same information as a gyro. A RLG is made from a single block of glass with three holes drilled through the glass to form a triangular path. Two of the openings are plugged with mirrors and the triangular tube is filled with helium neon or other lazing gas. When the gas is charged, the lazing gas

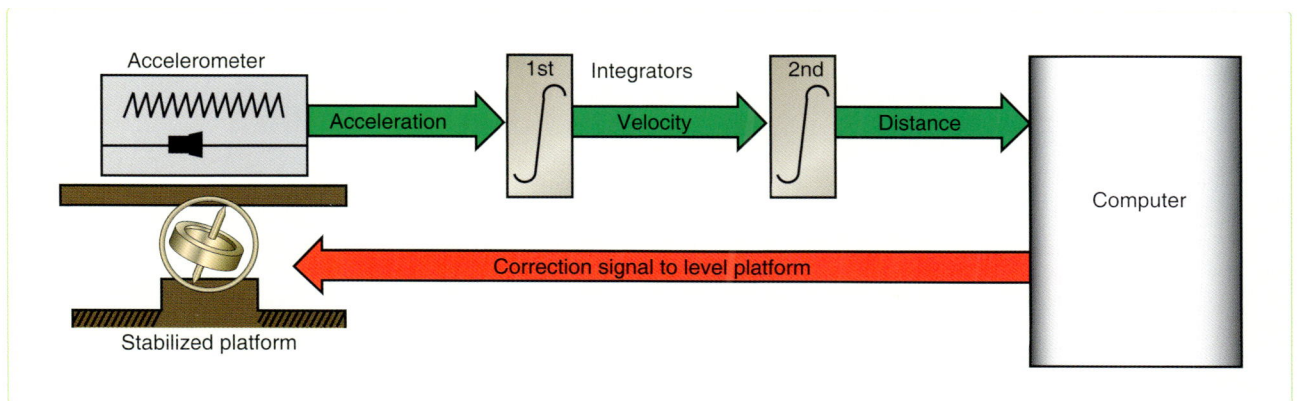

**Figure 16-1.** *Basic inertial system.*

produces two counter-rotating laser beams that are reflected around the path by the mirrors. Both laser beams emerge through the third hole in the glass and are superimposed upon each other to produce an interference pattern. As the RLG moves, one beam has a longer path to travel; the other a shorter path. This causes changes in the interference pattern that are detected by photocells. The angular rate and direction of motion are computed as accelerations.

*Acoustic Gyros*

Another recent development is the inertial sensor based on vibrating quartz crystal technology. Like the RLG, these are not true gyros. Acoustic gyros are manufactured from a single piece of microminiature quartz rate sensor. Angular accelerations affect the patterns produced by a vibrating tuning fork and result in torque on the fork proportional to the angular acceleration. These gyros appeared in inertial units in the late 1990s.

## Stable Platform

Autopilots and attitude indicators use gyrostabilized platforms. Inertial navigation simply requires a stable platform with higher specifications of accuracy. A gyro-stabilized platform on which accelerometers are mounted is called a stable element. It is isolated from the aircraft's angular motions by three concentric gimbals. The stable element is the mounting for the linear accelerometers, gyroscopes, and other supporting equipment. The supporting equipment includes torque motors, servo motors, pickoffs, amplifiers, and wiring. The effectiveness of the stable platform is determined by all parts of the platform, not just the accelerometers and gyros.

The linear accelerometers measure acceleration in all directions and the gyros control the orientation of the platform. The platform must contain at least two gyros with two degrees of freedom. A simple diagram of a two-degrees-

of-freedom gyro mounted on a single-axis platform is shown in *Figure 16-2*. If one-degree-of-freedom rate gyros are used, three units are needed, each gyro having its own independent feedback and control loop. The original gimbaled gyro was not very accurate by today's standards, producing sizeable amounts of gyroscopic precession. Recent developments such as the air-bearing gyro and the electronically suspended gyro have only 1/10,000,000 the friction of a standard gyro and negligible real precession. Today's gyros have real precession rates of less than 360° in 30 years.

The desired property of a gyro that we want to capitalize on is its stability in space. A spinning gyro tends to remain in its original position. A free spinning gyro aligned in space tends to remain pointed in the same direction unless a force acts on it. On a stable platform, any displacement of the stable element from its frame of reference is sensed by the electrical pickoffs in the gyroscopes. These electrical signals are amplified and used to drive the platform gimbals to realign the stable element in the original position. More advanced INS have a four-gimbal platform in a three-axis configuration. *[Figure 16-3]*

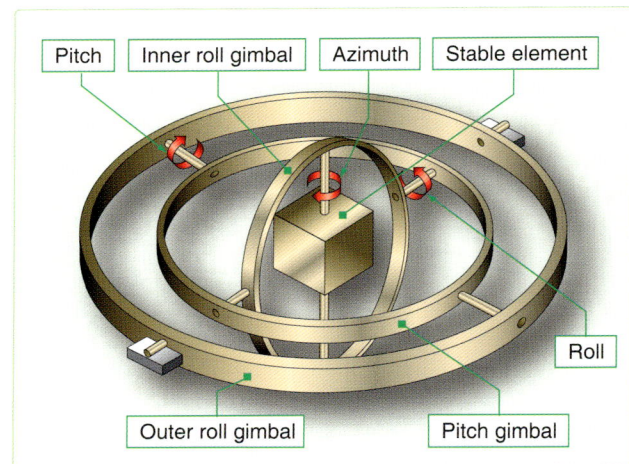

Figure 16-3. *Gimbal platform.*

The four-gimbal mounting provides a full 360° freedom of rotation about the stable element, thus allowing it to remain level with respect to local gravity and to remain oriented to true north. This is north as established by the gyros and accelerometers, regardless of the inflight attitude of the aircraft. The azimuth, pitch, and outer roll gimbals have a 360° freedom of rotation about their own individual axis. The fourth, or inner roll, gimbal has stops limiting its rotation about its axis. This gimbal is provided to prevent gimbal lock, which is a condition that causes the stable element to tumble. Gimbal lock can occur during flight maneuvers, such as a loop, when two of the gimbal axes become aligned parallel

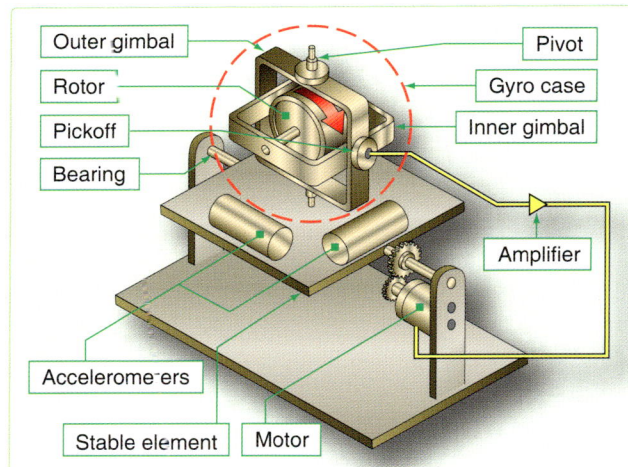

Figure 16-2. *Stable platform.*

to each other, causing the stable element to lose one of its degrees of freedom.

## Measuring Horizontal Acceleration

The key to a successful inertial system is absolute accuracy in measuring horizontal accelerations. A slight tilt of the stable platform introduces a component of earth's gravity as acceleration on the aircraft and results in incorrect distances and velocities. [Figure 16-4] Keeping the accelerometers level is the job of the feedback circuit. The computer

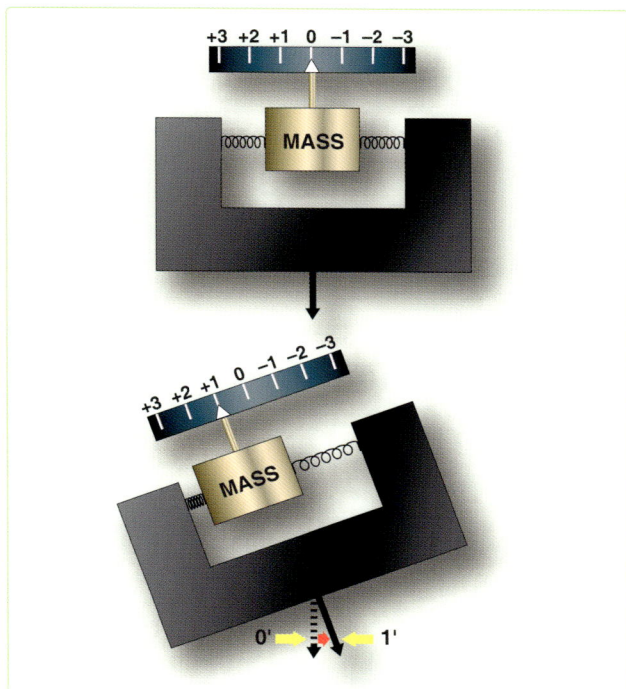

**Figure 16-4.** *Effect of accelerometer tilt.*

calculates distance traveled along the surface of the earth and moves the accelerometer through an equivalent arc. Several factors affect aligning the accelerometer using this method. The earth is not a sphere, but an oblate spheroid or geoid. Because the earth is not a smooth surface, there are local deviations in the direction of gravity. The feedback circuit operates on the premise that the arc traversed is proportional to distance traveled. Actually, the arc varies considerably because of the earth's shape; the variation is greatest at the poles. The computer must solve for this irregularity in converting distance to arc.

The accelerometers are kept level relative to astronomical rather than geocentric latitude. Using the astronomical latitude, the accelerometers are kept aligned with the local horizon and also with the earth's gravitational field. Feedback from the computer keeps the accelerometers level, correcting for two types of apparent precession. If the inertial unit were stationary at the equator, it would be necessary to rotate the accelerometers to maintain them level because of the earth's angular rotation of 15° per hour. Also, movement of the stabilized platform would require corrections to keep the accelerometers level. When using a local horizontal system, in which the accelerometers are maintained directly on the gyro platform, the gyro platform must be torqued by a signal from the computer to keep the platform horizontal. Apparent precession is illustrated in *Figure 16-5.*

A slight error in maintaining the horizontal would induce a major error in distance computation. If an accelerometer picked up an error signal of 1/100 of the G-force, the error on a 1-hour flight would be 208,000 feet (over 34 nautical miles (NM)). In 1923, Dr. Maxmillian Schuler showed a pendulum

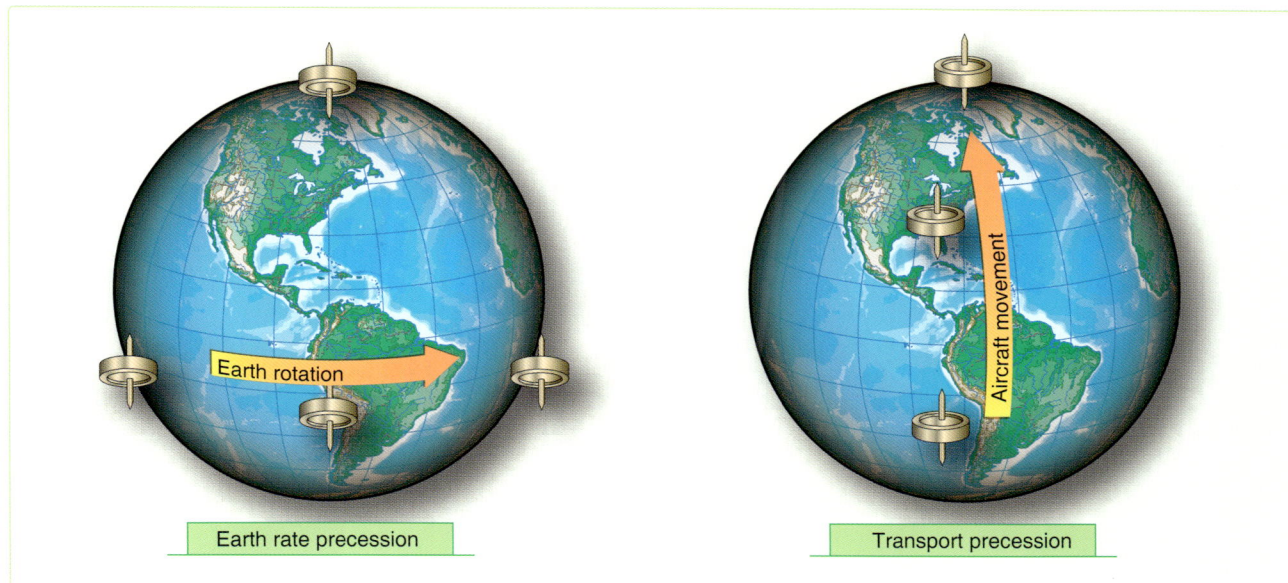

**Figure 16-5.** *Apparent precession.*

with a period of 84.4 minutes could solve the problem of eliminating inadvertent acceleration errors.

If a pendulum has a period of 84.4 minutes, it indicates the vertical, regardless of acceleration of the vehicle. He demonstrated that a device with a period of 84.4 minutes would remain vertical to the horizon despite any acceleration on the device. The fundamental principle of the 84.4-minute theorem is that if a pendulum had an arm equal in length to the radius of the earth, gravity would have no effect on the bob. This is because the center of the bob would be at the center of gravity (CG) of the earth, and the pendulum arm would always remain vertical for all motions of the pivot point. While it would be impossible to construct this pendulum, devices with an 84.4-minute cycle can be constructed using gyroscopes. The Schuler pendulum phenomenon prevents the accumulation of errors that would be caused by platform tilt and treating gravity as an acceleration. It does not compensate for errors in azimuth resulting from the precession of the steering gyro. The amplitude of the Schuler cycle depends upon the overall accuracy of the system. *Figure 16-6* shows the Schuler-tuned system.

A spinning, untorqued gyro is space-oriented and appears to move as the earth rotates underneath it. This is undesirable for older systems because the accelerometers are not kept perpendicular to the local vertical. To earth-orient the gyro, we control apparent precession. If a force is applied to the axis of a spinning gyro wheel that is free to move in a gimballing structure, the wheel moves in a direction at right angles to the applied force. This is called torquing a gyro and can be considered as mechanized or induced precession. A continuous torque, applied to the appropriate axis by electromagnetic elements called torques, reorients the gyro wheel to maintain the stable element level with respect to the earth and keeps it pointed north. An analog or digital computer determines the torque to be applied to the gyros through a loop that is tuned using the Schuler pendulum principle. The necessary correction for earth rate depends on the position of the aircraft; the correction to be applied about the vertical axis depends on the velocity of the aircraft.

It is important that the stable element be leveled accurately with respect to the local vertical and aligned in azimuth with respect to true north. Precise leveling of the stable element is accomplished prior to flight by the accelerometers that measure acceleration in the horizontal plane. The stable element is moved until the output of the X and Y accelerometers is zero, indicating that they are not measuring any component of gravity and that the platform is level. Azimuth alignment to true north is accomplished before flight by starting with the magnetic compass output and applying variation to roughly come up with true north reference. From this point, gyrocompassing is performed. This process makes use of the ability of the gyros to sense the rotation of the earth. If the stable element is misaligned in azimuth, the east gyro sees the wrong earth rate and causes a precession about the east axis. This precession causes the north accelerometer to tilt. The output of this accelerometer is then used to torque the azimuth and east gyro to ensure a true north alignment and a level condition.

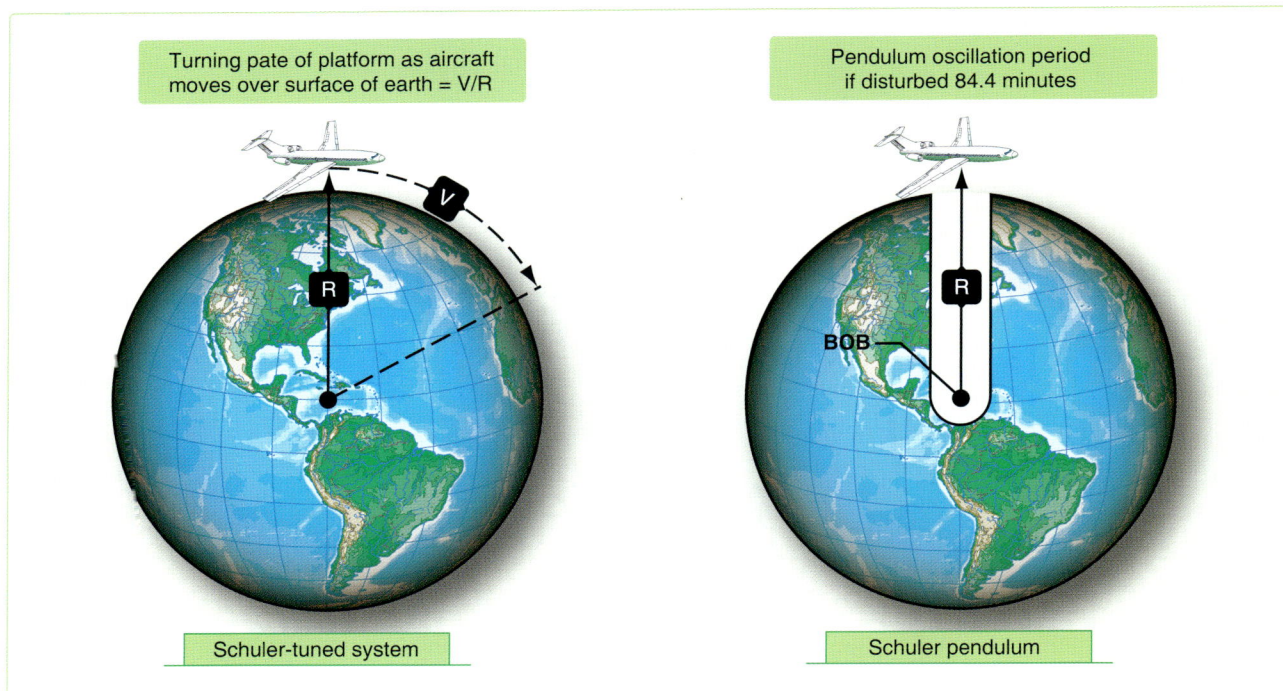

**Figure 16-6.** *Schuler pendulum phenomenon.*

In the more modern gyroscopes, the gyro cannot be physically torqued, because the gyro is either not moving or the gyro is electronically suspended. In these systems, the stable platform is leveled mathematically using gyro data. The precise orientation of the X and Y accelerometers on the stable platform is less critical since the computer can mathematically correct for any orientation. The next generation of INS may work without a stable platform, with orientation and stability maintained mathematically from accelerometer inputs.

## Integrator

Simply stated, the processing of acceleration is done with an integrator. An integrator integrates the input to produce an output: it multiplies the input signal by the time it was present. Accurate navigation demands extremely accurate integration of both acceleration and velocity. One of the most used analog integrators is the DC amplifier, which uses a charging current stabilized to a specific value proportional to an input voltage. Another analog integrator is the AC tachometer-generator that uses an input to turn a motor, which physically turns the tachometer-generator, producing an output voltage. The rotation of the motor is proportional to an integral of acceleration.

## Computer

The computer changes the integrator's outputs into useful navigation information. To do this, one accelerometer is mounted aligned to north and another is mounted 90° to the first, to sense east-west accelerations. Any movement of this system indicates distance traveled east-west and north-south. The INS maintains a local vertical reference and measures distance traveled over a reference spheroid perpendicular to the local vertical. On this spheroid, the latitude and longitude of the present position are continuously measured by the integration of velocity. In *Figure 16-7*, Θ represents latitude

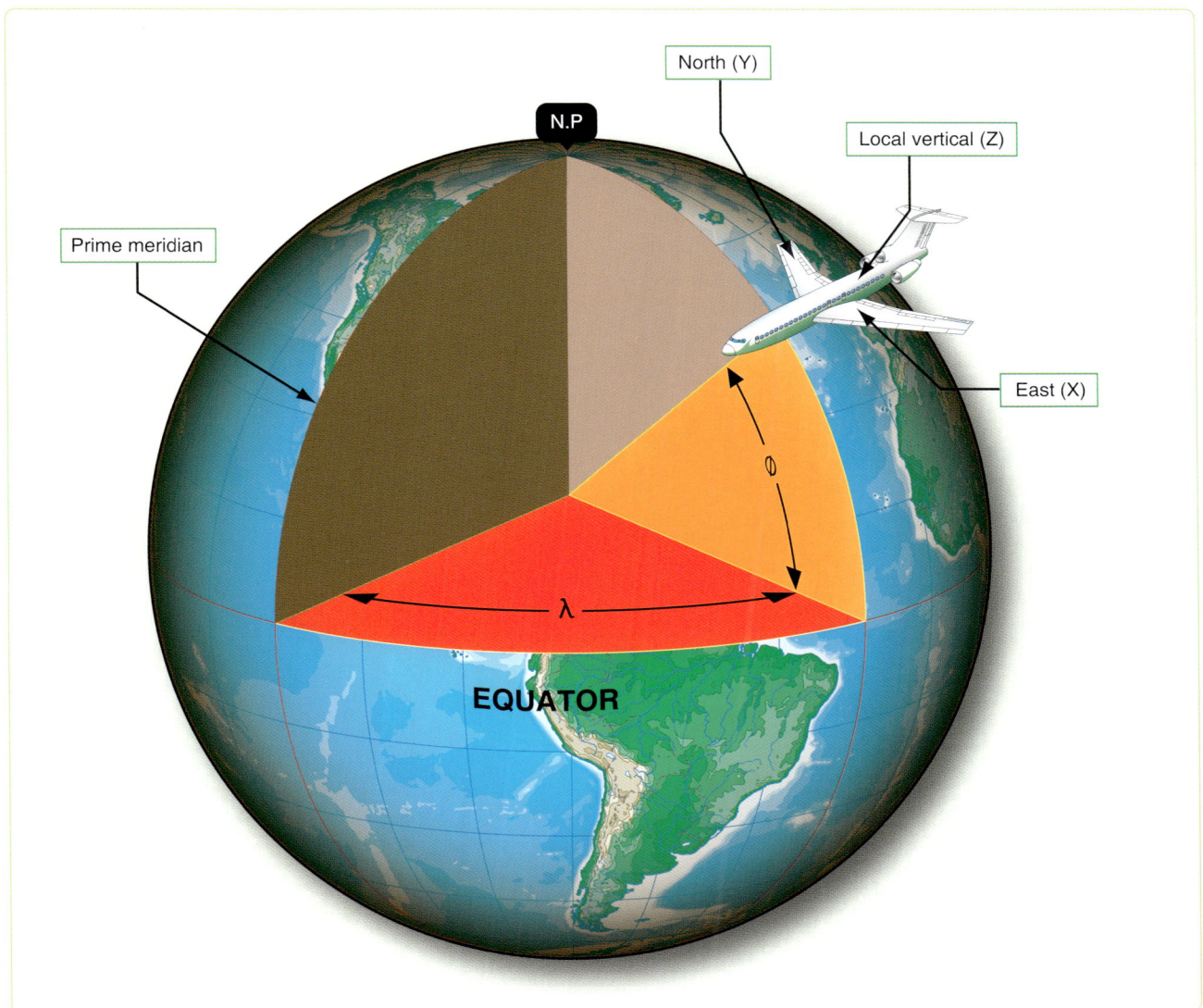

**Figure 16-7.** *Geographic references.*

and λ represents longitude. The axes are designated X, Y, and Z, corresponding to east, north, and local vertical. This defines their positive directions. References to velocities, attitude angles, and rotation rates are about the X, Y, and Z axes. The local vertical (Z) is established by platform leveling. This is the most fundamental reference direction. To complete platform alignment, the INS uses gyrocompassing to establish true north (Y). Gyrocompassing establishes platform alignment to the earth's axis of revolution or North Pole. The INS is capable of doing this to an accuracy of 10 minutes of arc or less. After alignment, the platform remains oriented to true north and the local vertical, regardless of the maneuvers of the aircraft.

Groundspeed components of velocity in track (V) are measured by the system along the X and Y axes. *[Figure 16-8]* These components, $V_X$ and $V_Y$, include all effects on the aircraft, such as wind, thermals, engine accelerations, and speed brake decelerations. Some form of digital readout usually displays the groundspeed (V).

The angles between the aircraft attitude and the platform reference attitude are continuously measured by synchros. The aircraft yaws, rolls, and pitches about the platform in a set of gimbals, each gimbal being rotated through some component of attitude. TH is measured as the horizontal angle between the aircraft's longitudinal axis and platform north. Roll and pitch angles are measured by synchro transmitters on the platform roll and pitch gimbals.

INS technology has advanced very rapidly within the past few years. Advanced navigation systems are commonly designed with INS as an essential component. INS reliability is exceptional and INS accuracies are second only to GPS.

Traditional INS design has capitalized on the advances of the digital computer to increase system responsiveness.

## NAVSTAR Global Positioning System

Space-based GPS, such as the U.S. GPS, the Russian GLONAS, or the forthcoming EU Galileo system all function on the same principles. In fact since the three systems use different frequencies and algorithms, in general receivers of all three systems can be more accurate than a receiver of just one system, since the system errors can be canceled out.

Deployment of the NAVSTAR GPS constellation of satellites began with the first launch in 1977. The satellites were launched into precisely controlled orbits, allowing users with GPS equipment to receive data to determine their position. The phenomenal accuracy of GPS was its major selling point, but its many different applications were a close second. GPS determines a position referenced to a common grid known as the World Geodetic System 1984 (WGS 84). The WGS 84 grid is based upon a mathematical model and compensates for the fact that the earth is not a perfect sphere. *[Figure 16-9]* Derived using precise satellite measurements, it creates an accurate model of the earth's surface. As a consequence, WGS 84 provides extremely accurate information when compared to older traditional datum references. The value of the WGS 84 grid is that positional data can be standardized worldwide. Many receiver sets are capable of converting WGS 84 data into these other commonly used references.

### General System Description
The GPS system is made up of three segments—space, user, and control.

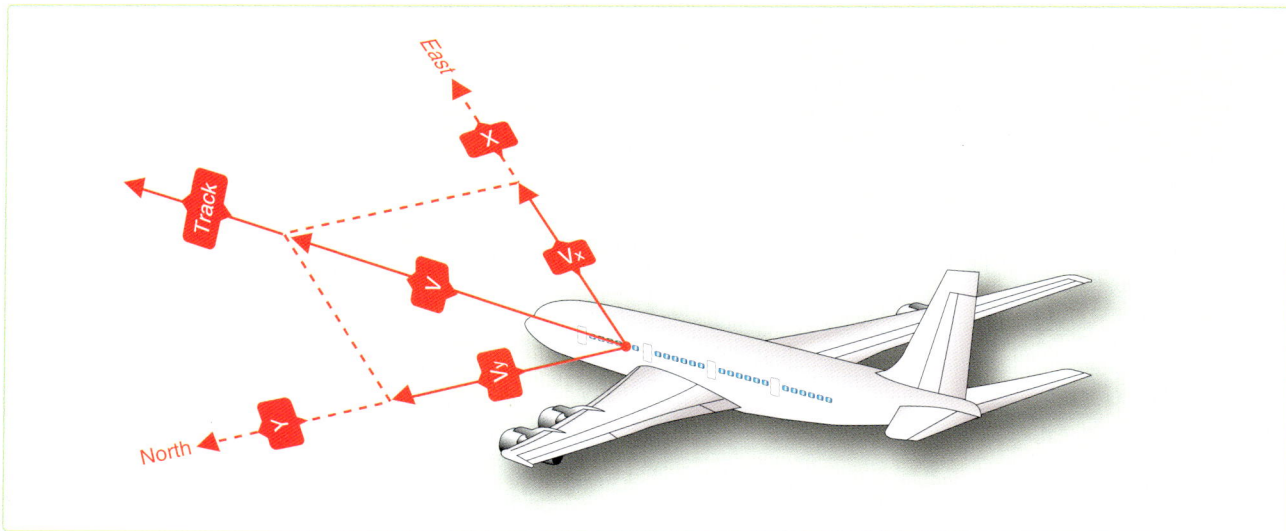

**Figure 16-8.** *Measurement of aircraft groundspeed.*

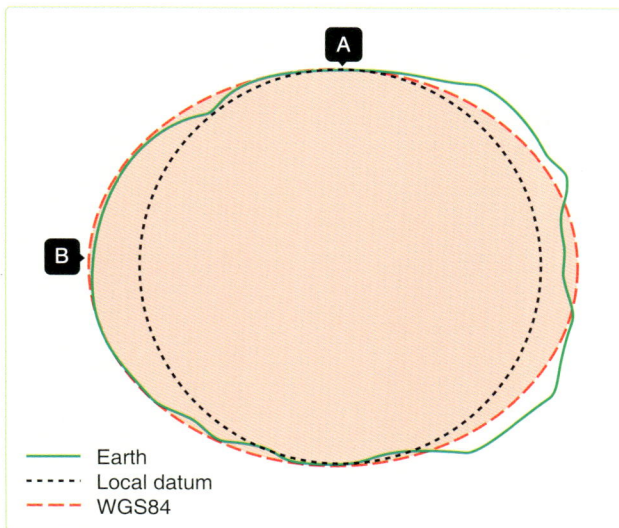

**Figure 16-9.** *World Geodetic System 1984.*

## Space Segment

The space segment is deigned to have 24 satellites plus spares in 6 orbital planes. *[Figure 16-10]* The orbits are arranged precisely such that a minimum of four satellites are in view at all times worldwide.

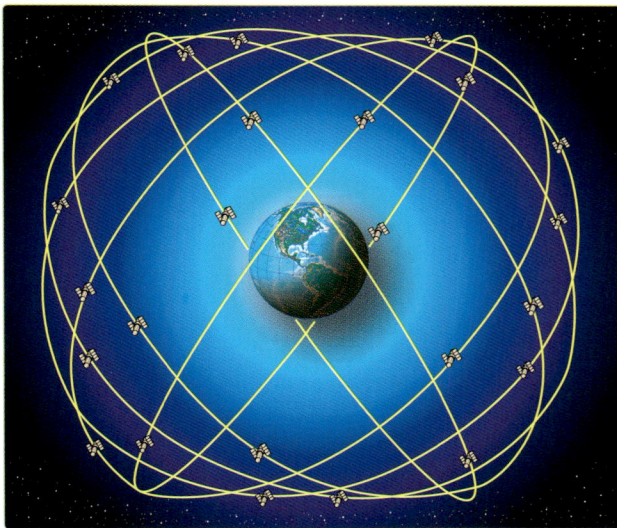

**Figure 16-10.** *GPS constellation.*

## User Segment

The user segment consists of user equipment (UE) sets, test equipment, and associated support equipment. *[Figure 16-11]* The UE set, using data transmitted by the satellites, determines the user's position, altitude, and velocity. Transmissions from the satellites also allow the UE set to evaluate the accuracy of the navigational information being received. This is based on built-in checks of its own performance, the configuration of the satellite constellation in view, and the jamming-to-signal ratios being experienced by the set.

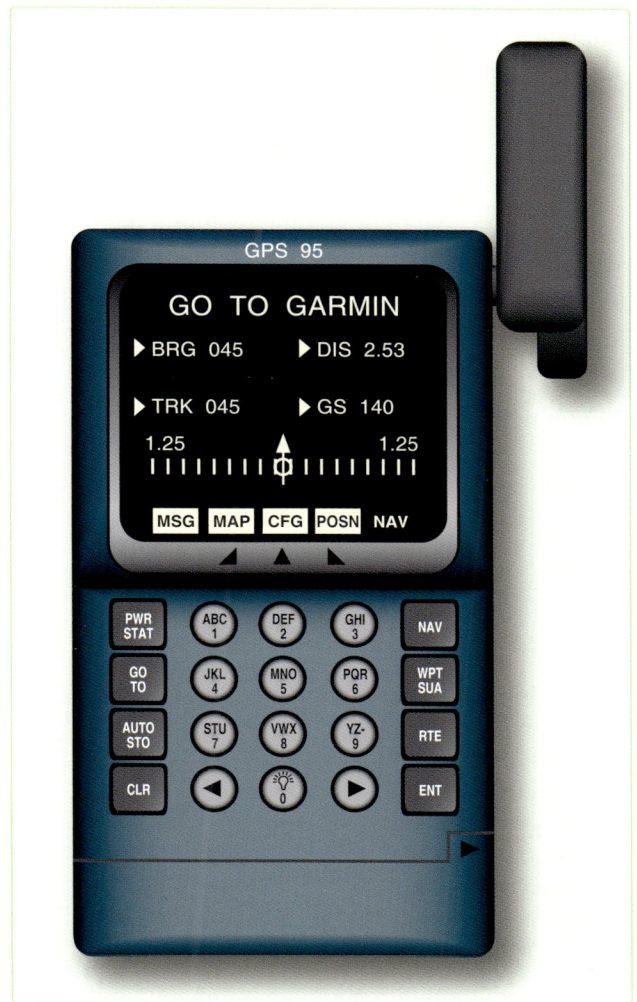

**Figure 16-11.** *Handheld user equipment set.*

## Control Segment

The control segment includes a network of monitor stations and ground antennas placed throughout the world. *[Figure 16-12]* The monitor stations track all satellites in view and monitor general health of the system. Data from the monitor stations is sent to and processed at the Master Control Station (MCS). This data is then used to refine and update the satellite's navigational signals. These corrections are transmitted to the individual satellites via ground antennas. The operational master control station is collocated with the Consolidated Space Operations Center at Peterson Field, Colorado. Three ground antenna stations are located at Diego Garcia, Ascension Island, and Kwajalein. Five monitor stations are positioned in Hawaii, Colorado, and at the three ground antenna locations.

## Theory of Operation

A UE set is capable of determining position, velocity, and time information by receiving ranging signals from a number of satellites. By measuring the difference between signal

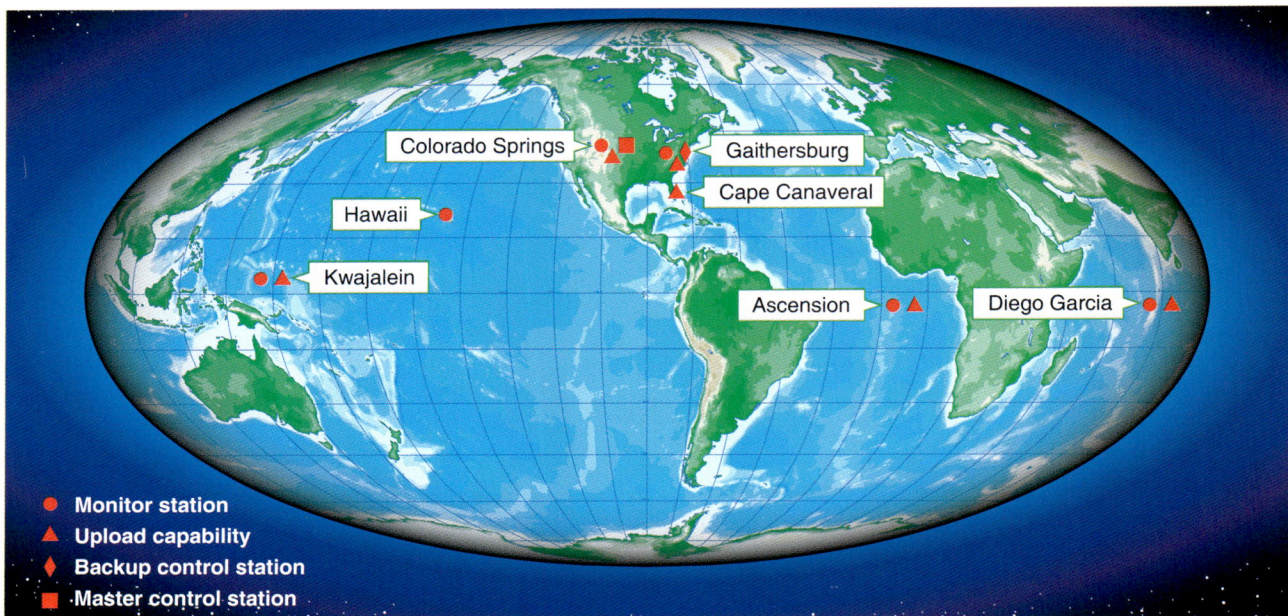

**Figure 16-12.** *GPS control segment.*

transmission and reception times and multiplying that time interval (Dt) by the speed of light, range to the satellite can be determined. In a general sense, this is very similar to the way TACAN DME functions, with one important difference. TACAN DME is an active system in that a signal must be sent from the aircraft to the selected TACAN ground station. The ground station in turn sends a reply signal back to the aircraft. The TACAN set then measures the Dt and then computes and displays the range to the station. GPS is a passive system; no signal is transmitted by the UE set to the satellite. How does the user or receiver determine when the signal was transmitted by the satellite? The solution is to encode the satellite signal so the receiver knows when it was transmitted.

In order to encode the signal with its transmission time, the satellite generates what is known as a pseudorandom noise (PRN) sequence or code. This code is broadcast continuously from each satellite. At the same time, the UE set simultaneously generates an identical code. When the set receives the satellite's signal, it compares it with the code that

it has been generating. If a signal arrives at the receiver with the same code generated two seconds ago, we know that the satellite's signal took 2 seconds to reach us. *[Figure 16-13]*

If we know the satellite's location in space and our distance from it, we know we are somewhere on the sphere having the satellite as its center. With two satellites in view, the user's position is somewhere on the circle representing the intersection of two spheres. A third satellite provides an additional sphere of position whose intersection with the other two defines a three-dimensional navigation fix with timing errors. *Figure 16-14* illustrates this concept in two dimensions for clarity. A fourth allows us to eliminate most of the timing errors. The accuracy of the navigation fix would be dependent on the accuracy of the measurement process (how accurately is the digital signal processed), the accuracy of the satellite positions, and the accuracy and stability of the satellite's clocks and the receiver clock. The user equipment should be able to track the satellite's signal to within 3 nanoseconds ($3 \times 10-9$ seconds). This is equivalent

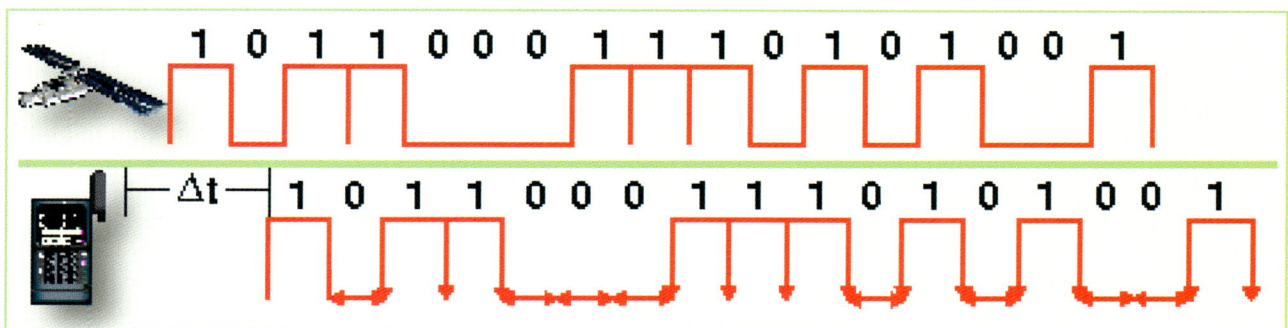

**Figure 16-13.** *PRN code comparison.*

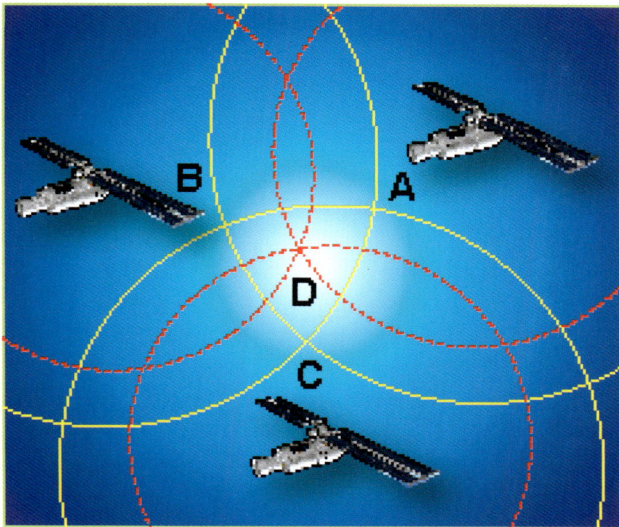

**Figure 16-14.** *Resolution of position.*

to a 1-meter error in position. If navigational accuracy on the order of 10 meters is desired, we must be able to establish satellite position at a particular time to within at least 10 meters. This is not a trivial problem. Since the satellite is moving and is subject to complex gravitational attractions and solar winds, measuring and predicting its position within 10 meters as a function of time is quite difficult. Fortunately, this ephemeris (orbital) data is transmitted to the receiver in the form of almanac data.

Ground stations continuously monitor each satellite so its position can be corrected and passed to the other satellites in the network. Each satellite in reception range transmits its coordinates, a time factor correction, and other data. The receiver solves simultaneous equations for the unknown receiver coordinates and the correct time. Four channel GPS receivers can use one channel per satellite to maintain continuous lock on and update. Receivers with less than four channels must continuously switch frequencies and hunt for new satellites that can limit the system responsiveness.

The satellites transmit two code signals: the precision (P) code on 1227.6 MHz and the coarse acquisition (C/A) code on 1575.42 MHz. *[Figure 16-15]* Both codes carry the same types of information. The C/A code is transmitted with intentional errors to deny the highly accurate position from unauthorized users. The P code, like its relative, the encrypted Y code, does not include these intentional errors. To circumvent the Y code encryption, differential GPS (DGPS) receivers are being designed that receive general GPS signals, as well as a fifth signal from a ground-based transmitter. These differential transmitters can easily determine the intentional error by comparing their GPS position to the surveyed coordinates of the transmitter. The difference is the intentional error. The ground transmitters compute and relay the amount of intentional bias in the C/A code so receivers can remove the position error without use of the P or Y codes.

## Clock Error and Pseudo Range

We assumed in the previous discussion that both the satellite and the UE set were generating identical pseudo codes at exactly the same time. Practically speaking, this is not the case. Each satellite carries an atomic clock accurate to 10-9 seconds. Achieving maximum accuracy in synchronizing the codes would require all users to carry atomic clocks with comparable accuracies, significantly increasing both the size and cost of each receiver set. As a compromise, each UE set is equipped with a quartz crystal clock.

Since the accuracy of a quartz crystal clock cannot approach that of an atomic clock, there is a difference between satellite GPS system time and UE set time. As a result, the generation of the two pseudo codes is not perfectly synchronized and a ranging error is induced. Instead of determining actual range, we measure the apparent, or pseudo range, to the satellite. This particular problem area is known as clock bias. Clock bias affects all range measurements equally. The problem is determining the amount of bias error. Using three satellites allows us to determine our position in three dimensions. By using a fourth satellite and comparing pseudo codes, the UE

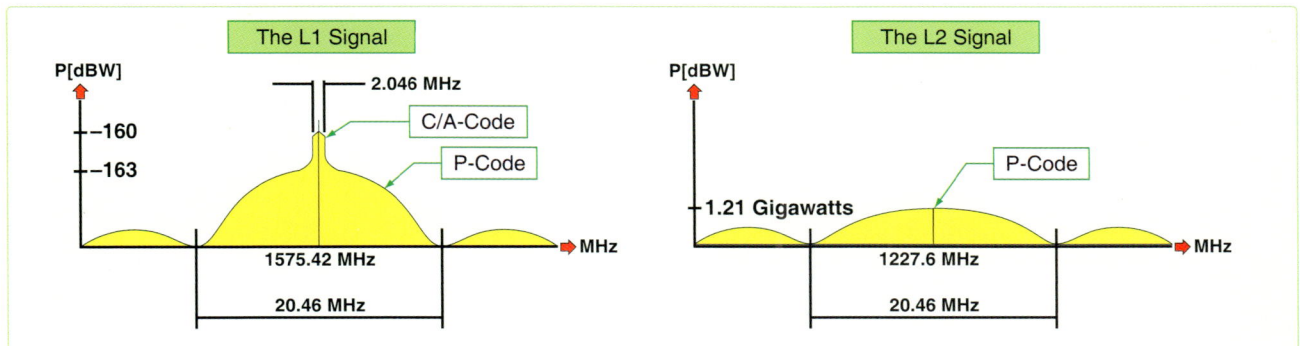

**Figure 16-15.** *Signal bandwidth.*

set internally determines the amount of adjustment necessary to make all of the measurements agree.

## Satellite Clock Error

It might be safe to assume that since each satellite carries an atomic clock, it would keep extremely accurate time. Since the compact dimensions of an orbiting satellite limit the clock size, its accuracy does not approach that of ground based atomic clocks. As a consequence, there is some error in each satellite's clock when compared with master GPS system time. The satellite's generation of the pseudo code is slightly out of synch and some ranging error is induced. This problem is known as satellite clock error. To compensate for this type of error, the GPS control segment comes into play. Monitor stations evaluate the accuracy of the satellite's clock and its pseudo code generation. This information is then relayed to the master control station where the necessary corrections to the satellite's transmissions are computed. Updated information is then uploaded to the satellite via the ground antennas.

## Ephemeris Error

Ephemeris is the ability to determine the location of a celestial body (in this case a satellite) at regular time intervals. Ephemeris error then is caused by the satellite not being exactly where we thought it was. By using estimation theory techniques, the computers at the master control station predict what the satellite's position should be at a specific time. This predicted position is then compared with the actual position as determined by the monitor stations. Updated information on the satellite's future position is then uploaded to each satellite on a regular basis via the ground antennas. Each satellite then continuously transmits these corrections to all users. In this way, ranging error caused by uncertainty as to the satellite's exact position is minimized.

## Atmospheric Propagation Error

We assumed that the satellite's RF signal traveled at the speed of light, as it does in a vacuum like space. But just as light is refracted through a prism, the RF signal is bent and slowed down as it enters the ionosphere. The degree to which the signal is affected depends on the atmospheric conditions between the satellite and receiver and on the signal's angle as it passes through the ionosphere. Atmospheric propagation error can cause position uncertainties up to 40 meters. By noting the time delay between the two L-Band signals, much of the effect caused by atmospheric propagation can be removed internally by the UE set. Since only the military is capable of simultaneously monitoring both of the frequencies, civilian users are forced to live with this error.

## Chapter Summary

GPS accuracy is unquestionable, producing accuracies within meters. An INS backup cannot be jammed or intercepted and provides acceptable accuracy in all weather. Combination INS/GPS units provide unmatched navigation accuracy and are not the backbone of many FMS (Flight Management Systems). Most FMSs maintain a position using the GPS and navigation beacons available. This adds resilience to the system in the case of prankster jamming of the GPS signals. These FMSs process multiple inputs and provide wind, heading, course, distance, and time information accurately and automatically.

# Appendix A
# References and Supporting Information

## References

AFPAM11-216 Air Navigation
NV Pub 9, Volumes 1 and 2, The American Practical Navigator
Department of Defense (DoD) Bulletin Digest
Global Navigation Chart (GNC) 9 available for sale at http://naco.faa.gov
Journal of The Institute Of Navigation
Nautical Almanac
Air Almanac
SR Pub 249, Volumes 1 through 3, Sight Reduction Tables for Air Navigation
Dutton's Nautical Navigation (formerly Dutton's Navigation and Piloting by Elbert S. Maloney) by Thomas J. Cutler

## Abbreviations and Acronyms

AD F—automatic direction finder
AGL—above ground level
AM—amplitude modulation
AP—air position
ARA—airborne radar approach
ARCP—air refueling control point
ARCT—air refueling control time
AREX—air refueling exit
ARIP—air refueling initial point
ARTCC—Air Route Traffic Control Center
AWAD S—Adverse Weather Aerial Delivery System
BAS—basic airspeed
BDHI—bearing direction heading indicator
CA—convergence angle
CARP—computed air release point
CAS—calibrated airspeed
CDI—course deviation indicator
CF—convergence factor
CH—compass heading
CHUM—Chart Updating Manual
Co-alt—co-altitude
CONUS—continental United States
CPC—constant pressure chart
CPU—central processing unit
CRT—cathode ray tube
CW—continuous wave
d—correction to tabulated altitude for minutes of declination

D—D-readings, difference between TA and PA
D1, D2—successive D readings
DA—density altitude, draft angle
DAS—density airspeed
Dec—declination
dev—deviation
DF—direction finder
DG—directional gyro
DGPS—differential GPS
DME—distance measuring equipment
DR—dead reckoning
E—East
EAD—effective air distance
EAP—effective airpath
EAS—equivalent airspeed
EMP—electromagnetic pulse
ERP—earth rate precession
ETA—estimated time of arrival
ETAS—effective true airspeed
ETP—equal time point
FAA—Federal Aviation Administration
FAR—Federal Aviation Regulations
FCG—Foreign Clearance Guide
FIH—Flight Information Handbook
FLIP—flight Information Publications
FSS—flight service station
FTD—forward travel distance

GC—grid course
GEOREF—geographic reference
GHA—Greenwich hour angle
GMT—Greenwich Mean Time
GN—grid north
GP—general planning
GPS—global positioning system
GS—groundspeed
GST—Greenwich sidereal time
GZD—grid zone designation
Hc—computed altitude
HF—high frequency
Hg—mercury
Ho—height observed
Hs—sextant altitude
Hz—cycles per second
IAF—initial approach fix
IAS—indicated airspeed
IAT—indicated air temperature
IAW—in accordance with
ICAO—International Civil Aviation Organization
ICE-T—method used to compute TAS using the ICE-T method on the DR computer. For each type of airspeed, solve in the order of I, C, E, and T.
IFF—identification friend or foe
IFR—instrument flight rules
ILS—instrument landing system
INS—inertial navigation system
IP—initial point
I/O—input/output
IRB—inverse relative bearing
ITA—indicated true altitude
JNC—jet navigation chart
JOG—Joint Operations Graphics
K—constant
KHz—kilocycles (1,000 cycles per second)
km—kilometer
L—trail
LAP—Longitude Adjustment Principle
LHA—local hour angle
LMT—local mean time
Long—longitude
LOP—line of position
LOS—line of sight
LST—local sidereal time
LZT—local zone time
MAP—missed approach point
MB—magnetic bearing
MC—magnetic course
MC&G—mapping, charting, and geodetic
MC S—master control station
MDA—minimum descent altitude
MEA—minimum en route altitude

MH—magnetic heading
mHz—megacycles per second
MN—magnetic north
MPP—Most probable position
MSL—mean sea level
N—North
NAVAID—navigation aids
NCP—north celestial pole
NDB—nondirectional radio beacon
NGA—National Geo-Spatial Intelligence Agency
NM—nautical mile
NOTAM—Notices to Airmen
NRB—nondirectional radio beacon
OAT—outside air temperature
ONC—Operational Navigation Chart
PA—pressure altitude
PCN—planning change notice
PGF—pressure gradient force
PLOP—pressure line of position
PPI—planned position indicator
PRN—pseudorandom noise
PW—pulse wave
RADAR—radio detection and reading
RB—relative bearing
RF—radio frequency
RLG—ring laser gyro
RMI—radio magnetic indicator
R/T—receiver and transmitter
S—south
SAD—sight angle drop
SAT—static air temperature
SCP—south celestial pole
SD—semidiameter
SHA—sidereal hour angle
SID—standard instrument departure
SIF—selective identification feature
SKE—station-keeping equipment
SSB—single sideband
SIOP—Single Integrated Operations Plan
STAR—standard terminal arrival route
STC—sensitivity time constant
TA—true altitude
TACAN—tactical air navigation
TAMPA—true altitude minus pressure altitude
TAS—true airspeed
TAT—true air temperature
TB—true bearing
TC—true course
TCN—terminal change notice
TF—time of fall
TFC—time of fall constant
TH—true heading

# Appendix B
# Mathematical Formulas

## Computers

Technological advances have made programmable handheld computers readily available at a reasonable cost. A variety of computers are acceptable for use in flight. Using the formulas in this chapter, the navigator increases calculating power for both preflight and in-flight situations. In addition to ease of operation, both speed and accuracy will improve significantly over manual and MB–4 computations. Also, the new handheld computers have capabilities never before available without expensive avionics. Use the following formulas as an aid in both preflight and in-flight computations.

NOTES:

1. All formulas are valid when trigonometric functions compute in degrees (not radians).

2. Unless otherwise indicated, velocities are in knots, temperatures are in degrees Celsius, and directions and angles are in degrees.

3. North, West, and Left are represented by positive values while South, East, and Right are represented by negative values.

4. Sq rt denotes the Square Root function; sin, cos, tan, asin, acos, atan are the standard trigonometric functions; and the ^ symbol represents the power of function.

## Flight Planning

### Variables

TC = True course

TAS = True airspeed

W = Wind direction

DCA = Drift correction angle

TW = Tailwind component

TH = True heading

GS = Groundspeed

V = Wind velocity

VAR = Variation

CW = Crosswind component

$$DCA = sin^{-1}\left[\frac{V\ sin(W-TC)}{TAS}\right]$$

or Drift Angle = $asin$((Wind Speed ÷ True Airspeed) × $sin$(Wind Direction – True Course))

$$GS = \frac{TAS\ sin(W-TC-DCA)}{sin(TC-W+180)}$$

or Groundspeed = $sin$(Wind Direction + 360 – True Course – Drift Angle) × True Airspeed ÷ $sin$(True Course – Wind Direction – 180)

or Groundspeed = True Airspeed × $cos$(Drift Angle) – Wind Speed × $cos$(Wind Direction – True Course)

GS = TAS $cos$(TH – TC) – V $cos$(W – TC)

or Groundspeed = True Airspeed × $cos$(True Heading – True Course) – Wind Speed × $cos$(Wind Direction – True Course)

TW = V $cos$(W – TH)
or Tailwind Component = Wind Speed × $cos$(Wind Direction – True Heading)

CW = V $sin$(W – TH)
or Crosswind Component = Wind Speed × $sin$(Wind Direction – True Heading)

## Inflight Wind Determination
### Variables

TC = True Course

TAS = True Airspeed

D = Drift Angle

GS = Groundspeed

TH = True Heading

DCA = Drift Correction Angle = –D

$$V = \sqrt{GS^2 = TAS^2 - 2(GS)(TAS)(cosDCA)}$$

or Wind Speed = sq rt (Groundspeed$^2$ + True Airspeed$^2$ – 2 × Groundspeed × True Airspeed × $cos$(Drift Correction Angle))

$$W = TC + sin^{-1}\left[\frac{TAS\ sinD}{V}\right] \qquad TAS > GS$$

or Wind Direction = True Course + $asin$(True Airspeed × $sin$(Drift) ÷ Wind Speed)

$$W = TC - sin^{-1}\left[\frac{TAS\ sinD}{V}\right] + 180 \qquad TAS \leq GS$$

or Wind Direction = True Course – $asin$(True Airspeed × $sin$(Drift) ÷ Wind Speed) + 180

## Pressure Pattern
### Variables

K = Constant

ETAS = Effective True Airspeed

T = Time between DRs

$D_2$ = Second D reading

BD = Bellamy Drift

ML = Mid Latitude between DRs

AD = Air Distance

ZN = Crosswind Displacement

$D_1$ = First D reading

GD = Ground Distance

$$K = \left[\frac{21.49}{sin\ ML}\right] \qquad ETAS = \frac{AD}{T}$$

or K = 21.49 ÷ $sin$(Latitude)
or Equivalent True Airspeed = Air Distance ÷ Elapsed Time

$$ZN = \frac{K\ (D_2 - D_1)}{ETAS} \qquad BD = \frac{ZN\ (57.3)}{GD}$$

or ZN = K × ($D_2$ – $D_1$) ÷ Equivalent True Airspeed or Bellamy Drift = ZN × 57.3 ÷ Ground Distance

## TAS÷Mach
### Variables

CAS = Calibrated Airspeed

IT = Indicated Air Temperature

TAT = True Air Temperature (° Celsius)

PA = Pressure Altitude

M = Mach

CT = Temperature Rise (+1 for most aircraft)

DA = Density Altitude

TAS = True Airspeed

$$M = \sqrt{5\left[\left(\chi\left\{\left[1+\left(\frac{CAS}{661.5}\right)^2\right]^{3.5}-1\right\}+1\right)^{2.86}-1\right]} \quad \text{where } \chi = \frac{1}{\left|\dfrac{518.67-(3.566\times 10^{-3})(PA)}{518.67}\right|^{5.2563}}$$

or MACH = sq rt(5 × (((1 ÷ (((518.67 − (0.003566 × Pressure Altitude)) ÷ 518.67)$^{5.2563}$ )) × ((1 + 0.2 × (Calibrated Airspeed ÷ 661.5)$^2$)$^{3.5}$ − 1) + 1)$^{.286}$) − 1)

or MACH = True Airspeed ÷ (39 × sq rt (Temperature Celsius + 273))

or Calibrated Airspeed = 661.5 × sq rt((((((( 1 + (MACH$^2$) ÷ 5)$^{(1 \div 0.286)}$) − 1) × (((518.67 − (0.003566 × Pressure Altitude)) ÷ 518.67)$^{5.2563}$) + 1)$^{(1 \div 3.5)}$) − 1) ÷ 0.2)

Indicated Airspeed = Calibrated Airspeed × F Factor

$$TAS = 39M\sqrt{TAT+273} = 39M\sqrt{(IT+273)\left[CT\left(\frac{1}{1+(.2)(M^2)}-1\right)+1\right]}$$

or True Airspeed = 39 × MACH × sq rt((Indicated Air Temperature + 273) × ((Temperature Rise × ((1 ÷ (1 + .2. × MACH$^2$)) − 1)) + 1))

NOTE: Temperature rise is generally +1 degree for most aircraft.

or True Airspeed = 39 × MACH × sq rt(True Air Temperature)

NOTE: True Air Temperature is in degrees Kelvin (Celsius + 273.15).

**Turn Performance**
*Variables*
BANK = Bank Angle
DIAM = Turn Diameter in NM
T = Time to Complete 360° turn

$$DIAM = \frac{TAS^2}{34208\,tan BANK}$$

or Turn Diameter = True Airspeed$^2$ ÷ (34208 × $tan$(Bank Angle))

$$T = \frac{.0055\,TAS}{tan BANK}$$

or Minutes for Complete 360 = .0055 × True Airspeed ÷ $tan$(Bank Angle)

G Force = 1 ÷ $cos$(Bank Angle)

Turn Stall Speed = Normal Stall Speed × G Force

## Celestial Precomputations

### *Variables*

LAT = Latitude of Assumed Position or DR      LONG = Longitude of Assumed Position or DR
Zn = True Azimuth      Z = Azimuth Angle
$H_C$ = Height Computed      DEC = Declination of the Body from Air Almanac
SHA = Sidereal Hour Angle      LHA = Local Hour Angle = GHA – W (+E) Long + SHA + corrections

NOTE: $H_C$ and Z are displayed in degrees and decimal degrees. You must convert the decimal degrees to minutes. Once the azimuth angle (Z) has been determined by computation, the ambiguity caused by LAT and LHA can be resolved by the following:

Zn = Z      $sin(LHA) < 0$
Zn = 360 – Z      $sin(LHA) \geq 0$

$$H_C = sin^{-1}[(sinLAT)(sinDEC) + (cosLHA)(cosDEC)(cosLAT)]$$

or Height Computed (Hc) = $asin(sin(\text{DR Latitude}) \times sin(\text{Body Declination}) + cos(LHA) \times cos(\text{Body Declination}) \times cos(\text{DR Latitude}))$

$$Z = cos^{-1}\left|\frac{(sinDEC) - (sinLAT)(sinH_C)}{(cosH_C)(cosLAT)}\right|$$

or Azimuth Angle = $acos((sin(\text{Body Declination}) - sin(\text{DRL altitude}) \times sin(\text{Height Computed})) \div (cos(\text{Height Computed}) \times cos(\text{DRL altitude})))$

Zn = Z      N LAT and LHA > 180°
Zn = 360 – Z      N LAT and LHA < 180°
Zn = 180 – Z      S LAT and LHA > 180°
Zn = 180 + Z      S LAT and LHA < 180°

## Motions

The formula for combined 1-minute motion can be separated as follows:

or Motion of Body = 15 $(cos(LAT)(sin Zn)$
or Motion of the Body (for 1 minute) = $15 \times cos(\text{Latitude}) \times sin(\text{Azimuth})$

$$\text{Motion of Observer} = \left(\frac{GS}{60}\right)cos(TC - Zn)$$

or Motion of the observer (for 1 minute) = $(\text{Groundspeed} \div 60) \times cos(\text{True Course} - \text{Azimuth})$

These quantities, whether combined or used separately, must be added algebraically to the Ho and subtracted from the Hc. To apply Coriolis/rhumb line correction to Ho, multiply Coriolis/rhumb line by $sin(ZN-TC)$. Note that P and N adjustments are not necessary with these computer applications since Hc is correct for fix time, not Pub. No. 249 EPOCH year time.

$$\text{1-Minute Motion} = [15(cosLAT)(sinZn)] - \left[cos(TC - Zn)\left(\frac{GS}{60}\right)\right]$$

or Motion of the Body (for 1 minute) = $15 \times cos(\text{Latitude}) \times sin(\text{Azimuth}) - (cos(\text{True Course} - \text{Azimuth}) \times \text{Groundspeed} \div 60)$

$$\text{1-Minute Motion} = [15(cosLAT)(cos(270 - Zn))] - \left[cos(TC - Zn)\left(\frac{GS}{60}\right)\right]$$

or Motion of the Body (for 1 minute) = 15 × $cos$(Latitude) × $cos$(270–Azimuth) – ($cos$(True Course – Azimuth) × Groundspeed ÷ 60)

$$\text{Rhumb line} = .146\left(\frac{GS}{100}\right)^2 (sinTC)(tanLAT)$$

or Rhumb Line = 0.146 × (Groundspeed ÷100)$^2$ × $sin$(True Course) × $tan$(Latitude)

$$\text{Coriolis} = (.0265)(GS)(sinLAT)$$

or Coriolis = 0.02625 × Groundspeed × $sin$(Latitude)

$$\text{Coriolis} = cos(90 - TC - Zn)(.0265)(GS)(sinLAT)$$

or Coriolis = $cos$(90 – True Course – Azimuth) × 0.02625 × Groundspeed × $sin$(Latitude)

$$\text{Coriolis/rhumb line} = [(.0265)(GS)(sinLAT)] + \left[(.146)\left(\frac{GS}{100}\right)^2(sinTC)(tanLAT)\right]$$

or Coriolis = 0.02625 × Groundspeed × $sin$(Latitude) + [0.146 × (Groundspeed ÷100)$^2$ × $sin$(True Course) × $tan$(Latitude)]

## Great Circle Planning
### Variables

$L_1$ = Departure Latitude (N and W = +)     $L_2$ = Destination Latitude (S and E = –)
$\lambda_1$ = Departure Longitude     $\lambda_2$ = Destination Longitude
$L_i$ = Intermediate Latitude     $\lambda_i$ = Intermediate Longitude
$H_i$ = Initial True Heading     D = Distance
H = Heading Angle     $\Delta t$ = Time between positions
GS = Groundspeed     TC = True Course

$$D = 60cos^{-1}[(sinL_1)(sinL_2) + (cosL_1)(cosL_2)cos(\lambda_2 - \lambda_1)]$$

Distance=60 × $acos$(($sin$(Departure Latitude) × $sin$(Destination Latitude)) + ($cos$(Departure Latitude) × $cos$(Destination Latitude) × $cos$(Destination Longitude – Departure Longitude)))

$$H = cos^{-1}\left[\frac{sinL_2 - sinL_1 cos\left(\frac{D}{60}\right)}{sin\left(\frac{D}{60}\right)cosL_1}\right]$$

or Heading Angle = $acos$(($sin$(Destination Latitude) – $sin$(Departure Latitude) × $cos$(Distance ÷ 60)) ÷ ($sin$(Distance ÷ 60) × $cos$(Departure Latitude)))

$H_i$ = H          $sin(\lambda_2 - \lambda_2) < 0$
$H_i$ = 360 – H     $sin(\lambda_2 - \lambda_1) > 0$

This formula computes the latitude of $L_i$ where $\lambda_i$ intersects the great circle defined by $(L_1, \lambda_1)$ and $(L_2, \lambda_2)$. This formula can be very useful when matching charts of different projections or scales.

$$L_i = tan^{-1}\left[\frac{(tanL_2)sin(\lambda_i - \lambda_2) - (tanL_1)sin(\lambda_i - \lambda_2)}{sin(\lambda_i - \lambda_2)}\right]$$

or Intermediate Latitude = $atan$((tan(Destination Latitude) × sin(Intermediate Longitude – Departure Longitude) – tan(Departure Latitude) × sin(Intermediate Longitude – Destination Longitude)) ÷ sin(Destination Longitude – Departure Longitude))

## Computing Position By Dead Reckoning:

$$L_2 = \left( \frac{(\Delta t)(GS)(cos TC)}{60} \right) + L_1$$

or DEST Latitude = (Elapsed Time × Groundspeed × $cos$(True Course)) ÷ 60 + Departure Latitude

$$\lambda_i = \lambda_2 - \left( \frac{(\Delta t)(GS)(sin TC)}{60\ cos L_1} \right) \qquad\qquad TC = 90°, 270°$$

or DEST Longitude = Departure Longitude – ((Elapsed Time × Groundspeed × $sin$(True Course)) ÷ (60 × $cos$(Departure Latitude)))

Otherwise

$$\lambda_2 = \lambda_1 - \frac{180}{\pi}\{(tan TC)[L_n\ tan(45 + \tfrac{1}{2}L_2)) - (L_n\ tan(45 + \tfrac{1}{2}L_2))]\}$$

or DR Longitude = Departure Longitude – (180 ÷ 3.14159) × ($tan$(True Course) × Ln ($tan$(45 + 0.5 × Destination Latitude)) – Ln ($tan$(45 + 0.5 × Departure Latitude)) )

NOTE: The flightpath may not cross either pole.

For long distances, use formula below:
DR Latitude = 90.0 – $acos$($sin$(– Departure Latitude) × $cos$(Distance ÷ 60.0) + $cos$(– Departure Latitude) × $sin$(Distance ÷ 60.0) × $cos$(True Course))

DR Longitude = Departure Longitude +/– $acos$(($cos$(Distance ÷ 60.0) –$sin$(– DR Latitude) × $sin$(– Departure Latitude)) ÷ ($cos$(– DR Latitude) × $cos$(– Departure Latitude)))

NOTE: Distance can be replaced with (Groundspeed × Elapsed Time) where Elapsed Time is in hours.

## Rhumb Line Planning
### Variables

| | |
|---|---|
| $\Delta t$ = Time between positions | D = Rhumb line Distance |
| C = Rhumb line True Course | $\pi$ = Pi (»3.14159) |

$$C = tan^{-1}\left[ \frac{\pi\,(\lambda_2 - \lambda_1)}{180 L_n\ tan(45 + \tfrac{1}{2}L_2) - (L_n\ tan(45 + \tfrac{1}{2}L_1)}\right]$$

or True Course = $atan$((3.14159 × (Departure Longitude – Destination Longitude)) ÷ (180 × Ln ($tan$(45 + 0.5 × Destination Latitude)) – Ln($tan$(45 + 0.5 × Departure Latitude)))

$$D = 60(\lambda_2 - \lambda_1)cos L_1 \qquad\qquad C = 0$$

or Distance = 60 × (Destination Longitude – Departure Longitude) × $cos$(Departure Latitude)

$$D = \frac{60\,(L_2 - L_1)}{cos C} \qquad\qquad C = 0$$

or Distance = 60 × (Destination Latitude – Departure Latitude) × $cos$(Rhumb Line True Course)

## Course Correction to Destination
### *Variables*

$D_1$ = Distance flown to current position

$D_2$ = Distance from start to destination

$D_3$ = Distance from start to checkpoint

$D_4$ = Distance from current position to destination

$M_1$ = Intended magnetic course

$M_2$ = Magnetic course flown to current position

$M_3$ = Magnetic course to fly from current position to destination

DOC = Distance off course (+ left, – right)

$$M_1 = M_2 + tan^{-1}\left(\frac{DOC}{D_3}\right)$$

or Intended Magnetic Course = Current Magnetic Course + $atan$(Distance Off Course ÷ Leg Distance To Checkpoint)

$$M_3 = M_1 + sin^{-1}\left(\frac{DOC}{D_4}\right)$$

Required Magnetic Course = Intended Magnetic Course + $asin$(Distance Off Course ÷ Distance To Go)

$$D_3 = \sqrt{D_1{}^2 - DOC^2}$$

or Distance To Checkpoint = sq rt(Distance Flown² – Distance Off Course²)

$$D_4 = \sqrt{(D_2 - D_3)^2 + DOC^2} \quad 0 \leq M_3 < 360°$$

or Distance To Go = sq rt((Total Distance – Distance Flown)² + Distance Off Course²)

## Point to Point
### *Variables*

$D_1$ = DME #1

$R_1$ = Radial #1

HA = Altitude MSL

T = Time between positions

GS = Groundspeed

$D_2$ = DME #2

$R_2$ = Radial #2

HT = Elevation of the TACAN

A = Altitude above TACAN (in NM)

MC = Magnetic Course

NOTE: These formulas are useful when flying from one radial and DME to another.

$$A = \left(\frac{HA - HT}{6076}\right)$$

$$GS = \frac{\sqrt{D_1{}^2 + D_2{}^2 - 2A^2 - 2\{\sqrt{D_1{}^2 - A^2}\,[\sqrt{D_2{}^2 - A^2}(R_1 - R_2))]\}}}{T}$$

or Groundspeed = sq rt(First DME² ≠ 2 × (((Aircraft Altitude MSL – Target Altitude MSL) ÷ 6076)²) – 2 × sq rt(First DME² – (((Aircraft Altitude MSL – Target Altitude MSL) ÷ 6076)²)) × sq rt(Second DME² – (((Aircraft Altitude MSL – Target Altitude MSL) ÷ 6076)²)) × $cos$(First Radial – Second Radial)) ÷ Elapsed Time

NOTE: If you do not divide by Elapsed Time, the result is Distance instead of Groundspeed

$$MC = R_2 - sin^{-1}\left[\frac{(D_1)\,(sin(R_1 - R_2))}{(GS)(T)}\right]$$

or Magnetic Course Flown = Second Radial – $asin$((First DME × $sin$(First Radial – Second Radial)) ÷ (Groundspeed × Elapsed Time))

NOTES:

1.  This assumes that the magnetic variation of the TACAN is the same as the one affecting the aircraft over the distance flown.

2.  This assumes that if using a RADAR target, the radials are taken as magnetic readings (or converted to magnetic)

## Rate of Climb

TAS = True Airspeed

ROC = Rate of climb in feet ÷ min

D = Ground distance over which the altitude change (Alt) occurs Alt = Change in Altitude

$$ROC = \frac{TAS(\Delta\ Alt)}{60\sqrt{\left[D^2 + \left(\frac{\Delta\ Alt}{6076}\right)^2\right]}}$$

or Required Rate of Climb = True Airspeed × Altitude Change ÷ (60 × sq rt(Distance Desired for Change$^2$ + (Altitude Change ÷ 6076)$^2$))

# Chart and Navigation Symbols

## Chart and Navigation Symbols

| Symbol | Description | Symbol | Description |
|---|---|---|---|
| ——— | Course line | ◎ | Alternate/emergency airfield (red) |
| —→— | True heading (air vector) | ▭ | Highest obstacle (red) |
| —⇒— | Track (ground vector) | + | Air position |
| —⇛— | Wind vector | ⊙ | Dead reckoning (DR) position |
| ←——→ | Line of position (LOP) | ⊙ MPP | Most probable position (MPP) |
| ⇐——⇒ | Advanced or retarded LOP | ⊡ | Computer position |
| ◄——► | Average LOP or pressure LOP (PLOP) | ⚠ | Fix |
| ⟍⟋ LOP #1 | LOP #1 | Λ̇ | Celestial assumed position |
| ⟍⟍⟋⟋ LOP #2 | LOP #2 | Λ̈ | Advanced or retarded celestial assumed position |
| ⟍⟍⟍⟋⟋⟋ LOP #3 | LOP #3 | ↓, √, or " | No change from previous log entry |
| ⊙ | Checkpoint/navigation point | ⬭ | Orbit pattern |
| ⊡ | Refueling action point or initial point | | |

# Celestial Computation Sheet

## CELESTIAL PRECOMPUTATION

| | SHEET NUMBER |
|---|---|

### PRECOMPUTATION—PERISCOPIC SEXTANT

| NAVIGATOR | | | ALT MSL | | DATE(Z) | FIX TIME |
|---|---|---|---|---|---|---|

| STAR SELECTOR BY AZIMUTH | TRACK | | BODY | | Spica | Pollux | |
|---|---|---|---|---|---|---|---|

Compass rose: 0, 30, 60, 90, 120, 150, 180, 210, 240, 270, 300, 330

| | | | | | | |
|---|---|---|---|---|---|---|
| | GS | | BASE GHA | | | |
| | CORIOLIS | R / L | CORR | | | |
| | PREC/NUT | | +360 | | | |
| | DR LAT | N / S | GHA | | | |
| | DR LONG | E / W | ASSUM LONG W / +E | | | |

|  | | | | | | |  | | | | | |
|---|---|---|---|---|---|---|---|---|---|---|---|---|
| MOTION OF OBSERVER | | | | | | | LHA | | | | | |
| MOTION OF BODY | | | | | | | ASSUM LAT | N / S | | | | |
| 4 MIN ADJUST | | | | | | | DEC | N/S | N/S | N/S | N/S | N/S |
| OFF-FIX TIME | E/L | E/L | E/L | E/L | E/L | | PLANNED MID-TIME | | | | | |
| TOTAL MOT. ADJUST | | | | | | | ACTUAL MID-TIME | | | | | |
| POLARIS Q / MOON PA SD | | | | | | | TAB Hc | | | | | |
| REF (−) | | | | | | | D / DEC CORR | | | | | |
| PERS/SEXT | | | | | | | CORR Hc | | | | | |
| TOTAL → ADJ | | | | | | | TOTAL → ADJ | | | | | |
| TH/GH | | | | | | | ADJ Hc | | | | | |
| Zn/GZa(−) | | | | | | | OFF TIME MOTION | | | | | |
| IRB | | | | | | | Hc | | | | | |
| IRB | | | | | | | Ho | | | | | |
| Za/GZa(+) | | | | | | | INT | T/A | T/A | T/A | T/A | T/A |
| TH/GH | | | | | | | Zn | | | | | |
| TRACK T/G | | | | | | | CONV +W ANGLE −E | | | | | |
| Za | | | | | | | GRID Zn | | | | | |

REFRACTION TABLE (condensed)

| Ro | Altitude MSL (thousands of feet) | | | | | |
|---|---|---|---|---|---|---|
| | 0 | 20 | 25 | 30 | 35 | 40 |
| 1 | 53 | 46 | 41 | 36 | 33 | 26 |
| 2 | 33 | 19 | 16 | 14 | 11 | 9 |
| 3 | 23 | 12 | 10 | 8 | 7 | 5 |
| 4 | 16 | 8 | 7 | 6 | 5 | 3–10 |
| | 12 | 7 | 5 | 4 | 3–10 | 2–10 |

| REL Za | | TK | DC | TH | VAR | MH | DEV | CH |
|---|---|---|---|---|---|---|---|---|

# Glossary

**Absolute altimeter.** The absolute or radar altimeter indicates the altitude above terrain, land, or water directly below the aircraft.

**Absolute altitude.** The height of an aircraft directly above the surface or terrain over which it is flying.

**Acceleration error.** An error caused by the deflection of the bubble due to any change in acceleration of the aircraft.

**Aeronautical chart.** A specialized representation of mapped features of the earth, produced to show selected terrain, cultural and hydrographic features, and supplemental information required for air navigation, pilotage or for planning air operations.

**Agonic line.** A line drawn on a map or chart joining points of zero magnetic declination for a specified year date.

**Air almanac.** A joint publication of the National Almanac Office of the United States Naval Observatory and Her Majesty's Nautical Almanac Office. It provides coordinates of celestial bodies and additional data required for celestial navigation.

**Air distance (AD).** Distance that is measured relative to the mass of air through which an aircraft passes; the no-wind distance flown in a given time (true airspeed X time).

**Air plot.** A continuous plot used in air navigation of a graphic representation of true headings and air distances flown.

**Air position.** The calculated position of an aircraft assuming no wind effect at a given time.

**Airspeed.** The speed of an aircraft relative to its surrounding airmass.

**Airspeed indicator.** An instrument that displays the indicated airspeed of the aircraft derived from inputs of pitot and static pressures.

**Airway.** A control area or portion thereof established in the form of a corridor marked with radio navigational aids.

**Alter course (A/C).** A change in course to a destination or a turn point.

**Alter heading (A/H).** The change in heading to make good the intended course.

**Altimeter.** A flight instrument that indicates the altitude above a given reference point.

**Altimeter setting.** The pressure datum in millibars or inches of mercury set on the altimeter subscale.

**Altitude.** The vertical distance of a level, a point, or an object considered as a point, measured from mean sea level.

**Altitude delay.** Synchronization delay introduced between the time of transmission of the radar pulse and the start of the trace on the indicator for the purpose of eliminating the altitude hole on the plan position indicator-type display.

**Altitude hole.** The blank area at the center of a radar display, the center of the periphery of which represents the point on the ground immediately below the aircraft.

**Apparent precession.** The apparent deflection of the gyro axis, relative to the earth, due to the rotating effect of the earth and not due to any applied force.

**Apparent time.** Time measured with reference to the true sun. The interval that has elapsed since the last lower transit of a given meridian by the true sun.

**Assumed position.** The geographic position upon which a celestial solution is based.

**Astronomical triangle.** A triangle on the celestial sphere bounded by the observer's celestial meridian, the vertical circle, and the hour circle through the body, and having as its vertices the elevated pole, the observer's zenith, and the body.

**Astronomical twilight.** That period which ends in the evening and begins in the morning when the sun reaches 18° below the horizon.

**Azimuth angle (Z).** The interior angle of the astronomical triangle at the zenith measured from the observer's meridian to the vertical circle through the body.

**Azimuth stabilization.** Orientation of the picture on a radarscope so as to place true north at the top of the scope.

**Barometric altimeter.** An instrument that displays the height of the aircraft above a specified pressure datum. The datum may be varied by setting the specified pressure on a subscale on the instrument.

**Barometric altimeter reversionary.** An altimeter in which the indication is normally derived electrically from an external source (central air data computer or altitude computer) but which, in case of failure or by manual selection, can revert to a pneumatic drive.

**Basic air temperature (BAT).** Indicated air temperature corrected for the instrument error.

**Beacon.** A light or electronic source that emits a distinctive signal used to determine bearings, courses, or location.

**Beam width.** The angle between the directions, on either side of the axis, at which the intensity of the radio frequency field drops to one-half the value it has on the axis.

**Beam-width error.** The effective width in azimuth of radiation from an antenna.

**Bearing.** The horizontal angle at a given point measured clockwise from a specific datum to a second point.

**Bellamy drift.** The net drift angle of the aircraft calculated between any two pressure readings.

**Blip.** The display of a received pulse on a cathode-ray tube; a spot of light representing a target.

**Cabin pressure altimeter.** An instrument that measures the pressure within an aircraft cabin and gives an indication in terms of height according to the chosen standard atmosphere.

**Calibrated airspeed (CAS).** Indicated airspeed corrected for instrument installation error.

**Calibrated altitude.** Indicated altitude corrected for instrument and installation errors.

**Celestial altitude.** The angular distance of a celestial body above or below the horizon, measured along the great circle passing through the body and the zenith. Altitude is 90° minus zenith distance.

**Celestial equator.** The great circle formed by the intersection of the plane of the earth's equator with the celestial sphere. Also known as Equinoctial.

**Celestial intercept.** The difference in minutes of arc between computed and observed altitudes or between precomputed and sextant altitudes. It is labeled T (toward) or A (away) as the observed (or sextant) altitude.

**Celestial meridian.** A great circle on the celestial sphere formed by the intersection of the celestial sphere and any plane passing through the North and South Poles. Any great circle on the celestial sphere which passes through the celestial poles.

**Celestial navigation.** The determination of position by reference to celestial bodies.

**Celestial poles.** The points where the earth's axis, if produced, would intersect the celestial sphere.

**Celestial sphere.** An imaginary sphere of infinite radius concentric with the earth, on which all celestial bodies, except the earth, are imagined to be projected.

**Checkpoint.** A geographical reference point used for checking the position of an aircraft in flight. Normally, well-defined and selected in preflight planning, a checkpoint can usually be easily identified from the air.

**Circle of equal altitude.** A circle on the earth that is the focus of all points equidistant from the substellar point of a celestial body. The altitude of a celestial body measured from any point on the circle is the same. Also called circle of position.

**Civil day.** The interval of time between two successive lower transits of a meridian by the mean (or civil) sun.

**Civil twilight.** That period which ends in the evening and begins in the morning when the sun reaches 6° below the horizon.

**Co-altitude (co-alt).** The small arc of a vertical circle, between the observer's position and the body (90° altitude).

**Co-declination (co-dec).** See Polar Distance.

**Co-latitude (co-lat).** The small arc of the observer's celestial meridian between the elevated pole and the body (90° latitude).

**Collimation.** The correct alignment of the images of the bubble of a sextant and the object being observed.

**Compass.** An instrument that indicates direction measured clockwise from true north or grid north.

**Compass direction.** The horizontal direction expressed as an angular distance measured clockwise from compass north.

**Compass heading (CH).** The reading taken directly from the compass.

**Compass North.** The uncorrected direction indicated by the north-seeking end of a compass needle.

**Compass rose.** A graduated circle, usually marked in degrees, indicating directions and printed or inscribed on an appropriate medium.

**Computed altitude (Hc).** Celestial altitude of a body calculated mathematically for a given position on the earth at a given time. Also called calculated altitude.

**Constellation.** A recognizable group of stars by means of which individual stars may be identified.

**Contour line.** A line on a map or chart connecting points of equal elevation.

**Controlled time of arrival.** A method of arriving at a destination at a specified time by changing direction and/or speed of an aircraft.

**Control point.** The position an aircraft must reach at a predetermined time.

**Coordinates.** Linear or angular quantities that designate the position that a point occupies in a given reference frame or system. Also used as a general term to designate the particular kind of reference frame or system, such as plane rectangular coordinates or spherical coordinates.

- **Celestial 1.** The equinoctial system involves the use of sidereal hour angle and declination to locate a point on the celestial sphere with reference to the first point of Aries and the equinoctial.

- **Celestial 2.** The horizon system involves the use of azimuth and altitude to locate a point on the celestial sphere for an instant of time from a specific geographical position on the earth.

- **Celestial 3.** The Greenwich system involves the use of Greenwich hour angle and declination to locate a point on the celestial sphere with reference to the Greenwich meridian and the equinoctial for a given instant of time.

**Coriolis error.** The error introduced in a celestial observation taken in flight resulting from the deflective force on the liquid in the bubble chamber, as caused by the path of the aircraft in counteracting the earth's rotation.

**Coriolis force.** An apparent force due to the rotation of the earth that causes a moving body to be deflected to the right in the Northern Hemisphere and to the left in the Southern Hemisphere.

**Course.** The direction of the intended path of an aircraft over the earth.

**Corrected mean temperature (CMT).** The average between the target temperature and the true air temperature at flight level.

**Course line.** A line of position that is parallel or approximately parallel to the track of the aircraft. A line of position used to check aircraft position relative to intended course.

**Crab angle.** The angle between the aircraft longitudinal axis and ground track used to correct for wind drift and maintain an intended track.

**Cruise control.** The operation of an aircraft to obtain the maximum efficiency on a particular flight (most miles per amount of fuel).

**D Reading.** The difference between pressure altitude and true altitude at a different time in flight (true altitude minus pressure altitude).

**Datum.** Any numerical or geometrical quantity or set of such quantities that may serve as reference or base for other quantities.

**Dead reckoning (DR).** Finding ones position by means of a compass and calculations based on speed, time elapsed, effect of wind, and direction from a known position.

**Declination (Dec).** The angular distance to a body on the celestial sphere measured north or south through 90° from the celestial equator along the hour circle of the body. Comparable to latitude on the terrestrial sphere.

**Density altitude.** Pressure altitude corrected for temperature. Pressure and density altitudes are the same when conditions are standard. As the temperature rises above standard, the density of the air decreases, hence an increase in density altitude.

**Deviation (dev).** The angular difference between magnetic and compass headings.

**Deviation correction.** The correction applied to a compass reading to correct for deviation error. The numerical equivalent of deviation with the algebraic sign added to magnetic heading to obtain compass heading.

**Direct-indicating compass.** A magnetic compass in which the dial, scale, or index is carried on the sensing element.

**Diurnal circle.** The daily apparent path of a body on the celestial sphere caused by the rotation of the earth.

**Dog leg.** A temporary divergence from the desired track to adjust your course timing (preplanned or spontaneous).

**Drift.** The rate of lateral displacement of the aircraft by wind, generally expressed in degrees.

**Drift angle (DA).** The angle between true heading and true course, expressed in degrees right or left according to the way the aircraft has drifted.

**Drift correction (DC).** Correction for drift, expressed in degrees (plus or minus) and applied to true course to obtain true heading.

**Ecliptic.** The great circle on the celestial sphere along which the sun, by reason of the earth's annual revolution about the sun, appears to move. The plane of the ecliptic is tilted to the plane of the equator at an angle of 23° 27'.

**Effective air distance (EAD).** The distance measured along the effective air path.

**Effective air path (EAP).** A straight line on a navigation chart connecting two air positions or a fix position and an air position. Commonly used between the last known position (fix) and the next air position associated with a fix position.

**Effective true airspeed (ETAS).** The effective air distance divided by the elapsed time between two pressure readings.

**Elevated poles.** Celestial pole that is on the same side of the equinoctial as the position of the observer.

**Equal Altitude.** See Circles.

**Equation of time.** The amount of time by which the mean sun leads or lags behind the true sun at any instant. The difference between mean and apparent times expressed in units of solar time with the algebraic sign so that, when added to mean time, it gives apparent time.

**Equator.** The great circle on the earth's surface equidistant from the poles. Latitude is measured north and south from the equator.

**Equinoctial.** See Celestial Equator.

**Equinox.** The Autumnal Equinox is the point on the equinoctial when the sun, moving along the ecliptic, passes from north to south declination. The Vernal Equinox is the point of intersection of the ecliptic and the celestial equator (equinoctial) when the sun is moving from south to north declination. Also called First Point of Aries.

**Equivalent airspeed (EAS).** Calibrated airspeed corrected for compressibility-of-air error.

**Field-elevation pressure.** The existing atmospheric pressure in inches of mercury at the elevation of the field. Also known as station pressure.

**First point of Aries.** The point of intersection of the ecliptic and the celestial equator (equinoctial) when the sun is moving from south to north declination. Also called vernal equinox.

**Fix.** A geographic position of an aircraft determined from terrestrial, electronic, or astronomical data for a specific time.

**Flight levels (FL).** Surfaces of constant atmospheric pressure that are related to a specific pressure datum, 1013.2 millibar (29.92 inches), and are separated by specific pressure intervals. Flight levels are expressed in three digits that represent hundreds of feet. For example, flight level 250 represents a barometric altimeter indication of 25,000 feet, and flight level 260 is an indication of 26,000 feet.

**Flight plan.** Predetermined information for the conduct of a flight. That portion of a flight log that is prepared before the mission.

**Geographical coordinates.** The latitude and longitude used to locate any given point on the surface of the earth.

**GEOREF.** A worldwide position reference system that may be applied to any map or chart graduated in latitude and longitude, regardless of projection. It is a method of expressing latitude and longitude in a form suitable for rapid reporting and plotting. (This term is derived from the words The World Geographic Reference System.)

**Geostrophic wind.** The mathematically calculated wind that theoretically blows parallel to the contour lines, in which only pressure-gradient force and Coriolis force are considered.

**Gradient wind.** Generally accepted as the actual wind above the friction level, influenced by Coriolis force, pressure gradient, and centrifugal force.

**Graticule.** In cartography, a network of lines representing the earth's parallels of latitude and meridians of longitude.

**Great circle.** Any circle on a sphere whose plane passes through the center of that sphere.

**Great circle course.** The route between two points on the earth's surface measured along the shorter segment of the circumference of the great circle between the two points. A great circle course establishes the shortest distance over the surface of the earth between any two terrestrial points.

**Greenwich hour angle (GHA).** The angular distance measured from the upper branch of the Greenwich meridian westward through 360° to the upper branch of the hour circle passing through a body.

**Greenwich mean time (GMT).** Local time at the Greenwich meridian measured by reference to the mean sun. It is the angle measured at the pole or along the equator (and converted from time) from the lower branch of Greenwich meridian westward through 360° to the upper branch of the hour circle through the mean sun.

**Greenwich meridian.** The prime meridian that passes through Greenwich, England, and from which longitude is measured east or west.

**Greenwich sidereal time (GST).** Local sidereal time at Greenwich. It is equivalent to the Greenwich hour angle of Aries converted to time.

**Grid coordinates.** Coordinates of a grid coordinate system to which numbers and letters are assigned for use in designating a point on a grid map, photograph, or chart. Also a rectangular grid of fictitious chart graticule which is oriented with grid north.

**Grid course (GC).** The horizontal angle measured clockwise from grid north to the course line. The course of an aircraft measured with reference to the north direction of a polar grid.

**Grid heading (GH).** The heading of an aircraft with reference to grid north.

**Grid navigation.** A method of navigation using a grid overlay for direction reference.

**Grid north (GN).** An arbitrarily selected direction of a rectangular grid. In grid navigation, the direction to the 180° geographical meridian from the pole is almost universally used as standard grid north.

**Grivation (GRIV).** Angular difference in direction between grid north and magnetic north. It is measured east or west from grid north.

**Ground plot.** A graphic representation of track and groundspeed.

**Ground range.** The horizontal distance from the subpoint of the aircraft to an object on the ground.

**Ground return.** The reflection from the terrain as displayed on a CRT.

**Groundspeed (GS).** The horizontal component of the speed of an aircraft relative to the earth's surface.

**Ground wave.** A radio wave that is propagated over the surface of the earth and tends to parallel the earth's surface.

**Heading.** The direction in which the longitudinal axis of an aircraft is pointed, usually expressed in degrees clockwise from north (true, magnetic, compass or grid).

**Heat of compression error.** The error caused by the increase in the indication of the free air temperature gauge due to air compression and friction on the case around the sensitive element.

**Hertz (Hz).** The standard unit notation for measure of frequency in cycles per second. Sixty cycles per second is 60 Hz.

**Homing.** The technique whereby an aircraft is directed toward a specific point, keeping the aircraft pointed toward the point by visual, radio, radar, or similar references.

**Horizon.** The apparent or visible junction of the earth and sky as seen from any specific position. Also called the apparent, visible, or local horizon. A horizontal plane passing through a point of vision or perspective center. The apparent or visible horizon approximates the true horizon only when the point of vision is very close to sea level.

- **Bubble horizon.** An artificial horizon parallel to the celestial horizon established by means of a bubble level.

- **Celestial horizon.** The great circle on the celestial sphere whose plane passes through the center of the earth and is parallel to the plane tangent to the earth at the observer's position.

- **Visible horizon.** The circle around the observer where earth and sky appear to meet. Also called natural horizon or sea horizon.

**Hour circle.** A great circle on the celestial sphere passing through the celestial poles and a given celestial body.

**Index error.** An error caused by the misalignment of the measurement mechanism of an instrument.

**Indicated air temperature (IAT).** The uncorrected reading from the free air temperature gauge.

**Indicated airspeed (IAS).** The airspeed shown by an airspeed indicator.

**Indicated altitude.** Altitude displayed on the altimeter.

**Induced (real) precession.** A precession resulting from a torque, deliberately applied to a gyro. The gyro precesses 90° from the point of applied pressure in the direction of rotation.

**Inherent distortion.** The distortion of the display of a received radar signal caused by the design characteristics of a particular set.

**International date line.** The line coinciding approximately with the antimeridian of Greenwich, modified to avoid certain habitable land. In crossing this line, there is a date change of 1 day (gain 1 day heading west, lose 1 day heading east).

**Isobar.** A line joining points of equal pressure.

**Isogonic line (Isogonal).** A line drawn on a chart joining points of equal magnetic variation.

**Isogriv.** A line on a map or chart that joins points of equal angular difference between grid north and magnetic north.

**Isotach.** A line drawn on a chart joining points of equal windspeed.

**Isotherm.** A line drawn on a chart joining points of equal temperature.

**Knots (k).** Nautical miles per hour.

**Lateral axis.** The straight line passing through the center of gravity of an aircraft perpendicular to the fuselage.

**Latitude (lat).** Angular distance measured north or south of the equator along a meridian, 0° through 90°.

**Line of constant bearing.** A line from a fixed or moving point to a moving object or fixed point that retains a constant angular value with respect to a reference line.

**Line of position (LOP).** A line indicating a series of positions in which the observer is estimated to be at the time of the observation.

**Local hour angle (LHA).** The angular distance measured from the upper branch of the observer's meridian westward through 360° to the upper branch of the hour circle passing through a body.

**Local mean time (LMT).** The time interval elapsed since the mean suns transit of the observer's antimeridian.

**Local sidereal time (LST).** Local time at the observer's meridian measured by reference to the first point of Aries. It is equivalent to the local hour angle of Aries converted to time.

**Log.** A written record of computed or observed flight data generally applied to the written navigational record of a flight.

**Longitude (long).** The angular distance east or west of the Greenwich meridian, measured in the plane of the equator or of a parallel from 0° to 180°.

**Longitudinal axis.** A straight line through the center of gravity of an aircraft parallel to the fuselage.

**Lower branch.** Half of an hour circle opposite from upper branch, defined below.

**Lubber line.** A reference mark representing the longitudinal axis of an aircraft.

**Mach number.** The ratio of the velocity of a body to that of sound in the surrounding medium.

**Magnetic bearing (MB).** The direction to an object from a point, expressed as a horizontal angle, measured clockwise from magnetic north.

**Magnetic compass.** An instrument containing a freely suspended magnetic element that displays the direction of the horizontal component of the earth's magnetic field at the point of observation.

**Magnetic course (MC).** The horizontal angle measured from the direction of magnetic north clockwise to a line representing the course of the aircraft. The aircraft course measured with reference to magnetic north.

**Magnetic dip.** The vertical displacement of the compass needle from the horizontal caused by the earth's magnetic field.

**Magnetic direction.** A direction measured clockwise from the magnetic meridian.

**Magnetic heading (MH).** The heading of an aircraft with reference to magnetic north.

**Magnetic north (MN).** The direction indicated by the north-seeking pole of a freely suspended magnetic needle, influenced only by the earth's magnetic field.

**Map reading.** The determination of position by identification of landmarks with their representations on a map or chart.

**Map symbols.** Figures and designs used to represent topographical, cultural, and aeronautical features on a map or chart.

**Maximum elevation figure (MEF).** Depicted on a sectional chart, the MEF shows the height of the highest feature within a quadrangle and is printed in thousands and hundreds of feet above mean sea level (MSL).

**Maximum range.** The maximum distance a given aircraft can cover under given conditions by flying at the economical speed and altitude at all stages of the flight.

**Mean sea level (MSL).** The average height of the surface of the sea for all stages of the tide, used as a reference for elevations.

**Mean sun.** An imaginary sun traveling around the equinoctial at the average annual rate of the true sun.

**Mean time.** Time measured by reference to the mean sun.

**Meridional part.** A unit of measurement equal to 1 minute of longitude at the equator.

**Most probable position (MPP).** The most accurate estimate of position where an element of doubt exists as to the true position.

**Nadir.** That point on the celestial sphere directly beneath the observer and directly opposite the zenith.

**Nautical mile (NM).** A measure of distance equal to 1 minute of arc on the earth's surface. The United States has adopted the international nautical mile equal to 1,852 meters or 6,076 feet.

**Nautical twilight.** That period which ends in the evening and begins in the morning when the sun reaches 12° below the horizon.

**Navigational Aids.** Any means of obtaining a fix or line of position as an aid to dead reckoning.

**NAVSTAR global positioning system.** A space-based navigation, three-dimensional positioning, and time distribution system. It provides precise, continuous, all-weather, common grid, worldwide navigation, positioning, and timing information to land, sea, air, and space-based users. The joint program, with the US Air Force the lead service, allows distances to be measured instantly rather than measuring angles as in celestial navigation.

**Observed altitude (Ho).** The sextant altitude corrected for sextant and observation errors.

**Parallax error.** The difference between a body's altitude above an artificial or visible horizon and above the celestial horizon. The error is present because the body is not at an infinite distance.

**Personal error.** Differences in observations caused by sighting limitations of an observer.

**Pitch.** The rotation of an aircraft about its lateral axis.

**Pitot-static tube.** A device that consists of a pitot tube and a static port that measures pressures in such a way that the relative airspeed of an aircraft may be determined.

**Polar coordinates.** Coordinates derived from the distance and angular measurements from a fixed point (pole).

**Polar distance.** Angular distance from a celestial pole or the arc of an hour circle between the celestial pole or the arc of an hour circle between the celestial pole and a point on the celestial sphere. It is measured along an hour circle and may vary from 0° to 180° since either pole may be used as the origin of measurement.

**Precession of the equinox.** The average yearly apparent movement of the first point of Aries to the west.

**Precomputed curve.** A graphical representation of the azimuth or altitude of a celestial body plotted against time for a given assumed position (or positions) and computed for subsequent use for celestial observations. Used in celestial navigation to determine position or to check a sextant.

**Pressure altimeter.** An instrument that measures and displays vertical distance above a selected pressure datum based on a standard atmosphere.

**Pressure altitude.** The altitude above the standard datum plane. The standard datum plane is where the air pressure is 29.92 inches of mercury (corrected to plus 15 °C).

**Pressure altitude variation.** The pressure difference, in feet or meters, between mean sea level and the standard datum plane.

**Pressure differential.** The determination of the average drift or the crosswind component of the wind effect on the aircraft for a given period by taking D readings and applying the formula:

**Pressure line of position (PLOP).** A line of position determined by pressure pattern formulas, plotted parallel to the effective heading of the aircraft, and indicating the net crosswind displacement.

**Procedure turn.** An aircraft maneuver in which a turn is made away from a designated track followed by a turn in the opposite direction, both turns being executed at a constant rate so as to permit the aircraft to intercept and proceed along the reciprocal of the designated track.

**Projection (Chart, MAP).** Any systematic arrangement of meridians and parallels portraying the curved surface of the earth upon a plane.

**Pulse duration.** In radar, measurement of pulse transmission time in microseconds; that is, the time the radars transmitter is energized during each cycle. Also called pulse length and pulse width.

**Pulse-length error.** A range distortion of a radar return caused by the duration of the pulse.

**Pulse recurrence time (PRT) or rate (PRR).** The interval of time, in microseconds, between the transmission of two successive radar or radio pulses.

**Pulse repetition frequency.** In radar, the number of pulses that occur each second. Not to be confused with transmission frequency which is determined by the rate at which cycles are repeated within the transmitted pulse.

**Radar and/or radio altimeter.** An instrument that displays the absolute altitude or vertical distance between the aircraft and the surface directly below the aircraft.

**Radar beacon (RACON).** A receiver-transmitter combination that sends out a coded signal when triggered by the proper type of pulse, enabling determination of range, and bearing information by the interrogating station or aircraft.

**Radar beam.** A directional concentration of radio energy.

**Radar nautical mile.** The time required for a radar pulse to travel out 1 nautical mile and the echo pulse to return (12.4 microseconds).

**Radar navigation.** The determination of position by obtaining bearing and range information (or a combination of each) from a radar scope.

**Radio compass (ADF).** A radio receiver antenna that is used to measure the bearing to a radio transmitter.

**Radio frequency (RF).** Any frequency of electrical energy above the audio range that is capable of being radiated into space.

**Radio navigation.** Radio location intended for the determination of position or direction or for obstruction warning in navigation.

**Radome.** A bubble-type cover for a radar antenna.

**Range control.** The operation of an aircraft to obtain the optimum flying time.

**Range definition.** The accuracy with which a radar set can measure range, usually a function of pulse shape.

**Rectangular coordinates.** A system of coordinates based on a rectangular grid, sometimes referred to as grid coordinates.

**Refraction error.** An error caused by the bending of light rays in passing through the various layers of the atmosphere.

**Relative bearing (RB).** The direction expressed as a horizontal angle, normally measured clockwise from the true heading or forward point of the longitudinal axis of the aircraft to an object or body.

**Remote indicating compass.** A magnetic compass, the magnetic detecting element of which is installed in an aircraft in a position as free as possible from causes of deviation. A transmitter system is included to enable compass indications to be read on a number of repeater dials suitably positioned in the aircraft.

**Revolution (of the earth).** The earth's elliptical path about the sun that determines the length of the year and causes the seasons.

**Rhumb line.** A line on the surface of a sphere that makes equal oblique angles with all meridians. A loxodromic curve.

**Rhumb line correction.** The correction applied for the bubble-acceleration error caused by the rhumb line path of the aircraft.

**Running fix.** A fix determined from a series of lines of position, based on the same object or body and resolved for a common time.

**Scan.** The motion of an electronic beam through space searching for a target. Scanning is produced by the motion of the antenna or by lobe switching.

**Semidiameter (SD).** The value in minutes of arc of the radius of the sun or the moon.

**Sextant.** An optical instrument normally containing a two-power telescope with a 15° field of vision. It also contains a series of prisms geared to an altitude scale permitting altitude measurement of a celestial body's altitude from –10° below the horizon to 92° above the horizon.

**Sextant altitude (Hs).** A celestial altitude measured with a sextant. The angle measured in a vertical plane between an artificial or sea horizon and a celestial body without application of any corrections.

**Sidereal day.** The interval of time between two successive upper transits of a meridian by the first point of Aries (23 hours, 56 minutes).

**Sidereal hour angle (SHA).** Angular distance measured from the upper branch of the hour circle of the first point of Aries westward through 360° to the upper branch of the hour circle passing through a body.

**Sidereal time.** Time measured by reference to the upper branch of the first point of Aries.

**Slant range.** The line of sight distance between two points, not at the same level relative to a specific datum.

**Small circle.** Any circle on a sphere whose plane does not pass through the center of that sphere.

**Solar day.** The interval of time between two successive lower transits of a meridian by the true (apparent) sun.

**Solstice.** Those points on the ecliptic where the sun reaches its greatest northern or southern declination. Also the times when these phenomena occur.

**Speed line.** A line of position that intersects the track at an angle great enough to be used as an aid in determining groundspeed.

**Spot-size error.** A distortion of a radar return caused by the size of the electron spot in a cathode-ray tube.

**Standard lapse rate.** A temperature decrease of approximately 2° centigrade for each 1,000 feet increase in altitude.

**Standard time.** An arbitrary time, usually fixed by the local mean time of the central meridian of the time zone.

**Star magnitude.** A measure of the relative apparent brightness of stars.

**Statute mile.** 5,280 feet or .867 nautical miles.

**Subpoint.** That point on the earth's surface directly beneath an object or celestial body.

**Summer.** That point on the ecliptic where the sun reaches its greatest declination having the same name as the latitude.

**Sun line.** A line of position obtained by computation based on observation of the altitude of the sun for a specific time.

**Sweep.** The luminous line produced on the screen of a cathode-ray tube by deflection of the electron beam. Also called time base line.

**Sweep delay.** The electronic delay of the start of the sweep used to select a particular segment of the total range.

**Target-timing wind.** A wind determined from a series of ranges and bearings on the same target taken within a relatively short period of time.

**Time zone.** A band on the earth approximately 15° of longitude wide, the central meridian of each zone generally being 15° or a multiple removed from the Greenwich meridian so that the standard time of successive zones differs by 1 hour.

**Track (Tr).** The actual path of an aircraft above the surface of the earth.

**True airspeed (TAS).** Equivalent airspeed corrected for error due to air density (altitude and temperature).

**True air temperature (TAT).** Basic air temperature corrected for the heat of compression error. Also known as outside air temperature.

**True altitude.** The actual height above mean sea level.

**True Azimuth (Zn).** The angle at the zenith measured clockwise from true north to the vertical circle passing through the body.

**True bearing (TB).** The direction to an object from a point, expressed as a horizontal angle measured clockwise from true north.

**True course (TC).** The angle measured clockwise from true north to the line representing the intended path of the aircraft.

**True heading (TH).** The heading of an aircraft with reference to true north.

**True North (TN).** The direction from an observer's position to the geographic North Pole. The north direction of any geographic meridian.

**Twilight.** The periods of incomplete darkness following sunset and preceding sunrise. Twilight is designated as civil, nautical, or astronomical, as the darker limit occurs when the center of the sun is below the celestial horizon.

**Upper branch.** That half of an hour circle or meridian that contains the celestial body or the observer's position.

**Variable range marker (VRM).** An electronic marker, variable in range, displayed on a CRT for purposes of accurate ranging.

**Variation (var).** The angle difference at a given point between true north and magnetic north, expressed as the number of degrees that magnetic north is displaced east or west from true north. The angle to be added to true directions to obtain magnetic directions.

**Vertical circle.** A great circle on the celestial sphere joining the observer's zenith and nadir.

**Wander error.** The bubble-acceleration error caused by a change of track during the celestial shooting period.

**Wind.** Moving air, especially a mass of air having a common direction or motion. The term is generally limited to air moving horizontally or nearly so; vertical streams of air are usually called currents.

**Wind direction and velocity.** The horizontal direction and speed of air motion. Windspeed is generally expressed in nautical miles or statue miles per hour.

**Winter.** That point on the ecliptic where the sun reaches its greatest declination having the opposite name as the latitude.

**Z or Zulu time. (NATO).** Greenwich mean time. (DOD: Coordinated Universal Time (UTC).) Expressed in four digits.

**Zenith.** The point on the celestial sphere directly above the observer's position.

**Zenith distance (ZD).** Angular distance on the celestial sphere measured along the great circle from the zenith to the celestial object. Zenith distance is 90° minus celestial altitude.

**ZN (pressure pattern displacement).** In pressure pattern flying, the displacement in nautical miles, at right angles to the effective airpath, due to the crosswind component of the geostrophic wind.

**Zone time.** The time used through a 15° band of longitude. The time is based on the local mean time for the center meridian of the zone.